Vibroacoustic Simulation

Vibroacoustic Simulation

An Introduction to Statistical
Energy Analysis and
Hybrid Methods

Alexander Peiffer

WILEY

This edition first published 2022

Registered Office
John Wiley & Sons, Inc., 111 River Street, Hoboken, NJ 07030, USA

Editorial Office
111 River Street, Hoboken, NJ 07030, USA

For details of our global editorial offices, customer services, and more information about Wiley products visit us at www.wiley.com.

Wiley also publishes its books in a variety of electronic formats and by print-on-demand. Some content that appears in standard print versions of this book may not be available in other formats.

A catalogue record for this book is available from the Library of Congress

Hardback ISBN: 9781119849841; ePub ISBN: 9781119849865; ePDF ISBN: 9781119849858;
Obook ISBN: 9781119849872

Cover image: © visibleimpression/Shutterstock; Courtesy of Alexander Peiffer
Cover design by Wiley

Set in 9.5/12.5pt STIXTwoText by Integra Software Services Pvt. Ltd, Pondicherry, India

10 9 8 7 6 5 4 3 2 1

To my parents and Ivonne my love who always supported me

Contents

Preface

The reason and stimulation for this book was an application for a lecturer position several years ago. I slightly panicked due to the fact that in case of success, I would be totally unprepared in terms of lecturing material. So, I decided to write a script in order to have at least something prepared. But, the more I got into the text and the more I tried to collect content for a script on vibroacoustics, I came to the conclusion that there is a need for a modern text book on the subject of vibroacoustic simulation focussing on statistical energy analyses and hybrid methods.

There are many excellent books on acoustics and vibration, but what is missing to my opinion is an overall treatment of vibroacoustic simulation methods. Especially when we are talking about statistical energy analysis (SEA) and the combination of finite element methods (FEM) and SEA, the hybrid FEM/SEA method. In addition, the hybrid FEM/SEA method allows a much clearer and more systematic approach to SEA compared to the original literature and might help to impart the knowledge to students and professionals. It is my persuasion, that every acoustic simulation engineer shall master these simulation techniques to be prepared for vibroacoustic prediction of the full audible frequency range.

What is so special about vibroacoustics that so many methods are required? One answer is that the dynamic properties of structure and fluid systems are so different. This leads to distinct dynamic behavior. There may fit a lot of wavelengths of acoustic waves into a chamber of a machine or a passenger compartment filled with air, whereas the surrounding structure is often stiff and robust, and only a few wavelengths of the structural bending waves fit into the area of the surrounding walls. This strongly influences how energy is transmitted via the walls into the cavity and how small uncertainties affect the system response.

Additionally, there is often a great variety of materials, like foams, fibers, rubbers, etc. in the structure or applied as noise and vibration control, all having different orders of magnitude in wavelengths or even completely different modes of wave propagation.

As a consequence, vibroacoustics is a complex engineering discipline or science because the engineer has to master all those modes of wave propagation in the different systems and media as far as the coupling between those waves for connected subsystems. A thorough treatment of all wave types, couplings, and properties is not possible in a typical lecture or textbook, but it is possible to explain the main idea of how to deal with vibroacoustic phenomena and which means are required to solve the engineering problem. This book tries to extract the basic concepts, so that candidates are in a position to determine, investigate and categorize vibroacoustic systems and make the right decision on how to simulate them.

The frequency range of interest is covering four orders of magnitude from 20 to 20 000 Hz. That is one further reason why various methods for the description of these phenomena are required. At low frequencies it makes sense to investigate the modal behavior of a structure like the first modes of a string. In contrast to this, calculating all standing waves at high frequencies for a large room is not reasonable, as small changes at the boundaries or even temperature will lead to totally different wave forms in the room. Both regimes are addressed by different approaches categorized as (i) deterministic or (ii) statistic methods. The first occurs normally at lower frequencies, whereas the latter is valid at high frequencies. Because of the different wavelengths, it often appears that both cases occur in one vibroacoustic system, and both approaches are necessary. The combination of the two methods is called hybrid FEM/SEA method.

As there are many books on the subjects of deterministic acoustics and vibration available, this book focuses on SEA and hybrid methods. However, as FEM systems of equations are involved in the hybrid method, a minimum understanding of deterministic systems is required.

How is the book organized? It starts with a simple but excellent example for a vibrating system: the harmonic oscillator. In chapter 1 phenomena such as resonances, off resonance dynamics, and numerous damping mechanisms are explained based on this test case. A first step towards complex and FEM systems is made by introducing multiple coupled oscillators as an example for multiple degree of freedom systems. Real excitations often are of random nature. Hence, this chapter ends with tools and methods to describe random signals and processes as far as the response of linear systems to such signals.

Chapters 2 and 3 deal with wave motion in fluids and structures, respectively. Both chapters bring into focus the physics of sources, because the source mechanisms reveal how energy is introduced into the wavefields and how the feedback to the excitation can be characterized. Furthermore, the source dynamics are required when systems are coupled. The dynamics of acoustic and structure systems are shown in chapters 4 and 5. This includes the natural resonances of such systems that will become important for the classification of random systems. Based on analytical models, the low and high frequency behavior of such systems is presented. One aim of the various examples is to illustrate that when sources are exciting those systems, the high frequency dynamics become similar to the free field results from chapters 2 and 3. Chapter 6 deals with the random description of systems. The concept of ensemble average and diffuse fields is applied to typical example systems by using Monte Carlo simulations. Based on such randomized systems and averaged values, it is shown that we get similar results to those you would get from deterministic methods when the uncertainty of dynamically complex systems is considered. This opens the door to the statistical energy analysis (SEA). Some typical one-, two-, and three-dimensional systems are presented in the very detail, so that the reader gets a feeling when and under which conditions the SEA assumptions are valid. The idea is to provide comprehensive examples for the rules of thumb usually used to determine if random methods are valid or not.

In chapter 7 methods for coupling deterministic (FEM) and random (SEA) systems are presented, and the hybrid FEM/SEA method is introduced by describing the coupling between FEM and SEA systems. Based on this, the effect of random on deterministic systems as far as the impact of deterministic on random subsystems is presented. The chapter closes with the global procedure of hybrid FEM/SEA modelling that calculates the joint response of both types of systems.

Chapter 8 applies these coupling formulas to several options of connections. Especially the coupling sections are often missing in text books on SEA for a certain reason: the calculation of coupling loss factors is not easy. However, as it is important to understand the assumptions and limits, the coupling loss factors of point, area, and line junctions are systematically derived. Since junctions are nothing else than noise paths, this chapter is also useful for practical applications, for example the acoustic transmission loss of plates that is an important quantity for airborne acoustic isolation.

The following chapters apply the theory to pure deterministic (chapter 9), pure random (chapter 10), and hybrid FEM/SEA examples (chapter 11). All examples are worked out in detail and show real engineering systems such as mufflers. In chapter 9 the transfer matrix method is introduced as an example of deterministic methods. This allows the simulation of complex lay-ups of noise control treatments applied in chapters 10 and 11.

The presented theory and the examples are calculated using Python scripts. The scripts and the related toolbox are made available as open source code. The author hopes that this toolbox helps to understand and to apply the presented topics. Further contributions to the code of the toolbox are very welcome. The documentation of the toolbox and the GIT repository can be found on the authors website www.docpeiffer.com.

As an acoustic engineer, I am in the somehow unique situation that I had the chance to work on several means of transportation: trains, aircraft, helicopters, launchers, satellites, and finally cars (mainly electric). Because of this experience I am convinced that a deep knowledge of vibroacoustic simulation methods is mandatory to create excellent *and* low-noise products. This know-how puts the acoustic engineer in the position to apply the right method in the right situation and frequency range. To underline this fact, chapter 12 presents models, basic ideas, pitfalls, and results of some industrial examples from aerospace, automotive, and train industries. Special thanks goes to my former colleague Ulf Orrenius who wrote the train and motivation section of chapter 12. His great experience and knowledge strongly enriched the content of this chapter.

This book is about simulation, but simulation is nothing without validation based on tests. In my view both – simulation and tests – are required to perform a good acoustic design and noise control engineering. Thus, chapter 13 briefly summarizes test and correlation methods together with an outlook to further topics of simulation and ongoing research in the field of acoustic simulation.

In most cases the life of an acoustic engineer means solving the target conflict between the acoustic performance and costs, weight, and space requirements. This is the reason why design engineers are not always the best friends of acousticians during the design phase. The more important it is that you are able to calculate the effect of your decisions for efficient application of the sometimes rare acoustic resources. I hope that this book provides some support for this demanding task.

Coming back to my initial motivation: If I would have to hold a lecture on vibroacoustic simulation now, I would sleep much better.

Alexander Peiffer

Planegg, Germany
15 October 2021

Acknowledgments

Special thanks goes to my former colleague Ulf Orrenius who wrote the train- and motivation section of chapter 12. His great experience and knowledge strongly enriched the content of this chapter.

I would also like to thank the Audi AG who provided the models to present the simulation strategy of automotive systems.

Acronyms

2DOF	Two degrees of freedom
ms	Mean square
rms	Root mean square
CFD	Computational fluid dynamics
CFRP	Carbon fiber reinforced plastic
DFT	Discrete Fourier transform
DVA	Dynamic vibration absorber
DOF	Degree of freedom
EMA	Experimental modal analysis
FEM	Finite element method
FFT	Fast Fourier transform
FT	Fourier transform
GFRP	Glass fiber reinforced plastic
HVAC	Heating, ventilation, and air conditioning
MAC	Modal assurance criterion
MDOF	Multiple degrees of freedom
MIMO	Multiple input multiple output
NVH	Noise, vibration, and harshnes
PAX	Passenger
SDOF	Single degree of freedom
SEA	Statistical Energy Analysis
SISO	Single input single output
TBL	Turbulent boundary layer
TPA	Transfer path analysis
TVA	Tuned vibration absorber

1

Linear Systems, Random Process and Signals

Simple systems with properties constructed by lumped elements as masses, springs and dampers are a good playground to understand and investigate the physics of dynamic systems. Many phenomena of vibration as resonance, forced vibration and even first means of vibration control can be explained and visualized by these lumped systems.

In addition, a basic knowledge of signal and system analysis is required to put the principle of cause and effect in the right context. Every vibroacoustic system response depends on excitation by random, harmonic or specific signals in the time domain and we need a mathematical tool set to describe this.

An excellent test case to demonstrate and define the principle effects of vibration is the harmonic oscillator. It consists of a point mass, a spring and a damper. The combination of many point masses connected via simple springs and dampers provides some further insight into dynamic systems.

As those systems are described by components that have no dynamics in themselves they are called lumped systems. In principle all vibroacoustic systems can by modelled and approximated by this simplified approach.

1.1 The Damped Harmonic Oscillator

A realization of the harmonic oscillator is given by a concentrated point mass m fixed at massless spring with stiffness k_s as in Figure 1.1. The static equilibrium is assumed at $u = 0$ being the displacement in x-direction. A damper connecting mass and fixation creates dissipation.

1.1.1 Homogeneous Solutions

Without external excitation as shown in Figure 1.1 a) the motion depends on the initial conditions at time $t = 0$ with the displacement $u(0) = u_0$ and velocity $v_x(0) = v_{x0}$. The damping is supposed to be viscous, thus proportional to the velocity $F_{xv} = -c_v\dot{u}$. The equation of motion

$$m\ddot{u} + c_v\dot{u} + k_s u = 0 \qquad (1.1)$$

Vibroacoustic Simulation: An Introduction to Statistical Energy Analysis and Hybrid Methods, First Edition. Alexander Peiffer.

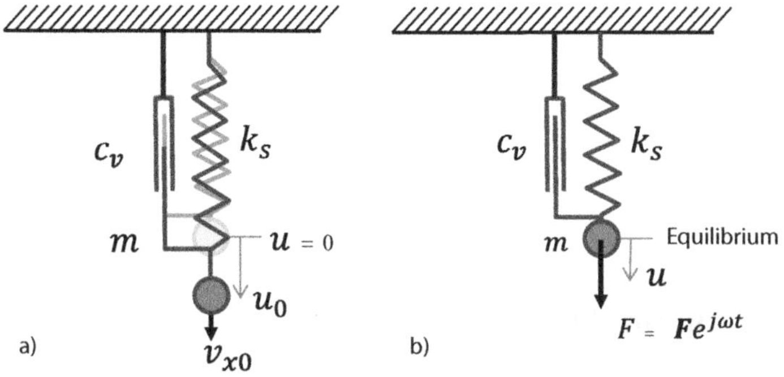

Figure 1.1 Damped harmonic oscillator with initial conditions a) and external force excitation b). *Source:* Alexander Peiffer.

is a homogeneous second order equation with a solution of the form $u = Ae^{st}$. Entering this into Equation (1.1) leads to the characteristic equation

$$ms^2 + c_v s + k_s = 0 \tag{1.2}$$

with the two solutions

$$s_{1/2} = -\frac{1}{2m}\left[-c_v \pm \sqrt{c_v^2 - 4mk_s^{1/2}}\right] \tag{1.3}$$

Hence,

$$u(t) = B_1 e^{s_1 t} + B_2 e^{s_2 t} \tag{1.4}$$

with B_1 and B_2 depending on the initial conditions. The root in Equation (1.3) is zero when c_v equals $\sqrt{4mk_s}$. This specific value is called the critical viscous damping

$$c_{vc} = \sqrt{4mk_s} \tag{1.5}$$

We use the following definitions:

$$\omega_0^2 = \frac{k_s}{m} \qquad \frac{c_v}{m} = 2\zeta\omega_0 \qquad \zeta = \frac{c_v}{\sqrt{4mk_s}} = \frac{c_v}{c_{vc}} \tag{1.6}$$

ω_0 is the natural angular frequency, ζ is ratio of the viscous-damping to the critical viscous-damping. There are additional expressions for the period and frequency

$$f_0 = \frac{\omega_0}{2\pi} \qquad T_0 = \frac{1}{f_0} \tag{1.7}$$

where f_0 is the natural frequency and T_0 the oscillation period. Equations (1.1)–(1.3) can now be written as

$$\ddot{u} + 2\zeta\omega_0\dot{u} + \omega_0^2 u = 0 \tag{1.8}$$

$$s^2 + 2\zeta\omega_0 s + \omega_0^2 = 0 \tag{1.9}$$

$$s_{1/2} = -\zeta\omega_0 \pm \omega_0\sqrt{\zeta^2 - 1} \tag{1.10}$$

The problem falls into three cases:

$\zeta > 1$ overdamped
$\zeta < 1$ underdamped
$\zeta = 1$ critically damped.

The first case leads to two real roots, and no oscillation is possible. The second case gives two complex roots, which means that (damped) oscillation occurs. The third case is a transition case between the two other. Subsections 1.1.2–1.1.4 deal with each case in detail.

1.1.2 The Overdamped Oscillator ($\zeta > 1$)

Both roots in Equation (1.10) are real, distinct and negative. The motion is called overdamped because introducing this into Equation (1.4) gives a sum of decaying exponential functions:

$$u(t) = B_1 e^{(-\zeta+\sqrt{\zeta^2-1})\omega_0 t} + B_2 e^{(-\zeta-\sqrt{\zeta^2-1})\omega_0 t} \tag{1.11}$$

The movement of such a system is illustrated in Figure 1.2. Using the above solution and applying the initial conditions u_0 and v_{x0} we get for B_i:

$$B_{1/2} = \pm\frac{u_0\omega_0(\zeta \pm \sqrt{\zeta^2-1}) + v_{x0}}{2\omega_0\sqrt{\zeta^2-1}} \tag{1.12}$$

1.1.3 The Underdamped Oscillator ($\zeta < 1$)

Here, the roots are complex conjugates and the solution of Equation (1.10) becomes:

$$u(t) = e^{-\zeta\omega_0 t}\left(B_1 e^{j(1-\zeta^2)^{1/2}\omega_0 t} + B_2 e^{-j(1-\zeta^2)^{1/2}\omega_0 t}\right) \tag{1.13}$$

$$= \hat{u}_0 e^{-\zeta\omega_0 t} cos((1-\zeta)^{1/2}\omega_0 t + \phi_0) \tag{1.14}$$

The motion is oscillatory with a frequency that is lower than in the undamped configuration:

$$\omega_d = \omega_0\sqrt{1-\zeta^2} = \omega_0\gamma \tag{1.15}$$

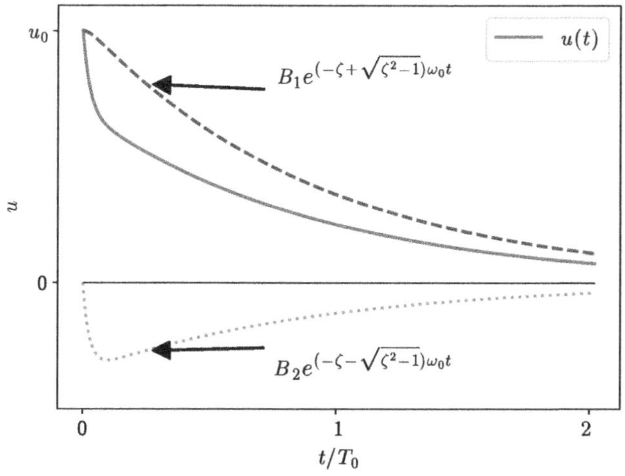

Figure 1.2 Decaying components of the overdamped oscillator.
Source: Alexander Peiffer.

Introducing the initial conditions u_0 and v_{x0} at $t = 0$ the solution for the initial amplitude $\hat{u}_0$ and phase ϕ_0 reads as:

$$\hat{u}_0 = \frac{\sqrt{u_0^2\omega_d^2 + (v_{x0} + \zeta\omega_0 u_0)^2}}{\omega_d} \tag{1.16}$$

$$\phi_0 = -\arctan\left(\frac{v_{x0} + \zeta\omega_0 u_0}{u_0\omega_d}\right) \tag{1.17}$$

The damped oscillatory motion is illustrated in Figure 1.3. It shows a decreasing motion that never approaches the equilibrium.

1.1.4 The Critically Damped Oscillator ($\zeta = 1$)

The last case is a transition between both systems. There is only one root $s = -\omega_0$, and the solution in Equation (1.4) becomes:

$$u(t) = (B_1 + B_2)e^{-\omega_0 t} \tag{1.18}$$

This solution does not provide enough constants to fulfil the initial conditions, so that we need an extra term $te^{-\omega_0 t}$:

$$u(t) = (B_3 + B_4 t)e^{-\omega_0 t} \tag{1.19}$$

Introducing the initial conditions again, the constants are:

$$B_3 = u_0 \tag{1.20}$$

$$B_4 = v_{x0} + \omega_0 u_0 \tag{1.21}$$

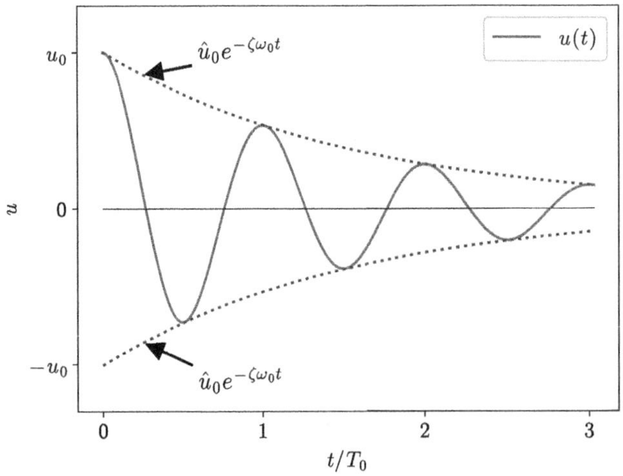

Figure 1.3 Damped, sinusoidal motion of the underdamped oscillator. *Source:* Alexander Peiffer.

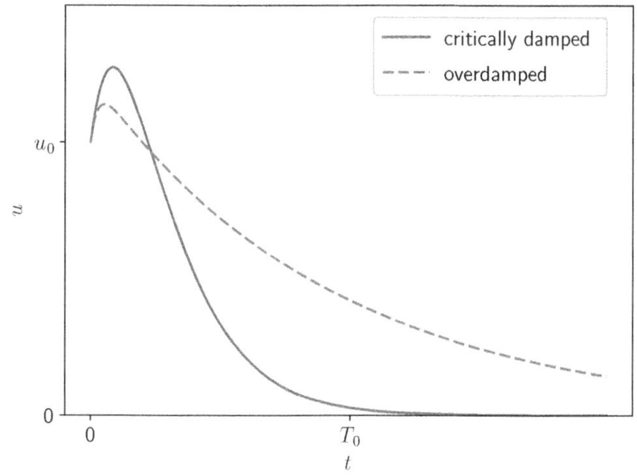

Figure 1.4 Motion of the critically damped oscillator. *Source:* Alexander Peiffer.

Critically damped systems can be of practical relevance, because the motion returns to rest in the shortest possible time, which is useful if periodic motion shall be prevented. In contrast to the overdamped oscillator the equilibrium is reached as can be seen in Figure 1.4.

Let us summarize some facts and observations about free damped oscillators:

1. Oscillation occurs only if the system is underdamped.
2. ω_d is always less than ω_0.
3. The motion will decay.
4. The frequency ω_d and the decay rate are properties of the system and independent from the initial conditions.
5. The amplitude of the damped oscillator is $\hat{u}(t) = \hat{u}_0 e^{-\beta t}$ with $\beta = \zeta\omega_0$. β is called the decay rate of the damped oscillator.

The decay rate is related to the decay time τ. This is the time interval where the amplitude decreases to e^{-1} of the initial amplitude. Thus, the decay time is:

$$\tau = \frac{1}{\beta} = \frac{1}{\zeta\omega_0} \tag{1.22}$$

1.2 Forced Harmonic Oscillator

When an external force $\hat{F}_x \cos(\omega t)$ is exciting the damped oscillator as shown in Figure 1.1 b), applying Newton's second law we get for the equation of motion:

$$m\ddot{u} + c_v\dot{u} + k_s u = \hat{F}_x \cos(\omega t) \tag{1.23}$$

This is an inhomogeneous, linear, second-order equation for u. The solution of this equation is given by a particular solution $u_P(t)$ and the solutions of the homogeneous

Equation (1.1) $u_H(t)$.

$$u(t) = u_H(t) + u_P(t) \tag{1.24}$$

Any linear combination of the homogeneous solution can be added to the particular solution because it equals zero.

1.2.1 Frequency Response

There are several methods to determine the particular solutions, like Laplace and Fourier transforms. Here, complex algebra will be used[1]. Amplitude and phase are given by a complex pointer denoted by bold italic type as depicted in Figure 1.5.

$$\hat{F}_x \cos(\omega t + \phi) = Re(\boldsymbol{F}_x e^{j\omega t}) \tag{1.25}$$

$\boldsymbol{F}_x$ is the complex amplitude of the force, and the $Re(\cdot)$ expression is usually omitted. The displacement and velocity response is then given by

$$u(t) = \boldsymbol{u}e^{j\omega t} \qquad v_x(t) = j\omega\boldsymbol{u}e^{j\omega t} = \boldsymbol{v}_x e^{j\omega t} \tag{1.26}$$

with $\boldsymbol{u}$ and $\boldsymbol{v}_x$ as complex amplitudes of the displacement and velocity, respectively. Introducing this into Equation (1.23).

$$-m\omega^2\boldsymbol{u}e^{j\omega t} + jc_v\omega\boldsymbol{u}e^{j\omega t} + k_s\boldsymbol{u}e^{j\omega t} = \boldsymbol{F}e^{j\omega t} \tag{1.27}$$

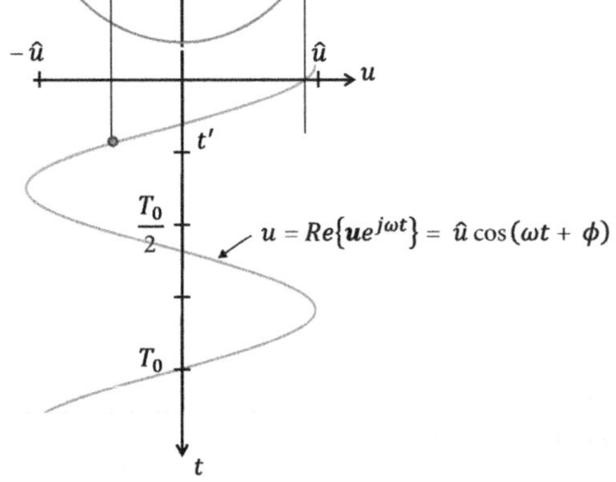

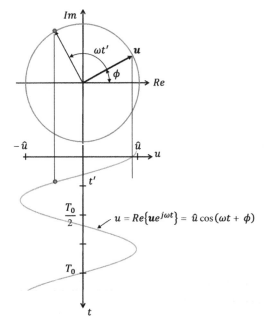

Figure 1.5 Complex pointer, amplitude and phase relationship. *Source:* Alexander Peiffer.

1 In this book the convention $e^{j\omega t}$ for the complex harmonic function is used. Literature that deals with wave propagation often use $e^{-j\omega t}$ to have positive wavenumber for positive wave propagation. However, as in every textbook in acoustics I denote the used convention on the first page to avoid confusion.

and solving this for $\boldsymbol{u}$ gives:

$$\boldsymbol{u} = \frac{\boldsymbol{F}_x}{k_s - m\omega^2 + jc_v\omega} \tag{1.28}$$

The magnitude $\hat{u}$ and the phase ϕ of $\boldsymbol{u}$ are:

$$\hat{u} = \frac{\hat{F}_x}{\sqrt{(k_s - m\omega^2)^2 + (c_v\omega)^2}} \tag{1.29}$$

$$\phi_0 = \arctan\left(\frac{c_v\omega}{k_s - m\omega^2}\right) \tag{1.30}$$

At $\omega = 0$ the static displacement amplitude is $\hat{u}_0 = \hat{F}_x/k_s$. Using the definitions from (1.6) and dividing $\hat{u}$ by $\hat{u}_0$ gives the normalized amplitude

$$\frac{\hat{u}}{\hat{u}_0} = \frac{1}{\sqrt{[1 - (\omega/\omega_0)^2]^2 + (2\zeta\omega/\omega_0)^2}} \tag{1.31}$$

and phase

$$\phi_0 = \arctan -\frac{2\zeta\omega/\omega_0}{1 - (\omega/\omega_0)^2} \tag{1.32}$$

It can be shown that the maximum of $\hat{u}$ is at

$$\frac{\omega_r}{\omega_0} = \sqrt{1 - 2\zeta^2} \tag{1.33}$$

and the maximum value is

$$\hat{u}_r = \frac{\hat{u}_0}{2\zeta\sqrt{1 - \zeta^2}} \tag{1.34}$$

with the corresponding phase

$$\phi_r = \arctan\left(-\frac{\sqrt{1 - 2\zeta^2}}{\zeta}\right) \tag{1.35}$$

The evolution of $\hat{u}/\hat{u}_0$ and ϕ_0 is shown in Figures 1.6 and 1.7 for different ζ. One can see the resonance amplification at ω_r that would be infinite in case of $\zeta = 0$ and the decrease of the amplitude with increasing damping. For $\zeta > 1/\sqrt{2}$ the maximum value occurs at $\omega = 0$, so the displacement is just a *forced* movement without any resonance effect.

The frequency of highest amplitude is called the amplitude resonance and it is different from the so called phase resonance with $\phi = -\frac{\pi}{2}$, which corresponds to the resonance of the undamped oscillator.

1.2.2 Energy, Power and Impedance

It is helpful to investigate the ongoing processes from the energy perspective. For an undamped system $c_v = 0$ that oscillates with $u(t) = \hat{u}\cos(\omega_0 t)$ and absence of external

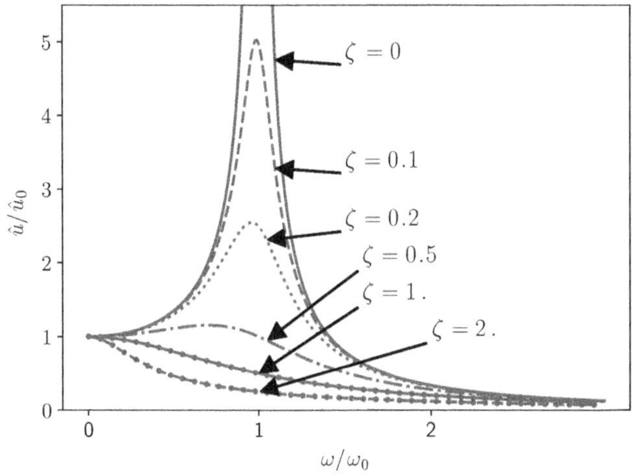

Figure 1.6 Normalized amplitude of forced harmonic oscillator. *Source:* Alexander Peiffer.

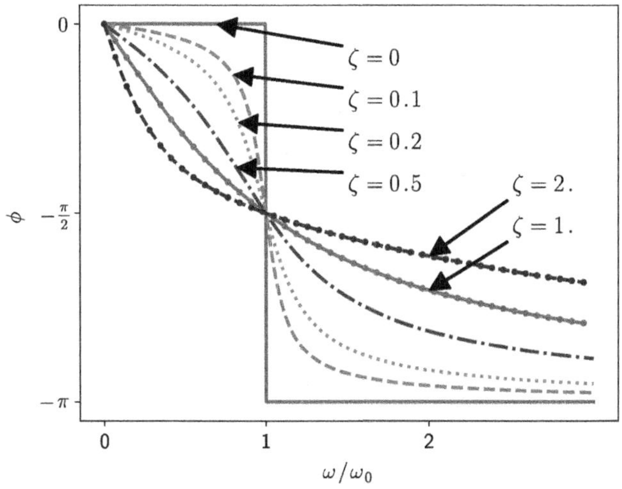

Figure 1.7 Phase of forced harmonic oscillator. *Source:* Alexander Peiffer.

forces the energy remains constant. The energy is the sum of kinetic E_{kin} and potential energy E_{pot}

$$
\begin{aligned}
E_{\text{kin}} + E_{\text{pot}} &= \frac{1}{2}m\dot{u}^2 + \frac{1}{2}k_s u^2 \\
&= \frac{1}{2}m\omega^2\hat{u}^2 \sin^2(\omega t) + \frac{1}{2}k_s\hat{u}\cos^2(\omega t) \\
&= \frac{1}{2}m\omega^2\hat{u}^2\left[\sin^2(\omega t) + \cos^2(\omega t)\right] = \frac{1}{2}m\omega^2\hat{u}^2
\end{aligned} \tag{1.36}
$$

and is constant, but spring and mass exchange energy twice over one period T_0. As energy quantities are not always constant, this motivates the definition of time average values. We introduce the mean-square (ms) and root-mean-square (rms) value for the

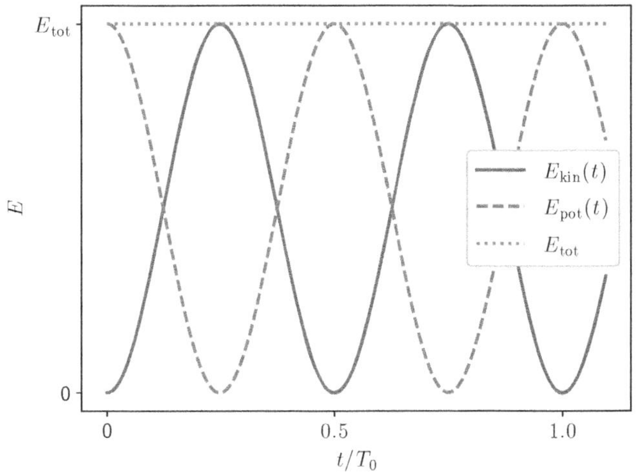

Figure 1.8 Kinetic and potential energy of the harmonic oscillator. *Source:* Alexander Peiffer.

time average for an arbitrary signal $x(t)$. The mean-square value over a time T is

$$u_{\text{ms}}^2 = \langle u^2 \rangle_T = \frac{1}{T} \int_0^T u^2(t) dt \qquad u_{\text{rms}} = \sqrt{\langle u^2 \rangle_T} \tag{1.37}$$

In the following $\langle \cdot \rangle_T = \frac{1}{T} \int_0^T \cdot dt$ denotes a time average. If the signal is harmonic with $u(t) = \hat{u} \cos(\omega_0 t)$ then

$$u_{\text{rms}} = \frac{\hat{u}}{\sqrt{2}} \qquad u_{\text{rms}}^2 = \frac{\hat{u}^2}{2} \tag{1.38}$$

1.2.3 Impedance and Response Functions

So far the frequency response of the oscillator was expressed as the relationship between displacement and force. The ratios $\boldsymbol{u}/\boldsymbol{F}_x$ and $\boldsymbol{D} = \boldsymbol{F}_x/\boldsymbol{u}$ are called *mechanical receptance* and *dynamic stiffness*, respectively. Often used force response relationships are the mechanical impedance (force/velocity= $\boldsymbol{F}_x/\boldsymbol{v}_x$) and the mobility (velocity/-force= $\boldsymbol{v}_x/\boldsymbol{F}_x$). The symbols and definitions are:

$$\text{Impedance: } \boldsymbol{Z} = \frac{\boldsymbol{F}_x}{\boldsymbol{v}_x} \quad \text{Mobility: } \boldsymbol{Y} = \frac{\boldsymbol{v}_x}{\boldsymbol{F}_x} \tag{1.39}$$

Considering the solution of the damped oscillator and $\boldsymbol{v}_x = j\omega \boldsymbol{u}$ both quantities become:

$$\boldsymbol{Z} = \frac{\boldsymbol{F}_x}{\boldsymbol{v}_x} = c_v + j\left(-\frac{k_s}{\omega} + m\omega\right) \tag{1.40}$$

The real and imaginary part have specific names

$$\boldsymbol{Z} = R + jX_Z \quad \text{resistance } + j \text{ reactance} \tag{1.41}$$

The meaning of this convention becomes clear when inspecting Figure 1.9. When the oscillator is excited in the mass- or stiffness-controlled regime below or above the resonance, the force introduces a reactive movement. The energy can be taken back from

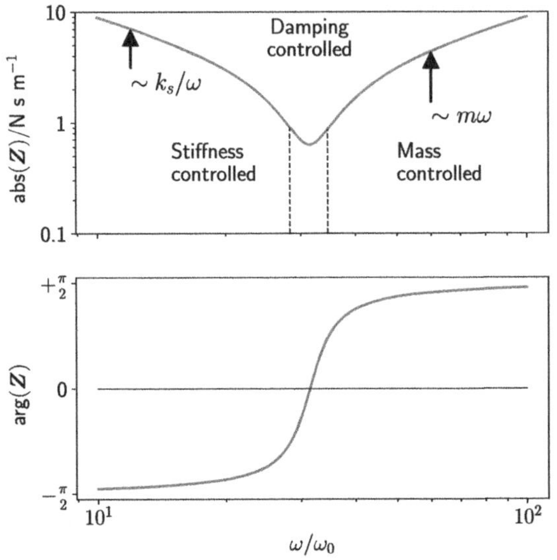

Figure 1.9 Magnitude and phase of oscillator impedance. *Source:* Alexander Peiffer.

the system and it is called *reactive*. Whether the system response is mass or stiffness controlled can be identified from the slope of the impedance magnitude and the phase. In the resonance regime the system is driven at resonance and the damper dissipates energy. The resistance of the system becomes important and is thus called *resistive*.

1.2.3.1 Power Balance

We multiply Equation (1.23) by $\dot{u}$

$$m\ddot{u}\dot{u} + c_v\dot{u}^2 + k_s\dot{u}x = F_x\dot{u} \tag{1.42}$$

The first and third term can be integrated

$$\frac{d}{dt}\left(\frac{1}{2}m\dot{u}^2 + \frac{1}{2}k_s u^2\right) + c_v\dot{u}^2 = F_x\dot{u}. \tag{1.43}$$

The terms in the parenthesis are kinetic and potential energy and known as constant. The expression $c_v\dot{u}^2$ is the dissipated power, because it is the damping force times velocity.

$$\Pi_{\text{diss}} = F_x\dot{u} = c_v\dot{u}^2 \tag{1.44}$$

And $F_x\dot{u}$ is the introduced power, thus

$$\Pi_{\text{in}} = F_x\dot{u} \tag{1.45}$$

So we get the power balance

$$\Pi_{\text{diss}} = \Pi_{\text{in}} \tag{1.46}$$

that is fluctuating for harmonic motion but with a net power flow.

For harmonic motion the power introduced into the system by a force $F_x(t) = \boldsymbol{F}_x e^{j\omega t}$ that generates the velocity response $v_x(t) = \boldsymbol{v}_x e^{j\omega t}$ is

$$\begin{aligned}\Pi(t) = F(t)v(t) &= Re\left(\boldsymbol{F}_x e^{j\omega t}\right) Re\left(\boldsymbol{v}_x e^{j\omega t}\right)\\ &= \frac{1}{2}Re\left(\boldsymbol{F}_x e^{j\omega t} + \boldsymbol{F}_x^* e^{-j\omega t}\right)\frac{1}{2}Re\left(\boldsymbol{v}_x e^{j\omega t} + \boldsymbol{v}_x^* e^{-j\omega t}\right)\\ &= \frac{1}{4}Re\left(\boldsymbol{F}_x\boldsymbol{v}_x^* + \boldsymbol{F}_x^*\boldsymbol{v}_x + \boldsymbol{F}_x\boldsymbol{v}_x e^{j2\omega t} + \boldsymbol{F}_x^*\boldsymbol{v}_x^* e^{-j2\omega t}\right)\\ &= \frac{1}{2}Re\left(\boldsymbol{F}_x\boldsymbol{v}_x^* + \boldsymbol{F}_x\boldsymbol{v}_x e^{j2\omega t}\right)\end{aligned} \tag{1.47}$$

The first term in the bracket is constant, the second oscillating with twice the excitation frequency. The first part is called active power and the second part the reactive. All introduced energy in one half cycle comes back in the next half cycle. The time average over one period leaves only the active part

$$\langle\Pi\rangle_T = \frac{1}{T}\int_0^T \frac{1}{2}Re\left(\boldsymbol{F}_x\boldsymbol{v}_x{}^* + \boldsymbol{F}_x\boldsymbol{v}_x e^{j2\omega t}\right)dt = \frac{1}{2}Re(\boldsymbol{F}_x\boldsymbol{v}_x{}^*) \tag{1.48}$$

The velocity can be expressed by the impedance $\boldsymbol{V} = \boldsymbol{Z}/\boldsymbol{F}$ or vice versa, so we get

$$\langle\Pi\rangle_T = \frac{1}{2}\hat{F}_x^2 Re(\boldsymbol{Z}^{-1}) = \frac{1}{2}\hat{F}_x^2 Re(\boldsymbol{Y}) \tag{1.49}$$

The power considerations further clarify the naming conventions for the real and imaginary parts of the impedance. With Equation (1.40) the power introduced into the system equals $\Pi = \frac{1}{2}\left|\boldsymbol{v}_x\right|^2 c_v$. Thus, the active power is controlled by the real part or resistance whereas the reactive part is determined by the imaginary component called reactance. The energy is dissipated in the *resistive* damping process, but power delivered to the *reactive* part goes into the kinetic and potential energy of mass and spring.

1.2.4 Damping

In many practical applications ζ is small and the amplitude can be estimated by linear expansion from (1.34)

$$\hat{u}_r \approx \frac{\hat{u}_0}{2\zeta}\left(1 + \frac{\zeta^2}{2}\right) \approx \frac{\hat{u}_0}{2\zeta} \tag{1.50}$$

with the corresponding phase angle

$$\phi_r \approx \arctan\left(-\frac{1}{\zeta}\right) \tag{1.51}$$

The amplitude- and phase resonances are assumed to be equal for systems with small damping. The magnification is thus $1/2\zeta$, and it is called the quality factor:

$$\frac{\hat{u}_r}{\hat{u}_0} = \frac{1}{2\zeta} = Q \tag{1.52}$$

This factor is a measure for the sharpness of the resonance peak or the *quality* of the resonator. In response diagrams when the shape of the amplitude over frequency is measured the half power bandwidth is used. This is the distance of the points where the amplitude is $1/\sqrt{2}$ of the peak value $\hat{u}_r$. Solving Equation (1.31) for $\hat{u}_r/\hat{u}_0 = 1/\sqrt{2}$ the frequencies of half power can be found:

$$\omega_{1/2} = (1 \pm \zeta)\omega_0 \tag{1.53}$$

and therefore

$$Q = \frac{1}{2\zeta} = \frac{\omega_0}{\omega_2 - \omega_1} \tag{1.54}$$

Obviously the decay time is also related to damping. If Equation (1.22) is considered we get

$$Q = \frac{\omega_0 \tau}{2} \tag{1.55}$$

We have presented several expressions that describe damping. Nevertheless, even more quantities for damping are used depending on the engineering discipline and will be shown in Section 1.2.5. Section 1.2.5.1 aims at sorting all those expressions and their relationships among each other.

1.2.5 Damping in Real Systems

Viscous damping is rare in real systems, it only exists if the surface that is connected to liquids moves so slow that no turbulent motion appears. Observation of experiments with damping normally doesn't show damping that increases with frequency as would be the case with viscous damping. Examples for damping processes are:

- structural or hysteretic damping
- coulomb or dry-friction damping
- velocity-squared or aerodynamic drag damping.

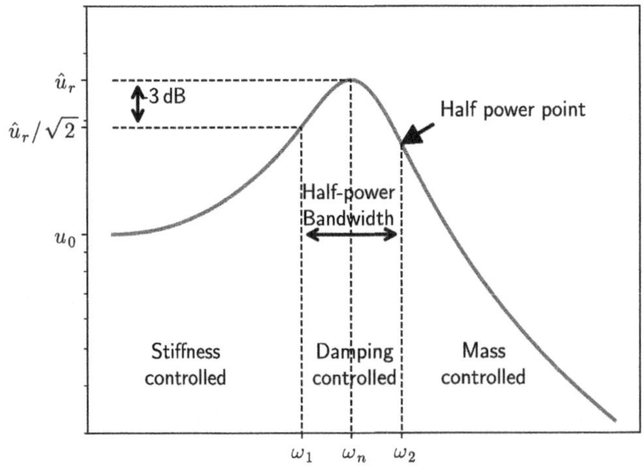

Figure 1.10 Half power bandwidth for harmonic oscillator. *Source:* Alexander Peiffer.

As most systems are lightly damped so that damping can be neglected except near resonance, it is sufficient to approximate the non-viscous damping in terms of equivalent viscous damping. A good way to formulate a criteria that works for all types of damping is to consider the dissipated energy per cycle of vibration. For viscous damping this reads as

$$\Delta E_{\text{cycle}} = \int_0^T c_v \dot{x} \frac{dx}{dt} dt = c_v \hat{u}^2 \omega^2 \int_0^{2\pi/\omega} sin^2(\omega t + \phi) dt = \pi c_v \omega \hat{u}^2$$

With the dissipated energy per cycle ΔE_{cycle} for any particular damping type the equivalent viscous damping $c_{v\text{eq}}$ can be determined from

$$\Delta E_{\text{cycle}} = \pi c_{v\text{eq}} \omega \hat{u}^2 \tag{1.56}$$

1.2.5.1 Hysteretic Damping

In many cases damping is caused by structural damping. If materials like aluminium or steel are cyclically stressed they form a hysteresis loop. Experimental observations show, that the energy dissipated per cycle is proportional to the square of the strain (displacement):

$$\Delta E_{\text{cycle}} = \alpha_x \hat{u}^2 \tag{1.57}$$

Comparing this with Equation (1.56) gives:

$$c_{v\text{eq}} = \frac{\alpha_x}{\pi \omega} \tag{1.58}$$

Entering this into the equation of motion in complex form (1.27)

$$-m\omega^2 \boldsymbol{u} e^{j\omega t} + j\frac{\alpha_x}{\pi\omega}\omega \boldsymbol{u} e^{j\omega t} + k_s \boldsymbol{u} e^{j\omega t} = \boldsymbol{F}_x e^{j\omega t} \tag{1.59}$$

and rewriting (1.59) leads to

$$-m\omega^2 \boldsymbol{u} e^{j\omega t} + k_s(1 + j\eta) \boldsymbol{u} e^{j\omega t} = \boldsymbol{F}_x e^{j\omega t} \tag{1.60}$$

with the structural loss factor

$$\eta = \frac{\alpha_x}{\pi k_s} \tag{1.61}$$

and

$$\boldsymbol{k}_s = k_s(1 + j\eta) \tag{1.62}$$

In this case the displacement response reads:

$$\boldsymbol{u} = \frac{\boldsymbol{F}_x}{k_s(1 + j\eta) - m\omega^2} = \frac{\boldsymbol{F}_x}{m(\omega_0^2(1 + \eta) - \omega^2)} \tag{1.63}$$

The structural loss factor is of great importance for structural dynamics not only because of the frequent occurrence in practical systems but also for numerical methods. The equation of motion remains similar to the equation for undamped systems. Additionally, the frequency of structurally damped systems does not change. This can be seen if the equivalent viscous damping of structurally damped systems is introduced in the solution for the magnification factor and the phase:

$$\frac{\hat{u}}{\hat{u}_0} = \frac{1}{\sqrt{[1-(\omega/\omega_0)^2]^2 + \eta^2}} \tag{1.64}$$

and

$$\phi_0 = \arctan\frac{-\eta}{1-(\omega/\omega_0)^2} \tag{1.65}$$

Amplitude and phase resonance occur at the same frequency ω_0. At resonance the viscously damped system amplification is $\hat{u}/\hat{u}_0 = 1/2\zeta$. Hence

$$\eta = 2\zeta = \frac{1}{Q} \tag{1.66}$$

There is a further interpretation of the loss factor. ΔE_{cycle} is the energy dissipated per cycle. The dissipated power given by $\Pi_{\text{diss}} = \Delta E_d/T = \Delta E_{\text{cycle}}\omega/2\pi$. Using equations (1.57) and (1.61) we get:

$$\Pi_{\text{diss}} = \eta\,\omega\frac{1}{2}k_s\hat{X}^2 = \eta\omega E \tag{1.67}$$

At the beginning of the cycle the total energy E is stored as potential energy in the spring. The dissipated power is a product of damping loss, frequency and the total energy of the system. This will be frequently used in the following sections, but particularly in Chapter 6 about statistical energy methods. The energy aspect leads to an equivalent definition of the loss factor:

$$\eta = \frac{1}{2\pi}\frac{\Delta E_{\text{cycle}}}{E} \tag{1.68}$$

Thus, the damping loss factor can be seen as the criteria, defining the relative loss of energy per cycle divided by 2π. In this zoo of damping criteria we have related c_v, ζ, c_{vc}, η, $\Delta\omega$, τ and Q. Table 1.1 summarizes those quantities and puts them in relation to the others.

In tools and software for vibroacoustic simulations many different quantities are used. The overview of all those different criteria shall help to avoid mistakes and confusion.

Table 1.1 Relation of important damping criteria.

Name	Symbol	c_v, c_{vc}	ζ	η	Q	$\Delta\omega$	τ
Viscous damping	c_v	1	ζc_{vc}				
Critical damping ratio	ζ	c_v/c_{vc}	1	$\eta/2$	$\frac{1}{2Q}$	$\Delta\omega/2\omega_0$	$1/\omega_0\tau$
Critical damping	c_{vc}	$\sqrt{4mk_s}$	$2\zeta m\omega_0$				
Damping loss	η	$2c_v/c_{\text{vc}}$	2ζ	1	$1/Q$	$\Delta\omega/\omega_0$	$2/\omega_0\tau$
Qualtity factor	Q	$\frac{c_{\text{cv}}}{2c_v}$	$\frac{1}{2\zeta}$	$1/\eta$	1	$\frac{\omega_0}{\Delta\omega}$	$\frac{\omega_0\tau}{2}$
3dB bandwidth	$\Delta\omega$		$2\zeta\omega_0$	$\eta\omega_0$	ω_0/Q	1	$\frac{1}{2\tau}$
Decay time	τ		$1/\zeta\omega_0$	$2/\omega_0\eta$	$2Q/\omega_0$	$2/\Delta\omega_0$	1

1.3 Two Degrees of Freedom Systems (2DOF)

The harmonic oscillator is also named as single degree of freedom (SDOF) system. Realistic systems consist of multiple degrees of freedom (MDOF). In order to keep things manageable we stay in a first step with two degrees of freedom. As for the SDOF case several phenomena can be treated exemplarily, especially the coupling effects. For presenting the idea of MDOF systems the start is done by two degrees of freedom (2DOF). The equations of motion for such a system as shown in Figure 1.11 are

$$m_1\ddot{u}_1 + c_{v1}\dot{u}_1 + k_{s1}u_1 + k_{sc}(u_1 - u_2) = F_{x1} \tag{1.69}$$

$$m_2\ddot{u}_2 + c_{v2}\dot{u}_1 + k_{s2}u_2 + k_{sc}(u_2 - u_1) = F_{x2} \tag{1.70}$$

By introducing harmonic motion for $u_1 = \boldsymbol{u}_1 e^{j\omega t}$ and $u_2 = \boldsymbol{u}_2 e^{j\omega t}$ we get

$$(-\omega^2 m_1 + j\omega c_{v1} + k_{s1} + k_{sc})\boldsymbol{u}_1 - k_{sc}\boldsymbol{u}_2 = \boldsymbol{F}_{x1} \tag{1.71}$$

$$-k_{sc}\boldsymbol{u}_1 + (-\omega^2 m_2 + j\omega c_{v2} + k_{s2} + k_{sc})\boldsymbol{u}_2 = \boldsymbol{F}_{x2} \tag{1.72}$$

neglecting the time dependence $e^{j\omega t}$. It is practical to write this in matrix form:

$$\begin{bmatrix} -\omega^2 m_1 + j\omega c_{v1} + k_{s1} + k_{sc} & -k_{sc} \\ -k_{sc} & -\omega^2 m_2 + j\omega c_{v2} + k_{s2} + k_{sc} \end{bmatrix} \begin{Bmatrix} \boldsymbol{u}_1 \\ \boldsymbol{u}_2 \end{Bmatrix} = \begin{Bmatrix} \boldsymbol{F}_{x1} \\ \boldsymbol{F}_{x2} \end{Bmatrix} \tag{1.73}$$

In the following, the square brackets and the curly brackets denote a coefficient matrix and vector, respectively.

1.3.1 Natural Frequencies of the 2DOF System

We start with a simplified system without damping and external forces in order to get the natural frequencies of the system.

$$\begin{bmatrix} -\omega^2 m_1 + k_{s1} + k_{sc} & -k_{sc} \\ -k_{sc} & -\omega^2 m_2 + k_{s2} + k_{sc} \end{bmatrix} \begin{Bmatrix} \boldsymbol{u}_1 \\ \boldsymbol{u}_2 \end{Bmatrix} = \begin{Bmatrix} 0 \\ 0 \end{Bmatrix} \tag{1.74}$$

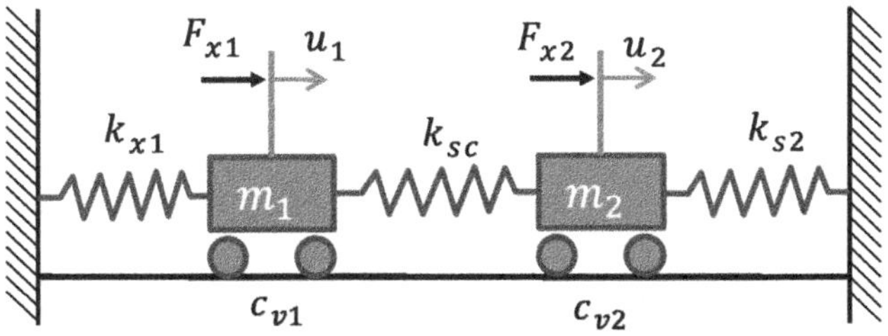

Figure 1.11 Two degrees of freedom system. *Source:* Alexander Peiffer.

This equation can be rearranged so that it corresponds to the general eigenvalue problem

$$\begin{bmatrix} k_{s1}+k_{sc} & -k_{sc} \\ -k_{sc} & k_{s2}+k_{sc} \end{bmatrix} \begin{Bmatrix} \boldsymbol{u}_1 \\ \boldsymbol{u}_2 \end{Bmatrix} - \omega^2 \begin{bmatrix} m_1 & 0 \\ 0 & m_2 \end{bmatrix} \begin{Bmatrix} \boldsymbol{u}_1 \\ \boldsymbol{u}_2 \end{Bmatrix} = \dots$$

$$[[K] - \omega^2 [M]]\{\boldsymbol{u}\} = \begin{Bmatrix} 0 \\ 0 \end{Bmatrix} \tag{1.75}$$

The non trivial solutions of this are given by:

$$\det\{[K] - \omega_n^2 [M]\} = 0 \tag{1.76}$$

This leads to the characteristic equation with $\lambda = \omega^2$

$$m_1 m_2 \lambda^2 - [k_{s1}m_2 + k_{s2}m_1 + k_{sc}(m_1+m_2)]\,\lambda + k_{s1}k_{s2} + (k_{s1}+k_{s2})k_{sc} = 0 \tag{1.77}$$

With $\omega_1^2 = k_{s1}/m_1$, $\omega_1^2 = k_{s1}/m_1$ and $\omega_c^2 = \frac{k_{sc}(m_1+m_2)}{m_1 m_2}$ the solutions are:

$$\omega_{n1/n2}^2 = \frac{\omega_1^2+\omega_2^2+\omega_c^2}{2} \pm \sqrt{\frac{(\omega_1^2+\omega_2^2+\omega_c^2)^2}{4} - \omega_1^2\omega_2^2 + \omega_c^2\frac{\omega_1^2 m_1 + \omega_2^2 m_2}{m_1+m_2}} \tag{1.78}$$

The eigenvalues shall be entered into the equations to solve for $\{\Psi_i\}$.

$$[[K] - \omega_{ni}^2 [M]]\{\Psi_i\} = \begin{Bmatrix} 0 \\ 0 \end{Bmatrix} \text{ with } i = 1,2 \tag{1.79}$$

leading to the surprisingly simple eigenvalues after some painful math

$$\{\Psi_i\} = \begin{Bmatrix} \frac{k_{s1}+k_{sc}-\omega_{ni}^2 m_1}{k_{sc}} \\ 1 \end{Bmatrix} \tag{1.80}$$

The results present the modes of the system or the shape of movement for this natural frequency. Let us simplify the above expression by additional conditions: $k_{s1} = k_{s2} = k_s$ and $m_1 = m_2 = m$.

So the modal frequencies read with $\omega_0^2 = k_s/m$

$$\omega_{n1/n2}^2 = \omega_0^2 + \frac{\omega_c^2}{2} \mp \frac{\omega_c^2}{2} \tag{1.81}$$

with the eigenvectors:

$$\Psi_1 = \begin{Bmatrix} 1 \\ 1 \end{Bmatrix} \quad \Psi_2 = \begin{Bmatrix} -1 \\ 1 \end{Bmatrix}$$

Thus, the first mode represents a uniform motion of both masses without any relative motion and thus no effect of the centre spring. The second mode is a symmetric resonance and the centre spring adds some extra stiffness leading to higher frequencies. See Figure 1.12 for both modes.

When we enter numerical figures with $m = 0.1$ kg, $k_s = 10$ N/m and $k_{sc} = 2$ N/m we get the modal frequencies $\omega_{n1} = 10.0\,\text{s}^{-1}(f_{n1} = 1.59\,\text{Hz})$ and $\omega_{n2} = 11.83\,\text{s}^{-1}(f_{n2} = 1.88\,\text{s}^{-1})$.

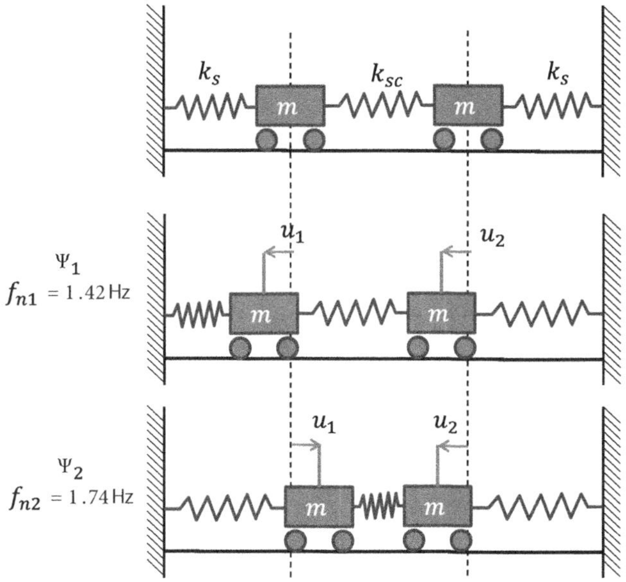

Figure 1.12 Mode shapes of the 2DOF example. *Source:* Alexander Peiffer.

1.3.1.1 Forced Vibration of the 2DOF System

If the 2DOF system is excited by a force or a combination of several forces, the system of equations (1.73) is solved for every frequency. Equation (1.73) can by written in a more generic way

$$[\boldsymbol{D}(\omega)]\{\boldsymbol{u}(\omega)\} = \{\boldsymbol{F}_x(\omega)\} \tag{1.82}$$

The solution can be written using the inverse of the stiffness matrix

$$\{\boldsymbol{u}(\omega)\} = [\boldsymbol{D}(\omega)]^{-1}\{\boldsymbol{F}_x(\omega)\} \tag{1.83}$$

This frequency response can be received for example by using numerical packages like MATLAB^{TM}, Python (with NumPy) for a set of frequencies. In Figure 1.13 the response of the system for unit excitation of $\boldsymbol{F}_{x1} = 1$ N and $\boldsymbol{F}_{x2} = 0$ N is shown.

1.3.1.2 Dynamic Vibration Absorber

The harmonic oscillator can be an anti-vibration device. This is called a tuned vibration absorber (TVA) or dynamic vibration absorber (DVA) and is a kind of multi-purpose tool whenever you have to combat resonance issues. It is a very useful device if vibration at a particular frequency must be reduced. Many applications are single frequency cases as for example propeller harmonics. In addition DVAs are used for reducing the resonance effects under broadband excitation. Usually real technical systems have multiple resonances but the principle can be shown with a SDOF system as master system. In Figure 1.14 such a setup is shown. The exciting force can be for example a rotating or vibrating machinery.

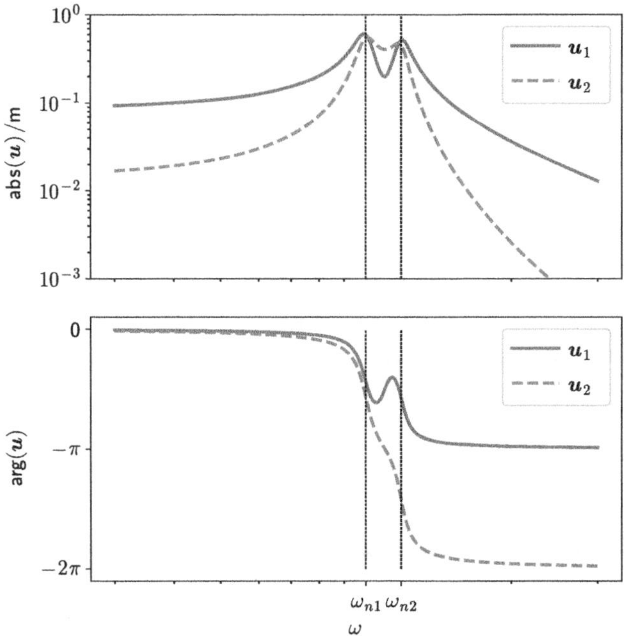

Figure 1.13 Magnitude and phase of response to unit force at mass 1. *Source:* Alexander Peiffer.

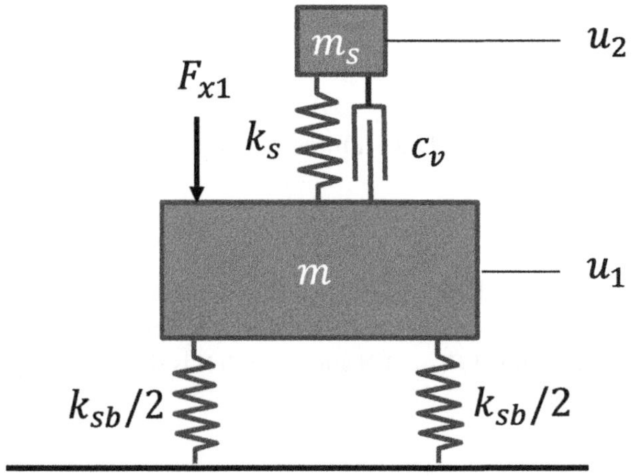

Figure 1.14 DVA mounted on resonant master system. *Source:* Alexander Peiffer.

The equation of motion is

$$\begin{bmatrix} -\omega^2 m + k_{sb} + k_s + j\omega c_v & -k_s - j\omega c_v \\ -k_s j\omega c_v & -\omega^2 m_s + j\omega c_v + k_s \end{bmatrix} \begin{Bmatrix} \boldsymbol{u}_1 \\ \boldsymbol{u}_2 \end{Bmatrix} = \begin{Bmatrix} \boldsymbol{F}_{x1} \\ 0 \end{Bmatrix} \tag{1.84}$$

with the following transfer function

$$\boldsymbol{u}_1 = -\frac{k\left(ic_v\omega + k_s - m_s\omega^2\right)}{\left(ic_v\omega + k_s\right)^2 - \left(ic_v\omega + k_s - m_s\omega^2\right)\left(ic_v\omega + k_{sb} + k_s - m\omega^2\right)}\boldsymbol{F}_{x1} \tag{1.85}$$

The result is non-dimensionalized by dividing through the static response $\boldsymbol{u}_1(0) = \boldsymbol{F}_{x1}/k_{sb}$.

Assuming zero damping gives the characteristic equation for the combined resonances

$$(k_s - \omega_i^2 m_s)(k_{sb} + k_s - \omega_i^2 m) - k_s^2 = 0 \tag{1.86}$$

With the resonance frequencies of each single system $\omega_0^2 = k_{sb}/m$, $\omega_s^2 = k_s/m_s$ and the mass ratio $\mu = m_s/m$ the resonance frequencies of the combined undamped system are given by

$$\omega_{1/2} = \frac{\omega_s^2(1+\mu) + \omega_0}{2} \mp \sqrt{\left(\frac{\omega_s^2(1+\mu) + \omega_0}{2}\right) + \omega_s^2\omega_0^2} \tag{1.87}$$

Figure 1.15 shows the result for a master system with $m = 0.1$ kg and $k_{sb} = 10$ N/m and an additional DVA tuned to the same frequency as the master system with $m_s = 0.02$ kg and $k_s = 2$ N/m. Several curves for different critical damping ζ are given. From the undamped case we learn that the response can theoretically be reduced to zero but implicating two resonances at different frequencies. With additional damping the response can be diminished for a broad frequency range. The design of the best DVA is an optimisation task depending on several constraints as discussed in detail by Harris and Crede (1976). In the optimisation procedures issues such as total mass, DVA mass displacement, linearity range of the spring, and the dynamic must be considered.

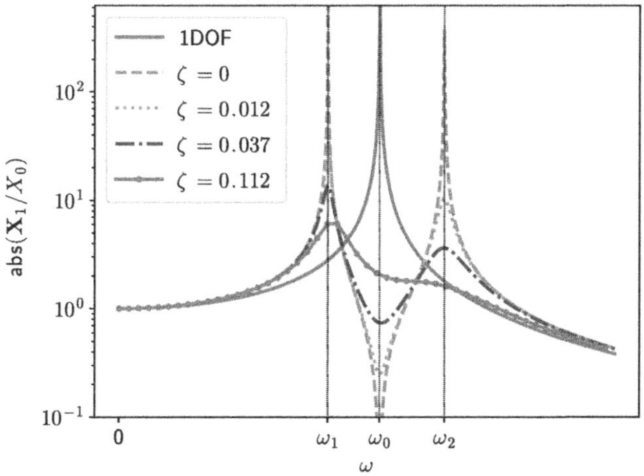

Figure 1.15 1DOF system with and without DVA. $m = 0.1$ kg, $m_s = 0.02$ kg, $k_{sb} = 10$ N/m, $k_s = 2$ N/m. *Source:* Alexander Peiffer.

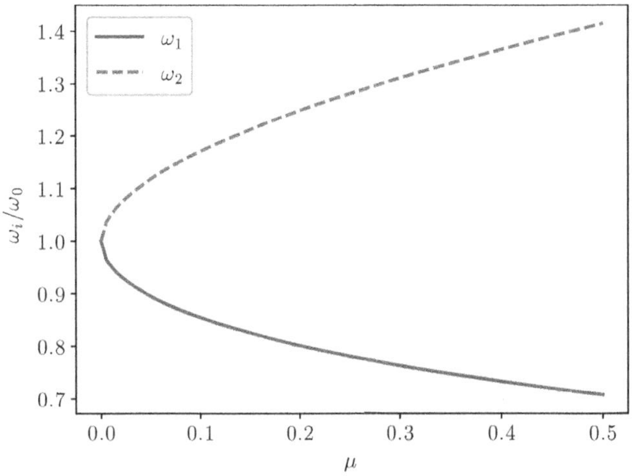

Figure 1.16 Frequency spread for DVA tuned to the same frequency depending on mass ratio. *Source:* Alexander Peiffer.

1.4 Multiple Degrees of Freedom Systems MDOF

The considerations above show the concept of how to write the equation of motion in matrix form. The matrices of the equation of motion in the frequency domain follow a certain convention in order to separate mass, stiffness and damping effects. As in Equation (1.73) the equation of motion of every discrete linear mechanical system can be approximated by the following form:

$$[M]\{\ddot{q}\} + [C]\{\dot{q}\} + [K]\{q\} = \{F(t)\} \tag{1.88}$$

or in the frequency domain as

$$[\boldsymbol{D}(\omega)]\{\boldsymbol{q}\} = (-\omega^2[M] + j\omega[C] + [K] + j[B])\{\boldsymbol{q}\} = \{\boldsymbol{F}\} \tag{1.89}$$

The coefficients $\boldsymbol{q}_i$ of $\{\boldsymbol{q}\}$ are generic displacement degrees of freedom, for example the displacements u, v, w in x-, y- and z-directions at different positions. The first matrix $[\boldsymbol{D}]$ is called the dynamic stiffness matrix. The matrices in the parentheses are called mass matrix, damping matrix, stiffness matrix and proportional damping matrix. The solution of equation 1.89 with regard to $\{\boldsymbol{q}\}$ is called frequency response.

The stiffness matrix must not be confused with the dynamic stiffness matrix. The dynamic stiffness matrix is frequency dependent and includes all matrices whereas the stiffness matrix includes the real and frequency independent stiffness part of the equation of motion. For specific lumped elements like springs and dampers there exist simple sub matrices that can be used to set up the global matrix of a more complex set-up. For illustration, see the example network of masses and springs in Figure 1.17.

We speak about *nodes* for the different locations in space running over index i. Each node may have different *degrees of freedom*(DOF). In our case, there are two translational coordinates u and v. So the displacement and force vector is running over all

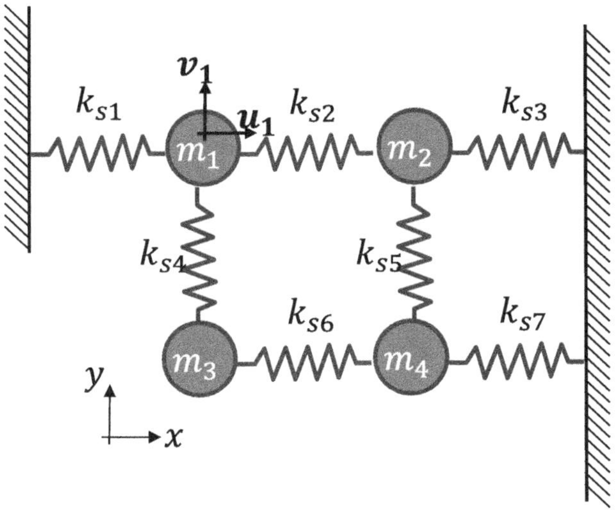

Figure 1.17 Multiple degrees of freedom network *Source:* Alexander Peiffer.

DOF of all N nodes:

$$\{\boldsymbol{q}\} = \begin{Bmatrix} \boldsymbol{u}_1 \\ \boldsymbol{v}_1 \\ \boldsymbol{u}_2 \\ \boldsymbol{v}_2 \\ \vdots \\ \boldsymbol{u}_N \\ \boldsymbol{v}_N \end{Bmatrix} \quad \{\boldsymbol{F}\} = \begin{Bmatrix} \boldsymbol{F}_{x1} \\ \boldsymbol{F}y_1 \\ \boldsymbol{F}_{x2} \\ \boldsymbol{F}y_2 \\ \vdots \\ \boldsymbol{F}_{xN} \\ \boldsymbol{F}y_N \end{Bmatrix} \tag{1.90}$$

A recipe can be used to assemble the equations of motion in the above given matrix form. For this purpose we derive a set of rules that allows for creating those matrices. This is practical for the understanding of complex mechanical networks but also a good start for understanding the basics of finite element simulation which is also based on mathematical methods to create mass and stiffness matrices from a discretized model. The concept of nodes and DOF is also kept in the finite element simulation.

1.4.1 Assembling the Mass Matrix

The mass matrix in multidimensional space follows from Newton's law of point masses in free space.

$$\{F\} = m\{\ddot{u}\} \tag{1.91}$$

Using the complex amplitude notation of harmonic motion this leads to:

$$\boldsymbol{F}_{ij} = -\omega^2 m_i \boldsymbol{q}_{ij} \text{ with } j = x, y, z \tag{1.92}$$

For every mass m_i at node i the *local* mass matrix is depending on the available degrees of freedom for each mass node. For a two-dimensional system with $\{\boldsymbol{q}\}_i = \{\boldsymbol{u}_i, \boldsymbol{v}_i\}^T$

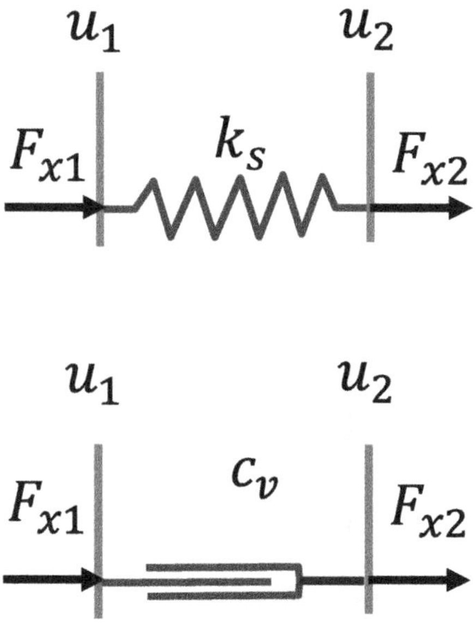

Figure 1.18 Spring and damper connection of two nodes. *Source:* Alexander Peiffer.

we get:

$$-\omega^2 \begin{bmatrix} m_i & 0 \\ 0 & m_i \end{bmatrix} \begin{Bmatrix} \boldsymbol{u}_i \\ \boldsymbol{v}_i \end{Bmatrix} = \begin{Bmatrix} \boldsymbol{F}_{xi} \\ \boldsymbol{F}_{yi} \end{Bmatrix} \tag{1.93}$$

The local matrices must by assembled for all degrees of freedom, so the system from Figure 1.17 will have the follwing mass matrix:

$$[M] = \begin{bmatrix} m_1 & 0 & \dots & & & & & \\ 0 & m_1 & 0 & \dots & & & & \\ & 0 & m_2 & 0 & & & & \\ & & 0 & m_2 & 0 & & & \\ & & & 0 & m_3 & 0 & & \\ & & & & 0 & m_3 & 0 & \\ & & & & & 0 & m_4 & 0 \\ & & & & & & 0 & m_4 \end{bmatrix} \tag{1.94}$$

1.4.2 Assembling the Stiffness Matrix

A similar procedure is applied to the springs and the related stiffness matrix. Every spring k_i connects different points in space, so there are at least two nodes involved, unless the spring is connected to a rigid wall. Inspecting the spring in Figure 1.18 the force balance is:

$$\begin{bmatrix} k_s & -k_s \\ -k_s & k_s \end{bmatrix} \begin{Bmatrix} \boldsymbol{u}_1 \\ \boldsymbol{u}_2 \end{Bmatrix} = \begin{Bmatrix} \boldsymbol{F}_{x1} \\ \boldsymbol{F}_{x2} \end{Bmatrix} \tag{1.95}$$

Springs connected to rigid walls are treated differently. In that case the DOF at the wall is constrained. Here for example $\boldsymbol{u}_1 = 0$, and we get for Equation (1.95)

$$\begin{Bmatrix} \boldsymbol{F}_{x1} \\ \boldsymbol{F}_{x2} \end{Bmatrix} = \begin{bmatrix} k_s & -k_s \\ k_s & k_s \end{bmatrix} \begin{Bmatrix} 0 \\ \boldsymbol{u}_2 \end{Bmatrix} = \begin{Bmatrix} -k_s \boldsymbol{u}_2 \\ k_s \boldsymbol{u}_2 \end{Bmatrix} \qquad \Rightarrow \boldsymbol{F}_{x1} = -\boldsymbol{F}_{x2} = -k_s \boldsymbol{u}_2 \tag{1.96}$$

If the spring is connected to the wall, only u_2 changes with applied forces, and the reaction force in the constraint is $-k_s u_2$. Practically that is the reaction force that must be provided by the wall to keep the node in place. In the finite element terminology these constraints are called *single point constraints* (SPC).

The local matrices must be distributed among the global stiffness matrix $[K]$. For better visibility this is done for springs 2 and 4. The place of the local matrix of spring 2 in the global stiffness matrix follows from the $\boldsymbol{u}_1$ and $\boldsymbol{u}_2$, the position of spring 4 from $\boldsymbol{v}_1$ and $\boldsymbol{v}_3$.

$$[K_2] = \begin{bmatrix} k_{s2} & 0 & -k_{s2} & 0 & & & & \\ 0 & 0 & 0 & 0 & & & & \\ -k_{s2} & 0 & k_{s2} & 0 & & & & \\ 0 & 0 & 0 & 0 & & & & \\ & & & & 0 & & & \\ & & & & & 0 & & \\ & & & & & & 0 & \\ & & & & & & & 0 \end{bmatrix} \quad [K_4] = \begin{bmatrix} 0 & 0 & & & 0 & 0 & & \\ 0 & k_{s4} & & & 0 & -k_{s4} & & \\ & & 0 & & & & & \\ & & & 0 & & & & \\ 0 & 0 & & & 0 & 0 & & \\ 0 & -k_{s4} & & & 0 & k_{s4} & & \\ & & & & & & 0 & \\ & & & & & & & 0 \end{bmatrix} \tag{1.97}$$

For the system from Figure 1.17 and putting all free and constraint springs together we get:

$$[K] = \begin{bmatrix} k_{s1}+k_{s2} & 0 & -k2 & 0 & & & & \\ 0 & k_{s4} & & & 0 & -k_{s1} & & \\ -k_{s2} & 0 & k_{s2}+k_{s3} & 0 & & & & \\ 0 & 0 & 0 & k_{s5} & & & & -k_{s5} \\ & & & & k_{s6} & & -k_{s6} & \\ 0 & -k_{s4} & & & & k_{s4} & & \\ & & & & -k_6 & & k_{s6}+k_{s7} & \\ & & & -k_{s5} & & & & k_{s5} \end{bmatrix} \tag{1.98}$$

With this procedure the matrix formulation of the equation of motion can be created. A similar approach can be used if other elements like dampers are involved. Generally the local elements can be everything that can be expressed by a dynamic stiffness matrix $[\boldsymbol{D}(\omega)]$ and can be added into a global matrix, independent from the fact if it comes from other models, simulation or test.

1.4.3 Power Input into MDOF Systems

The power introduced into the MDOF system is calculated according to (1.49).

$$\Pi = \frac{1}{2} Re\{\boldsymbol{v}^*\}\{\boldsymbol{F}\} = \frac{\omega}{2} Im\{\boldsymbol{q}^*\}\{\boldsymbol{F}\} = \frac{\omega}{2}\sum_i Im[\boldsymbol{q}_i^* \boldsymbol{F}_i] \tag{1.99}$$

The input power can be reconstructed from the dynamic stiffness matrix (1.89). We know that

$$\{\boldsymbol{F}\} = [\boldsymbol{D}]\{\boldsymbol{q}\} \tag{1.100}$$

hence

$$\Pi = \frac{\omega}{2}\sum_{i,j} Im[\boldsymbol{q}_i^* \boldsymbol{D}_{ij} \boldsymbol{q}_j] \tag{1.101}$$

or in matrix notation

$$\Pi = \frac{\omega}{2} Im\left[\{\boldsymbol{q}\}^H [\boldsymbol{D}]\{\boldsymbol{q}\}\right] \tag{1.102}$$

1.4.4 Normal Modes

Modes are natural shapes of vibration for a dynamic system. For a given excitation it would be of interest to see how well each mode is excited. In addition, these considerations lead to a coordinate transformation that simplifies the equations of motion.

We start with the discrete equation of motion in the frequency domain (1.89) and set the damping matrices $[C]$ and $[B]$ to zero

$$[[K] - \omega^2 [M]]\{\boldsymbol{q}\} = \{\boldsymbol{F}\} \tag{1.103}$$

Without external forces we get the equation for free vibrations, and we get the generalized eigenvalue problem

$$[[K] - \omega_n^2 [M]]\{\Psi_n\} = 0 \tag{1.104}$$

The non-trivial solutions of this are determined by zero determinants

$$\det\{[K] - \omega_n^2 [M]\} = 0 \tag{1.105}$$

providing the modal frequencies ω_n. Entering these frequencies and solving for Ψ_n provides the mode shapes of the dynamic system. These are the natural modes (shapes) of vibration that occur at the modal frequencies.

The mode shapes are orthogonal as can be derived by assuming two different solutions m, n

$$[[K] - \omega_m^2 [M]]\{\Psi_m\} = 0 \tag{1.106}$$

$$[[K] - \omega_n^2 [M]]\{\Psi_n\} = 0 \tag{1.107}$$

Multiplying (1.107) from the left with the transposed $\{\Psi_m\}^T$ gives

$$\{\Psi_m\}^T [[K] - \omega_n^2 [M]]\{\Psi_n\} = 0 \tag{1.108}$$

Transposing (1.106) and multiplying from the right with $\{\Psi\}_n$ reads as

$$\{\Psi_m\}^T \left[[K] - \omega_m^2 [M]\right]^T \{\Psi_n\} = 0 \tag{1.109}$$

The difference between (1.108) and (1.109) leads to

$$(\omega_n^2 - \omega_m^2)\{\Psi_m\}^T [M]\{\Psi_n\} = 0 \tag{1.110}$$

Since $\omega_n^2 \neq \omega_m^2$ this requires

$$\{\Psi_m\}^T [M]\{\Psi_n\} = 0 \text{ for } m \neq n \tag{1.111}$$

Using this in Equation (1.109) gives also

$$\{\Psi_m\}^T [K]\{\Psi_n\} = 0 \text{ for } m \neq n \tag{1.112}$$

Thus, the mode shapes are orthogonal to each other with respect to $[K]$ and $[M]$. For normalisation we multiply (1.106) from the left with $\{\Psi\}_m$

$$\{\Psi\}_m^T [K]\{\Psi\}_m = -\omega_m^2 \{\Psi\}_m^T [M]\{\Psi\}_m \; (m = 1, 2, \dots N) \tag{1.113}$$

and get

$$\{\Psi\}_n^T [M]\{\Psi\}_n = m_n \tag{1.114}$$

$$\{\Psi\}_n^T [K]\{\Psi\}_n = k_n \tag{1.115}$$

for the modal mass m_n and stiffness k_n with the following relation to the modal frequency

$$\omega_m^2 = \frac{k_m}{m_n} \quad (m = 1, 2, \dots N) \tag{1.116}$$

1.4.4.1 Equation of Motion in Modal Coordinates

The orthogonality of the mode shapes allows using them as a base for new coordinates that will simplify or *condense* the equation of motion. It is convenient to chose a normalisation with modal mass unity, thus $m_n = 1$, hence:

$$\{\Phi\}_n = \frac{1}{\sqrt{m_n}} \{\Psi\}_n \tag{1.117}$$

$\{\Phi\}_n$ is called the mass-normalized mode shape of the system. With Equation (1.116) and writing the mass normalized modes in matrix form

$$[\Phi] = \left[\{\Phi\}_1 \quad \{\Phi\}_1 \quad \cdots \quad \{\Phi\}_N\right] \tag{1.118}$$

it becomes clear that the normalisation from (1.114) and (1.115) now reads as follows:

$$[\Phi]^T [M][\Phi] = [I] = [M]' \tag{1.119}$$

$$[\Phi]^T [K][\Phi] = [K]' = \text{diag}(\omega_n^2) \tag{1.120}$$

$$\text{with diag}(\omega_n) = \begin{bmatrix} \omega_1^2 & 0 & \cdots & 0 \\ 0 & \omega_2^2 & \cdots & 0 \\ 0 & \cdots & \ddots & 0 \\ 0 & \cdots & 0 & \omega_N^2 \end{bmatrix} \tag{1.121}$$

The prime (′) denotes modal coordinates or matrices and $[I]$ is the unit matrix. Due to the orthogonality of the modes every solution $\boldsymbol{q}$ can be expressed as

$$\{\boldsymbol{q}\} = \sum_{n=1}^{N} \boldsymbol{q}'_n \{\Phi\}_n = [\Phi]\{\boldsymbol{q}\}' \qquad \text{with } \{\boldsymbol{q}\}' = \begin{Bmatrix} \boldsymbol{q}'_1 \\ \boldsymbol{q}'_2 \\ \vdots \\ \boldsymbol{q}'_N \end{Bmatrix} \tag{1.122}$$

and

$$[\Phi] = \begin{bmatrix} \Phi_1 & \Phi_2 & \cdots & \Phi_N \end{bmatrix} \tag{1.123}$$

The vector $\{\boldsymbol{q}'\}$ with components $\boldsymbol{q}'_n$ is the displacement in modal coordinates. Entering this into (1.103) and multiplication from the left with $[\Phi]^T$ provides

$$\left([\Phi]^T [K] [\Phi] - \omega^2 [\Phi]^T [M] [\Phi]\right) \{\boldsymbol{q}\}' = \left([K]' - \omega^2 [M]'\right) \{\boldsymbol{q}\}' = [\Phi]^T \{\boldsymbol{F}\} = \{\boldsymbol{F}\}' \tag{1.124}$$

$\boldsymbol{F}'_n$ are the components of the modal force vector $\boldsymbol{F}'$. Using the above defined orthogonality and normalisation we get:

$$\begin{bmatrix} \omega_1^2 - \omega^2 & & & \\ & \ddots & & \\ & & \omega_n^2 - \omega^2 & \\ & & & \omega_N^2 - \omega^2 \end{bmatrix} \begin{Bmatrix} \boldsymbol{q}'_1 \\ \boldsymbol{q}'_n \\ \vdots \\ \boldsymbol{q}'_N \end{Bmatrix} = \begin{Bmatrix} \boldsymbol{F}'_1 \\ \boldsymbol{F}'_n \\ \vdots \\ \boldsymbol{F}'_N \end{Bmatrix}' \tag{1.125}$$

or in another form:

$$\boldsymbol{q}'_n = \frac{\boldsymbol{F}'_n}{\omega_n^2 - \omega^2} \tag{1.126}$$

The response shape is reconstructed using (1.122). Thus, with modal coordinates expressed by independent degrees of freedom – the modes – the system decouples. This can by seen by the diagonal form of the matrix. For a system with N degrees of freedom, the modal solution is exact. The similarity to Equation (1.30) underlines that every mode can be interpreted as a single resonator. Thus, every dynamical system can be interpreted as a group of independent resonators of modal mass m_n and stiffness $\omega_n^2 m_n$ In accordance with Equation (1.63) we may introduce a modal damping η_n for every mode. In that case the response in modal coordinates is:

$$\boldsymbol{q}'_n = \frac{\boldsymbol{F}'_n}{\omega_n^2(1 + j\eta_n) - \omega^2} \tag{1.127}$$

In computational dynamics the solution of the homogeneous equation for normal modes and the calculation of the system response using the diagonal form is called *modal frequency response*. Finite element methods are frequently applying this method to simplify the solution and to *condense* the system equations to a reduced coordinate system. In addition modes are an excellent option to exchange model information or simulation results between different solvers. Usually, the modal base is truncated and not all modes are considered. In that case the modal solution is only an approximation of the direct solution. The denominator $\omega_n^2 - \omega^2$ reveals that the participation of every mode to the global solution decreases with the distance from the modal frequency. Thus, a reasonable set of modes *near* the frequency of interest can by sufficient.

Finally, we conclude that the change in the coordinate base to modal coordinates keeps the general format of the equation of motion.

$$\left([K]' - \omega^2 [M]'\right)\{q\}' = \{F\}' \tag{1.128}$$

1.5 Random Process

As clean harmonic signals are rare in technical systems we need a mathematical toolbox to deal with nonharmonic signals. Due to the Fourier theory (Nelson and Elliott, 1993) every deterministic time signal can be synthesized by a sum of harmonic components. This provides a theoretical link between periodic and harmonic but also transient signals. See appendix A.1 for a brief introduction.

The same theory can also be applied to stochastic or random signals but in a slightly different manner. Fourier analysis and methods for investigating the random processes and the description of mechanical systems by impulse response or frequency response functions is an important toolset for the description of vibroacoustic systems. In addition and due to digitalisation most signals or spectra are given as a discrete, digital set of values. This discrete formulation creates some pitfalls that may also lead to misinterpretations. Even though the signal analysis is not a major subject in vibroacoustic simulation it is a very important especially when acoustic experiments are performed.

Signals from technical processes are often not predictable and are generated randomly like pressure fluctuations in a turbulent flow, the impact of raindrops on a roof, or the stochastic combustion in a jet. For dealing with such signals the above described methods must be adapted. In addition we need formulations that allow for the definition of the statistics of random processes as there is no deterministic functionality between time or frequency and the physical quantity, for example force or displacement.

1.5.1 Probability Function

Imagine a random process creating signal sequences as shown in Figure 1.19. At each time the signal value $f(t)$ may be different and has a certain continuous value. One option to characterize this signal is to define the probability that the signal value is less or equal to a specific value f_k. Thus we define a probability for $f(t)$ to be less than or equal to f_k

$$P(f) = \text{Prob}\,[f \le f_k] \tag{1.129}$$

Next, we are interested in the probability that the value of $f(t)$ is in a range defined by $\Delta f = f_2 - f_1$ meaning the probability $\text{Prob}[f_1 < f \le f_2]$. Now we can define the probability density function as

$$p(f) = \lim_{\Delta f \to 0} \frac{\text{Prob}[f_1 < f \le f_2]]}{\Delta f} \tag{1.130}$$

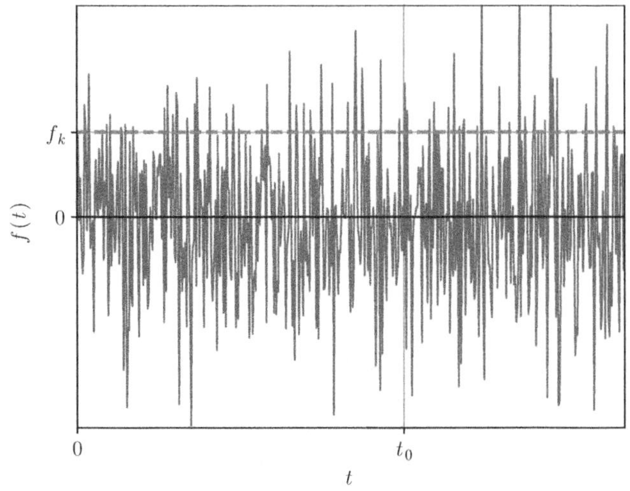

Figure 1.19 Stochastic fluctuation with time. *Source:* Alexander Peiffer.

Consequently in the limit $\Delta f \to 0$ the probability function density p is defined by

$$p(f) = \frac{dP(f)}{df} \tag{1.131}$$

On the other side we can reconstruct the probability to be in the range of f_1 to f_2 by integrating over the probability density function

$$\text{Prob}[f_1 < f \leq f_2] = \int_{f_1}^{f_2} p(f)df = P(f_2) - P(f_1) \tag{1.132}$$

In Figure 1.20 the examples for above defined functions are depicted. Those distinct ways of averaging reveal that the different averaging methods must be described in more detail. Until now averaging was performed over time intervals. This must not be confused with averaging over an ensemble. Ensemble averaging means averaging over an ensemble of experiments, systems, or even random signals. It will be denoted by $\langle \cdot \rangle_E$. Ensemble averaging can be similar to time averaging but this is only valid for specific time signals or random processes.

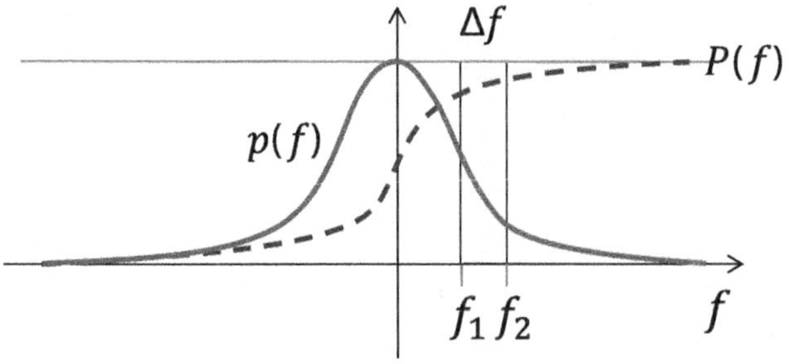

Figure 1.20 Probability and probability density function of a continuous random process. *Source:* Alexander Peiffer.

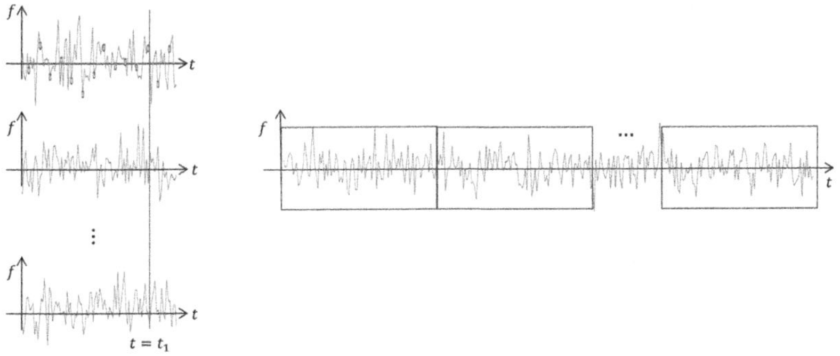

Figure 1.21 Ensemble and time averaging of signals from random processes. *Source:* Alexander Peiffer.

In Figure 1.21 the differences between time and ensemble averaging are shown. On the left hand side (ensemble) we perform a large set of experiments and take the value at the same time t_1, on the right hand side we perform one experiment but investigate sequent time intervals.

Consider now the mean value of an ensemble of N experiments. The mean value is defined by

$$\langle f\rangle_E = \frac{1}{N}\sum_{i=1}^{N} f_i \tag{1.133}$$

If we assume N_k discrete results f_k that occur with frequency $r_k = n_k/N$ we can also write

$$\langle f\rangle_E = \frac{1}{N_k}\sum_{k=1}^{N_k} f_k r_k$$

In a continuous form this can be expressed as $r_k = p(f_k)\Delta f_k$

$$\langle f\rangle_E = \sum_{k=1}^{N_k} f_k p(f_k)\Delta f_k \tag{1.134}$$

For $f_k \to f$ we get the definition of the *expected value* based on ensemble averaging and expressed as the integral over the probability density.

$$E[f] = \langle f\rangle_E = \int_{-\infty}^{\infty} f p(f) df \tag{1.135}$$

Similar to the expression for time signal rms-value, we define in addition the expected mean square value

$$E[f^2] = \langle f^2\rangle_E = \int_{-\infty}^{\infty} f^2 p(f) df \tag{1.136}$$

and the variance

$$\sigma_f^2 = E\left[(f - E[f])^2\right] = \int_{-\infty}^{\infty} (f - E[f])^2 p(f) df \tag{1.137}$$

We come back to the difference between ensemble and time averaging as shown in Figure 1.21. A process is called ergodic when the ensemble averaging can be replaced by time averaging, thus

$$E[f] = E[f(t)] = \lim_{T\to\infty} \frac{1}{T} \int_{-T/2}^{T/2} f(t)dt \tag{1.138}$$

$$E[f^2] = E[f^2(t)] = \lim_{T\to\infty} \frac{1}{T} \int_{-T/2}^{T/2} f^2(t)dt \tag{1.139}$$

We are usually not able to perform an experiment for an ensemble of similar but distinct experimental set-ups, but we can easily record the signals over a long time and take several separate time windows out of this signal.

1.5.2 Correlation Coefficient

Even more important than the key figures of one random process is the relationship between two different processes, the so-called correlation. It defines how much a random process is linearly linked to another process. Imagine two random processes $f(t)$ and $g(t)$. Without loss of generality we assume the mean values to be zero:

$$E[f] = E[g] = 0 \tag{1.140}$$

Let us consider that the process g is linked to f by a linear factor K:

$$g = Kf$$

The error or deviation of this assumption reads

$$\delta = g - Kf \tag{1.141}$$

or in terms of the mean square value

$$J = E[\delta^2] = E[(g - Kf)^2] \tag{1.142}$$

This can be rewritten as

$$J = E[g^2] + K^2E[f^2] - 2KE[fg] \tag{1.143}$$

This is a minimization problem and we search for the slope K that minimizes the sum of squared deviations J. This function has a quadratic dependence and can be rewritten as

$$J = K^2A + 2BK + C \tag{1.144}$$

where $A = E[f^2], B = -E[fg]$, and $C = E[g^2]$, with all three terms being real expected values from real random processes. Equation (1.143) is parabolic in shape, and the minimum is found by setting the first derivative with respect to K to zero.

$$\frac{dJ}{dK} = 2AK + 2B = 0 \tag{1.145}$$

Therefore, the point that minimizes J is given by $K_0 = -B/A$. In order to assure K_0 being a minimum we need $d^2J/dK^2 > 0$, meaning that A must be positive. This can be easily proven, as the expected value of the squared function $E[f^2]$ must be positive. If we substitute K_0 into Equation (1.143) we get the following relationships for K_0 and J_0

$$K_0 = -B/A = \frac{E[fg]}{E[f^2]} \tag{1.146}$$

$$J_0 = C - B^2/A = E[g^2] - \frac{E^2[fg]}{E[f]} \tag{1.147}$$

Using the definition of variances we can write J_0 in the case of zero mean processes in a non-dimensional form:

$$\frac{J_0}{\sigma_g^2} = 1 - \left(\frac{E^2[fg]}{\sigma_f \sigma_g}\right) = 1 - \rho_{fg} \tag{1.148}$$

The quantity $\rho_{fg} = E[fg]/\sigma_f\sigma_g$ is the normalized correlation coefficient between f and g. If both processes are perfectly correlated $\rho_{fg} = 1$. If they are fully uncorrelated $\rho_{fg} = 0$. In terms of the linear relationship from (1.141) all points would be perfectly on the line for full correlation and would be arbitrarily distributed for no correlation (Figure 1.22).

1.5.3 Correlation Functions for Random Time Signals

In the above considerations we have taken the values from an ensemble of random processes or signals taken at t_1. We can also define a correlation coefficient for values taken from two processes at different times t_1 and t_2:

$$\rho_{fg}(t_1, t_2) = \frac{E[f(t_1)g(t_2)]}{\sigma_f \sigma_g} \tag{1.149}$$

The numerator is called the *cross correlation function* :

$$R_{fg}(t_1, t_2) = E[f(t_1)g(t_2)]. \tag{1.150}$$

If the two processes are stationary the value of the cross correlation function depends only on the distance between the two times, i.e. $t_2 = t_1 + \tau$. τ is called the lag or

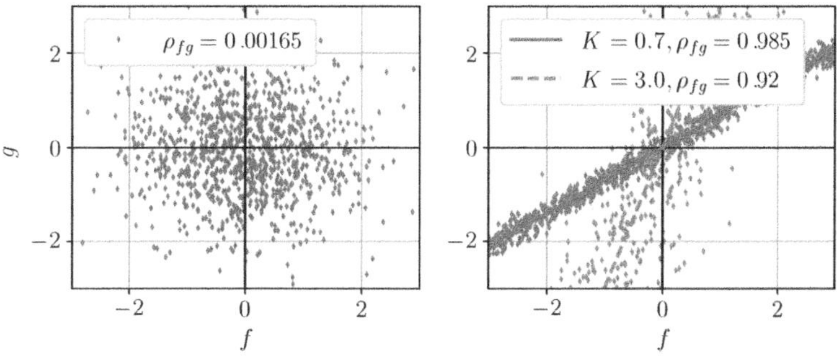

Figure 1.22 Example for correlation of random processes. No correlation (left) and different correlation values (right). *Source:* Alexander Peiffer.

separation between the two time samples and we can write:

$$R_{fg}(\tau) = E[f(t)g(t+\tau)] \tag{1.151}$$

It also makes sense to correlate the function $f(t)$ with itself at later moments $f(t+\tau)$. This is called the autocorrelation function defined by:

$$R_{ff}(\tau) = E[f(t)f(t+\tau)] \tag{1.152}$$

This function will later enable us to describe the spectrum of random functions. At $\tau = 0$ the value is known as variance of $f(t)$ as given by Equation (1.137):

$$R_{ff}(0) = E[f^2(t)] \tag{1.153}$$

The autocorrelation is symmetric in time, proven by:

$$R_{ff}(\tau) = E[f(t)f(t+\tau)] = E[f(t'-\tau)f(t')] = R_{ff}(-\tau) \tag{1.154}$$

In addition some useful properties can be derived for the cross correlation function

$$R_{fg}(\tau) = E[f(t)g(t+\tau)] = E[f(t'-\tau)g(t')] = R_{gf}(-\tau). \tag{1.155}$$

So we get finally

$$R_{fg}(\tau) = R_{gf}(-\tau) \tag{1.156}$$

For the stationary ergodic process we can replace the ensemble averaging by the average over time

$$R_{fg}(\tau) = \lim_{T\to\infty} \frac{1}{T} \int_{-T/2}^{T/2} f(t)g(t+\tau)dt \tag{1.157}$$

1.5.4 Fourier Analysis of Random Signals

The Fourier transform of a random signal would lead to infinite results because it is not approaching zero. Fourier series cannot be applied too, because there is no periodicity in the signal. A smart solution is to use the correlation function for the Fourier transform and not the random signal itself. The correlation is a decaying function that is suitable for infinite integration due to the $1/T$ factor in (1.157). We start with the pair of Fourier transforms of the autocorrelation:

$$S_{ff}(\omega) = \int_{-\infty}^{\infty} R_{ff}(\tau)e^{-j\omega\tau}d\tau \tag{1.158a}$$

$$R_{ff}(\tau) = \frac{1}{2\pi} \int_{-\infty}^{\infty} S_{ff}(\omega)e^{j\omega\tau}d\omega \tag{1.158b}$$

$S_{ff}(\omega)$ is called the *auto spectral density* of the signal $f(t)$. It is a measure of how the signal energy is distributed over the frequency range. This becomes quite obvious if we look at $\tau = 0$ and use (1.153):

$$R_{ff}(0) = \frac{1}{2\pi} \int_{-\infty}^{\infty} S_{ff}(\omega)d\omega = E[f^2(t)] \tag{1.159}$$

The autocorrelation is a real symmetric function, thus the auto spectrum as Fourier transform of this is a symmetric real valued function in frequency.

We define the cross spectral density function as the Fourier transform of the cross correlation function from Equation (1.151) and the back transformation:

$$\boldsymbol{S}_{fg}(\omega) = \int_{-\infty}^{\infty} R_{fg}e^{-j\omega\tau}d\tau \tag{1.160}$$

$$R_{fg}(\omega) = \frac{1}{2\pi}\int_{-\infty}^{\infty} \boldsymbol{S}_{fg}e^{j\omega\tau}d\tau. \tag{1.161}$$

Changing the integration constant $\tau = -\tau'$ and using symmetry (1.156) reveals

$$\boldsymbol{S}_{fg} = \boldsymbol{S}_{gf} \tag{1.162}$$

Mathematically, the expected value of an ergodic processes is derived by the investigation of infinite time periods. However, this can neither be realized in tests nor numerical simulations. We are restricted to specific time slots, so let us assume the Fourier transform of two random signals $f(t)$ and $g(t)$.

$$\boldsymbol{F}_k(\omega) = \int_0^T f_k(t)e^{-j\omega t}dt \qquad \boldsymbol{G}_k(\omega) = \int_0^T g_k(t)e^{-j\omega t}dt \tag{1.163}$$

It can be shown by a straightforward but lengthy proof by Bendat and Piersol (1980) that the cross spectral density is a complex function given by

$$\boldsymbol{S}_{fg}(\omega) = \lim_{T\to\infty} E\left[\frac{1}{T}\boldsymbol{F}_k^*(\omega)\boldsymbol{G}_k(\omega)\right] \tag{1.164}$$

Here, we take the signals of infinite duration from an ensemble of recordings. The same can be shown for the autospectrum.

$$S_{ff}(\omega) = \lim_{T\to\infty} E\left[\frac{1}{T}\boldsymbol{F}_k^*(\omega)\boldsymbol{F}_k(\omega)\right] \tag{1.165}$$

So these important relationships show, that if we have an infinite time period T and we average over a whole ensemble of such records we can determine the auto and cross spectra from this. In practical test and simulation situations we will see that for a given number of time signals of finite time T we can estimate the power- and cross spectral density. For convenience we abbreviate Equation (1.164)

$$\boldsymbol{S}_{fg}(\omega) = E\left[\boldsymbol{F}^*(\omega)\boldsymbol{G}(\omega)\right] \tag{1.166}$$

Note that the cross correlation $R_{fg}(\tau)$ of time signals $f(t)$ and $g(t)$ can be reconstructed from the inverse Fourier transform.

1.5.5 Estimation of Power and Cross Spectra

As shown in section 1.5.4 and Equation (1.164) the expected value of the Fourier transform of an ensemble of signals of infinite duration is given by:

$$\boldsymbol{S}_{fg}(\omega) = \lim_{T\to\infty} E\left[\frac{1}{T}F_k^*(\omega)G_k(\omega)\right]$$

It is helpful to convert this into an expression for a finite number of signals of finite duration. Usually, there is one signal of time length T available that is separated into

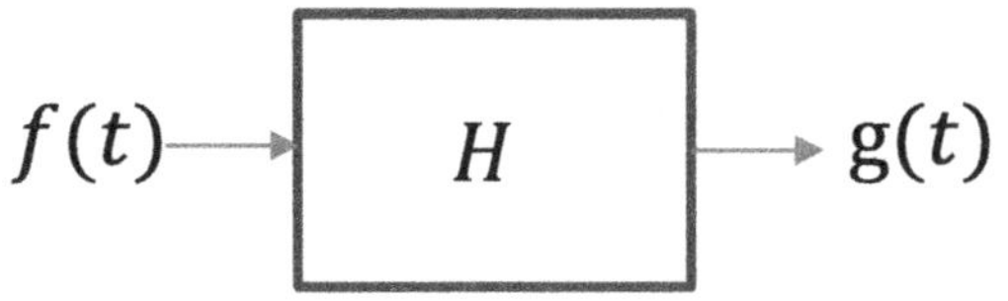

Figure 1.23 Sketch of a single-input–single-output system. *Source:* Alexander Peiffer.

M partitions of length T_w. So the estimate would be:

$$\hat{S}_{fg} = \frac{1}{M}\sum_{m=1}^{M} F_m^*(\omega)G_m(\omega) \tag{1.167}$$

In Bendat and Piersol (1980) it is shown that the relative error is given by:

$$\frac{E\left[[\hat{\boldsymbol{S}}_{fg}(\omega) - \boldsymbol{S}_{fg}(\omega)]^2\right]}{\boldsymbol{S}_{fg}^2(\omega)} = \frac{T_w}{T} = \frac{1}{M} \tag{1.168}$$

So for a given time length T the more we average, the more precise is the spectral estimate, but we sacrifice spectral resolution against statistical precision.

1.6 Systems

Any car, building, air plane or machine represents a dynamic system. This dynamic system is excited by source function $f(t)$, e.g. a force. The excitation is called the *input* of the system. This excitation leads to a specific response $g(t)$ called the output of the system. The simplest cases are systems with one input and output, e.g. the forced harmonic oscillator. Those systems are called single-input–single-output systems (SISO). Real systems are usually driven by multiple inputs or sources and have continuous or multiple responses. Those are called *multiple-input–multiple-output systems* (MIMO). Every system is described by its transfer function: i.e. a functional expression H that relates the inputs to the outputs.

$$g(t) = H[f(t)] \tag{1.169}$$

The practical advantage of such a formalism is that a complex realistic system is reduced to the quantities of interest. All intermediate steps of sound and vibration prediction are neglected. A practical example would be the force from the engine mount as input exciting the car system. A reasonable output could be the sound pressure at the drivers ear.

We restrict our considerations to linear systems. Thus, the response of the system to the sum of two signals $f_1(t)$ and $f_2(t)$ is given by the sum of each single response

$$H\left[f_1(t) + f_2(t)\right] = H[f_1(t)] + H[f_2(t)] \tag{1.170}$$

If the inputs are multiplied by constants a and b the responses scale linearly with the input

$$H\left[af_1(t) + bf_2(t)\right] = aH[f_1(t)] + bH[f_2(t)] \tag{1.171}$$

This is called the superposition principle.

1.6.1 SISO-System Response in Frequency Domain

The Equation (1.28) for the displacement response of the damped harmonic oscillator

$$\boldsymbol{u}(\omega) = \frac{1}{k_s - m\omega^2 + jc_v\omega}\boldsymbol{F}_x(\omega) \tag{1.172}$$

is an example for a system response in frequency domain. With $\boldsymbol{u}$ and $\boldsymbol{F}_s$ as complex amplitudes of harmonic signals we can express the system properties easily in the frequency domain

$$\boldsymbol{q}(\omega) = \boldsymbol{G}(\omega) = \boldsymbol{H}(\omega)\,\boldsymbol{F}(\omega) \text{ with } \boldsymbol{H}(\omega) = \frac{1}{k_s - m\omega^2 + jc_v\omega} \tag{1.173}$$

Thus, $\boldsymbol{H}(\omega)$ is the system transfer function in frequency domain. It is usually a complex function.

1.6.2 System Response in Time Domain

The product of Fourier transforms corresponds to a convolution in time domain (1.42), hence:

$$g(t) = \int_{-\infty}^{\infty} h(t-\tau)f(\tau)d\tau \tag{1.174}$$

or alternatively, as

$$g(t) = \int_{-\infty}^{\infty} h(\tau)f(t-\tau)d\tau. \tag{1.175}$$

A system is called causal (and this is definitely the case for any mechanical system) when the impulse response is zero at times smaller than zero (it can only react when it has experienced an excitation), so

$$h(t) = 0 \text{ for } t < 0 \tag{1.176}$$

That means we can rearrange the integration limits as follows:

$$h(t) = \int_{0}^{\infty} h(t-\tau)f(\tau)d\tau \tag{1.177}$$

All those forms are convolution integrals according to (1.46) and the response can be written as

$$g(t) = f(t) * h(t) = h(t) * f(t) \tag{1.178}$$

So, with this function we can express any response of a linear system using the above convolution integral. When considering a unit frequency excitation $\boldsymbol{F}(\omega) = 1$ Equation (1.173) shows that the response equals the transfer function.

$$\boldsymbol{G}(\omega) = \boldsymbol{H}(\omega) \cdot 1 \tag{1.179}$$

The inverse Fourier transform of the unit spectrum $e^{j\omega t_0} = 1$ for $t_0 = 0$ is the delta function $\delta(t)$ according to (1.32). In the time domain this reads as the following convolution integral

$$h(t) = \int_{-\infty}^{\infty} h(t-\tau)\delta(\tau)d\tau \tag{1.180}$$

We see that the transfer function in time domain is the response of the system to a delta function. Consequently, this function is called the impulse response of the system. When we calculate the inverse Fourier transform of the frequency response function of the damped harmonic oscillator we get as impulse response function

$$h(t) = \frac{1}{\sqrt{1-\zeta^2}\,\omega_n m} e^{-\zeta m \omega_n t} \sin\left(\sqrt{1-\zeta^2}\,\omega_n t\right) \tag{1.181}$$

This is the solution of the following equation of motion

$$m\ddot{u} + c_v\dot{u} + k_s u = \delta(t) \tag{1.182}$$

which corresponds to the solution of the homogeneous version in (1.1) for the following initial conditions

$$u(0) = 0 \qquad v_{x0} = \dot{u}(0) = 1/m \tag{1.183}$$

Finally, with the definition of system response functions in time and frequency domain, there is a powerful description available. This will be used in Section 1.6.3 in order to describe the system response to random signals.

1.6.3 Systems Excited by Random Signals

We know from section 1.5.4 that the random signal cannot be Fourier transformed. As a consequence we apply the correlation methods from above to the output of a system excited by random signals. We start with the autocorrelation of the system output g from excitation by random input f

$$\begin{aligned} R_{gg}(\tau) &= E\left[g(t)g(t+\tau)\right] \\ &= E\left[\int_{-\infty}^{\infty} h(\tau_1)f(t-\tau_1)d\tau_1 \int_{-\infty}^{\infty} h(\tau_2)f(t+\tau-\tau_2)d\tau_2\right] \end{aligned} \tag{1.184}$$

The impulse response can by taken out from the expected value operation, because it does not depend on the time. Hence we get

$$R_{gg}(\tau) = \int_{-\infty}^{\infty}\int_{-\infty}^{\infty} h(\tau_1)h(\tau_2)E\left[f(t-\tau_1)f(t+\tau-\tau_2)\right]d\tau_1 d\tau_2 \tag{1.185}$$

When we assume a retarded time argument of $\tau' = \tau + \tau_1 - \tau_2$ the expected value can be interpreted as the autocorrelation of $f(t)$. With this assumption we get:

$$R_{gg}(\tau) = \int_{-\infty}^{\infty} h(\tau_1)\left[\int_{-\infty}^{\infty} h(\tau_2)R_{ff}(t+\tau_1-\tau_2)d\tau_2\right]d\tau_1 \tag{1.186}$$

The term in the rectangular brackets can be seen as the convolution with argument $(\tau + \tau_1)$.

$$R_{gg}(\tau) = \int_{-\infty}^{\infty} h(\tau_1)\left[h(t+\tau_1) * R_{ff}(t+\tau_1)\right]d\tau_1 \tag{1.187}$$

By replacing τ_1 by $-u$ this reads as

$$R_{gg}(\tau) = \int_{-\infty}^{\infty} h(-u)\left[h(t-u) * R_{ff}(t-u)\right] du = h(\tau) * h(-\tau) * R_{ff}(\tau) \tag{1.188}$$

using that $R_{ff}(\tau)$ is symmetric in τ. This equation can be converted into frequency domain using the fact that time reversal in time corresponds to complex conjugate spectra:

$$S_{gg}(\omega) = \boldsymbol{H}(\omega)\boldsymbol{H}^*(\omega)S_{ff}(\omega) = |\boldsymbol{H}(\omega)|^2 S_{ff}(\omega) \tag{1.189}$$

So we know now the autospectrum of the system excited by random response. Next we investigate the cross correlation between input and output.

$$R_{fg} = E\left[f(t)g(t+\tau)\right] = E\left[f(t)\int_{-\infty}^{\infty} h(\tau_1)f(t+\tau-\tau_1)d\tau_1\right] \tag{1.190}$$

$$= \int_{-\infty}^{\infty} h(\tau_1)E\left[f(t)f(t+\tau-\tau_1)\right]d\tau_1 \tag{1.191}$$

This expression can be simplified by assuming the expected value to be the auto-convolution with argument $\tau - \tau_1$ to:

$$R_{fg} = \int_{-\infty}^{\infty} h(\tau_1)R_{ff}(\tau-\tau_1)d\tau_1 = h(\tau) * R_{ff}(\tau) \tag{1.192}$$

Converting this into the frequency domain gives:

$$\boldsymbol{S}_{fg}(\omega) = \boldsymbol{H}(\omega)S_{ff}(\omega) \tag{1.193}$$

This is a very important result: every transfer function (also using deterministic signals) can be determined by the ratio of the cross spectrum to auto spectrum.

$$\boldsymbol{H}(\omega) = \frac{\boldsymbol{S}_{fg}(\omega)}{S_{ff}(\omega)} \tag{1.194}$$

In principle this can also be done by using the Fourier transform (1.173), using time-limited excitation and dividing output and input FT, but the cross spectral variant is much more robust against measurement noise.

1.7 Multiple-input–multiple-output Systems

Mechanical set-ups with multiple degrees of freedom are multiple-input–multiple-output (MIMO) systems. The frequency response is determined by matrix inversion or solution of the matrix as shown in Equation (1.83)

$$\{\boldsymbol{q}(\omega)\} = \left[\boldsymbol{D}(\omega)\right]^{-1}\{\boldsymbol{F}(\omega)\}$$

In a more general form a MIMO system is defined as a system with N input signals $f_n(t)$ and M output signals $g_m(t)$.

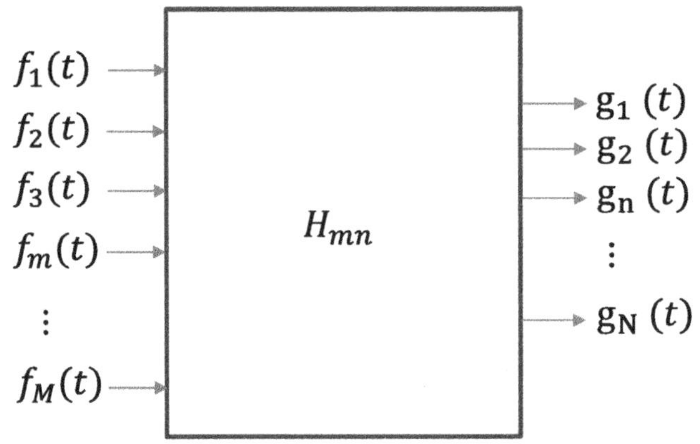

Figure 1.24 Multiple-input–multiple-output (MIMO) system. *Source:* Alexander Peiffer.

The linear system function is described by $M \times N$ functions $H_{mn}(\omega)$ in frequency domain or $h_{mn}(\tau)$ in time domain. The response of the mth output channel in time domain is given by

$$\begin{aligned} g_m(t) = \int_{-\infty}^{\infty} & h_{m1}(t-\tau)f_1(\tau) + h_{m2}(t-\tau)f_2(\tau) + \cdots \\ & \cdots h_{mn}(t-\tau)f_n(\tau) + h_{mN}(t-\tau)f_N(\tau)d\tau \\ & = h_{m1}(t) * f_1(t) + \cdots + h_{mn}(t) * f_n(t) + \cdots + h_{mN}(t) * f_N(t) \end{aligned} \tag{1.195}$$

In the frequency domain, in the response at output m the convolution is replaced by multiplication

$$\begin{aligned} \boldsymbol{G}_m(\omega) = {} & \boldsymbol{H}_{m1}(\omega)\boldsymbol{F}_1(\omega) + \boldsymbol{H}_{m2}(\omega)\boldsymbol{F}_2(\omega) + \cdots \\ & + \boldsymbol{H}_{mn}(\omega)\boldsymbol{F}_n(\omega) + \cdots + \boldsymbol{H}_{mN}(\omega)\boldsymbol{F}_N(\omega) \end{aligned} \tag{1.196}$$

Both equations can be written in matrix form. The frequency domain matrix reads as

$$\begin{Bmatrix} \boldsymbol{G}_1(\omega) \\ \vdots \\ \boldsymbol{G}_m(\omega) \\ \vdots \\ \boldsymbol{G}_M(\omega) \end{Bmatrix} = \begin{bmatrix} \boldsymbol{H}_{11}(\omega) & \cdots & \boldsymbol{H}_{1n}(\omega) & \cdots & \boldsymbol{H}_{1N} \\ \vdots & \ddots & \vdots & \ddots & \vdots \\ \boldsymbol{H}_{m1}(\omega) & \cdots & \boldsymbol{H}_{mn}(\omega) & \cdots & \boldsymbol{H}_{mN} \\ \vdots & \ddots & \vdots & \ddots & \vdots \\ \boldsymbol{H}_{M1}(\omega) & \cdots & \boldsymbol{H}_{Mn}(\omega) & \cdots & \boldsymbol{H}_{MN} \end{bmatrix} \begin{Bmatrix} \boldsymbol{F}_1(\omega) \\ \vdots \\ \boldsymbol{F}_n(\omega) \\ \vdots \\ \boldsymbol{F}_N(\omega) \end{Bmatrix} \tag{1.197}$$

or in short form

$$\{\boldsymbol{G}(\omega))\} = [\boldsymbol{H}(\omega)]\{\boldsymbol{F}(\omega)\} \tag{1.198}$$

The force excitation and displacement response matrix $[\boldsymbol{H}(\omega)]$ corresponds to the inverse dynamic stiffness matrix of Equation (1.83).

1.7.1 Multiple Random Inputs

The contemplations in section 1.6.3 dealt with a single random input of one system. The cross spectrum or correlation was only applied to input and output. Imagine a system that has many sources that have random nature. In the case of a car this could be the exhaust noise, the cooling ventilation, and the engine. The noise linked to the stroke movement will be correlated to the component that is transmitted via the muffler to exhaust, whereas the flow noise generated by the turbulent flow in the muffler is not correlated to the stroke harmonics. Thus, in practical vibroacoustic systems there is a mix of sources that are correlated to each other. Some are independent – thus uncorrelated – and some are a mix of both, as the discussed exhaust example.

Equations (1.196) and (1.195) can only be used in case of fully correlated input because each summand in those equations will have a different phase or time delay for each set of input signals taken from the ensemble of possible input or for each separate test.

Consider a set of N random input signals as given in Equation (1.195) or (1.196). For the description of the input we need the power spectral density or autocorrelation of all signals $f_n(t)$ given by

$$S_{ff,nn} = E\left[F_n^*(\omega)F_n(\omega)\right] \tag{1.199}$$

Here the index ff denotes that only the input is considered, and nn that it is the autocorrelation of the nth input. In addition the cross correlation between the input signals is given by:

$$S_{ff,mn} = E\left[F_m^*(\omega)F_n(\omega)\right] \tag{1.200}$$

This expression is called the cross spectral density matrix that looks in large form

$$[S_{ff}] = \begin{bmatrix} S_{ff,11}(\omega) & \cdots & S_{ff,1n}(\omega) & \cdots & S_{ff,1N} \\ \vdots & \ddots & \vdots & \ddots & \vdots \\ S_{ff,m1}(\omega) & \cdots & S_{ff,mn}(\omega) & \cdots & S_{ff,mN} \\ \vdots & \ddots & \vdots & \ddots & \vdots \\ S_{ff,N1}(\omega) & \cdots & S_{ff,Nn}(\omega) & \cdots & S_{ff,NN} \end{bmatrix} \tag{1.201}$$

This is a hermitian matrix due to the symmetry relationship of cross spectra also following from the expected value of each spectral product

$$S_{ff,nm} = E\left[F_n^*(\omega)F_m(\omega)\right] = E\left[F_n(\omega)F_m^*(\omega)\right]^* = S_{ff,mn}^* \tag{1.202}$$

For further considerations it is helpful to write the cross spectral matrix using the input spectra in vector form. Because of the matrix multiplication notation – row times column – the cross spectral density matrix can be written as

$$[S_{ff}] = E\left[\begin{Bmatrix} F_1^*(\omega) \\ F_2^*(\omega) \\ \vdots \\ F_N^*(\omega) \end{Bmatrix} \{F_1(\omega) \quad F_2(\omega) \quad \dots \quad F_N(\omega)\}\right] = E\left[\{F\}^* \{F\}^T\right] \tag{1.203}$$

In some literature on hybrid theory the complex conjugate of the cross spectral density matrix is used

$$[S_{ff}]^* = E\left[\begin{Bmatrix} F_1(\omega) \\ F_2(\omega) \\ \vdots \\ F_N(\omega) \end{Bmatrix} \{F_1^*(\omega), \quad F_2^*(\omega), \quad \dots, \quad F_N^*(\omega)\}\right] = E\left[\{F\}\{F\}^H\right] \quad (1.204)$$

It is helpful to understand how the matrix coefficients look for the following extreme cases:

- Fully uncorrelated signals
- Fully correlated signals.

The hermitian operator $\cdot^H$ is used, which combines the operations of complex conjugation and transposition.

1.7.1.1 Fully Uncorrelated Signals – Rain on the Roof Excitation

Fully uncorrelated input signals mean that the cross correlation is zero. A model of this is the "rain-on-the-roof" excitation because each drop falls fully independent from its brother drops on the roof. This means that the cross correlation matrix has only diagonal components. All off-diagonal components are zero.

1.7.1.2 Fully Correlated Signals

In this case the signals are correlated and have a clear phase relationship to a reference. Thus, all signals are linearly dependent to this reference. So every column of the cross spectral matrix can by derived by a linear combination of the other columns. So, we don't need the full matrix and Equation (1.198) can be applied.

1.7.2 Response of MIMO Systems to Random Load

When the input can only be defined by a cross spectral matrix $[S_{ff}]$, the same is true for the response $G_m(\omega)$.

$$[S_{gg}] = E\left[\{G\}^* \{G\}^T\right] \quad (1.205)$$

with

$$\{G^*(\omega)\} = [H^*(\omega)]\{F^*(\omega)\} \qquad \{G(\omega)\}^T = \{F(\omega)\}^T [H(\omega)]^T$$

and so

$$[S_{gg}] = E\left[[H^*]\{F^*\}([H]\{F\})^T\right] = E\left[[H^*]\{F^*\}\{F\}^T [H]^T\right] \quad (1.206)$$

The system matrix $\boldsymbol{H}$ can be removed from the expected value operator

$$[S_{gg}] = [H^*]E\left[\{F^*\}\{F\}^T\right][H]^T \quad (1.207)$$

and finally

$$[S_{gg}] = [H^*][S_{ff}][H]^T \quad (1.208)$$

This expression determines the cross spectral response of a linear MIMO system excited by random load. Typical numerical system equations have millions of degrees of freedom. Therefore, in many cases the above equation cannot by used straightforwardly, because this would lead to triple multiplication of Hermitian-but-huge matrices. Thus, for random excitation we need different approaches for the response calculation. The conjugate version of Equation (1.208) is

$$[\boldsymbol{S}_{gg}^{*}] = [\boldsymbol{H}]\,[\boldsymbol{S}_{ff}^{*}]\,[\boldsymbol{H}]^{H} \tag{1.209}$$

Bibliography

Julius S. Bendat and Allan G. Piersol. *Engineering Applications of Correlation and Spectral Analysis*. Wiley, New York, 1980. ISBN 978-0-471-05887-8.

Cyril M. Harris and Charles E Crede. *Shock and Vibration Handbook*. McGraw-Hill, New York, NY, U.S.A., second edition, 1976. ISBN 0-07-026799-5.

P. A. Nelson and S. J. Elliott. *Active Control of Sound*. Academic, London, 1993. ISBN 0-12-515426-7.

Bibliography

2

Waves in Fluids

2.1 Introduction

The acoustic wave motion is described by the equations of aerodynamics that are linearized because of the small fluctuations that occur in acoustic waves compared to the static state variables. The fluid motion is described generally by three equations:

- Continuity equation – conservation of mass
- Newton's law – conservation of momentum
- State law – pressure volume relationship.

For lumped systems the velocity is simply the time derivative of the point mass position in space. The same approach can be used in fluid dynamics, but here the continuous fluid is subdivided into several *cells* and their movement is described by trajectories. This is called the *Lagrange description* of fluid dynamics. Even if the equation of motions are simpler in that formulation it is quite complicated to follow all coordinates of fluid volumes in a complex flow. Thus, the *Euler description* of fluid dynamics is used. In this description the conservation equations are performed for a control volume that is fixed in space and the flow passes through this volume.

In this chapter, the three dimensional space is given by the Cartesian coordinates $\mathbf{x} = \{x, y, z\}^T$ and the velocity of the fluid $\mathbf{v} = \{v_x, v_y, v_z\}^T$.

2.2 Wave Equation for Fluids

2.2.1 Conservation of Mass

For simplicity we consider first the flow in the x-direction as in Figure 2.1. The mass flow balance contains the following quantities:

1. The elemental mass $m = \rho dV = \rho A$ with $A = dydz$.
2. Mass flow into the volume $(\rho v_x A)_x$.
3. Mass flow out of the volume $(\rho v_x A)_{x+dx}$.
4. Mass input from external sources $\dot{m}$.

Vibroacoustic Simulation: An Introduction to Statistical Energy Analysis and Hybrid Methods,
First Edition. Alexander Peiffer.

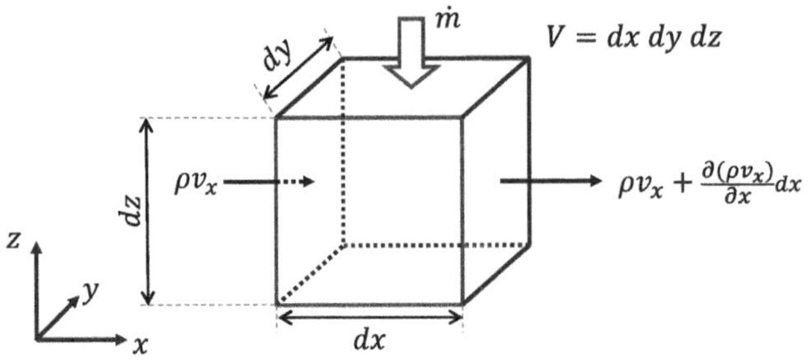

Figure 2.1 Mass flow in x-direction through control volume. *Source:* Alexander Peiffer.

leading to equation

$$\frac{\partial(\rho A dx)}{\partial t} = (\rho v_x A)_x - (\rho v_x A)_{x+dx} + \dot{m} \tag{2.1}$$

for mass conservation. Expanding the second term on the right hand side in a Taylor series gives

$$\frac{\partial(\rho A dx)}{\partial t} = \left[(\rho v_x A)_x - (\rho v_x A)_x - \frac{\partial(\rho v_x A)_x}{\partial x} dx\right] + \dot{m}$$

and finally

$$\frac{\partial \rho}{\partial t} + \frac{\partial(\rho v_x)}{\partial x} = \dot{\rho}_s \tag{2.2}$$

This one dimensional equation of mass conservation in x-direction can be extended to three dimensions:

$$\frac{\partial \rho}{\partial t} + \nabla(\rho \mathbf{v}) = \dot{\rho}_s \tag{2.3}$$

The second term of (2.3) may be confusing, but it says that the change of density is not only determined by a gradient in the velocity field but also by a gradient of the density.

2.2.2 Newton's law – Conservation of Momentum

The same procedure is applied to the momentum of the fluid. As shown in Figure 2.2 we get for flow in x-direction:

1. The momentum of the control volume is $\rho v_x dV = \rho v_x A dx$.
2. The momentum flow into the volume $(\rho v_x^2 A)_x$.
3. mass flow out of the volume $(\rho v_x^2 A)_{x+dx}$.
4. The force at position x is $(PA)_x$.
5. The force at position $x + dx$ is $(PA)_{x+dx}$.
6. External volume force density f_x.

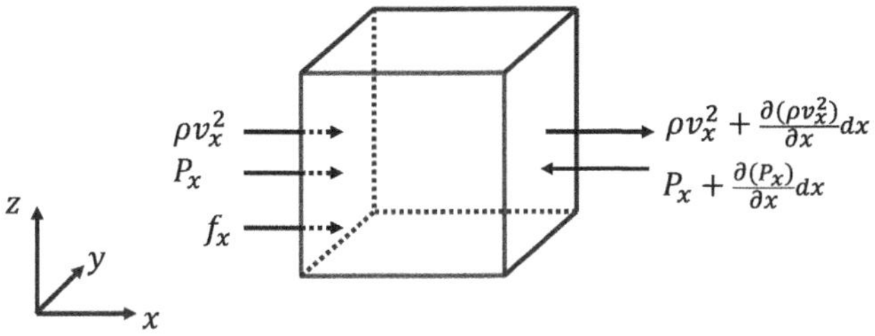

Figure 2.2 Momentum flow in x-direction through control volume.
Source: Alexander Peiffer.

Thus, the conservation of momentum in x reads

$$\frac{\partial(\rho v_x A dx)}{\partial t} = (\rho v_x^2 A)_x - (\rho v_x^2 A)_{x+dx} + (PA)_x - (PA)_{x+dx} + F_x \tag{2.4}$$

Using Taylor expansions for $(\rho v_x^2 A)_{x+dx}$ and $(PA)_{x+dx}$ gives

$$\frac{\partial(\rho v_x)}{\partial t} = -\frac{\partial(\rho u_x^2)}{\partial x} - \frac{\partial P}{\partial x} + f_x \tag{2.5}$$

Here, $f_x = F_x/(Adx)$ is the volume force density (force per volume). Using the chain law the partial derivatives of the first and second term lead to

$$\rho\frac{\partial v_x}{\partial t} + v_x\left\{\frac{\partial \rho}{\partial t} + v_x\frac{\partial \rho}{\partial x} + \rho\frac{\partial v_x}{\partial x}\right\} + \rho v_x\frac{\partial v_x}{\partial x} + \frac{\partial P}{\partial x} = f_x \tag{2.6}$$

The term in brackets is the homogeneous continuity Equation (2.2), and Equation (2.6) simplifies to

$$\rho\frac{\partial v_x}{\partial t} + \rho v_x\frac{\partial v_x}{\partial x} + \frac{\partial P}{\partial x} = f_x \tag{2.7}$$

As with the conservation of mass, this can be extended to three dimensions:

$$\rho\left\{\frac{\partial \mathbf{v}}{\partial t} + (\mathbf{v}\nabla)\mathbf{v}\right\} + \nabla P = \mathbf{f} \tag{2.8}$$

This equation is the non-linear, inviscid momentum equation called the Euler equation.

2.2.3 Equation of State

The above equations relate pressure, velocity and density. For further reducing this set we need a third equation. The easiest way would be to introduce the ideal gas law. Here we start with the first law of thermodynamics in order to show the difference between isotropic (or adiabatic) equation of state and other relationships.

$$dq = du + Pdv - dr \tag{2.9}$$

With the following specific quantities per unit mass

dq ... specific heat $q = q(T, \rho)$

dv ... specific volume $v = V/M$

Pdv ... specific expansion work

dr ... specific friction losses

With the specific entropy $ds = \frac{dq+dr}{T}$ we get:

$$
\begin{aligned}
ds &= \left(\frac{\partial u}{\partial T}\right)_T \frac{dT}{T} + \frac{P}{T} dV \\
&= \frac{c_v}{T} dT + \frac{P}{T} dV \\
&= \frac{c_v}{T} dT + \frac{P}{T} d\left(\frac{1}{\rho}\right) \\
&= \frac{c_v}{T} dT + \frac{P}{T\rho^2} d\rho
\end{aligned} \tag{2.10}
$$

The relation $dv = d(1/\rho)$ comes from the fact that v is a mass specific value and therefore the reciprocal of the density $\rho = 1/v$. For an ideal gas we have

$$P = \rho RT \text{ with } R = c_p - c_v \tag{2.11}$$

c_p and c_v are the specific thermal heat capacities for constant pressure and volume, respectively. That is the ratio of temperature change ∂T per increase of heat ∂q. From the total differential

$$dP(T, \rho) = \left(\frac{\partial p}{\partial T}\right)_\rho dT + \left(\frac{\partial p}{\partial \rho}\right)_T d\rho \tag{2.12}$$

we can derive

$$\frac{dT}{T} = \frac{dP}{P} - \frac{d\rho}{\rho} \tag{2.13}$$

Using all above relations the change in density $d\rho$ is:

$$d\rho = \frac{\rho}{\kappa P} dP - \frac{\rho}{c_p} ds \tag{2.14}$$

with $\kappa = c_v/c_p$. In most acoustic cases the process is isotropic: i.e. time scales are too short for heat exchange in a free gas; thus $ds = 0$, and the change of pressure per density is

$$\left(\frac{dP}{d\rho}\right)_s = \kappa \frac{P}{\rho} = c_0^2 \tag{2.15}$$

In case of constant temperature (isothermal) $dT = 0$ we get with (2.12) and the ideal gas law (2.11):

$$\left(\frac{dP}{d\rho}\right)_T = \frac{P}{\rho} = c_{0T}^2 \tag{2.16}$$

As we will later see, c_0 is the speed of sound. Newton calculated the wrong speed of sound based on the assumption of constant temperature that was later corrected by

Laplace by the conclusion that the process is adiabatic. For fluids and liquids like water a different quantity is used because there is no such expression as the ideal gas law. The bulk modulus is defined as:

$$K = \rho \left(\frac{\partial P}{\partial \rho} \right) \tag{2.17}$$

Due to (2.15) and (2.16) the relationship between the bulk modulus K and c_0 is:

$$c_0^2 = \frac{K}{\rho} \tag{2.18}$$

The bulk modulus can be defined for gases too, but we must distinguish between isothermal or adiabatic processes.

$$K_s = \rho \left(\frac{\partial P}{\partial \rho} \right)_s = \kappa P \qquad K_T = \rho \left(\frac{\partial P}{\partial \rho} \right)_T = P \tag{2.19}$$

2.2.4 Linearized Equations

Equations (2.3) and (2.8) can be linearized if small changes around a certain equilibrium are considered:

$$\rho = \rho_0 + \rho' \tag{2.20}$$

$$P = P_0 + p \tag{2.21}$$

$$\mathbf{v} = \mathbf{v}_0 + \mathbf{v}' \tag{2.22}$$

Inserting (2.22) into the equation of continuity (2.3), neglecting all second order terms as far as source terms, and setting[1] $\mathbf{v}_0 = 0$ the linear equation of continuity is:

$$\frac{\partial \rho'}{\partial t} + \rho_0 \nabla \mathbf{v}' = 0 \tag{2.23}$$

Doing the same for the equation of motion (2.8) leads to:

$$\rho_0 \frac{\partial \mathbf{v}'}{\partial t} + \nabla p = 0 \tag{2.24}$$

Using the curl ($\nabla \times$) of this equation it can be shown that the acoustic velocity $\mathbf{v}'$ can be expressed using a so-called velocity potential which will be useful for the calculation of some wave propagation phenomena.

$$\mathbf{v}' = \nabla \Phi \tag{2.25}$$

1 Keeping $v_0 \neq 0$ would lead to the convective wave equation that is used in the context of flow related acoustic problems, which is more a topic of aero acoustics

2.2.5 Acoustic Wave Equation

From the following operation

$$\frac{\partial}{\partial t}(2.23) - \nabla(2.24)$$

follows

$$\frac{\partial^2 \rho'}{\partial t^2} - \nabla^2 p = 0 \tag{2.26}$$

With the equation of state (2.15) for the density we get the linear wave equation for the acoustic pressure p

$$\frac{1}{c_0^2}\frac{\partial^2 p}{\partial t^2} - \nabla^2 p = 0 \tag{2.27}$$

Inserting the velocity $\mathbf{v}' = \nabla\Phi$ derived from the potential Φ into the linear equation of motion (2.24) provides the required relation between pressure and the velocity potential

$$\rho_0 \frac{\partial}{\partial t}\nabla\Phi + \nabla p = \nabla\left(\rho_0 \frac{\partial \Phi}{\partial t} + p\right) = 0 \tag{2.28}$$

Thus, the relationship between pressure p and the velocity potential Φ is

$$p = -\rho_0 \frac{\partial \Phi}{\partial t} \tag{2.29}$$

Entering this into the wave equation (2.27) and eliminating one time derivative gives:

$$\frac{1}{c_0^2}\frac{\partial^2 \Phi}{\partial t^2} - \nabla^2\Phi = 0 \tag{2.30}$$

The definition of the velocity potential (2.25) and equation (2.29) can be applied for the derivation of a relationship between acoustic velocity and pressure:

$$\mathbf{v}' = \nabla\Phi = -\frac{1}{\rho_0}\int \nabla p dt \tag{2.31}$$

2.3 Solutions of the Wave Equation

In acoustics we stay in most cases in the linear domain, so we change the notations from equations (2.20)–(2.22):

$$\mathbf{v}' \to \mathbf{v} \qquad \rho' \to \rho \tag{2.32}$$

Equations (2.27) and (2.30) define the mathematical law for the propagation of waves. For the explanation of basic concepts the wave equation is used in one dimensional form.

2.3.1 Harmonic Waves

According to D'Alambert every function of the form $p(x,t) = Af(x - c_0t) + Bg(x + c_0t)$ is a solution of the one-dimensional wave equation. In the following we will consider harmonic motion or waves so we replace the functions f and g by the exponential function with

$$f(x), g(x) = e^{j\omega x/c_0} = e^{jkx} \text{ with } k = \frac{\omega}{c_0} \tag{2.33}$$

and get

$$\boldsymbol{p}(x,t) = \boldsymbol{A}e^{j(\omega t - kx)} + \boldsymbol{B}e^{j(\omega t + kx)} \tag{2.34}$$

The first term of the right hand side of this equation is travelling in positive directions, the second in negative directions[2]. Harmonic waves are characterized by two quantities, the angular frequency ω and the wavenumber k. The first is the frequency (in time) as for the harmonic oscillator, and the second is a frequency in space. A similar relationship can be found between the time period T and the wavelength λ. Space and time domains are coupled by the sound velocity c_0 as shown in Table 2.1.

The time integration in Equation (2.31) corresponds to the factor $1/(j\omega)$ and reads in the frequency domain:

$$\mathbf{v} = -\frac{1}{j\omega\rho_0}\nabla\boldsymbol{p} \tag{2.35}$$

For one-dimensional waves in the x-direction this leads to:

$$\boldsymbol{v}_x = -\frac{1}{j\omega\rho_0}\frac{\partial\boldsymbol{p}}{\partial x} = \frac{1}{\rho_0 c_0}\left(\boldsymbol{A}e^{j(\omega t - kx)} - \boldsymbol{B}e^{j(\omega t + kx)}\right) \tag{2.36}$$

Depending on the wave orientation the ratio between pressure and velocity is given by:

$$\boldsymbol{v}_x = \pm\frac{1}{\rho_0 c_0}\boldsymbol{p} \tag{2.37}$$

In accordance with the impedance concept from section 1.2.3 we define the ratio of complex pressure and velocity as specific acoustic impedance $\boldsymbol{z}$

$$\boldsymbol{z} = \frac{\boldsymbol{p}}{\boldsymbol{v}} \tag{2.38}$$

also called acoustic impedance. For plane waves this leads to:

$$z_0 = \pm\rho_0 c_0 \tag{2.39}$$

Table 2.1 Quantities of wave propagation in time and space domains.

Name	Time		Space	
	Symbol	**Unit**	**Symbol**	**Unit**
Period	T	s	$\lambda = c_0 T$	m
Frequency	$f = \frac{1}{T}$	s^{-1}(Hz)	$(\cdot) = \frac{1}{\lambda}$	m^{-1}
Angular frequency	$\omega = 2\pi f = \frac{2\pi}{T}$	s^{-1}	$k = \frac{2\pi}{\lambda} = \omega/c_0$	m^{-1}

2 Negative k means propagation in positive x-direction, because of the $j\omega t$ convention. This is the reason why pure acoustic textbooks usually take the $-j\omega t$ convention in order to get positive wavenumber for positive propagation.

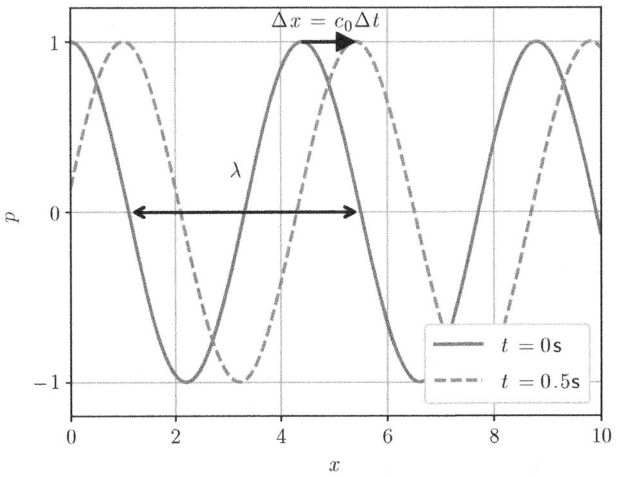

Figure 2.3 One-dimensional harmonic waves travelling in the positive x-direction ($c_0 = 2$m/s, $T = 2.2$s). *Source:* Alexander Peiffer.

$z_0 = \rho_0 c_0$ is called the characteristic acoustic impedance of the fluid. The specific acoustic impedance $\mathbf{z}$ is complex, because for waves that are not plane the velocity may be out of phase with the pressure. However, for plane waves the specific acoustic impedance is real and an important fluid property.

The above description of plane waves can be extended to three-dimensional space by introducing a wavenumber vector $\mathbf{k}$.

$$\boldsymbol{p} = Ae^{j(\omega t - \mathbf{kx})} \tag{2.40}$$

2.3.2 Helmholtz equation

Entering (2.40) into the wave Equation (2.27) provides

$$\left(\frac{1}{c_0^2}\frac{\partial^2}{\partial t^2} - \Delta\right)\boldsymbol{p}(\mathbf{x}, t) = -\left(\frac{\omega^2}{c_0^2} + \Delta\right)\boldsymbol{p}(\mathbf{x}, \omega)e^{j\omega t} = 0. \tag{2.41}$$

The $e^{j\omega t}$ term is often omitted and with $k = \omega / c_0$ we get the homogeneous Helmholtz equation.

$$\left(k^2 + \Delta\right)\boldsymbol{p}(\mathbf{x}, \omega) = 0 \tag{2.42}$$

2.3.3 Field Quantities: Sound Intensity, Energy Density and Sound Power

A sound wave carries a certain amount of energy that is moving with the speed of sound. We start with the instantaneous acoustic power Π:

$$\Pi(t) = \mathbf{Fv} \tag{2.43}$$

$\mathbf{F}$ is the force acting on a fluid particle and $\mathbf{v}$ the associated velocity. The acoustic intensity $\mathbf{I}$ is defined as the power per unit area $\mathbf{A} = A\,\mathbf{n}$ in the direction of the unit vector

n and with $\mathbf{F} = pA\mathbf{n}$ we get:

$$\mathbf{I}(t) = p\mathbf{v}\mathbf{n} \tag{2.44}$$

As in Equation (1.48) the time average is given by:

$$\langle \mathbf{I} \rangle_T = \frac{1}{T}\int_0^T p\mathbf{v}dt \Rightarrow \frac{1}{2}Re\left[\boldsymbol{p}\boldsymbol{v}_x^*\right] \tag{2.45}$$

Using the harmonic plane wave solutions for pressure (2.34) and velocity (2.36)

$$\begin{aligned}\boldsymbol{p}(x,t) &= \boldsymbol{A}e^{j(\omega t - kx)}\\ \boldsymbol{v}_x^*(x,t) &= \frac{\boldsymbol{A}^*}{\rho_0 c_0}e^{j(\omega t-kx)}\end{aligned}$$

the time averaged mean intensity yields:

$$\langle I\rangle_T = \frac{1}{T}\int_0^T \frac{|A|^2}{2\rho_0 c_0}Re\left[e^{j2(\omega t-kx)}\right] = \frac{1}{T}\int_0^T \frac{\hat{p}^2}{2\rho_0 c_0}\cos^2(\omega t - kx) \tag{2.46}$$

and finally:

$$\langle I\rangle_T = \frac{\hat{p}^2}{2\rho_0 c_0} = \frac{p_{\text{rms}}^2}{\rho_0 c_0} = \frac{1}{2}\rho_0 c_0 \hat{v}^2 \tag{2.47}$$

We see that the specific impedance $z_0 = \rho_0 c_0$ relates the intensity to the squared pressure.

The kinetic energy density e_{kin} in a control volume V_0 is written as

$$e_{\text{kin}} = \frac{E_{\text{kin}}}{V_0} = \frac{1}{2}\rho_0 v_{\text{x}}^2 = \frac{p^2}{2\rho_0 c_0^2} \tag{2.48}$$

The potential energy density e_{pot} follows from the adiabatic work integral as in equation (2.9)

$$e_{\text{pot}} = \frac{E_{\text{pot}}}{V_0} = -\frac{1}{V_0}\int_{V0}^{V} PdV \tag{2.49}$$

If we use Equation (2.15) we get the change in density as a start for the change in volume

$$d\rho = \frac{1}{c_0^2}dP$$

With unit mass M in the control volume V_0 it follows from $\rho = M/V_0$ that

$$dV = -\frac{V}{\rho_0}d\rho = -\frac{V}{\rho_0 c_0^2}dP$$

Finally we get:

$$e_{\text{pot}} = \frac{E_{\text{pot}}}{V_0} = \int_{P_0}^{P_0+p}\frac{P}{\rho_0 c_0^2}dP = \frac{p^2}{2\rho_0 c_0^2} \tag{2.50}$$

Pressure and velocity are in phase for plane waves; the same is true for the potential and kinetic energy density, so the total energy density is given by:

$$e(x,t) = e_{\text{kin}} + e_{\text{pot}} = \frac{p(x,t)^2}{\rho_0 c_0^2} \tag{2.51}$$

Using Equation (2.34) the time average over one period leads to:

$$\langle e \rangle_T = \frac{\hat{p}^2}{2\rho_0 c_0^2} = \frac{p_{\text{rms}}^2}{\rho_0 c_0^2} \tag{2.52}$$

Finally, we can see that the speed of sound relates energy density to the sound intensity.

$$e = \frac{I}{c_0} \tag{2.53}$$

All those above expressions are useful for the description and evaluation of sound fields. Especially in case of statistical methods that are based on the energy density of acoustic subsystems they link the wave fields to the energy in the systems and the power irradiating at the system boundaries.

If the intensity can be determined over a certain surface the source power is calculated by integrating the intensity component perpendicular to the surface

$$\Pi = \int_A \mathbf{I} d\mathbf{A} \tag{2.54}$$

2.3.4 Damping in Waves

There is no motion without damping, and a sound wave propagating over a long distance will vanish. This is considered by adding a damping component to the one-dimensional solution of the wave equation similar to the decay rate in (1.22)

$$\boldsymbol{p} = \boldsymbol{A} e^{-\alpha x} e^{j(\omega t - kx)} \tag{2.55}$$

Here α is the damping constant. There are several reasons for the attenuation of acoustic waves:

- Viscous damping due to inner viscosity.
- Thermal damping due to irreversible heat flow during wave propagation.
- Molecular damping due to excitation of degrees of freedom of molecules (for additional content of the gas, e.g. humidity in air).

The damping loss η as defined in (1.68) is based on the amount of energy dissipated during one cycle of wave motion. The harmonic pressure wave performs one cycle of oscillation in one period in time T or space λ. So we get for η:

$$\eta = \frac{1}{2\pi} \frac{\Delta E}{E} = \frac{1}{2\pi} \frac{\boldsymbol{A}^2 - \boldsymbol{A}^2 e^{-2\alpha\lambda}}{\boldsymbol{A}^2} \tag{2.56}$$

For small damping the exponential function can be approximated by $e^x \approx 1 + x - \ldots$ providing the relationship between damping loss and fluid wave attenuation.

$$\eta \approx \frac{1}{2\pi} 2\alpha\lambda \text{ with } \lambda = \frac{2\pi}{k} \tag{2.57}$$

Table 2.2 Field and energy properties of acoustic waves.

Quantity	Symbol	Formula	Units	Plane wave	Equation
Acoustic velocity	$\mathbf{v}$	$\frac{1}{j\omega c_0}\nabla \boldsymbol{p}$	m/s	$\frac{p}{\rho_0 c_0}$	(2.35)
Acoustic impedance	z	$\boldsymbol{p}/\mathbf{v}$	Pa s/m	$z_0 = \rho_0 c_0$	(2.38)
Intensity	$\mathbf{I}$	$\frac{1}{2}Re(\mathbf{pv}^*)$	Pa m/s	$\langle I \rangle_T = \frac{\hat{p}^2}{2\rho_0 c_0}$	(2.47)
Energy density	e		J/m^3	$\langle e \rangle_T = \frac{\hat{p}^2}{2\rho_0 c_0^2}$	(2.52)
Acoustic power	Π	$\Pi = \mathbf{IA}$	W	$\langle \Pi \rangle_T = \frac{A\hat{p}^2}{2\rho_0 c_0}$	(2.43)

Hence, the attenuation can be given by:

$$\alpha = \eta \frac{k}{2} \quad \eta = \frac{2\alpha}{k} \tag{2.58}$$

An appropriate way to consider this relationship in the solution of the wave equation is to include this into a complex wavenumber $\boldsymbol{k}$:

$$\boldsymbol{p} = \boldsymbol{A}e^{j(-\boldsymbol{k}x+\omega t)} = \boldsymbol{A}e^{-\frac{\eta x}{2}} e^{j(-kx+\omega t)} \text{ with } \boldsymbol{k} = k\left(1 - j\frac{\eta}{2}\right) \tag{2.59}$$

This complex wavenumber naturally impacts the speed of sound

$$\boldsymbol{c} = \frac{\omega}{\boldsymbol{k}} = \frac{c_0}{1 - j\frac{\eta}{2}} \tag{2.60}$$

and the acoustic impedance

$$\boldsymbol{z} = \rho_0 \boldsymbol{c} = \frac{z_0}{1 - j\frac{\eta}{2}} \tag{2.61}$$

The shown quantities of the plane wave field can also be applied in three-dimensional space and they are summarized in Table 2.2.

2.4 Fundamental Acoustic Sources

The radiation of sound is key to understanding how energy is introduced into wave fields. Depending on the wavelength, geometry, and dimension of the source the behavior varies. A detailed understanding of fundamental sources is helpful for the radiation of vibrating structures and thus, how they exchange acoustic energy.

2.4.1 Monopoles – Spherical Sources

The most simple geometry we might think of is a point in space. For simple derivation of the sound field of a point source the spherical coordinate system is introduced as

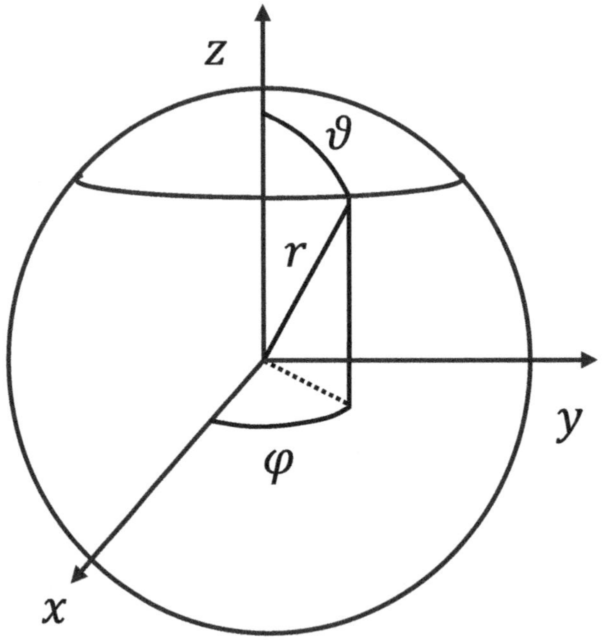

Figure 2.4 Definition of a spherical coordinate system. *Source:* Alexander Peiffer.

shown in Figure 2.4 and defined by the following coordinate transformation

$$x = r\sin\vartheta\cos\varphi \quad (2.62a)$$

$$y = r\sin\vartheta\sin\varphi \quad (2.62b)$$

$$z = r\cos\vartheta \quad (2.62c)$$

Using this coordinate system and neglecting the angular components the Laplace operator Δ reads as

$$\Delta = \frac{1}{r^2}\frac{\partial}{\partial r}\left(r^2\frac{\partial}{\partial r}\right) = \frac{2}{r}\frac{\partial}{\partial r} + \frac{\partial^2}{\partial r^2} \quad (2.63)$$

The wave equation for the velocity potential (2.30) becomes

$$\left(\frac{1}{c_0^2}\frac{\partial^2}{\partial t^2} - \frac{2}{r}\frac{\partial}{\partial r} - \frac{\partial^2}{\partial r^2}\right)\Phi = 0 \quad (2.64)$$

The two right terms can be written in a different form using $r\Phi$ as argument

$$\left(\frac{1}{c_0^2}\frac{\partial^2}{\partial t^2} - \frac{1}{r}\frac{\partial^2}{\partial r^2}\right)(r\Phi) = 0. \quad (2.65)$$

Equation (2.30) is the one-dimensional wave equation for the argument $r\Phi$, so we can use the D'Alambert solution

$$r\Phi(r,t) = f(r - c_0t) + g(r + c_0t) \quad (2.66)$$

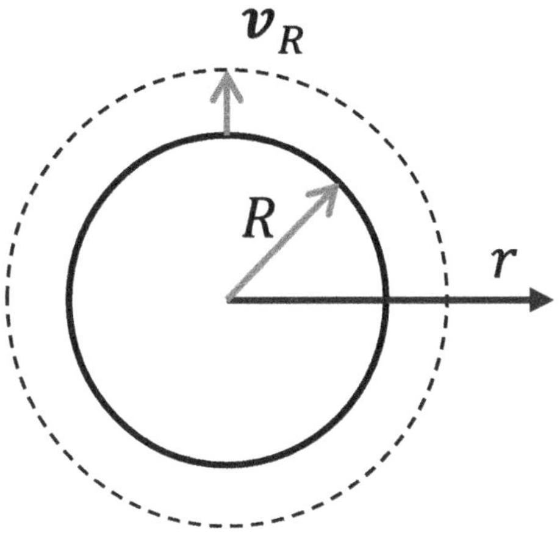

Figure 2.5 Breathing sphere as source model for a monopole. *Source:* Alexander Peiffer.

The first term represents an outgoing wave travelling away from the source, the second an incoming wave travelling to the source. As we are interested in sound being emitted from the source we consider the outgoing harmonic solution with complex amplitude $\boldsymbol{A}$

$$\Phi(r,t) = \frac{\boldsymbol{A}}{r} e^{j(\omega t - kr)} \tag{2.67}$$

Consider a pulsating sphere of radius R in the centre with normal surface velocity $\boldsymbol{v}_R$. With the velocity potential the radial velocity can be easily derived from the solution (2.67):

$$\frac{\partial \Phi}{\partial r} = v_r \tag{2.68}$$

Substituting Equation (2.67) into (2.68) and solving for $\boldsymbol{A}$ gives

$$\boldsymbol{A} = -\boldsymbol{v}_R \frac{R^2}{1 - jkR} e^{jka} \tag{2.69}$$

Hence,

$$\Phi(r,t) = -\frac{\boldsymbol{v}_R}{r} \frac{R^2}{1 - jkR} e^{j[\omega t - k(r-R)]} \tag{2.70}$$

The strength $\boldsymbol{Q}(t)$ of the source is defined by the volume flow rate. This is the surface of the sphere times normal velocity v_R

$$Q(t) = \dot{V} = 4\pi R^2 v_R(t) \tag{2.71}$$

With the harmonic source strength

$$\boldsymbol{Q}(t) = 4\pi R^2 \boldsymbol{v}_R e^{j\omega t} \qquad \boldsymbol{Q}(\omega) = 4\pi R^2 \boldsymbol{v}_R \tag{2.72}$$

the spherical wave solution is

$$\Phi(r,\omega) = -\frac{\boldsymbol{Q}(\omega)}{4\pi r} \left(\frac{1}{1 + jkR} \right) e^{-jk(r-R)} \tag{2.73}$$

Using equations (2.25) and (2.35) pressure and velocity are given by

$$\mathbf{p}(r,\omega) = -jk\rho_0 c_0 \Phi = \frac{\mathbf{Q}(\omega)}{4\pi r}\left(\frac{jk\rho_0 c_0}{1+jkR}\right)e^{-jk(r-R)} \tag{2.74}$$

and

$$\mathbf{v}_r(r,\omega) = -\left(\frac{1+jkr}{r}\right)\Phi = \frac{\mathbf{Q}(\omega)}{4\pi r^2}\left(\frac{1+jkr}{1+jkR}\right)e^{-jk(r-R)} \tag{2.75}$$

2.4.1.1 Field Properties of Spherical Waves

The acoustic impedance $\mathbf{z}$ is according to Equation (2.38)

$$\mathbf{z} = \frac{j\rho_0 c_0 kr}{1+jkr} = \rho_0 c_0 \left\{\frac{k^2r^2}{1+k^2r^2} + j\frac{kr}{1+k^2r^2}\right\} \tag{2.76}$$

In contrast to the plane wave, the specific acoustic impedance is not real. It contains a resistive and a reactive part. When the resistive part is dominant the pressure is in phase with the velocity. When the reactive part dominates, the velocity is out of phase to the pressure. The out of phase component does not generate any power in the sound field as it was the case for moving a mass or driving a spring. The motion is partly introduced into the local kinetic energy, and this part can be recovered as it is the case for an oscillating mass. For the acoustic field of a spherical source the reactive field represents the near-field fluid volume that is carried by the sphere motion but not emitting a wave.

There are two limit cases in Equation (2.76):

(i) $kr \ll 1$; the wave length λ is much larger than distance r.
(ii) $kr \gg 1$; the wave length λ is much smaller than distance r.

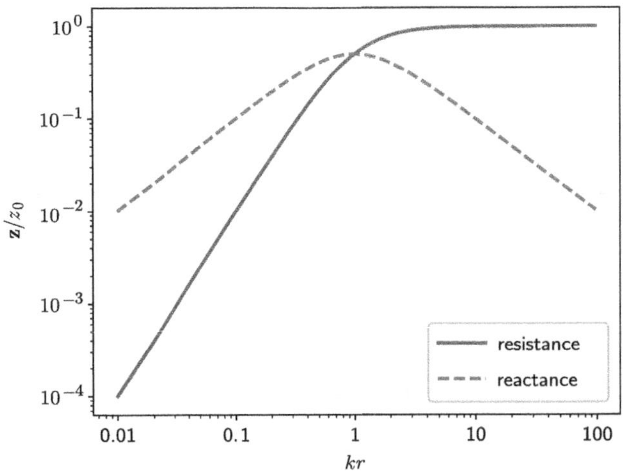

Figure 2.6 Reactance and resistance of specific acoustic impedance of a pulsating sphere. *Source:* Alexander Peiffer.

Introducing the above approximations into Equation (2.76) gives a fully reactive impedance for (i)

$$kr \ll 1 \Longrightarrow z = j\rho_0 c_0 kr = j\rho_0 r\omega \tag{2.77}$$

and a resistive part equal to plane waves for (ii)

$$kr >> 1 \Longrightarrow z = \rho_0 c_0 \tag{2.78}$$

2.4.1.2 Field Intensity, Power and Source Strength

The time averaged radiated intensity is

$$\langle I(r)\rangle_T = \frac{\hat{Q}^2 k^2 \rho_0 c}{32\pi^2 r^2(1+k^2R^2)} = \frac{Q_{rms}^2 k^2 \rho_0 c}{16\pi^2 r^2(1+k^2R^2)} \tag{2.79}$$

The total radiated power can now be evaluated from Equation (2.54) and the integration surface $4\pi r^2$

$$\langle \Pi\rangle_T = 4\pi r^2 \langle I(r)\rangle_T = \frac{Q_{rms}^2 k^2 \rho_0 c_0}{4\pi(1+k^2R^2)} \tag{2.80}$$

The mean square pressure can be derived from (2.74) and expressed by the intensity using (2.45)

$$p_{\text{rms}}^2 = \frac{1}{2} Re\left[\boldsymbol{p}\boldsymbol{p}^*\right] = \frac{Q_{rms}^2 k^2(\rho_0 c_0)^2}{16\pi^2 r^2(1+k^2R^2)} = \langle I(r)\rangle_T \, \rho_0 c_0. \tag{2.81}$$

Replacing the intensity in (2.81) gives the rms pressure in the spherical sound field due to power

$$p_{\text{rms}}^2 = \frac{\rho_0 c_0}{4\pi r^2} \langle \Pi\rangle_T \tag{2.82}$$

2.4.1.3 Power and Radiation Impedance at the Surface Sphere

The characteristic impedance of the sphere exactly at the surface at radius R can be translated into the *radiation impedance* of the sphere as a volume source. The radiation impedance is defined as the ratio of pressure to source strength at the vibrating surface

$$\boldsymbol{Z}_a = \frac{\boldsymbol{p}_{\text{surf}}}{\boldsymbol{Q}_{\text{surf}}} = \frac{\boldsymbol{p}_{\text{surf}}}{A_s \boldsymbol{v}_{\text{surf}}} \qquad A_s = 4\pi R^2 \tag{2.83}$$

If we assume a constant harmonic surface velocity $\boldsymbol{v}_R$ we get for the radiation impedance of the breathing sphere and according to the acoustic impedance (2.76)

$$\boldsymbol{Z}_a = \frac{j\rho_0 c_0 kR}{4\pi R^2(1+jkR)} \tag{2.84}$$

The acoustic radiation impedance is the ratio pressure and normal velocity at the sphere's surface

$$\boldsymbol{z}_a = 4\pi a^2 \boldsymbol{Z}_a = \frac{j\rho_0 c_0 kR}{1+jkR} \tag{2.85}$$

We can now use this impedance to eliminate either $\boldsymbol{p}$ or $\boldsymbol{v}_r$. The power transmitted by a vibrating sphere using Equation (2.54) over the surface of the sphere

$$\Pi = \frac{1}{2} A_s Re\left[\boldsymbol{p}\boldsymbol{v}_r^*\right] = \frac{1}{2} Re\left(\boldsymbol{p}\boldsymbol{Q}^*\right) \tag{2.86}$$

$$\Pi = \frac{1}{2}Re(\boldsymbol{p}\boldsymbol{Q}^*) = \frac{1}{2}Re(\boldsymbol{Z}_a\boldsymbol{Q}\boldsymbol{Q}^*) = Re(\boldsymbol{Z}_a)Q_{rms}^2 \tag{2.87}$$

or for constant source pressure

$$\Pi = \frac{1}{2}Re(\boldsymbol{p}\boldsymbol{Q}^*) = \frac{1}{2}Re\left(\boldsymbol{p}\boldsymbol{p}^*\frac{1}{\boldsymbol{Z}_a^*}\right) = Re\left(\frac{1}{\boldsymbol{Z}_a}\right)p_{rms}^2 \tag{2.88}$$

It is instructive to see the mechanical properties considering the limit cases from above and extract the mass that is moved by the surface. From Newtons's law a force given by $F = 4\pi R^2 p$ leads to an in-phase acceleration of $j\omega v_r$ of a mass m

$$\boldsymbol{F} = \boldsymbol{p}4\pi R^2 = mj\omega\boldsymbol{v}_r$$

hence

$$m = Re(\frac{1}{j\omega}\frac{\boldsymbol{p}}{\boldsymbol{Q}}) = \frac{1}{\omega}Im(\boldsymbol{Z}_a) \tag{2.89}$$

For $kR \ll 1$ we get: $m = 4\pi R^3\rho_0 = 3V_{\text{sph}}\rho_0$. Thus, at low frequencies the source surface motion carries three times the fluid volume of the sphere. This motion near the source is called an evanescent wave, because it is oscillatory motion of fluid that does not radiate.

2.4.1.4 Point Sources

A point source is a spherical source with an infinitely small radius. Performing the limit $kR \to 0$ for Equation (2.73) leads to the velocity potential for point sources of strength $\boldsymbol{Q}$

$$\Phi(r,\omega) = -\frac{\boldsymbol{Q}(\omega)}{4\pi r}e^{-jkr} \tag{2.90}$$

The pressure and velocity field of such a source is given by

$$\boldsymbol{p}(r,\omega) = -jk\rho_0 c_0\Phi = \frac{jk\rho_0 c_0\boldsymbol{Q}(\omega)}{4\pi r}e^{-jkr} \tag{2.91}$$

and

$$\boldsymbol{v}_r(r,\omega) = \frac{\partial\Phi}{\partial r}\Phi = \frac{\boldsymbol{Q}(\omega)}{4\pi r}jk\left(1+\frac{1}{jkr}\right)e^{-jkr} \overset{r\gg k}{=} \frac{jk\boldsymbol{Q}}{4\pi r}e^{-jkr} \tag{2.92}$$

All other relations regarding power and intensity expressions remain. We see that the limit is expressed for kR and not for the wavelength. The reason is that it is the ratio of a characteristic length (in this case the sphere radius) to the wavelength that determines if the geometrical details must be considered or not. In other words, a wave of a certain wavelength doesn't care about details that are much smaller.

With the D'Alambert solution for spherical waves (2.66) we can also derive a point source in time domain

$$\Psi(r,t) = -\frac{Q(t-c_0/r)}{4\pi r} \tag{2.93}$$

The point source is of great importance for the solution of the inhomogeneous wave equation in combination with complex boundary conditions. Any source can be reconstructed by a superposition of point sources as shown in Section 2.7.

Performing the limit process with $kR \to 0$ and taking the power from equation 2.86 we get the intensity of the point source:

$$\langle I(r)\rangle_T = \frac{Q_{\mathrm{rms}}^2 k^2 \rho_0 c}{16\pi^2 r^2} \tag{2.94}$$

and the total radiated power

$$\langle \Pi \rangle_T = \frac{Q_{\mathrm{rms}}^2 k^2 \rho_0 c_0}{4\pi} \tag{2.95}$$

with radiation impedance following from this

$$Z_a = \frac{k^2 \rho_0 c_0}{4\pi} \tag{2.96}$$

2.5 Reflection of Plane Waves

A plane wave striking a plane surface is a first example of interaction with obstacles. Imagine a configuration as shown in Figure 2.7. The impedance of the surface is $\boldsymbol{z}_2$, and it is given by using the velocity $\boldsymbol{v}_z$ perpendicular to the plane.

Without loss of generality the wave front is parallel to the y-axis and all properties are functions of x and z. The solution in the half space of $z > 0$ is the superposition of two plane waves.

$$\Phi_1(r) = \Phi e^{-j\mathbf{kr}} + \Phi^{(R)} e^{-j\mathbf{k}^{(R)}\mathbf{r}} \tag{2.97}$$

With the following arguments of the exponential function

$$\mathbf{kr} = k(\sin\vartheta x - \cos\vartheta z) \tag{2.98}$$

$$\mathbf{k}^{(R)}\mathbf{r} = k(\sin\vartheta^{(R)} x + \cos\vartheta^{(R)} z) \tag{2.99}$$

The pressure at the surface $z = 0$ is given by

$$\boldsymbol{p}(x, z=0) = j\omega\rho_0\Phi_1 = j\omega\rho_0\left(\Phi e^{-jk\sin\vartheta x} + \Phi^{(R)} e^{-jk\sin\vartheta^{(R)} x}\right) \tag{2.100}$$

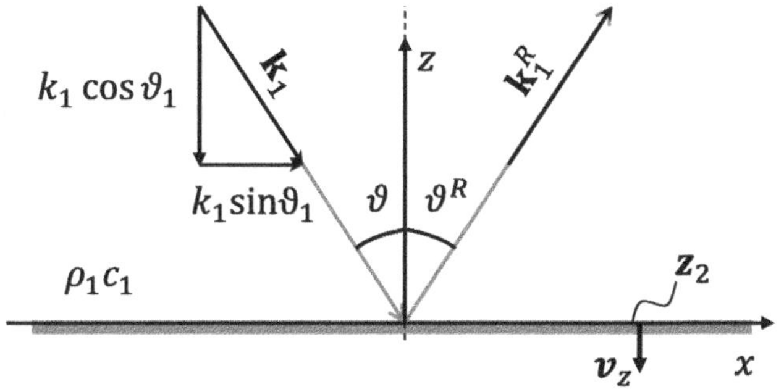

Figure 2.7 Reflection of a plane wave at an infinite surface with impedance $\boldsymbol{Z}_2$. *Source:* Alexander Peiffer.

and the velocity in z-direction reads

$$\boldsymbol{v}_z(x, z=0) = -\frac{\partial \Phi}{\partial z} = jk\left(\cos\vartheta \Phi e^{-jk\sin\vartheta x} - \cos\vartheta^{(R)}\Phi^{(R)}e^{-jk\sin\vartheta^{(R)}x}\right) \tag{2.101}$$

We certainly shall not be able to match the impedance $\boldsymbol{z}_2 = \boldsymbol{p}/\boldsymbol{v}_z$ at every surface position unless the arguments of the exponential functions are equal, hence

$$\vartheta = \vartheta^{(R)}$$

So, we get from the surface impedance condition

$$\boldsymbol{z}_2(x, z=0) = \frac{\boldsymbol{p}(x, z=0)}{\boldsymbol{v}_z(x, z=0)} = \frac{z_0(\Phi + \Phi^{(R)})}{\cos\vartheta(\Phi - \Phi^{(R)})} \tag{2.102}$$

With $z_0 = \rho_0 c_0$ and rearranging the above equation, the reflection factor is given by

$$\boldsymbol{R} = \frac{\Phi^{(R)}}{\Phi} = \frac{\boldsymbol{z}_2\cos\vartheta - z_0}{\boldsymbol{z}_2\cos\vartheta + z_0} \tag{2.103}$$

The ratio between irradiated power to reflector power is the squared reflection factor called the reflection coefficient.

$$r_s(\vartheta) = R^2(\vartheta) = \frac{(\boldsymbol{z}_2\cos\vartheta - z_0)^2}{(\boldsymbol{z}_2\cos\vartheta + z_0)^2} \tag{2.104}$$

$$\alpha_s(\vartheta) = 1 - r_s(\vartheta) = (1 - |R(\vartheta)|^2) \tag{2.105}$$

Note that those coefficients are exclusively described by the impedance of fluid and surface and not density or speed of sound. Thus, the impedance is the relevant quantity here.

2.6 Reflection and Transmission of Plane Waves

A plane wave passing a flat interface between two infinite fluid volumes with different density and sound velocity as shown in Figure 2.8 is a first example of continuous systems exchanging acoustic energy. Applications of such a system could be for example the interface between a liquid (water) and a gas (air) or just different gases.

Region 1 of the incoming wave has two wave components, the incoming and the reflected wave, and region 2 the transmitted wave. Thus, both velocity potentials read

$$\Phi_1(\mathbf{r}_1) = \Phi_1 e^{-j\mathbf{k}_1\mathbf{r}_1} + \Phi_1^{(R)}e^{-j\mathbf{k}_1^{(R)}\mathbf{r}_1} \tag{2.106}$$

$$\Phi_2(\mathbf{r}_2) = \Phi_2 e^{-j\mathbf{k}_2\mathbf{r}_2}$$

Using the given angles as sketched in Figure 2.8 the wavenumber space vector products are given by

$$\mathbf{k}_1\mathbf{r}_1 = k_1(\sin\vartheta_1 x - \cos\vartheta_1 z) \tag{2.107}$$

$$\mathbf{k}_1\mathbf{r}_1^{(R)} = k_1(\sin\vartheta_1^{(R)}x + \cos\vartheta_1^{(R)}z) \tag{2.108}$$

$$\mathbf{k}_2\mathbf{r}_2 = k_2(\sin\vartheta_2 x - \cos\vartheta_2 z) \tag{2.109}$$

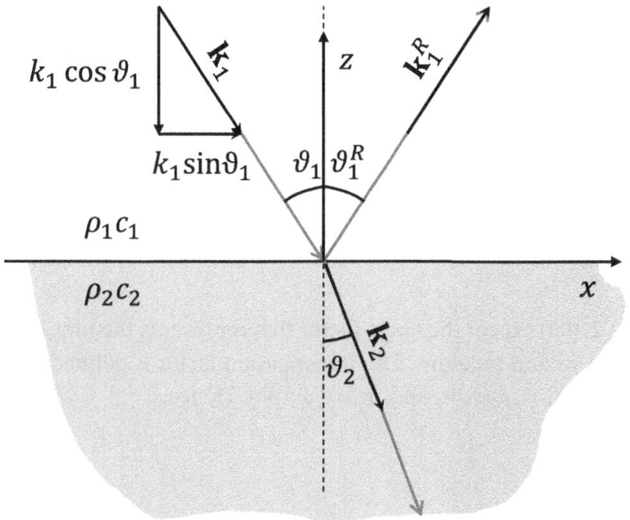

Figure 2.8 Transmission and reflection of a plane wave at the interface of two fluids. *Source:* Alexander Peiffer.

with $k_1 = \omega/c_1$ and $k_2 = \omega/c_2$. The contact face between the fluid requires the continuity of pressure and velocity in the z-direction. We start with the pressure $p_{1/2} = j\omega\rho_{1/2}\Phi_{1/2}$. Entering equations (2.107)–(2.109) into (2.106) and determining the pressure relation

$$p_1(x, z = 0) = p_2(x, z = 0)$$

gives

$$\rho_1\Phi_1 e^{jk_1 \sin\vartheta_1 x} + \rho_1\Phi_1^{(R)} e^{jk_1 \sin\vartheta_1^{(R)} x} = \rho_2\Phi_2 e^{jk_2 \sin\vartheta_2 x} \tag{2.110}$$

A solution for any x is only possible if the arguments of the exponential functions are equal.

$$k_1 \sin\vartheta_1 = k_1 \sin\vartheta_1^{(R)} \tag{2.111}$$

So also in the transmission case the incident angle equals the angle of the reflected wave. Additionally we have

$$k_1 \sin\vartheta_1 = k_2 \sin\vartheta_2 \tag{2.112}$$

This represents the acoustic equivalent of Snell's law of transmission:

$$\frac{\sin\vartheta_1}{c_1} = \frac{\sin\vartheta_2}{c_2} \tag{2.113}$$

With these conditions we can factor out the exponential function

$$\rho_1(\Phi_1 + \Phi_1^{(R)}) = \rho_2\Phi_2 \tag{2.114}$$

After clarifying the angles of reflection and transmission the next point is to assess the fraction of transmitted and reflected wave. From the continuity condition for the velocity in the z-direction we get with $\boldsymbol{v}_z = -\partial\Phi/\partial z$ when going through the algebra

$$\frac{\sin\vartheta_1}{c_1}(\Phi_1 - \Phi_1^{(R)}) = \frac{\sin\vartheta_2}{c_2}\Phi_2 \tag{2.115}$$

Rearranging equations (2.114) and (2.115) the reflection factor is

$$\boldsymbol{R} = \frac{\Phi_1^{(R)}}{\Phi_1} = \frac{\dfrac{z_2}{\cos\vartheta_2} - \dfrac{z_1}{\cos\vartheta_1}}{\dfrac{z_2}{\cos\vartheta_2} + \dfrac{z_1}{\cos\vartheta_1}} \tag{2.116}$$

This expression is similar to (2.103) except the angle factor that represents the physics of the wave propagation in the second medium. The transmission factor is defined by the ratio of pressure amplitudes $\boldsymbol{p}_1 = j\omega\rho_1\Phi_1$ and $\boldsymbol{p}_2 = j\omega\rho_2\Phi_2$. Hence,

$$\boldsymbol{T} = \frac{\dfrac{2z_2}{\cos\vartheta_2}}{\dfrac{z_1}{\cos\vartheta_1} + \dfrac{z_2}{\cos\vartheta_2}} \tag{2.117}$$

The transmitted acoustic power follows from the square of the amplitudes. We introduce a transmission coefficient τ by

$$\tau = \frac{\Pi_{\text{trans}}}{\Pi_{\text{in}}} \tag{2.118}$$

Without loss of generality we assume $\vartheta = 0$ so each power is given by

$$\Pi_{\text{in}} = \frac{A}{2}Re\left(\frac{\hat{p}_1^2}{\boldsymbol{z}_1}\right) \qquad \Pi_{\text{trans}} = \frac{A}{2}Re\left(\frac{\hat{p}_2^2}{\boldsymbol{z}_2}\right) = \frac{A}{2}Re\left(\frac{\boldsymbol{T}^2\hat{p}_1^2}{\boldsymbol{z}_2}\right) \tag{2.119}$$

With (2.117) this reads as:

$$\tau = \frac{4Re(\boldsymbol{z}_1)Re(\boldsymbol{z}_2)}{|\boldsymbol{z}_1 + \boldsymbol{z}_2|^2} \tag{2.120}$$

It should be noted that the transmission coefficient of the flat interface between two fluids is determined by the impedance of each half space 'seen' from the other side. This is the first indication for the coupling of subsystems determined by the radiation impedance into the free fields of each subsystem. Similar expressions will be found in Sec. 8.2.4.1 when transmission is dealt with in the context of coupled random subsystems.

2.7 Inhomogeneous Wave Equation

In the considerations in this chapter so far, we neglected the source terms related to the conservation of mass and momentum. All sources discussed until now are caused by vibrating surfaces. For establishing a physical link between the source term and the specific mass flow $\dot{q}_s$ in Equation (2.3) and force density term $\mathbf{f}$ in Equation (2.8) we keep the terms this time. The source terms are not influenced by the linearization procedure; thus, the inhomogeneous and linear equations of momentum (2.24) and continuity (2.23) read as

$$\frac{\partial \rho'}{\partial t} + \rho_0 \nabla \mathbf{v}' = \dot{\rho}_s \qquad \rho_0 \frac{\partial \mathbf{v}'}{\partial t} + \nabla p = \mathbf{f} \tag{2.121}$$

Repeating the steps of section 2.2.5 we finally get the inhomogeneous wave equation

$$\frac{1}{c_0^2}\frac{\partial^2 p}{\partial t^2} - \nabla^2 p = \ddot{\rho}_s - \nabla \mathbf{f} \tag{2.122}$$

The density source is converted into a volume source strength density by $\dot{\rho}_s = \rho_0 q_s(t)$. The above equation can also be converted into the frequency domain and hence to the inhomogeneous Helmholtz equation

$$\left(k^2 + \Delta\right) \boldsymbol{p}(\mathbf{x}, \omega) = -j\omega\rho_0 \boldsymbol{q}_s + \nabla \mathbf{f} = -\boldsymbol{f}_q(\mathbf{r}) \tag{2.123}$$

2.7.1 Acoustic Green's Functions

This section presents the concept of the Green's function that uses a formalism to calculate the sound field for arbitrary source and boundary configurations as shown for example by Morse and Ingard (1968).

The Green's function is defined as the solution of the following inhomogeneous wave equation.

$$\Delta g(\mathbf{r}, \mathbf{r}_0) + k^2 g(\mathbf{r}, \mathbf{r}_0) = -\delta(\mathbf{r} - \mathbf{r}_0) \tag{2.124}$$

The inhomogeneous part is the delta function which allows for this elegant derivation of the Kirchhoff integral. The delta function is introduced in the appendix A.1.3 in the time domain. However, it can also be applied in space. The multidimensional delta function is simply the product of three Dirac delta functions in space

$$\delta(\mathbf{r} - \mathbf{r}_0) = \delta(x - x_0)\delta(y - y_0)\delta(y - y_0) \text{ with } \mathbf{r} = \{x, y, z\}^T \tag{2.125}$$

The sifting properties and the value of the integration is defined by volume integral

$$f(\mathbf{r}) = \int_V f(\mathbf{r}_0)\delta(\mathbf{r} - \mathbf{r}_0) d\mathbf{r}_0 \text{ and } \int_V \delta(\mathbf{r}) d\mathbf{r} = 1 \tag{2.126}$$

The solution of Equation (2.124) is the point source (2.91)[3].

$$g(\mathbf{r}|\mathbf{r}_0) = \frac{1}{4\pi l} e^{-jkl} \quad \text{with} \quad l = |\mathbf{r} - \mathbf{r}_0| \tag{2.127}$$

In order to achieve a common formulation we add an arbitrary solution χ of the homogeneous wave equation

$$\Delta\chi(\mathbf{r}|\mathbf{r}_0) + k^2\chi(\mathbf{r}|\mathbf{r}_0) = 0 \tag{2.128}$$

to the Green's function to get the generalized Green's function

$$G(\mathbf{r}, \mathbf{r}_0) = g(\mathbf{r}, \mathbf{r}_0) + \chi(\mathbf{r}, \mathbf{r}_0) \tag{2.129}$$

3 This can be proven using the relationship between the δ and exponential functions

The purpose of the additional homogeneous solution is to create freedom to fulfill boundary conditions that do not occur in the free sound field. The task is to find the solution for the inhomogeneous wave equation

$$\Delta \boldsymbol{p}(\mathbf{r}) + k^2 \boldsymbol{p}(\mathbf{r}) = -\boldsymbol{f}_q(\mathbf{r}) \tag{2.130}$$

The generalized Green's function must be a solution of the following equation for the special case with $\mathbf{r}, \mathbf{r}_0 \in V$ and the boundary ∂V as shown in Figure 2.9.

$$\Delta G(\mathbf{r}, \mathbf{r}_0) + k^2 G(\mathbf{r}, \mathbf{r}_0) = -\delta(\mathbf{r} - \mathbf{r}_0) \tag{2.131}$$

In order to receive a global solution we perform the operation

$$G(\mathbf{r}, \mathbf{r}_0) \cdot (2.130) - \boldsymbol{p}(\mathbf{r}) \cdot (2.131) \tag{2.132}$$

This leads to

$$G(\mathbf{r}, \mathbf{r}_0)\Delta \boldsymbol{p}(\mathbf{r}) - \boldsymbol{p}(\mathbf{r})\Delta G(\mathbf{r}, \mathbf{r}_0) = -[G(\mathbf{r}, \mathbf{r}_0)\boldsymbol{f}_q(\mathbf{r}) - \boldsymbol{p}(\mathbf{r})\delta(\mathbf{r}, \mathbf{r}_0)] \tag{2.133}$$

Exchanging $\mathbf{r}$ and $\mathbf{r}_0$ and integrating $\mathbf{r}_0$ over the volume V gives

$$\begin{aligned} &\int_V G(\mathbf{r}_0, \mathbf{r})\Delta \boldsymbol{p}(\mathbf{r}_0) - \boldsymbol{p}(\mathbf{r}_0)\Delta G(\mathbf{r}_0, \mathbf{r}) d\mathbf{r}_0 = \\ &\qquad -\int_V G(\mathbf{r}_0, \mathbf{r})\boldsymbol{f}_q(\mathbf{r}_0) d\mathbf{r}_0 + \underbrace{\int_V \boldsymbol{p}(\mathbf{r}_0)\delta(\mathbf{r} - \mathbf{r}_0) d\mathbf{r}_0}_{=\boldsymbol{p}(\mathbf{r})} \end{aligned} \tag{2.134}$$

The last term on the RHS follows from the sifting property of the delta function

$$\begin{aligned} \boldsymbol{p}(\mathbf{r}) = &\int_V G(\mathbf{r}_0, \mathbf{r})\boldsymbol{f}_q(\mathbf{r}) d\mathbf{r}_0 \\ &+ \int_V G(\mathbf{r}_0, \mathbf{r})\Delta \boldsymbol{p}(\mathbf{r}_0) - \boldsymbol{p}(\mathbf{r}_0)\Delta G(\mathbf{r}_0, \mathbf{r})\, d\mathbf{r}_0. \end{aligned} \tag{2.135}$$

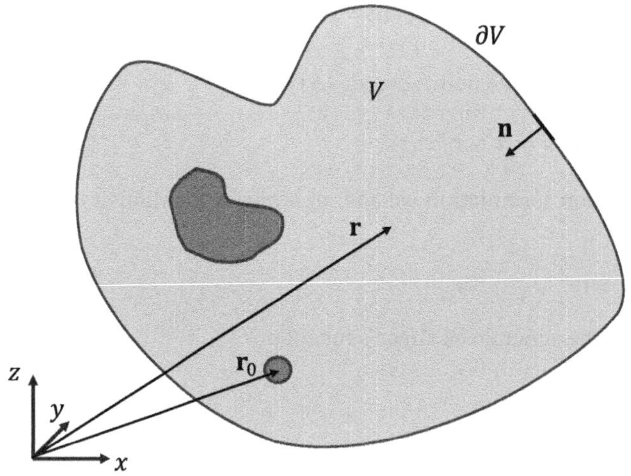

Figure 2.9 Solution volume and boundaries. *Source:* Alexander Peiffer.

With Green's law of vector analysis

$$\int_V (\Phi\Delta\Psi - \Psi\Delta\Phi)dV = \int_{\partial V} (\Phi\nabla\Psi - \Psi\nabla\Phi)dS \tag{2.136}$$

some volume integrals can be transferred into surface integrals and we get finally

$$\begin{aligned} \boldsymbol{p}(\mathbf{r}) &= \int_V G(\mathbf{r}_0, \mathbf{r})\boldsymbol{f}_q(\mathbf{r}_0)d\mathbf{r}_0 \\ &+ \int_{\partial V} G(\mathbf{r}_0, \mathbf{r})\nabla \boldsymbol{p}(\mathbf{r}_0) - \boldsymbol{p}(\mathbf{r}_0)\nabla G(\mathbf{r}_0, \mathbf{r})d\mathbf{r}_0 \end{aligned} \tag{2.137}$$

The first term on the right-hand side is the volume integral over all sources $\boldsymbol{f}_q(\mathbf{r})$ in the volume. So given a known source distribution we can calculate the according sound field. The two terms in the surface integral take care of the boundary condition. The pressure gradient in the first can be converted into the normal velocity using (2.35). The second surface integral allows establishing the correct surface impedance. Equation (2.137) is called the constant frequency version of the Kirchhoff-Helmholtz integral.

2.7.2 Rayleigh integral

The Rayleigh integral is a special solution of the Kirchhoff-Helmholtz integral applied to flat and infinite surfaces. We assume a configuration as shown in Figure 2.10. The integration volume is the right half space for $z > 0$ closed by a half sphere of infinite radius. The Green's function of any source at $\mathbf{r}_0 = (x_0, y_0, z_0)$ with $z_0 > 0$ is as defined in equation (2.127). The rigid surface acts as a reflector. Thus, there is a mirror source located at $\mathbf{r}_0' = (x_0, y_0, -z_0)$. This source is not in volume V, and the added wave field is

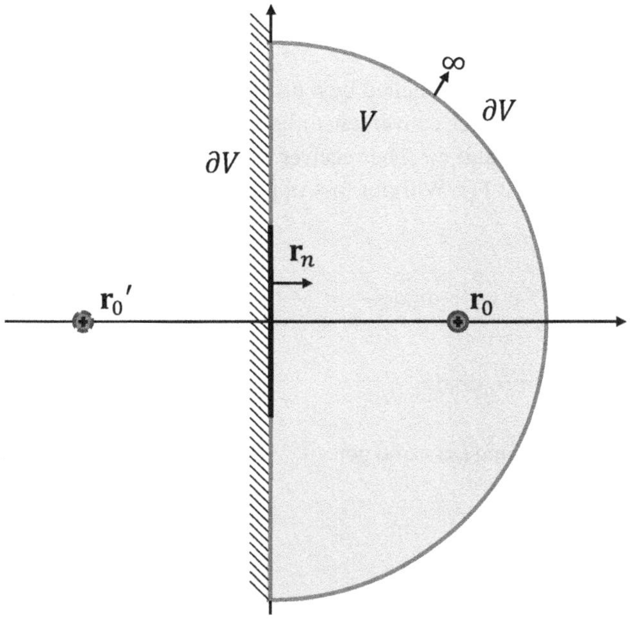

Figure 2.10 Half space in front of a rigid wall. *Source:* Alexander Peiffer.

therefore considered as a homogeneous solution in the volume. Hence, we get for the generalized Green's function

$$G(\mathbf{r}, \mathbf{r}_0) = \frac{1}{4\pi l} e^{-jkl} + \frac{1}{4\pi l'} e^{-jkl'} \text{ with } l = |\mathbf{r} - \mathbf{r}_0| \quad l' = |\mathbf{r} - \mathbf{r}_0{}'| \tag{2.138}$$

We enter this version of the Green's function in Equation (2.137) and we get

$$\begin{aligned} \mathbf{p}(\mathbf{r}) &= \int_V G(\mathbf{r}_0, \mathbf{r}) f_q(\mathbf{r}_0) d\mathbf{r}_0 \\ &+ \int_{-\infty}^{\infty} \int_{-\infty}^{\infty} G(\mathbf{r}_0, \mathbf{r}) \frac{\partial \mathbf{p}(\mathbf{r}_0)}{\partial z} - \mathbf{p}(\mathbf{r}_0) \frac{\partial G(\mathbf{r}_0, \mathbf{r})}{\partial z} dx_0 dy_0. \end{aligned} \tag{2.139}$$

We assume a source-free half space so $f_q(\mathbf{r}) = 0$, and due to the mirror source symmetry $\frac{\partial G(\mathbf{r}_0,\mathbf{r})}{\partial z} = 0$ is also true. By clever selection of the Green's function we fulfilled the boundary condition automatically. For the surface integral the contributions from the half sphere with infinite radius are supposed to be zero. From Equation (2.35) the first expression can be converted into an expression for the surface velocity $\mathbf{v}_z$. Performing the limit process $z_0 \to 0$ we get

$$G(\mathbf{r}_0, \mathbf{r}) = \frac{2}{4\pi l} e^{-jkl} \quad l = \sqrt{z^2 + (x - x_0)^2 + (y - y_0)^2} \tag{2.140}$$

and with this Green's function we can derive the Rayleigh integral that allows the calculation of infinite half space sound fields excited by a rigid vibrating plane with arbitrary velocity distribution $\mathbf{v}_z(x_0, y_0)$.

$$\mathbf{p}(\mathbf{r}) = \int_{-\infty}^{\infty} \int_{-\infty}^{\infty} \frac{j\omega\rho_0}{2\pi l} e^{-jkl} \mathbf{v}_z(x_0, y_0) dx_0 dy_0 \tag{2.141}$$

2.7.3 Piston in a Wall

A cylindrical loudspeaker in a wall can be modelled by a piston of radius R vibrating with velocity $\mathbf{v}_z$ located in a rigid wall. For convenience the surface integral will be expressed in cylindrical coordinates r_0 and φ_0. The receiver coordinates are given as spherical coordinates r and ϑ(Figure 2.11). Without loss of generality the azimuthal angle φ is set to zero.

$$\begin{aligned} \mathbf{p}(\mathbf{r}, \vartheta) &= \int_0^{2\pi} \int_0^{\infty} \frac{j\omega\rho_0}{2\pi l} e^{-jkl} \mathbf{v}_z(r_0) r_0 dr_0 d\varphi_0 \\ &= \int_0^{2\pi} \int_0^{R} \frac{j\omega\rho_0}{2\pi l} e^{-jkl} \mathbf{v}_z r_0 dr_0 d\varphi_0 \end{aligned} \tag{2.142}$$

In the far field approximation we assume $l \approx r$ and get

$$l = r + r_0 \sin\vartheta \cos\varphi_0. \tag{2.143}$$

So, the approximate result is

$$\mathbf{p}(r, \vartheta) = \frac{j\omega\rho_0}{2\pi r \mathbf{v}_z} \int_0^{2\pi} \int_0^{R} e^{jkr_0 \sin\vartheta\cos\varphi_0} d\varphi_0 r_0 dr_0 \tag{2.144}$$

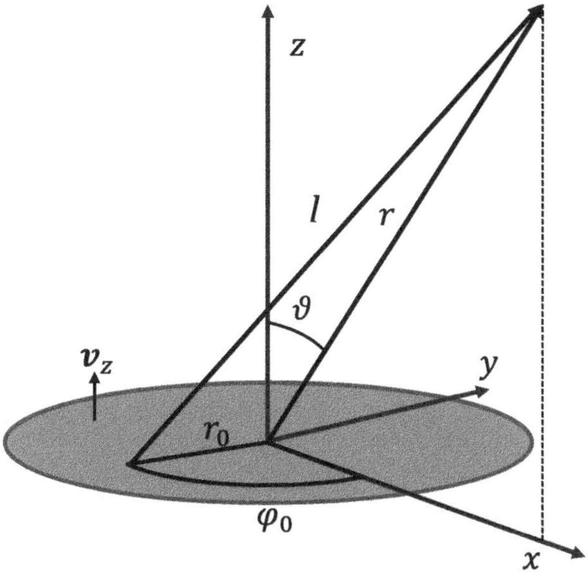

Figure 2.11 Coordinate definitions for the piston in the wall. *Source:* Alexander Peiffer.

The integral is the Bessel function of first order

$$\mathbf{p}(r,\vartheta) = \frac{j\omega\rho_0}{2\pi r \mathbf{v}_z}\left(\frac{2J_1(kR\sin\vartheta)}{kR\sin\vartheta}\right). \tag{2.145}$$

The results for some values of kR are shown in Figure 2.12 over the angular range of $\pm\pi/2$. For a piston size small compared to the wavelength ($kR \leq 1$) the radiation pattern is similar to a point source. The smaller the wavelength gets in relation to the piston radius R the more a specific radiation pattern develops.

2.7.3.1 Impedance Concept

The radiation impedance of the piston is calculated from the pressure averaged over the surface related to the piston velocity $\mathbf{v}_z$. As shown by Lerch and Landes (2012) the mechanical impedance of the piston due to radiation is given by

$$\mathbf{Z} = \frac{1}{\mathbf{v}_z}\int_S p(r)dS = \frac{2\pi}{\mathbf{v}_z}\int_0^R \mathbf{p}(r)rdr \tag{2.146}$$

According to equation (2.141) assuming a constant velocity $\mathbf{v}_z$ over the surface A the pressure is

$$\mathbf{p}(r) = \frac{j\omega\rho_0\mathbf{v}_z}{2\pi}\int_A \frac{e^{-jks}}{s}dA. \tag{2.147}$$

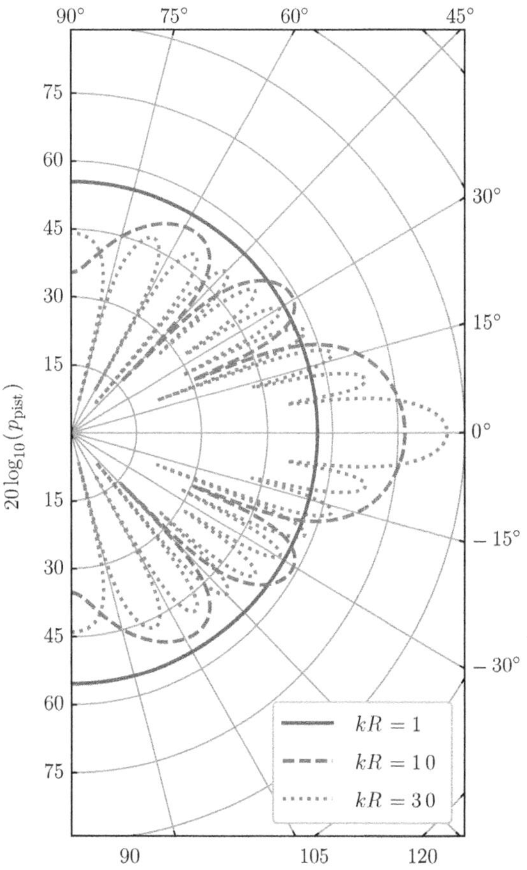

Figure 2.12 Angular distribution (radiation pattern) of the pressure field of the piston. *Source:* Alexander Peiffer.

Thus, we get the pressure at r from integrating the contribution from the rest of the piston in circles of radius s. The angle integration over φ_0 runs from 0 to 2π. From every angle φ_0 follows the integration limits s_{max} of the second integral.

$$s_{max} = r\cos\varphi + \sqrt{R^2 - r^2\sin^2\varphi} \tag{2.148}$$

Using those limits gives

$$\begin{aligned} \boldsymbol{p}(r) &= \frac{j\omega\rho_0\boldsymbol{v}_z}{2\pi}\int_0^{2\pi}\int_0^{r\cos\varphi+\sqrt{R^2-r^2\sin^2\varphi}} \frac{e^{-jks}}{s} s\,ds\,d\varphi \\ &= \frac{\rho_0 c_0\boldsymbol{v}_z}{2\pi}\int_0^{2\pi}\left(1 - e^{-jkr\cos\varphi - jk\sqrt{R^2-r^2\sin^2\varphi}}\right)d\varphi \\ &= \rho_0 c_0\boldsymbol{v}_z\left(1 - \frac{1}{2\pi}\int_0^{2\pi} e^{-jkr\cos\varphi - jk\sqrt{R^2-r^2\sin^2\varphi}}d\varphi\right). \end{aligned} \tag{2.149}$$

Inserting equation (2.149) into (2.146) leads to the expression

$$\boldsymbol{Z} = \rho_0 c_0\pi R^2\left(1 - \frac{1}{\pi R^2}\int_{\varphi=0}^{2\pi}\int_{r=0}^{R} e^{-jkr\cos\varphi - jk\sqrt{R^2-r^2\sin^2\varphi}} r\,dr\,d\varphi\right). \tag{2.150}$$

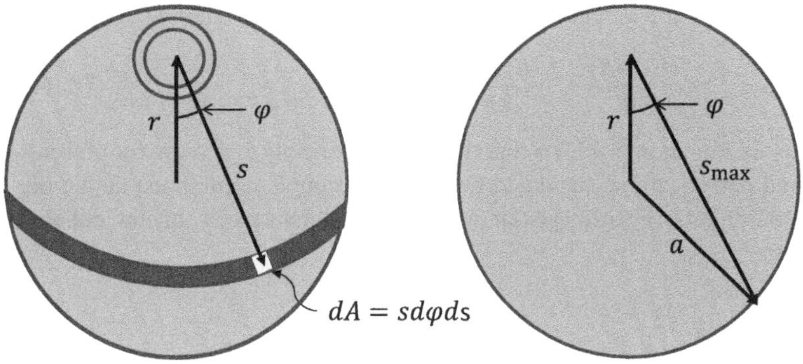

Figure 2.13 Surface integration over piston for radiation impedance. *Source:* Alexander Peiffer.

Running through quite a lot of algebraic modifications we get the expression for the impedance of a piston

$$\boldsymbol{Z} = \rho_0 c_0 \pi R^2 \left(1 - \frac{J_1(2kR)}{kR} + j\frac{H_1(2kR)}{kR}\right) \tag{2.151}$$

or

$$z_a = \rho_0 c_0 \left(1 - \frac{J_1(2kR)}{kR} + j\frac{H_1(2kR)}{kR}\right) \tag{2.152}$$

$H_1(z)$ is the Hankel function of first order. In Figure 2.14 the real and imaginary parts of the acoustic radiation impedance are compared to those of the pulsating sphere. Both sources have a similar shape except some waviness for the piston resulting from interference effects from the integration over the piston surface. For large kR the impedance is real for both radiators and approaches the acoustic impedance of a plane wave $z_0 = \rho_0 c_0$.

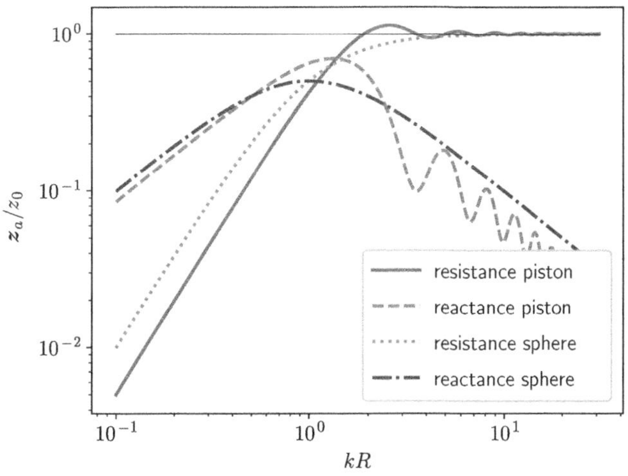

Figure 2.14 Acoustic radiation impedance of the piston. *Source:* Alexander Peiffer.

With Equation (2.87) the radiated power of a piston of source strength $\boldsymbol{Q} = \pi R^2 \boldsymbol{v}_z$ is

$$\langle \Pi \rangle_T = \rho_0 c_0 \left(1 - \frac{J_1(2kR)}{kR} + j \frac{H_1(2kR)}{kR} \right) Q_{\text{rms}}^2 \tag{2.153}$$

The main use of Equation (2.153) is that the required velocity to achieve (or prevent) a certain sound power can be calculated from it, for example if one must define the boundary condition for a radiating piston in simulation software and only the radiated power is known.

2.7.3.2 Inertia Effects

The Bessel functions can be approximated by a series in $2kR$ taking the first series term of both functions (Jacobsen, 2011)

$$\boldsymbol{Z} = \rho_0 c_0 \pi R^2 \left(\frac{1}{2}(2kR)^2 + j \frac{8}{3\pi} kR \right) \tag{2.154}$$

This expression is valid for $ka < 0.5$. From the imaginary part we get for the mass

$$m = \frac{Im(\boldsymbol{Z})}{\omega} = \frac{8R^3 \rho_0}{3} \tag{2.155}$$

Assuming a cylindrical volume $V = \pi R^2 l_c$ of the fluid above the piston we can calculate the length of the moving mass cylinder to be

$$l_c = \frac{8R}{3\pi} \approx 0.85R \tag{2.156}$$

meaning that at low frequencies the piston is moving a fluid layer of 0.85 times the radius acting as an inertia without radiation.

2.7.4 Power Radiation

For the radiated power calculation of the piston we took the pressure at the piston surface and integrated the pressure–velocity product over the surface. Due to the fact that the velocity is constant the surface integral involves mainly the pressure as a space-dependent property. In case of vibrating structures with complex shapes of vibration the velocity distribution over the surface is not homogeneous, and we need a more detailed approach.

$$\boldsymbol{p}(\mathbf{r}) = \iint_{-\infty}^{\infty} \frac{j\omega\rho_0}{2\pi l} e^{-jkl} \boldsymbol{v}_z(\mathbf{r}_0) d\mathbf{r}_0 \text{ with } l = |\mathbf{r} - \mathbf{r}_0| \tag{2.157}$$

In the above equation a function with argument $(\mathbf{r} - \mathbf{r}_0)$ is multiplied by the velocity function for $\mathbf{r}_0$ and integrated over the two-dimensional space. Mathematically, this can be interpreted as a two-dimensional convolution in space

$$\boldsymbol{p}(\mathbf{r}) = \frac{j\omega\rho_0}{2\pi|\mathbf{r}|} e^{jk|\mathbf{r}|} * \boldsymbol{v}_z(\mathbf{r}) \tag{2.158}$$

Thus, when we apply the two-dimensional Fourier transform to the Rayleigh integral the result is the product of the Fourier transform of the vibration shape $\boldsymbol{v}_z(\mathbf{r}_0)$ and the

Green's function in wavenumber space leading to

$$\boldsymbol{p}(\mathbf{k}) = \frac{1}{4\pi}\frac{\rho_0\omega}{\sqrt{k_a^2 - k^2}}\boldsymbol{v}_z(\mathbf{k}) \text{ with } k = |\mathbf{k}| \tag{2.159}$$

So, we have replaced the expensive convolution operation by a multiplication. This simplification is at the cost of two-dimensional Fourier transforms that are required to get the expressions in wavenumber domain.

The time averaged intensity of a sound field is given by the product of pressure and velocity (2.45). As the velocity is not uniform over the surface we perform a surface integration over the vibrating area to get the total radiated power

$$\begin{aligned}\Pi &= \iint\limits_{-\infty}^{\infty} \frac{1}{2} Re(\boldsymbol{p}(\mathbf{r})\boldsymbol{v}_z^*(\mathbf{r}))d\mathbf{r}\\ &= \iint\limits_{-\infty}^{\infty}\iint\limits_{-\infty}^{\infty} \frac{j\omega\rho_0}{4\pi l} Re(e^{-jkl}\boldsymbol{v}_z(\mathbf{r}_0)\boldsymbol{v}_z^*(\mathbf{r}))d\mathbf{r}_0 d\mathbf{r} \text{ with } l = |\vec{r} - \vec{r}_0|\end{aligned} \tag{2.160}$$

Thus, for the determination of radiated power a double area integral is required that may become computationally expensive.

In the above expression we can also switch to the wavenumber domain. In this case the area integration is replaced by an integration over the two-dimensional wavenumber space.

$$\begin{aligned}\Pi &= \iint\limits_{-\infty}^{\infty} \frac{1}{2} Re(\boldsymbol{p}(\mathbf{k})\boldsymbol{v}_z^*(\mathbf{k}))d\mathbf{k}\\ &= \iint\limits_{-\infty}^{\infty} \frac{1}{2} Re\left(\frac{1}{4\pi}\frac{\rho_0\omega}{\sqrt{k_a^2 - k^2}}\boldsymbol{v}_z(\mathbf{k})\boldsymbol{v}_z^*(\mathbf{k})\right)d\mathbf{k}\\ &= \iint\limits_{-\infty}^{\infty} \frac{1}{2} Re\left(\frac{1}{4\pi}\frac{\rho_0\omega}{\sqrt{k_a^2 - k^2}}v_z^2(\mathbf{k})\right)d\mathbf{k}\end{aligned} \tag{2.161}$$

The double integral is replaced by a single two-dimensional wavenumber integration. Thus, once the shape function is available the power calculation in wavenumber space is much faster than in real space (Graham, 1996).

2.7.4.1 Radiation Efficiency

The radiation efficiency is a quantity that relates the power of a plane wave to the radiated power of a vibrating surface with same surface averaged velocity. The definition of the radiation efficiency was motivated by experimental procedures because it allows the estimation of the radiated power from the measurements of the vibration velocity.

The squared average velocity of a vibrating surface is

$$\langle \hat{v}_z^2 \rangle_S = \frac{1}{S} \int_S \mathbf{v}_z(\mathbf{r}) \mathbf{v}_z^*(\mathbf{r}) d\mathbf{r} \tag{2.162}$$

and the power radiated by a plane wave through the same area S is given by (2.47)

$$\Pi_0 = \frac{1}{2} S \rho_0 c_0 \langle \hat{v}_z^2 \rangle_S \tag{2.163}$$

The radiation efficiency is defined as the ratio between the radiated power of a velocity profile $\mathbf{v}_z(\mathbf{r})$ of a surface S and the standardized power of the plane wave:

$$\sigma_{\text{rad}} = \frac{\Pi}{\Pi_0} = \frac{\Pi}{\frac{S}{2} \rho_0 c_0 \langle \hat{v}_z^2 \rangle_S} \tag{2.164}$$

The radiation efficiency is used to determine the radiated power of vibrating structures from calculated, estimated, or measured radiation efficiency of specific surfaces

$$\Pi = \frac{S}{2} \sigma_{\text{rad}} \rho_0 c_0 \langle \hat{v}_z^2 \rangle_S = S \sigma_{\text{rad}} \rho_0 c_0 \langle v_{z,\text{rms}}^2 \rangle_S \tag{2.165}$$

2.8 Units, Measures, and Levels

The dynamic range of acoustic quantities can be very high; thus, a logarithmic scale is well established for the quantification of acoustic signals. For convenience a certain time averaged quantity, for example the mean square pressure $\langle p \rangle_T^2 = p_{\text{rms}}^2$, is compared to a mean square reference value p_{ref}^2. The pressure level in decibels is defined as follows:

$$L_p = 10 \log_{10} \frac{p_{\text{rms}}^2}{p_{\text{ref}}^2} = 20 \log_{10} \frac{p_{\text{rms}}}{p_{\text{ref}}} \text{ with } p_{\text{ref}} = 20 \mu\text{Pa} \tag{2.166}$$

The factor of 10 is introduced to spread the scale. Linear quantities such as pressure, velocity, or displacement use the mean square values. As level and decibel are used on time signals too, one should not apply the decibel scale to amplitudes. This may lead to confusion, as it is not clear if the mean square values of the amplitude is meant. This makes even more sense when the energy and power levels are defined. The energy must be averaged, as there is no constant value over the period – see Equation (2.46). Energy quantities such as energy, intensity, or power are compared with mean values and not mean square values; hence:

$$L_w = 10 \log_{10} \frac{\Pi_{\text{mean}}}{\Pi_{\text{ref}}} \text{ with } \Pi_{\text{ref}} = 10^{-12} \text{W} = 1 \text{pW} \tag{2.167}$$

Table 2.3 Field and energy properties of acoustic waves.

Source	Source strength	Impedance / Velocity	Pressure / Radiated power
Mono pole	$Q, j\omega V$	$Z_{\rm rad} = \frac{k^2\rho_0 c_0}{4\pi}$	$p = \frac{jk\rho_0 c_0 Q(\omega)}{4\pi r}e^{-jkr}$
		$v_r = \frac{Q(\omega)}{4\pi r}jk\left(1+\frac{1}{jkr}\right)e^{-jkr}$	$\frac{Q_{rms}^2 k^2\rho_0 c_0}{4\pi}$
Breath. sphere	$Q = 4\pi R^2 v_r$	$z_R = \frac{j\rho_0 c_0 kR}{1+jkR}$	$\frac{Q}{4\pi r}\left[\frac{jk\rho_0 c_0}{1+jkR}\right]e^{-jk(r-R)}$
		$\frac{Q}{4\pi r^2}\left[\frac{1+jkr}{1+jkR}\right]e^{-jk(r-R)}$	$\frac{Q_{rms}^2 k^2\rho_0 c_0}{4\pi(1+k^2R^2)}$
Piston	$Q = \pi R^2 v_z$	$z_R = \rho_0 c_0\left(1-\frac{J_1(2kR)}{kR}+j\frac{H_1(2kR)}{kR}\right)$	$\frac{j\omega\rho_0}{2\pi r v_z}\left[\frac{2J_1(kR\sin\vartheta)}{kR\sin\vartheta}\right]$
			$\rho_0 c_0\left(1-\frac{J_1(2kR)}{kR}+j\frac{H_1(2kR)}{kR}\right)Q_{rms}^2$

In addition, the decibel scale is used for ratios of similar quantities. A typical example is the transmission loss that is the decibel scale of the transmission factor from Equation (2.118) that relates the transmitted to the radiated power. The definition of the transmission loss (TL) is:

$$\mathrm{TL} = 10\log_{10}\frac{1}{\tau} = -10\log_{10}\tau \tag{2.168}$$

The reciprocal definition was chosen in order to get positive values for losses. When linear quantities are compared, for example the pressure at two locations, the mean square values are related. When the squared pressure is compared to the situation with and without a specific equipment or installation, this is called insertion loss (IL)

$$\mathrm{IL} = 10\log_{10}\frac{p_{\rm out}^2}{p_{\rm in}^2} = 20\log_{10}\frac{p_{\rm out}}{\mathrm{p}_{\rm in}} \tag{2.169}$$

The reference quantities for power and pressure are chosen conveniently to simplify the calculations with levels. Taking the equation for the spherical source (2.82) and dividing it by the squared reference value for the pressure yields

$$\frac{p_{\rm rms}^2}{p_{\rm ref}^2} = \frac{\rho_0 c_0}{p_{\rm ref}^2 A_{\rm ref}}\frac{A_{\rm ref}}{4\pi r^2}\langle\Pi\rangle_T \tag{2.170}$$

and taking the decibel of this

$$10\log_{10}\left\{\frac{p_{\rm rms}^2}{p_{\rm ref}^2}\right\} = 10\log_{10}\left\{\frac{\langle\Pi\rangle_T}{\frac{A_{\rm ref}p_{\rm ref}^2}{\rho_0 c_0}}\right\} + 10\log_{10}\left\{\frac{A_{\rm ref}}{4\pi r^2}\right\}$$

yields

$$L_p = L_w + 10\log_{10}\left\{\frac{A_{\rm ref}}{4\pi r^2}\right\}\ \text{with}\ \Pi'_{\rm ref} = \frac{A_{\rm ref}p_{\rm ref}^2}{\rho_0 c_0}. \tag{2.171}$$

Entering typical values for air with $\rho_0 = 1.23$ kg/m^3 and $c_0 = 343$ m/s using $A_{\text{ref}} = 1$ m^2 we get

$$\Pi'_{\text{ref}} \approx 10^{-12}\text{W} = \Pi_{\text{ref}}$$

matching well to the reference value of acoustic power.

Bibliography

W.R. Graham. BOUNDARY LAYER INDUCED NOISE IN AIRCRAFT, PART I: THE FLAT PLATE MODEL. *Journal of Sound and Vibration*, 192(1):101–120, April 1996. ISSN 0022-460X.

Finn Jacobsen. PROPAGATION OF SOUND WAVES IN DUCTS. Technical Note 31260, Technical University of Denmark, Lynby, Denmark, September 2011.

Reinhard Lerch and H. Landes. Grundlagen der Technischen Akustik, September 2012.

P.M.C. Morse and K.U. Ingard. *Theoretical Acoustics*. International Series in Pure and Applied Physics. Princeton University Press, 1968. ISBN 978-0-691-02401-1.

3

Wave Propagation in Structures

3.1 Introduction

In comparison with the wave motion in structures the acoustic wave motion is simple. The equations are isotropic, the speed of sound is (in most cases) not dependent on frequency, and even for complex shapes the Green's function provides a powerful tool to calculate the wave field. For structural waves the situation is different. In structures there is a variety of wave types described by displacements and rotations in several space dimensions or degrees of freedom. The chance to find a practicable analytical solution is low and solutions are only available for simple systems like straight bars, rectangular thin plates, or membranes. Thus, wave propagation in structures is a natural field for numerics: i.e. finite element methods (FEM) that discretize the real system into many small and simple elements that have an analytical solution or at least an approximation. The dynamics of the full system or the full mesh are defined by the complete set of all these elements.

This book is not about FEM, but we will often use a discrete form of the equation of motion or wave equation. There are two reasons for this:

Complex systems	The easiest way to describe the dynamics of a realistic technical system is by numerical methods; thus, there will always by a matrix description of the deterministic system. This is a standard approach in the industrial simulation of structural dynamics.
Formulation	There are many ways to describe dynamic wave coupling of different systems, but the discrete formulation is simple, structured, and straightforward.

The various theories to derive the discrete equation are described for example by Bathe (1982). However, the result of this dicretisation process – the matrix presentation of complex structures – will be frequently used. The global structure of the matrices is the same as described in section 1.4 or in Equation (1.89).

There is a large amount of literature dealing with technical mechanics and the statics and dynamics of solid systems. The following deductions are taken from Szabo's text books Szabo (2013a,b) or from Lerch and Landes (2012) who used the reference from Cremer et al. (2005). This chapter can only summarize these studies in such a way that the concepts of wave propagation in typical structural systems can be applied

Vibroacoustic Simulation: An Introduction to Statistical Energy Analysis and Hybrid Methods,
First Edition. Alexander Peiffer.

for the wave field descriptions in later chapters. Please refer to those publications if more details are required for specific dynamic systems. In order to clarify the principle behavior of waves in representative structures we deal with the wave equations in structures like bars, beams, plates, shells, membranes, and the bulk material. Even if there are not many solutions for real systems available these equations will form the base for all random modelling approaches and provide understanding of the specific characteristics of wave propagation.

3.2 Basic Equations and Definitions

In fluids the state variable of the wave equation was the pressure. In solids there is a similar quantity but with different sign and spatial dependence: the stress. Let us imagine a small block of isotropic material as shown in Figure 3.1. The left hand side depicts a force F_z that pulls a perfectly stiff area connection on top of the block. The force per area is called stress. The nomenclature is such that positive stress is related to an elongation of the block. The normal stress σ_{zz} defines the force in the z-direction pulling at the free surface A in the direction normal to it.

$$\sigma_{zz} = \frac{F_z}{A_z} \tag{3.1}$$

We use double indices zz because both the force and the normal vector of the surface point in the z-direction. The ratio of the elongation dz to the unloaded length l_z is called the strain in z-direction

$$\varepsilon_{zz} = \frac{dz}{l_z} \tag{3.2}$$

The relationship between the stress and strain defines the Young's modulus E with

$$\varepsilon_{zz} = E\sigma_{zz}. \tag{3.3}$$

The stretching in z is accompanied by a contraction perpendicular to it. The Poisson ratio ν_{xz} is defined as the ratio between the strain orthogonal to the stretching e.g. x and the strain in the direction of the elongation.

$$\nu_{xz} = -\frac{\varepsilon_{xx}}{\varepsilon_{zz}} \tag{3.4}$$

The surface of the block from Figure 3.1 on the right hand side can alternatively be loaded by a force F_x transverse to the surface normal of A_z. This transversal or shear

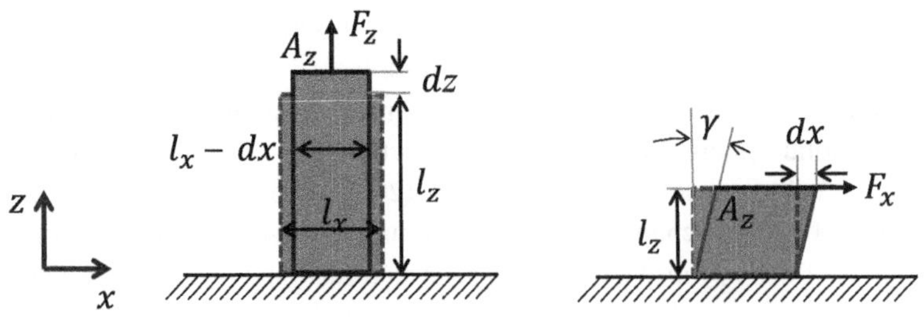

Figure 3.1 Strain and shear of a small solid block. *Source:* Alexander Peiffer.

stress is often denoted by τ instead of σ. We stay with σ for all stresses

$$\tau_{xz} = \sigma_{xz} = \frac{F_x}{A_z} \tag{3.5}$$

This shear stress leads to a shear strain γ

$$\gamma = \tan\frac{dx}{lx} \approx \frac{dx}{lx}. \tag{3.6}$$

The shear modulus is defined as the ratio between the shear strain and shear stress

$$\gamma = G\sigma_{xz} \tag{3.7}$$

In this chapter, we will further derive the elastodynamic theory and the related vector conventions to describe the solid wave motion in three-dimensional space.

The displacement $\mathbf{d}$ of a point is defined by its coordinates (x, y, z) in the three-dimensional space and its displacement components related to a system of unit vectors $\mathbf{e}_x$, $\mathbf{e}_y$ and $\mathbf{e}_z$.

$$\mathbf{d}(x,y,z) = \begin{Bmatrix} u(x,y,z)\,\mathbf{e}_x \\ v(x,y,z)\,\mathbf{e}_y \\ w(x,y,z)\,\mathbf{e}_z \end{Bmatrix} \tag{3.8}$$

A point located at $\{x, y, z\}^T$ that is displaced by $\{u, v, w\}^T$ will be located at $\{x + u, y + v, z + w\}^T$. In fluid dynamics this would correspond to the Lagrange description (chapter 2). There, the control volume is fixed and the mass and impulse flow are balanced based on this volume. In solid mechanics flow does not occur and we may assume small displacements, e.g. $\varepsilon_{zz} \ll 1$, and stay with the space-fixed convention.

3.2.1 Mechanical Strain

We consider a thin layer of thickness dx in the yz-plane. The displacement of the left point of position x is u. But the displacement of the right point originally located at $x + dx$ is determined by $u + \frac{\partial u}{\partial x}dx$. Thus the strain in the x-direction is given by:

$$\varepsilon_{xx} = \frac{\partial u}{\partial x} \tag{3.9}$$

A similar relationship can be found for y and z-directions.

$$\varepsilon_{yy} = \frac{\partial v}{\partial y} \tag{3.10}$$

$$\varepsilon_{zz} = \frac{\partial w}{\partial z} \tag{3.11}$$

The deformation of a solid can be described by the superposition of the following particular deformations:

dilatation	or global volume change into the orthogonal dimensions of space, i.e. strain in ε_{ii} in dimension $i = x, y, z$. (Figure 3.3a)
shear	deformation with angle $\gamma_{ij} = 2\varepsilon_{ij}/2 = \varepsilon_{ij}$. (Figure 3.3b)
rotations	or rigid body motion with rotation angle β_i with i being the normal vector to the plane of rotation. (Figure 3.3c)

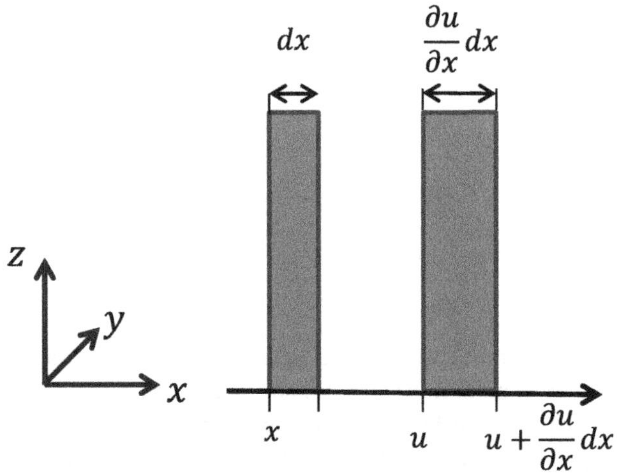

Figure 3.2 One dimensional displacement and strain of a thin layer of thickness dx.
Source: Alexander Peiffer.

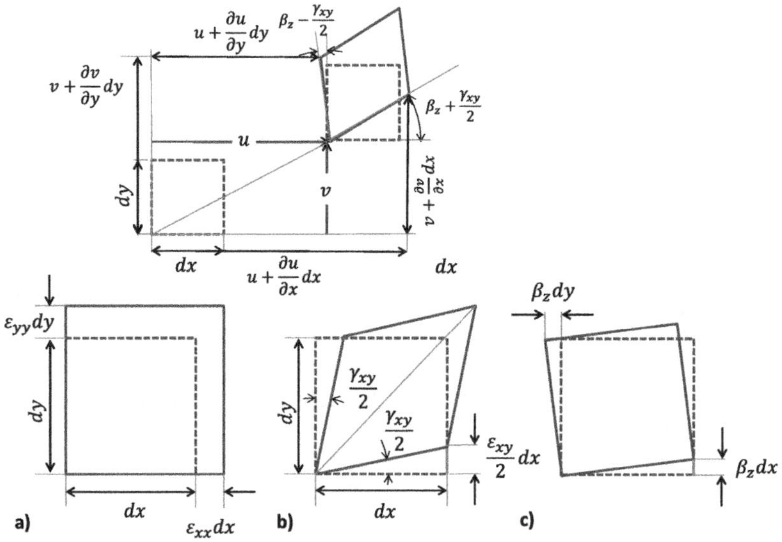

Figure 3.3 Deformation of a two-dimensional element of edge length dx and dy.
Source: Alexander Peiffer.

With those considerations we get the relationship between the shear angle $\gamma/2$ and the rotation angle β to the mechanical displacement. Assuming very small angles so that $\tan\gamma = \tan\varepsilon \approx \gamma$ this leads to

$$\frac{\partial u}{\partial y} = \frac{1}{2}\gamma_{xy} - \beta_z = \frac{1}{2}\varepsilon_{xy} - \beta_z \tag{3.12}$$

$$\frac{\partial v}{\partial x} = \frac{1}{2}\gamma_{xy} + \beta_z = \frac{1}{2}\varepsilon_{xy} + \beta_z \tag{3.13}$$

The difference of equations (3.12) and (3.13) gives

$$2\beta_z = \frac{\partial v}{\partial x} - \frac{\partial u}{\partial y} \tag{3.14}$$

and the sum of both equations leads to an expression for the deformation angle $\gamma_{xy} = \varepsilon_{xy}$

$$\varepsilon_{xy} = \gamma_{xy} = \frac{\partial v}{\partial x} + \frac{\partial u}{\partial y} \tag{3.15}$$

A similar expression can be found for the other shear angles ε_{yz} and ε_{zx}

$$\varepsilon_{yz} = \gamma_{yz} = \frac{\partial v}{\partial z} + \frac{\partial w}{\partial y} \tag{3.16}$$

$$\varepsilon_{zx} = \gamma_{zx} = \frac{\partial w}{\partial x} + \frac{\partial u}{\partial z} \tag{3.17}$$

3.2.1.1 Mechanical Strain - Voigt Notation

The mechanical strain ε_{ij} in three-dimensional space is the relative displacement of component i along dimension j. This is given by a symmetrical tensor of second order

$$\varepsilon_{kl} = \begin{pmatrix} \varepsilon_{xx} & \varepsilon_{xy} & \varepsilon_{xz} \\ \varepsilon_{yx} & \varepsilon_{yy} & \varepsilon_{yz} \\ \varepsilon_{zx} & \varepsilon_{zy} & \varepsilon_{zz} \end{pmatrix} = \begin{pmatrix} \varepsilon_{xx} & \gamma_{xy} & \gamma_{xz} \\ \gamma_{yx} & \varepsilon_{yy} & \gamma_{yz} \\ \gamma_{zx} & \gamma_{zy} & \varepsilon_{zz} \end{pmatrix} \tag{3.18}$$

We separate out the normal strain in the diagonal ε_{ii} that is defined as the extension into dimension i related to unit length. The shear strain is given be the shearing angle $\gamma_{ij} = \varepsilon_{ij}$. Thus we can define for $i, j = x, y, z, \eta_i = u, v, w$ and $\partial_i = \partial x, \partial y, \partial z$

$$\varepsilon_{ij} = \frac{\partial \eta_i}{\partial j} + \frac{\partial \eta_j}{\partial i} \tag{3.19}$$

Due to its symmetry the tensor in equation 3.18 has only six independent components, so the tensor can be alternatively represented in the *Voigt notation*

$$\varepsilon = \begin{Bmatrix} \varepsilon_1 \\ \varepsilon_2 \\ \varepsilon_3 \\ \varepsilon_4 \\ \varepsilon_5 \\ \varepsilon_6 \end{Bmatrix} = \begin{Bmatrix} \varepsilon_{xx} \\ \varepsilon_{yy} \\ \varepsilon_{zz} \\ \varepsilon_{yz} \\ \varepsilon_{xz} \\ \varepsilon_{xy} \end{Bmatrix} \tag{3.20}$$

3.2.1.2 Dilatation – Relative Change in Volume

The relative change ΔV of a volume $V = \Delta x \Delta y \Delta z$ can be found from the linear strains:

$$\frac{\delta V}{V} = \varepsilon_{xx} + \varepsilon_{yy} + \varepsilon_{zz} = \nabla \mathbf{d} \tag{3.21}$$

3.2.2 Mechanical Stress

According to the introduction of this section every force vector can be separated into two components (Figure 3.4):

1. **normal stress:** A normal force F_n in the direction of the normal vector of the surface.
2. **shear stress:** A tangential force perpendicular to the normal vector.

We consider the surfaces of a cube as shown in Figure 3.5 with the axes parallel to the coordinate axes. The symbol for stress is σ; the first index denotes the direction normal vector, and the second index the direction of the force. Thus, σ_{xx}, σ_{yy} and σ_{zz} are the normal stresses. σ_{ij} with $i \neq j$ and $i, j = x, y, z$ are the shear stresses. The definition is

$$\sigma_{ij} = \frac{dF_j}{dA_i} \tag{3.22}$$

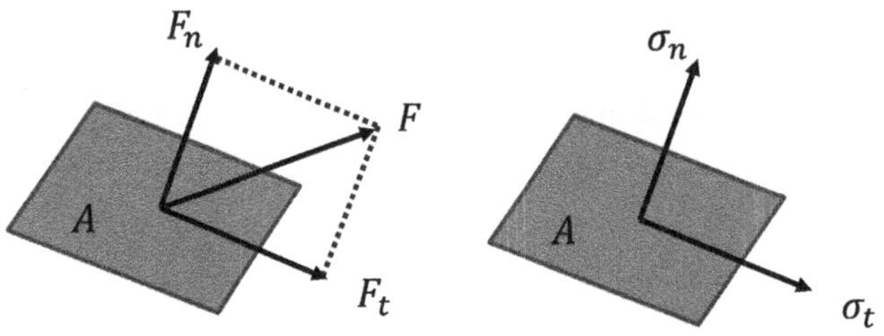

Figure 3.4 Force and unit area A. *Source:* Alexander Peiffer.

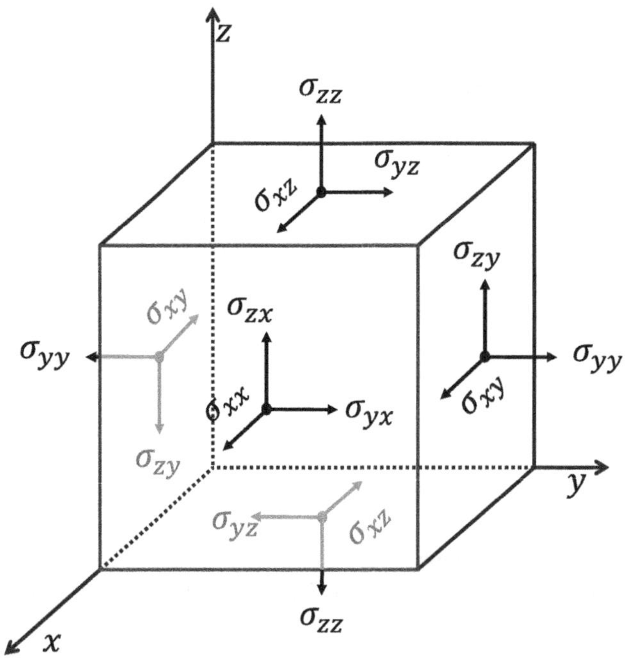

Figure 3.5 Stresses at the surfaces of a control volume *Source:* Alexander Peiffer.

Because of the moment of equilibrium the stresses of exchanged indices must be equal

$$\sigma_{xy} = \sigma_{yx} \quad \sigma_{yz} = \sigma_{zy} \quad \sigma_{xz} = \sigma_{zx} \tag{3.23}$$

Finally σ_{ij} is a symmetric tensor of second order.

$$\sigma_{xy} = \begin{bmatrix} \sigma_{xx} & \sigma_{xy} & \sigma_{xz} \\ \sigma_{yx} & \sigma_{yy} & \sigma_{yz} \\ \sigma_{zx} & \sigma_{zy} & \sigma_{zz} \end{bmatrix} \tag{3.24}$$

So we have six independent components of the stress tensor, and we can apply the Voigt notation

$$\sigma = \begin{Bmatrix} \sigma_{xx} \\ \sigma_{yy} \\ \sigma_{zz} \\ \sigma_{yz} \\ \sigma_{xz} \\ \sigma_{xy} \end{Bmatrix} = \begin{Bmatrix} \sigma_1 \\ \sigma_2 \\ \sigma_3 \\ \sigma_4 \\ \sigma_5 \\ \sigma_6 \end{Bmatrix} \tag{3.25}$$

3.2.3 Material Laws

The relationship between the strain and stress tensors is given by the material laws called Hooke's law for linear elastic media. These laws in their most common form are given by:

$$\sigma_{ij} = c_{ijkl}\, \epsilon_{kl} \tag{3.26}$$

$$\epsilon_{kl} = s_{klij}\, \sigma_{ij} \tag{3.27}$$

The first tensor is called the stiffness tensor or elasticity tensor and the second is named compliance tensor. They have 81 components that can be reduced to 36 because of symmetry. Using the Voigt notation the stress tensor can be written as a matrix

$$\begin{Bmatrix} \varepsilon_1 \\ \varepsilon_2 \\ \varepsilon_3 \\ \varepsilon_4 \\ \varepsilon_5 \\ \varepsilon_6 \end{Bmatrix} = \begin{bmatrix} s_{11} & s_{21} & s_{31} & s_{41} & s_{51} & s_{61} \\ s_{12} & s_{22} & s_{32} & s_{42} & s_{52} & s_{62} \\ s_{13} & s_{23} & s_{33} & s_{43} & s_{53} & s_{63} \\ s_{14} & s_{24} & s_{34} & s_{44} & s_{54} & s_{64} \\ s_{15} & s_{25} & s_{35} & s_{45} & s_{55} & s_{65} \\ s_{16} & s_{26} & s_{36} & s_{46} & s_{56} & s_{66} \end{bmatrix} \begin{Bmatrix} \sigma_1 \\ \sigma_2 \\ \sigma_3 \\ \sigma_4 \\ \sigma_5 \\ \sigma_6 \end{Bmatrix} \tag{3.28}$$

Those complex notations and equations are fortunately not that often necessary. For most engineering applications two special cases are applied: (1) Isotropic material, for example (steel, rubber, aluminium) and (2) orthotropic material (carbon fibre reinforced plastic (CFRP), steel enforced concrete, wood, plywood, etc.)

3.2.3.1 Isotropic Materials

Isotropy means, there is no dependency of the Hooke's law on space dimension or orientation. For the normal stress in the x-direction (or any other) we have

$$\sigma_1 = E\varepsilon_1 \tag{3.29}$$

This behavior is valid if the solid may perform lateral contraction. That means the elongation of a rod leads to lateral contraction. If this contraction is not constrained by surrounding solid material the elongation is reduced by the Poisson ratio ν, the negative ratio of transverse motion to axial strain. The strain due to stress in one direction must be corrected by the strain from other normal directions and the Poisson number.

$$E\varepsilon_1 = \sigma_1 - \nu(\sigma_2 + \sigma_3) \tag{3.30}$$

$$E\varepsilon_2 = \sigma_2 - \nu(\sigma_1 + \sigma_3) \tag{3.31}$$

$$E\varepsilon_3 = \sigma_3 - \nu(\sigma_1 + \sigma_2) \tag{3.32}$$

We can also apply shear stresses. There is a similar relationship to Equation (3.29)

$$\sigma_i = G\varepsilon_i \text{ for } i = 4, 5, 6 \tag{3.33}$$

Those three material constants are not independent, and they are linked by

$$E = 2G\,(1 + \nu) \tag{3.34}$$

Summarising all properties we can write the compliance and stiffness matrices:

$$[c] = \frac{E}{(1+\nu)(1-2\nu)} \begin{bmatrix} (1-\nu) & \nu & \nu & & & \\ \nu & (1-\nu) & \nu & & 0 & \\ \nu & \nu & (1-\nu) & & & \\ & & & \frac{1-2\nu}{2} & & \\ & 0 & & & \frac{1-2\nu}{2} & \\ & & & & & \frac{1-2\nu}{2} \end{bmatrix} \tag{3.35}$$

$$[s] = [c]^{-1} = \frac{1}{E} \begin{bmatrix} 1 & -\nu & -\nu & & & \\ -\nu & 1 & -\nu & & 0 & \\ -\nu & -\nu & 1 & & & \\ & & & 2(1+\nu) & & \\ & 0 & & & 2(1+\nu) & \\ & & & & & 2(1+\nu) \end{bmatrix} \tag{3.36}$$

3.2.3.2 Ortotropic Solids

This class of material occurs for example in composites with fibres or layered microstructures. The properties change with the normal axes of the three-dimensional coordinate system. Thus, orthotropic materials have three orthogonal planes of symmetry. If they coincide with the coordinate system we get for the stiffness matrix a common notation.

$$\begin{Bmatrix} \sigma_1 \\ \sigma_2 \\ \sigma_3 \\ \sigma_4 \\ \sigma_5 \\ \sigma_6 \end{Bmatrix} = \begin{bmatrix} C_{11} & C_{12} & C_{13} & 0 & 0 & 0 \\ C_{12} & C_{22} & C_{23} & 0 & 0 & 0 \\ C_{13} & C_{23} & C_{33} & 0 & 0 & 0 \\ 0 & 0 & 0 & C_{44} & 0 & 0 \\ 0 & 0 & 0 & 0 & C_{55} & 0 \\ 0 & 0 & 0 & 0 & 0 & C_{66} \end{bmatrix} \begin{Bmatrix} \epsilon_1 \\ \epsilon_2 \\ \epsilon_3 \\ \epsilon_4 \\ \epsilon_5 \\ \epsilon_6 \end{Bmatrix} \tag{3.37}$$

This matrix can be derived from the inversion of the compliance matrix. However, this leads to lengthy expressions that will not provide any additional insight, whereas the compliance matrix can be presented in a compact form

$$\begin{Bmatrix} \epsilon_1 \\ \epsilon_2 \\ \epsilon_3 \\ \epsilon_4 \\ \epsilon_5 \\ \epsilon_6 \end{Bmatrix} = \begin{bmatrix} \frac{1}{E_x} & -\frac{\nu_{yx}}{E_y} & -\frac{\nu_{zx}}{E_z} & 0 & 0 & 0 \\ -\frac{\nu_{xy}}{E_x} & \frac{1}{E_y} & -\frac{\nu_{zy}}{E_z} & 0 & 0 & 0 \\ -\frac{\nu_{xz}}{E_x} & -\frac{\nu_{yz}}{E_y} & \frac{1}{E_z} & 0 & 0 & 0 \\ 0 & 0 & 0 & \frac{1}{G_{yz}} & 0 & 0 \\ 0 & 0 & 0 & 0 & \frac{1}{G_{zx}} & 0 \\ 0 & 0 & 0 & 0 & 0 & \frac{1}{G_{xy}} \end{bmatrix} \begin{Bmatrix} \sigma_1 \\ \sigma_2 \\ \sigma_3 \\ \sigma_4 \\ \sigma_5 \\ \sigma_6 \end{Bmatrix} \tag{3.38}$$

where E_i is the Young's modulus along axis i, G_{ij} is the shear modulus in direction j on the plane whose normal is in direction i. ν_{ij}, is the Poisson's ratio that corresponds to a contraction in direction j when an extension is applied in direction i.

3.3 Wave Equation

In solids there are several modes of wave propagation. In addition the type of wave mode depends on the structure geometry, if it is infinite bulk material, a flat or curved plate, a beam or rod. In the following sections we will go step by step through the different wave equations for several structural configurations. For fluids the wave equation was derived from the linearized versions of the equation of motion or Newton's law and a combination of the continuity equations and the state law. For solids the wave equation is determined from the equations of motion and the material laws. In this section we will only deal with isotropic materials.

3.3.1 The One-dimensional Wave Equation

The field quantity for solids is the displacement $\mathbf{d} = \{u, v, w\}^T$. We start from the control volume as shown in Figure 3.5. If we assume a cut-free cube there is a stress balance at each cube surface: i.e. that the stresses are in balance at each interface and have opposite direction (Figure 3.6). From the stress balance over the cube surfaces the resultant

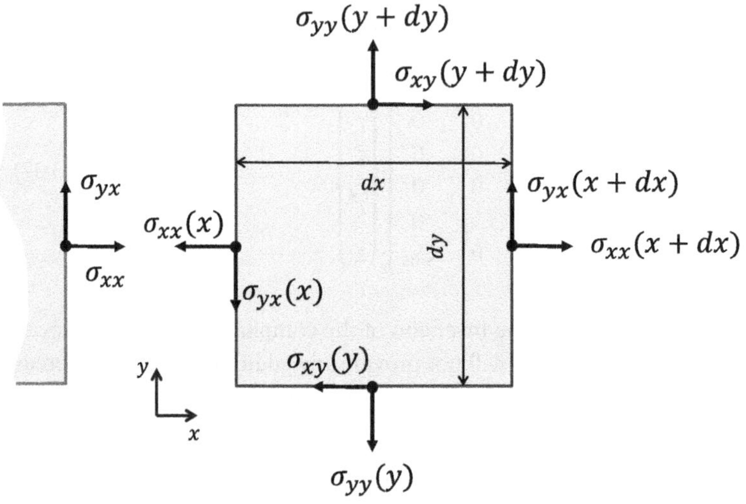

Figure 3.6 Stresses at the surfaces of a control volume (top view) *Source:* Alexander Peiffer.

force in the x-direction is:

$$\begin{aligned} dF_x &= (\sigma_{xx}(x+dx) - \sigma_{xx}(x))dxdy \\ &+ (\sigma_{xy}(y+dy) - \sigma_{xy}(y))dxdz \\ &+ (\sigma_{xz}(z+dz) - \sigma_{xz}(z))dxdy \end{aligned} \tag{3.39}$$

From Newton's law $dF_x = dm\,\ddot{u}$ and with $dm = \rho_0 dxdydz$ we get after division with $dV = dxdydz$ and the series expansion for $\sigma_{xx}(x+dx)$

$$\begin{aligned} \frac{\sigma_{xx}(x+dx) - \sigma_{xx}(x)}{dx} &\approx \frac{\sigma_{xx}}{dx} + \frac{\partial \sigma_{xx}}{\partial x} - \frac{\sigma_{xx}}{dx} \\ &= \frac{\partial \sigma_{xx}}{\partial x} \end{aligned} \tag{3.40}$$

the equation of motion in the x-direction

$$\frac{\partial \sigma_{xx}}{\partial x} + \frac{\partial \sigma_{xy}}{\partial y} + \frac{\partial \sigma_{xz}}{\partial z} = \rho_0 \frac{\partial^2 u}{\partial t^2} \tag{3.41}$$

$$\tag{3.42}$$

Similar treatment leads to the two equations of motion in y and z-directions

$$\frac{\partial \sigma_{xy}}{\partial x} + \frac{\partial \sigma_{yy}}{\partial y} + \frac{\partial \sigma_{yz}}{\partial z} = \rho_0 \frac{\partial^2 v}{\partial t^2} \tag{3.43}$$

$$\frac{\partial \sigma_{xz}}{\partial x} + \frac{\partial \sigma_{yz}}{\partial y} + \frac{\partial \sigma_{zz}}{\partial z} = \rho_0 \frac{\partial^2 w}{\partial t^2} \tag{3.44}$$

Thus, for deriving the wave equation in displacement coordinates the stresses must be expressed in terms of the displacement making use of the stress–strain relationship of isotropic solids. We start with the sum of equations (3.30)–(3.32).

$$E\,(\varepsilon_{xx} + \varepsilon_{yy} + \varepsilon_{zz}) = \sigma_{xx} + \sigma_{yy} + \sigma_{zz} - 2\nu(\sigma_{xx} + \sigma_{yy} + \sigma_{zz}) \tag{3.45}$$

With the definition of the dilatation $\nabla\mathbf{d}$ (3.21) and (3.34) this gives:

$$2G(1+\nu)\nabla\mathbf{d} = (1-2\nu)(\sigma_{xx} + \sigma_{yy} + \sigma_{zz}) \tag{3.46}$$

The goal is to find an expression for σ_{xx} depending only on displacements not linked to other stresses. We start with (3.30) solving for σ_{xx} and adding $\nu\sigma_{xx}$ on both sides.

$$\sigma_{xx}(1+\nu) = E\varepsilon_{xx} + \nu(\sigma_{xx} + \sigma_{yy} + \sigma_{zz}) \tag{3.47}$$

With (3.45) this leads to

$$\sigma_{xx}(1+\nu) = E\varepsilon_{xx} + \frac{\nu 2G(1+\nu)}{1-2\nu}\nabla\mathbf{d}. \tag{3.48}$$

Eliminating E yields

$$\sigma_{xx} = 2G\varepsilon_{xx} + \frac{2G}{1-2\nu}\nu\nabla\mathbf{d} \tag{3.49}$$

or

$$\sigma_{xx} = 2G\left(\frac{\partial u}{\partial x} + \frac{\nu}{1-2\nu}\nabla\mathbf{d}\right) \tag{3.50}$$

We introduce the Lamé constants

$$\lambda = 2G\frac{\nu}{1-2\nu} = \frac{E}{1+\nu}\frac{\nu}{1-2\nu} \tag{3.51}$$

$$\mu = G \tag{3.52}$$

λ is a measure for the extension perpendicular to the stress. Applying these constants we get from (3.50)

$$\sigma_{xx} = 2\mu\frac{\partial u}{\partial x} + \lambda\nabla\mathbf{d} \tag{3.53}$$

The equations for the y and z-directions can be derived similarly:

$$\sigma_{yy} = 2\mu\frac{\partial v}{\partial y} + \lambda\nabla\mathbf{d} \tag{3.54}$$

$$\sigma_{zz} = 2\mu\frac{\partial w}{\partial z} + \lambda\nabla\mathbf{d} \tag{3.55}$$

From the definition of the isotropic shear modulus (3.33) and the shear strains (3.15) the equations for the other components are found:

$$\sigma_{yz} = \mu\left(\frac{\partial v}{\partial z} + \frac{\partial w}{\partial y}\right) \tag{3.56}$$

$$\sigma_{xz} = \mu\left(\frac{\partial u}{\partial z} + \frac{\partial w}{\partial x}\right) \tag{3.57}$$

$$\sigma_{xy} = \mu\left(\frac{\partial u}{\partial y} + \frac{\partial v}{\partial x}\right) \tag{3.58}$$

Using equations (3.53)–(3.58) we get after some algebraic modifications the wave equation for the one-dimensional case in the x-direction

$$\mu \Delta u + (\mu + \lambda) \frac{\partial}{\partial x} (\nabla \mathbf{d}) = \rho_0 \frac{\partial^2 u}{\partial t^2} \tag{3.59}$$

Similar and equivalent derivation can be done for the remaining two dimensions.

$$\mu \Delta v + (\mu + \lambda) \frac{\partial}{\partial y} (\nabla \mathbf{d}) = \rho_0 \frac{\partial^2 v}{\partial t^2} \tag{3.60}$$

$$\mu \Delta w + (\mu + \lambda) \frac{\partial}{\partial z} (\nabla \mathbf{d}) = \rho_0 \frac{\partial^2 w}{\partial t^2} \tag{3.61}$$

3.3.2 The Three-dimensional Wave Equation

Collecting the one-dimensional wave equations and putting them together for the displacement vector $\mathbf{d}$ leads to

$$\mu \Delta \mathbf{d} + (\mu + \lambda) \nabla (\nabla \mathbf{d}) = \rho_0 \frac{\partial^2 \mathbf{d}}{\partial t^2} \tag{3.62}$$

With the identity $\nabla(\nabla \mathbf{d}) = \Delta \mathbf{d} + \nabla \times \nabla \mathbf{d}$ we get the wave equation in a different form

$$(2\mu + \lambda) \Delta \mathbf{d} + (\mu + \lambda) \nabla \times (\nabla \times \mathbf{d}) = \rho_0 \frac{\partial^2 \mathbf{d}}{\partial t^2} \tag{3.63}$$

This wave equation can only be solved in closed form for simple geometries. In most cases numerical methods like the finite element method must be applied to solve this equation for realistic systems.

There are two differential operators in space on the LHS of Equation (3.63) indicating two wave types in solids. Any vector field can be separated into two independent components. The irrotational and the source-free or zero-divergence component

$$\mathbf{d} = \mathbf{d}_L + \mathbf{d}_T \tag{3.64}$$

An irrotational field is characterized by

$$\nabla \times \mathbf{d}_L = \mathbf{0} \tag{3.65}$$

and a source free field by

$$\nabla \mathbf{d}_T = 0 \tag{3.66}$$

Thus

$$\nabla \times \mathbf{d}_T = \nabla \times \mathbf{d} \tag{3.67}$$

$$\nabla \mathbf{d}_L = \nabla \mathbf{d} \tag{3.68}$$

These relationships show that there are two independent wave types in solids, called longitudinal waves (L) and shear or transversal waves (S).

3.4 Waves in Infinite Solids

In order to understand the principle dynamics of both waves we consider an unbounded infinite solid. Practical applications can be found in geophysics or ultrasonics for non-destructive testing. Engineering acoustics and vibroacoustics usually deal with vehicles. Thus, the construction must be lightweight, and bulky solid materials are rarely found. With one exception, some materials for acoustic treatment such as foams have soft bulk material with wavelengths small enough that those wave types occur in practical technical systems.

3.4.1 Longitudinal Waves

Longitudinal waves are rotation free, thus $\nabla \times \mathbf{d} = \mathbf{0}$ and the wave equation (3.63) simplifies to

$$\Delta \mathbf{d} = \frac{1}{c_L^2}\frac{\partial^2 \mathbf{d}}{\partial t^2} \text{ with } c_L = \sqrt{\frac{2\mu + \lambda}{\rho_0}} = \sqrt{\frac{2G(1-\nu)}{\rho_0(1-2\nu)}} = \sqrt{\frac{E(1-\nu)}{\rho_0(1+\nu)(1-2\nu)}} \tag{3.69}$$

c_L is called the longitudinal wave speed. In isotropic materials this is the wave with the highest speed. When we calculate c_L for typical materials like iron, aluminium, wood etc., it becomes obvious why the practical relevance in vibroacoustics is quite low. For example, for iron the wavelength at $f = 1$ kHz equals 5 m, extending by far beyond the typical dimension of bulk parts in a car. In order to illustrate the propagation of a longitudinal wave we calculate the solution for propagation into the x-direction. That means $v = w = 0$ and $\frac{\partial}{\partial y} = \frac{\partial}{\partial z} = 0$. The longitudinal wave equation reduces to

$$\frac{\partial^2 u}{\partial x^2} - \frac{1}{c_L^2}\frac{\partial^2 u}{\partial t^2} = 0 \tag{3.70}$$

The wave motion is characterized by particles oscillating in the direction of propagation. All shear stresses are zero, and a plane of constant dilatation propagates in the wave direction. Thus, those waves are also called dilatation or density waves. Equation (2.34) from chapter 2 can be applied, and the harmonic solution is

$$u(x) = \boldsymbol{u} e^{j(\omega t - k_L x)} \text{ with } k_L = \frac{\omega}{c_L}. \tag{3.71}$$

In the upper part of Figure 3.7 the particle motion is shown for this wave type.

3.4.2 Shear waves

In contrast to the longitudinal wave the shear wave is source free, thus $\nabla \mathbf{d} = 0$. In that case the wave Equation (3.63) leads to

$$\nabla \times \mathbf{d} = \frac{1}{c_S^2}\frac{\partial^2 \mathbf{d}}{\partial t^2} \text{ with } c_S = \sqrt{\frac{\mu}{\rho_0}} = \sqrt{\frac{G}{\rho_0}} \tag{3.72}$$

The wave motion is iso-voluminous. It can be shown by the introduction of a rotation vector that the particle motion is orthogonal or *transversal* to the wave propagation

(Graff, 1991). This is the reason why this wave type is also called transversal wave. A simple plane wave equation for w is

$$\frac{\partial^2 w}{\partial x^2} - \frac{1}{c_S^2}\frac{\partial^2 w}{\partial t^2} = 0 \tag{3.73}$$

A solution in the positive x-direction is for example

$$w(x) = \boldsymbol{w} e^{j(\omega t - k_S x)} \text{ with } k_S = \frac{\omega}{c_S} \tag{3.74}$$

In the lower part of Figure 3.7 the particle motion for one fixed time is shown. From (3.34) it follows that shear waves are slower than longitudinal waves.

3.5 Beams

Beams are one-dimensional structures that have a cross section small compared to their length. Due to the large wavelength derived for the bulk material we can assume that there is no wave propagation perpendicular to the main axis of the beam. There are a variety of wave types, but we will deal only with two in this section.

3.5.1 Longitudinal Waves

Longitudinal waves are associated with normal stresses in the direction of the beam axis. We consider a free cut small layer of the beam and take the stress balance (3.32).

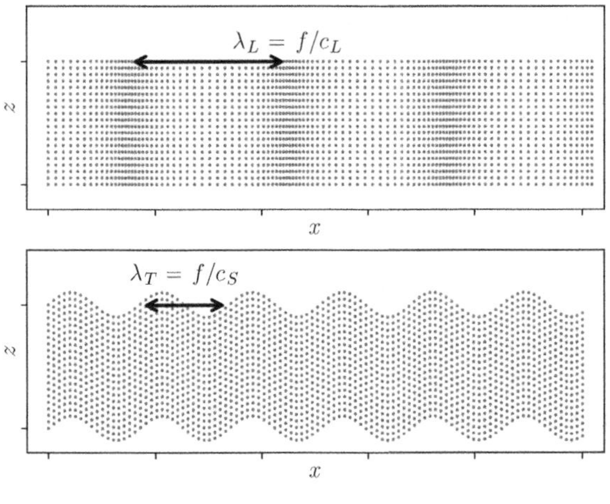

Figure 3.7 Displacement of solid particles for the different wave types. *Source:* Alexander Peiffer.

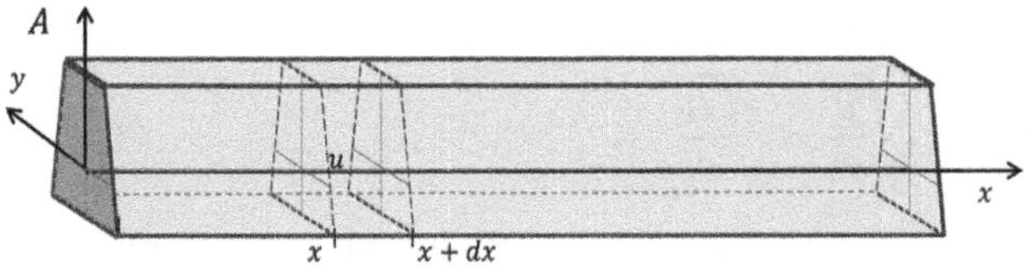

Figure 3.8 Beam in x-direction with cross section A. *Source:* Alexander Peiffer.

There are no constraints for y and z surfaces and therefore $\sigma_{yy}, \sigma_{zz}, \sigma_{yz}, \sigma_{xz} = 0$. Thus, the stress strain relationship in the x-direction reads

$$\sigma_{xx} = E\frac{\partial u}{\partial x} \tag{3.75}$$

Using this we get with Newton's law from Equation (3.41)

$$\frac{\partial^2 u}{\partial x^2} - \frac{1}{c_{LB}}\frac{\partial^2 u}{\partial t^2} = 0 \quad \text{with} \quad c_{LB} = \sqrt{\frac{E}{\rho_0}} \tag{3.76}$$

or in harmonic form for $u(x,t) = \boldsymbol{u}e^{j\omega t}$

$$\frac{\partial^2 \boldsymbol{u}}{\partial x^2} + k_{LB}^2\boldsymbol{u} = 0 \quad \text{with} \quad k_{LB} = \sqrt{\frac{\omega^2\rho_0}{E}} \tag{3.77}$$

The solution of this one-dimensional wave equation is

$$\boldsymbol{u}(x,t) = \boldsymbol{u}^+e^{-jkx} + \boldsymbol{u}^-e^{+jkx} \tag{3.78}$$

Due to the fact that the bar is stress free in the direction normal to the bar axis, the wave speed in bars c_{LB} is slower than the longitudinal wave speed in infinite solids. This lead to displacement in y and z that can be calculated from the first column of (3.36):

$$\varepsilon_{yy} = -\frac{\nu}{E}\sigma_{xx} \tag{3.79}$$

$$\varepsilon_{zz} = -\frac{\nu}{E}\sigma_{xx} \tag{3.80}$$

The only information related to the cross section that is required for the determination of the longitudinal wave speed is that the dimensions of the cross sections must be small compared to the wave length.

3.5.2 Power, Energy, and Impedance

When exciting a half beam as shown in Figure 3.9 we create a wave propagating in the positive direction leaving out the e^{-jkx} from (3.78)

$$\boldsymbol{u}(x,t) = \boldsymbol{u}^+e^{-jkx} \tag{3.81}$$

The stress in x is with (3.75):

$$\begin{aligned}\sigma_{xx}(x) &= E\frac{\partial \boldsymbol{u}}{\partial x} = -jEk_{LB}\boldsymbol{u}^+e^{-jkx} \\ &= -j\omega\boldsymbol{u}^+\rho_0 c_{LB}e^{-jkx} = -\boldsymbol{v}^+z_{LB}e^{-jkx}\end{aligned} \tag{3.82}$$

So, there is a longitudinal characteristic impedance $z_{LB} = \rho_0 c_{LB}$ of the longitudinal wave in beams. Assuming a force $\boldsymbol{F}_x$ exciting the beam in the x-direction via a stiff plate

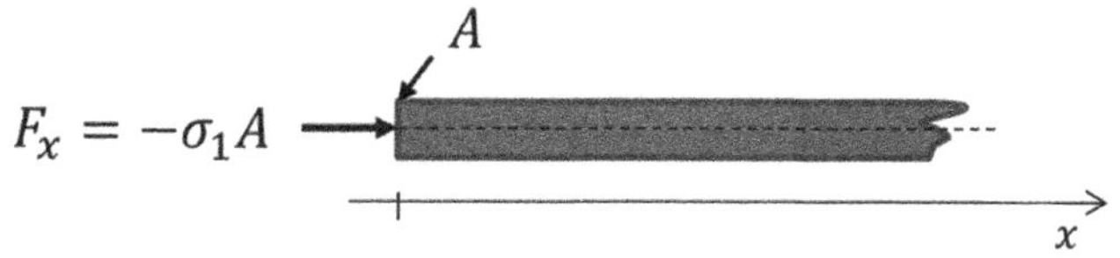

Figure 3.9 Excitation of longitudinal waves in half infinite beam. *Source:* Alexander Peiffer.

at the beam cross section leads to the boundary condition at $x = 0$ with $\boldsymbol{F}_z = -A\boldsymbol{\sigma}_1(x)$

$$\boldsymbol{\sigma}_1(0) = -j\omega\boldsymbol{u}^+ z_{LB} = -\boldsymbol{v}^+ z_{LB} \quad \Rightarrow \quad \boldsymbol{u}^+ = -\frac{\boldsymbol{\sigma}(0)}{j\omega z_{LB}} = \frac{\boldsymbol{F}_x}{j\omega A z_{LB}} \tag{3.83}$$

The mechanical impedance follows from this

$$\boldsymbol{Z} = \frac{\boldsymbol{F}_x}{\boldsymbol{v}} = \frac{\boldsymbol{F}_x}{j\omega\boldsymbol{u}^+} = A z_{LB} \tag{3.84}$$

The power introduced into the system is, when we assume without loss of generality $\boldsymbol{F}_x = \hat{F}_x$

$$\Pi = \frac{1}{2}\frac{\hat{F}_x^2}{A\, z_{LB}} \tag{3.85}$$

This power propagates into the beam and creates an energy density per length $E' = \Pi/c_{LB}$, because the power per time unit is distributed with velocity c_{LB} over the related length unit. This can also be proven in a more complicated way by deriving the kinetic and potential energy density per length

$$E'_{\text{kin}} = \frac{1}{2}A\rho_0 Re(\boldsymbol{v}_x(x)\boldsymbol{v}_x^*(x)) = \frac{1}{2}\frac{\hat{F}_x^2}{A z_{LB} c_{LB}}\cos^2(kx) \tag{3.86}$$

For the potential energy we have to integrate the stress over the strain

$$E'_{\text{pot}} = A\int_0^{\varepsilon_{xx}} \sigma_{xx} d\varepsilon_1 = \frac{A}{2}E\varepsilon_1^2 = \frac{A}{2E}\sigma_1^2 \tag{3.87}$$

with $\sigma_{xx}(x) = Re(-\frac{\hat{F}_x}{A})$ we get for the potential energy density per length

$$E'_{\text{pot}} = \frac{A}{2E}\frac{\hat{F}_x^2}{A^2}\cos^2(kx) = \frac{1}{2}\frac{\hat{F}_x^2}{A z_{LB} c_{LB}}\cos^2(kx) \tag{3.88}$$

So we find $E'_{kin} = E'_{pot}$. With (3.83) we get for the total energy expressed for the velocity

$$E'_{\text{tot}} = \frac{1}{2}A\left(\rho_0 + \frac{z_{LB}^2}{E}\right) Re(\boldsymbol{v}_x(x)\boldsymbol{v}^*(x)) = A\rho_0 Re(\boldsymbol{v}_x(x)\boldsymbol{v}^*(x)). \tag{3.89}$$

Integrating over one wavelength λ gives the total length specific energy

$$E'_{\text{tot}} = \frac{\hat{F}_x^2}{A z_{LB} c_{LB}}\int_0^{\lambda} \cos^2(kx)dx = \frac{1}{2}\frac{\hat{F}_x^2}{A z_{LB} c_{LB}} \tag{3.90}$$

proving the above arguments regarding power and speed of sound.

3.5.3 Bending Waves

The bending stiffness depends on area and shape of the cross section. In Figure 3.11 the cross section, forces, and moments of a cut-free slice of the beam are shown. We assume that there is a neutral axis without any dilation and pure bending. That means that each cross section of the un-loaded straight beam remains orthogonal to the neutral axis as shown in Figure 3.12. This is called the Bernoulli assumption.

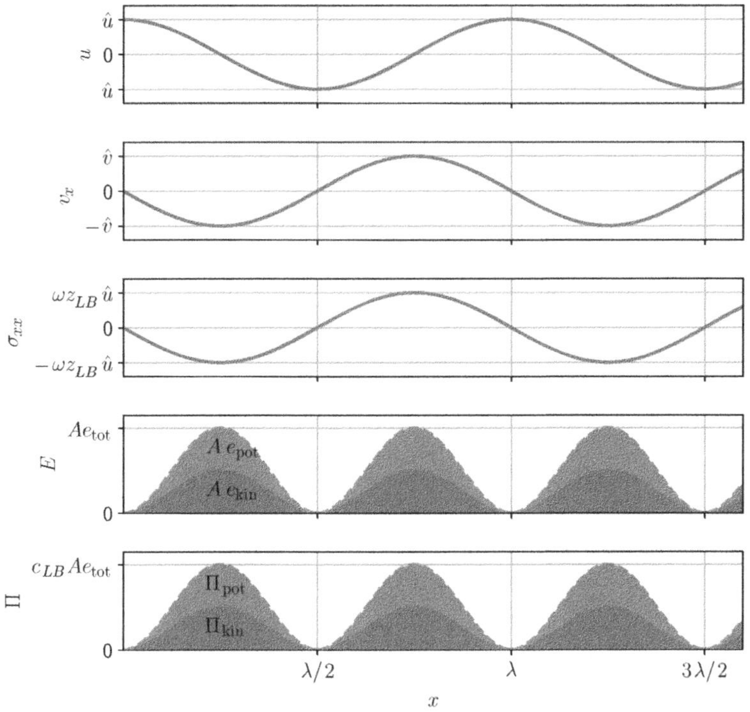

Figure 3.10 Various quantities for longitudinal wave propagation in structures. *Source:* Alexander Peiffer.

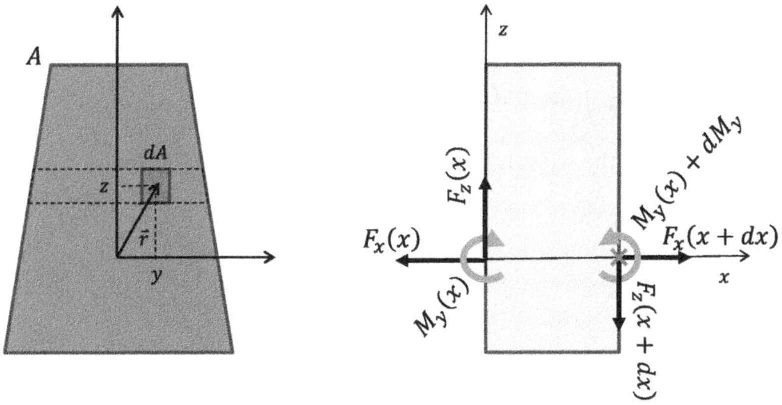

Figure 3.11 Cross section and segment of a beam. *Source:* Alexander Peiffer.

Geometrically this assumption means that all fibres stay orthogonal to the cross section. The stress is linear with the bending radius R as depicted in Figure 3.12. The stress in the direction of the neutral axis vanishes at the neutral axis[1]. The strain out of

1 The direction convention differs from the plate theory. The main reason for that is that the moments are defined in such a way that the curvature for positive moment is also positive

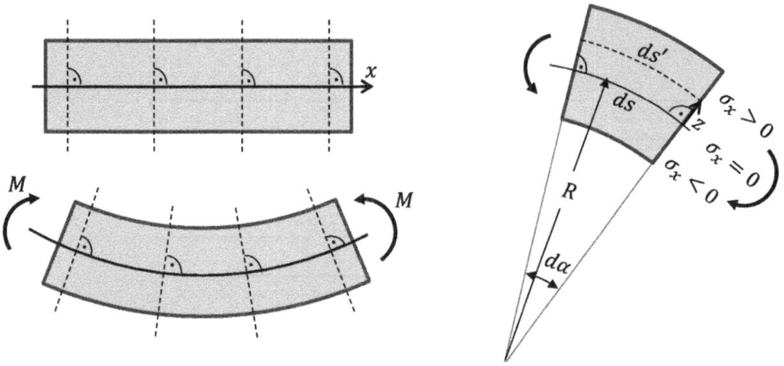

Figure 3.12 Bernoulli bending of beams. *Source:* Alexander Peiffer.

the neutral axis can be derived from Figure 3.12.

$$ds' = (R + z)d\alpha = Rd\alpha + zd\alpha = ds + zd\alpha$$

and therefore

$$\varepsilon_{xx}(z) = \frac{ds' - ds}{ds} = z\frac{d\alpha}{ds} = \frac{z}{R}$$

Due to Equation (3.29) it can be shown, that the stress is a linear function of the distance from the neutral axis z and the curvature $1/R$.

$$\sigma_{xx}(z) = E\frac{z}{R(x)} \tag{3.91}$$

The moment vector $\mathbf{M} = (0, M_y, M_y)$ is given by the integral over all small moment contributions at $\mathbf{r}$ with $d\mathbf{M} = \mathbf{r} \times \sigma_{xx} dA$. Thus:

$$M_y = \int_A \sigma_{xx} z dA = \frac{E}{R}\int_A \sigma_{xx} z^2 dA \tag{3.92}$$

and due to the assumptions the stress is symmetrical to the z-plane

$$M_z = \int_A \sigma_{xx} y dA = 0 \tag{3.93}$$

We call

$$I_{yy} = \int_A z^2 dA > 0 \tag{3.94}$$

the second moment of area, in most literature called moment of inertia[2]. The moment

$$I_{yz} = \int_A yzdF \tag{3.95}$$

that vanishes in our special case is called the product moment of area. There are axes of the cross section where this product moment is zero. They are called the main axes

2 There are many wordings used for the second moment of area: moment of inertia of plane area, area moment of inertia, or second area moment. There is no mass involved in the definition, so the expression "inertia" might be misleading.

and are for example the symmetry axes of a symmetrical section. Using (3.94) we can write (3.92) as

$$M_y(x) = \frac{E}{R(x)} I_{yy}(x) \tag{3.96}$$

Using the following differential equation for the curvature $1/R$

$$1/R(x) = \frac{\frac{\partial w^2(x)}{\partial x^2}}{-\left[1 + \left(\frac{\partial w(x)}{\partial x}\right)^2\right]^{3/2}} = \frac{M_y(x)}{EI_{yy}(x)} \tag{3.97}$$

We assume small displacements $w \ll 1$, and we get approximately

$$1/R(x) \approx \frac{\partial w^2(x)}{\partial x^2} \tag{3.98}$$

and with that

$$EI_{yy} \frac{\partial^2 w}{\partial x^2} = -M_y \tag{3.99}$$

For the z-axis we get accordingly

$$EI_{zz} \frac{\partial^2 v}{\partial x^2} = -M_z \tag{3.100}$$

The force in the z-direction can be derived from the moment equilibrium at the position $x + dx$ in Figure 3.11.

$$M_y(x) + F_z dx = M_y(x) + dM_y \quad \Rightarrow \quad F_z(x) = -\frac{dM_y}{dx} \tag{3.101}$$

Now, we set up the force balance at the control segment from Figure 3.11.

$$\rho_0 dx A \frac{\partial^2 w}{\partial t^2} = F_z(x) - F_z(x + dx) \tag{3.102}$$

Performing the limiting process for $dx \to 0$ gives

$$m' \frac{\partial^2 w}{\partial t^2} = -\frac{\partial F_z}{\partial x} = +\frac{\partial^2 M_y}{\partial x^2} = -EI_{yy} \frac{\partial^4 w}{\partial x^4} \tag{3.103}$$

with m' being the mass per length or length specific density. The prime ($'$) specifies units per length in this section. Using the harmonic wave solution $w(x,t) = \boldsymbol{w} e^{j(\omega t - k_B x)}$ we get the bending wave equation in the frequency domain

$$-\omega^2 m' \boldsymbol{w} + B_y \frac{\partial^4 \boldsymbol{w}}{\partial x^4} = 0 \tag{3.104a}$$

$$-\omega^2 m' \boldsymbol{v} + B_z \frac{\partial^4 \boldsymbol{v}}{\partial x^4} = 0 \tag{3.104b}$$

with the beam bending stiffness $B_y = EI_{yy}$ and $B_z = EI_{zz}$ for the principle y and z-axes of the cross section. Neglecting the time dependence and performing the spatial derivatives we get for the bending wavenumber

$$k_B^4 = \frac{m'}{B_y}\omega^2 \tag{3.105}$$

with two real and two complex solutions

$$k_{B1/2} = \pm\sqrt[4]{\frac{\omega^2 m'}{B_y}} \quad k_{B3/4} = \pm j\sqrt[4]{\frac{\omega^2 m'}{B_y}} \tag{3.106}$$

Thus the solution includes two propagating waves and two decaying

$$\boldsymbol{w}(x,\omega) = \boldsymbol{w}_1 e^{jk_B x} + \boldsymbol{w}_2 e^{-jk_B x} + \boldsymbol{w}_{1D} e^{k_B x} + \boldsymbol{w}_{2D} e^{-k_B x} \tag{3.107}$$

We neglect the complex wave numbers leading to exponential decaying waves and calculate the wave speed from the fourth root of (3.105)

$$c_B = \frac{\omega}{k_B} = \sqrt{\omega}\sqrt[4]{\frac{B_y}{m'}} \tag{3.108}$$

Here, we have the effect of dispersion, and the wave speed depends on the frequency. The main effect is that the propagation of energy is different from the propagation of constant phase in the wave. This was not the case for fluids where the speed of sound is constant over frequency. The consequences of this will be discussed in section 3.8.

If we calculate the bending wave speed for a typical system we see that we have a wave speed of similar order of magnitude as in air. Let us consider a rectangular cross section as shown in Figure 3.13. The moments are

$$I_{yy} = \frac{h^3 b}{12} \quad I_{zz} = \frac{hb^3}{12} \tag{3.109}$$

Assuming aluminium with isotropic material properties $E = 71 \times 10^9$ Pa, $\rho_0 = 2700\,\text{kg/m}^3$ and $\nu = 0.34$ the rectangular cross section with $b = 3\,cm$ and $h = 2\,cm$ yields the following wave speed for bending in the z and y-directions (Figure 3.14).

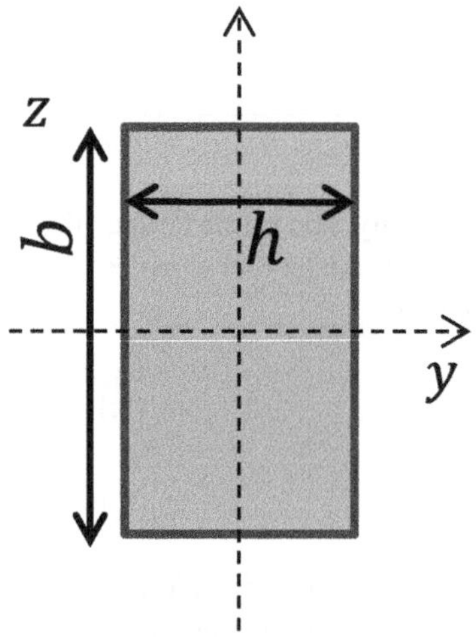

Figure 3.13 Rectangular cross section. *Source:* Alexander Peiffer.

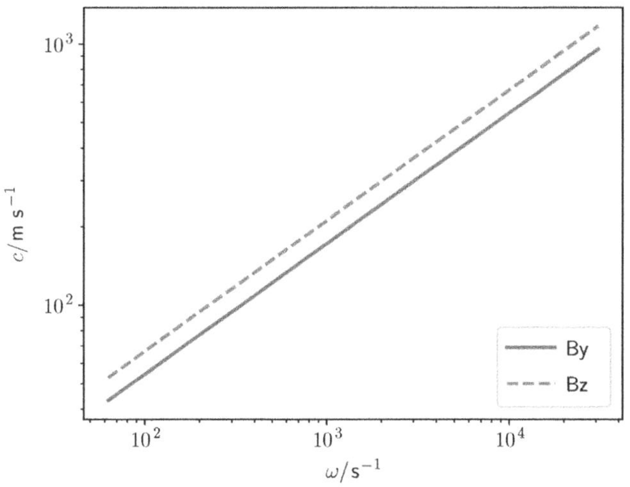

Figure 3.14 Bending wave speed of rectangular aluminium beam (b = 3 cm,h = 2 cm). *Source:* Alexander Peiffer.

3.5.4 Power, Energy, and Impedance

The frequency dependence of the wave speed leads to a different propagation speed, point of constant phase, and energy stored in the wave. For the power considerations we summarize the following relationships from the above calculations

$$\boldsymbol{M}_y = -B\frac{\partial^2 \boldsymbol{w}}{\partial x^2} \quad \boldsymbol{F}_z = -\frac{\partial \boldsymbol{M}_y}{\partial x} \quad -\omega^2 m' \boldsymbol{w} = \frac{\partial \boldsymbol{F}_z}{\partial x} \tag{3.110}$$

One power component in beams is the force–velocity product as in (1.47). In addition the product of moment M_y and the rotational speed $\dot{\beta}_y$ must be taken into consideration

$$\Pi_{\text{beam}} = \Pi_M + \Pi_F \tag{3.111}$$

with

$$\Pi_F = \frac{1}{2} Re\left(\boldsymbol{F}_z \boldsymbol{v}_z{}^* + \boldsymbol{F}_z \boldsymbol{v}_z e^{j2\omega t}\right) \tag{3.112}$$

$$\Pi_M = \frac{1}{2} Re\left(\boldsymbol{M}_y \dot{\boldsymbol{\beta}}_y{}^* + \boldsymbol{M}_y \dot{\boldsymbol{\beta}}_y e^{j2\omega t}\right) \tag{3.113}$$

We keep the oscillating part of the power in order to show a characteristic effect for wave propagation in beams. We will use the positively propagating part of the harmonic solution (3.107) and neglect the evanescent wave that is not relevant for power transport. For further simplifications of the expressions we assume a real displacement amplitude $\boldsymbol{w}_2 = \hat{w}_2$ and we get for the force and the velocity

$$\boldsymbol{F}_z = j\boldsymbol{w}_2 B k_B^3 e^{-jk_B x} \tag{3.114}$$

$$\boldsymbol{v}_z = j\omega \boldsymbol{w}_2 e^{-jk_B x} \tag{3.115}$$

$$\begin{aligned} \Pi_F &= \frac{1}{2}\omega \hat{w}_2 B k_B^3 \left[1 - Re\left(e^{-j2(k_B x - \omega t)}\right)\right] \\ &= \omega \hat{w}_2 B k_B^3 \sin^2(\omega t - k_B x) \end{aligned} \tag{3.116}$$

Force and velocity are in phase and the same is true for moment and rotational velocity, but with a phase shift of $\pi/2$

$$\boldsymbol{M}_y = \boldsymbol{w}_2 B k_B^2 e^{-jk_B x} \tag{3.117}$$

$$\boldsymbol{\beta}_y = \omega k_B \boldsymbol{w}_2 e^{-jk_B x} \tag{3.118}$$

$$\begin{aligned} \Pi_M &= \frac{1}{2}\omega \hat{w}_2 B k_B^3 \left[1 + Re\left(e^{-j2(k_B x - \omega t)}\right)\right] \\ &= \omega \hat{w}_2 B k_B^3 \cos^2(\omega t - k_B x) \end{aligned} \tag{3.119}$$

Thus, both contributions sum up to the constant value

$$\Pi_{\text{beam}} = \omega \boldsymbol{w}_2 B k_B^3 \tag{3.120}$$

The same must be true for the energy. If we multiply the kinetic energy density e_{kin} by the beam area we get for the translational energy per length $E' = Ae$

$$\begin{aligned} E'_{\text{kin}} &= Ae_{\text{kin}} = \frac{m'}{2} Re(\boldsymbol{v}_z)^2 = \frac{A\rho}{2} Re(j\omega \boldsymbol{w}_2 e^{-jk_B x})^2 \\ &= \frac{A\rho}{2}\omega^2 \hat{w}_2 \sin^2(\omega t - k_b x) \end{aligned} \tag{3.121}$$

The rotational component can be neglected, as we can see from

$$\begin{aligned} E'_{\text{rot}} &= \frac{\rho I_{yy}}{2} Re(\dot{\boldsymbol{\beta}}_y)^2 = \frac{\rho I_{yy}}{2} Re(\omega k_B \boldsymbol{w}_2 e^{-jk_B x})^2 \\ &= \frac{\rho I_{yy}}{2}\omega^2 k_B^2 \hat{w}_2 \sin^2(\omega t - k_B x) \end{aligned} \tag{3.122}$$

The averaged ratio between both kinetic energies is

$$\frac{E'_{\text{rot}}}{E'_{\text{kin}}} = \left(\frac{2\pi\sqrt{I_{yy}/A}}{\lambda_B}\right) \tag{3.123}$$

The so called radius of gyration $\sqrt{I_{yy}/A}$ is the perpendicular position of a point mass which gives the same rotational inertia. As we stated in the beginning, one main assumption was that the cross sectional dimensions must be small compared to the wavelength λ_B. Consequently we can neglect the rotational part of the energy, too.

Due to the Bernoulli assumption only in-plane stress and strain must be considered. Here the stress and strain in the x-direction. Therefore, the potential energy per length is given by

$$\begin{aligned} E'_{\text{pot}} &= Ae_{\text{pot}} = \frac{1}{2}\int_A \sigma_{xx}\varepsilon_x dA = \frac{E}{2}\int_A \left(\frac{\partial^2 w}{\partial x^2}\right)^2 z^2 dA \\ &= \frac{EI_{yy}}{2}\left(\frac{\partial^2 w}{\partial x^2}\right)^2 = \frac{B_y k_B^4}{2}\hat{w}_2 \cos^2(\omega t - k_B x). \end{aligned} \tag{3.124}$$

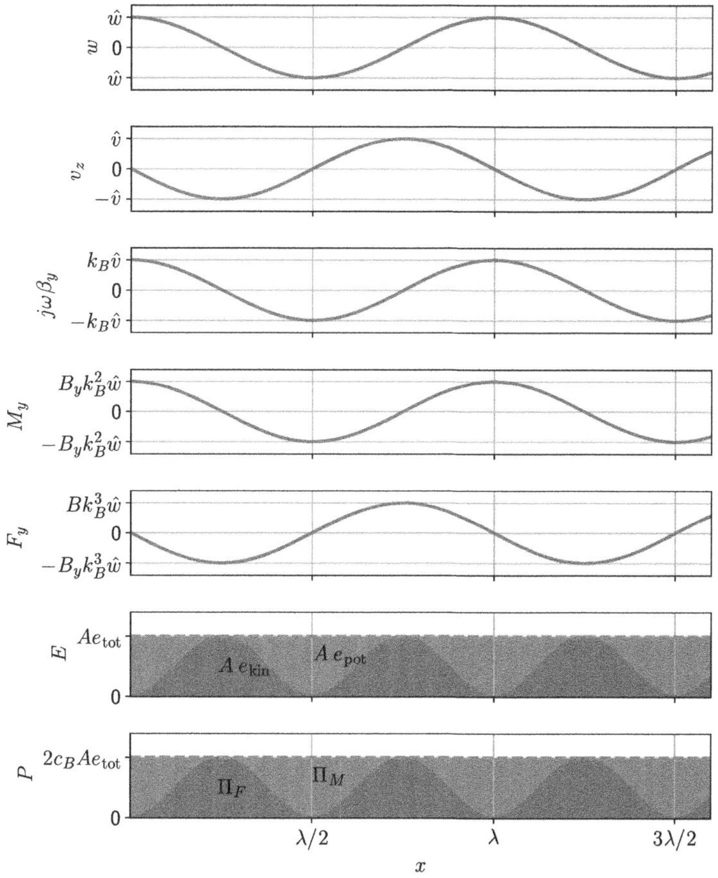

Figure 3.15 Various quantities for bending wave propagation in beams. *Source:* Alexander Peiffer.

Both kinds of energy oscillate between 0 and the maximum value similar to the power. Using the dispersion relation (3.105) the sum of both gives

$$E'_{\text{tot}} = A(e_{\text{kin}} + e_{\text{pot}}) = \frac{\hat{w}_2^2}{2}(m'\omega^2 \sin^2 k_B x + B_y k_B^4 \cos^2 k_B x)$$
$$= \frac{B_y k_B^4 \hat{w}_2^2}{2} = \frac{\rho_0 h \omega^2 \hat{w}_2^2}{2} \tag{3.125}$$

So, we have a constant energy distribution in the wave. There is no energy oscillation as was the case for fluid waves. If we relate the power transport (3.120) to the above energy E'_{tot} we get

$$\Pi = 2\frac{\omega}{k_B} E'_{tot} = 2c_B A e_{tot} \tag{3.126}$$

The power transportation is given by twice the phase speed c_B, which is fine in this case, because the uniform distribution of power and energy allows this effect. This is not an error and comes from the dispersion of the bending waves as we will see in section 3.8. The power transport in dispersive media is not given by the phase velocity but by the group velocity $c_{B,\text{gr}} = 2c_B$ for bending waves in beams.

A force exciting the beam radiates waves into the beam. In order to describe the source performance we come back to our impedance concept. As the force shall excite waves in the direction the bending occurs we apply a harmonic force $F_z(\omega) = \boldsymbol{F}_z e^{j\omega t}$ in the z-direction.

We start with a half infinite beam that is excited at $x = 0$, Figure 3.16. From the boundary condition it follows that $M_y = 0$ as it is an open end. From the origin only waves that propagate in the positive x-direction occur. Thus, we use the following part of Equation (3.107)

$$\boldsymbol{w} = \boldsymbol{w}_2 e^{-jk_B x} + \boldsymbol{w}_{2D} e^{-k_B x} \tag{3.127}$$

So we get for for M_y and F_z

$$\boldsymbol{M}_y = -B_y \frac{\partial^2 \boldsymbol{w}}{\partial x^2} = B_y k_b^2 (\boldsymbol{w}_2 e^{-jk_B x} + \boldsymbol{w}_{2D} e^{-k_B x}) \tag{3.128}$$

$$\boldsymbol{F}_z = B_y \frac{\partial \boldsymbol{w}^3}{\partial x^3} = B_y (\boldsymbol{w}_2 j k_B^3 e^{-jk_B x} - \boldsymbol{w}_{2D} k_B^3 e^{-k_B x}) \tag{3.129}$$

At $x = 0$ the first condition leads to $w_2 = w_{2D}$ and with this we get for the second equation balancing the force $F_z(x = 0)$:

$$\boldsymbol{w}_2 = \frac{\boldsymbol{F}_z}{k_B^3 B_y (j - 1)} \tag{3.130}$$

and hence

$$\boldsymbol{v}_z = j\omega \boldsymbol{w} = \frac{\boldsymbol{F}_z \omega}{k_B^3 B_y (1 + j)} (e^{-jk_B x} + e^{-k_B x}) \tag{3.131}$$

For the mechanical impedance at $x = 0$ we get now:

$$\boldsymbol{Z}_z = \frac{\boldsymbol{F}_z}{\boldsymbol{v}_z(0)} = \frac{Bk_B^3}{2\omega}(1 + j) \tag{3.132}$$

For the infinite beam we assume a mirrored wave at $x = 0$ propagating away from the point of excitation, but now the moment does not need to be zero. For symmetry reasons the the rotation β_y must vanish at the exciting force position. Thus, with Equation (3.16)

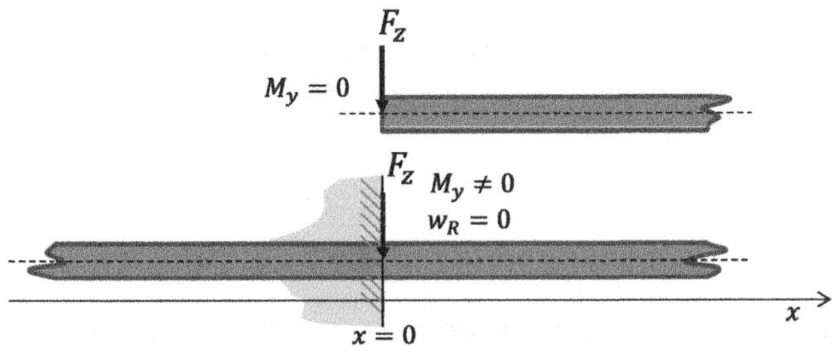

Figure 3.16 Force exciting a half infinite (top) and infinite beam at $x = 0$.
Source: Alexander Peiffer.

used in the y-direction and vanishing displacement the condition is

$$\boldsymbol{\beta}_y(0) = \frac{\partial \boldsymbol{w}}{\partial x}(0) = 0 \quad \Rightarrow \quad \boldsymbol{w}_2(0) = -j\boldsymbol{w}_{2D} \tag{3.133}$$

The force is now driving two half beams, thus

$$\frac{\boldsymbol{F}_z}{2} = (B_y k_B^3)(j\boldsymbol{w}_2 - \boldsymbol{w}_{2D}) = B_y k_B^3 2j\boldsymbol{w}_2 \tag{3.134}$$

and

$$\boldsymbol{v}_z = j\omega\boldsymbol{w} = \frac{\omega \boldsymbol{F}_z}{4B_y k_B^3}(e^{-jk_B|x|} - je^{-k_B|x|}) \tag{3.135}$$

This gives for the impedance of the full beam:

$$\boldsymbol{Z}_z = \frac{\boldsymbol{F}_z}{\boldsymbol{v}_z(0)} = \frac{2B_y k_B^3}{\omega}(1+j) \tag{3.136}$$

The full beam impedance is four times higher than the impedance of the half infinite beam. With growing frequency the impedance decreases, but we must keep in mind that for higher frequencies the Bernoulli beam assumptions are no longer valid. The impedance is complex with a phase of $\pi/4$. The driving force experiences a reactive and a resistive part. In Figure 3.17 three time or phase slots of an infinite and half infinite beam excited by a force are shown in combination with the solution without the evanescent part. We see that this part contributes only to the near field at the point of excitation and therefore contributes only to the impedance.

3.6 Membranes

Membranes don't have an inert bending stiffness from their mechanical properties. The stiffness comes solely from the tension T_{mem} in the membrane. Though membranes don't occur frequently in technical systems, it helps predefining the effect of pressurisation in a closed cabin, e.g. in an aircraft. The pressurisation leads to an additional bending stiffness coming from the tension.

If a small plate in the xy-plane is displaced in the z-direction by w the force due to the tension T'_{mem} per length is only determined by the slope θ of the membrane as shown in Figure 3.18. For small displacement w we get for the force Q'_x per length in the z-direction acting on the membrane element edge of length dy

$$Q'_x = T'_{x,\text{mem}} sin(\theta) \approx T'_{x,\text{mem}}\theta \tag{3.137}$$

The total force from tension acting on a membrane element along dy and due to the slope in x is

$$F_z = dy\left[T'_{x,\text{mem}}\left(\theta + \frac{\partial\theta}{\partial x}dx\right) - T'_{x,\text{mem}}\theta\right] = dy\left[T'_{x,\text{mem}}\frac{\partial\theta}{\partial x}dx\right] \approx dxdyT'_{x,\text{mem}}\frac{\partial^2 w}{\partial x^2} \tag{3.138}$$

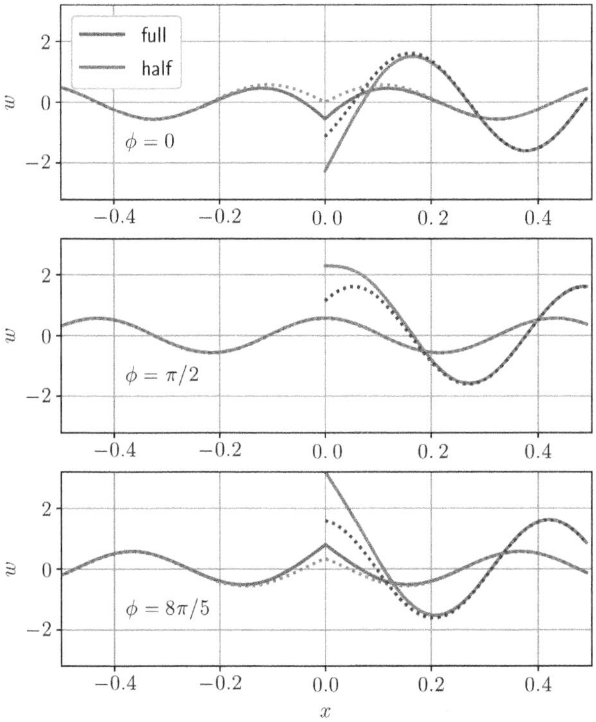

Figure 3.17 Beam displacement w for certain phases ($\phi = 0, \pi/2, 8\pi/5$). The dotted line is without the evanescent part e^{-kx}. *Source:* Alexander Peiffer.

Figure 3.18 Membrane forces in the x-direction. *Source:* Alexander Peiffer.

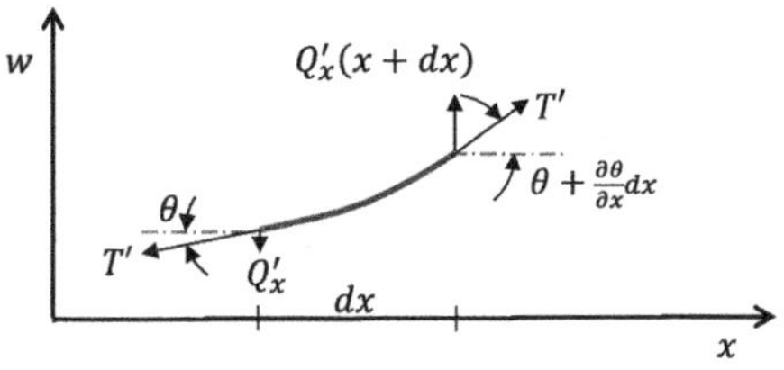

Using the similar equation in the y-direction we get with Newton's law the membrane wave equation

$$\Delta w - \frac{1}{c_M^2}\frac{\partial^2 w}{\partial t^2} = \frac{F''(x,y,t)}{T_{\text{mem}}} \quad \text{with} \quad c_M = \sqrt{\frac{T_{\text{mem}}}{m''}} \tag{3.139}$$

Here, m'' is the mass per area for the membrane, and we have assumed $T'_{x,\text{mem}} = T'_{y,\text{mem}} = T'_{\text{mem}}$. The membrane wave equation does not have any dispersion effects. This is the reason why drums should be made from membranes and not plates. Plates have dispersion effects as with beams, and this does not allow the wanted multiple harmonics as desired for musical instruments.

3.7 Plates

The theory of thin plates is summarized here according to Ventsel and Krauthammer (2001). The usual nomenclature distinguishes between plates that are flat, and thin structures and shells that can be curved. In terms of technical relevance plates and shells are the most important structure in many vibroacoustic investigations. Plates provide high stiffness at low weight. Thus, most means of transportation vehicles are created by enclosures made of plates and shells. For example the body of a car is made out of several sheets of steel that are stamped and point welded. The following assumptions for thin plates are made:

1. The plate is thin: the thickness h is much smaller than the other physical dimensions.
2. The in-plain strains are small compared to unity.
3. Transverse shear strains ε_{xz} and ε_{yz} are negligible.
4. Tangential displacements u and v are linear functions of the z coordinate.
5. The transverse shear stresses vanish at the surfaces at $z = \pm h/2$ ($\sigma_{zz} = \sigma_{zy} = \sigma_{xz} = 0$).

3.7.1 Strain–displacement Relations

Using the fourth assumption the tangential displacements are of the form

$$u = u_0(x, y, t) + z f_1(x, y, t) \tag{3.140}$$

$$v = v_0(x, y, t) + z f_2(x, y, t) \tag{3.141}$$

u_0 and v_0 are the tangential displacements of the middle plane. Entering these equations into the related strain–displacement relationships (3.16) and (3.17) and using assumption three yields

$$\varepsilon_{xz} = f_1(x, y, t) + \frac{\partial w}{\partial x} = 0 \tag{3.142}$$

$$\varepsilon_{yz} = f_2(x, y, t) + \frac{\partial w}{\partial y} = 0 \tag{3.143}$$

Therefore,

$$f_1(x, y, t) = -\frac{\partial w}{\partial x} \quad f_2(x, y, t) = -\frac{\partial w}{\partial y} \tag{3.144}$$

This result is illustrated in Figure 3.19 for a pure bending configuration.

The strain displacement relation (3.19) leads to

$$\varepsilon(z)_{xx} = \frac{\partial u}{\partial x} = \frac{\partial u_0}{\partial x} - z\frac{\partial^2 w}{\partial x^2} \tag{3.145}$$

$$\varepsilon(z)_{yy} = \frac{\partial v}{\partial y} = \frac{\partial v_0}{\partial x} - z\frac{\partial^2 w}{\partial y^2} \tag{3.146}$$

$$\varepsilon(z)_{xy} = \frac{1}{2}\left(\frac{\partial u}{\partial y} + \frac{\partial v}{\partial x}\right) = \frac{1}{2}\left(\frac{\partial u_0}{\partial y} + \frac{\partial v_0}{\partial x}\right) - 2\frac{\partial^2 w}{\partial x \partial y} \tag{3.147}$$

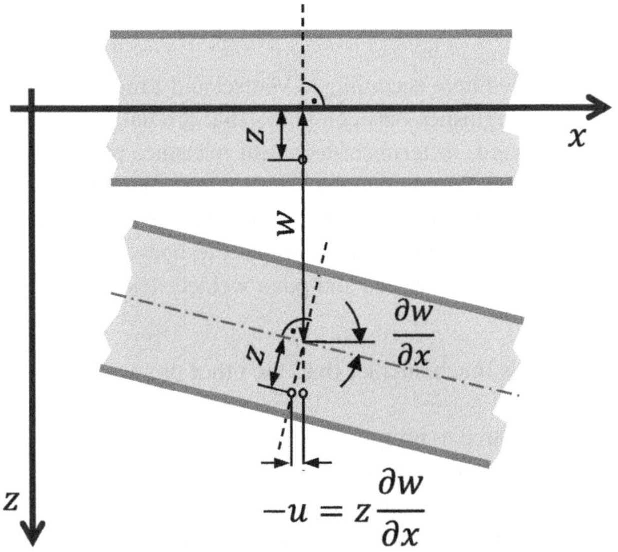

Figure 3.19 Deformation of a small plate element in the xz-plane. *Source:* Alexander Peiffer.

We consider an infinitesimally and cut-free small plate element as shown in figure 3.20. For the plate dynamics it is convenient to integrate over the plate thickness to get quantities defined per length instead of area. The integration over the normal stresses reads:

$$\begin{Bmatrix} N'_x \\ N'_y \\ N'_{xy} \end{Bmatrix} = \int_{-h/2}^{h/2} \begin{Bmatrix} \sigma_{xx} \\ \sigma_{yy} \\ \sigma_{xy} \end{Bmatrix} dz \tag{3.148}$$

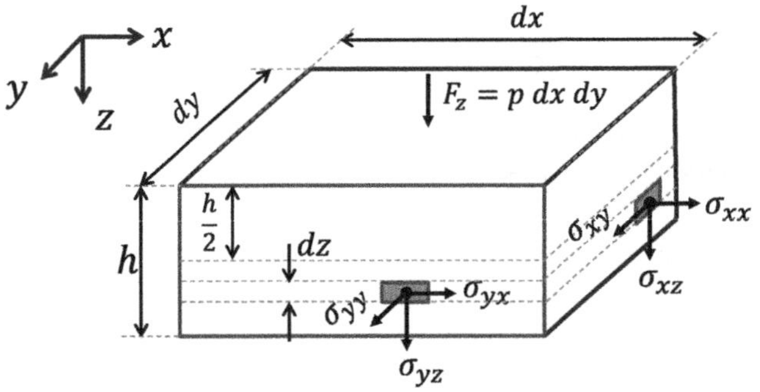

Figure 3.20 Plate element with stresses on the free faces. *Source:* Alexander Peiffer.

3.7.2 In-plane Wave Equation

From Figure 3.19 it becomes obvious that the bending is decoupled from the in-plane motion of shear and stress in case of homogeneous thin plates. Bending leads to symmetric stress and strain, and the integrals in Equation (3.148) are thus decoupled from the in-plane motion.

From assumption five it follows that the strains ε_x and ε_y can be given according to equations (3.30) and (3.31)

$$\varepsilon_x = \frac{\partial u}{\partial x} = \frac{1}{E}(\sigma_{xx} - \nu\sigma_{yy}) \tag{3.149}$$

$$\varepsilon_y = \frac{\partial v}{\partial y} = \frac{1}{E}(\sigma_{yy} - \nu\sigma_{xx}) \tag{3.150}$$

With the above equations and component xy in Voigt notation from (3.33) corresponding to the xy index in (3.19) we get

$$\sigma_{xx} = \frac{E}{1-\nu^2}\left(\frac{\partial u}{\partial x} + \nu\frac{\partial v}{\partial y}\right) \tag{3.151}$$

$$\sigma_{yy} = \frac{E}{1-\nu^2}\left(\frac{\partial v}{\partial y} + \nu\frac{\partial u}{\partial x}\right) \tag{3.152}$$

$$\sigma_{xy} = G\left(\frac{\partial u}{\partial y} + \frac{\partial v}{\partial x}\right) \tag{3.153}$$

Entering equations (3.151)–(3.153) into (3.148) provides

$$N'_x = \frac{E}{1-\nu^2}\int_{-h/2}^{h/2}\left(\frac{\partial u(z)}{\partial x} + \nu\frac{\partial v(z)}{\partial y}\right)dz = \frac{Eh}{1-\nu^2}\left(\frac{\partial u_0}{\partial x} + \nu\frac{\partial v_0}{\partial y}\right) \tag{3.154}$$

because the last term in Equation (3.145) vanishes due to symmetry. We get for N'_y and N'_{xy} accordingly

$$N'_y = \frac{Eh}{1-\nu^2}\left(\frac{\partial v_0}{\partial y} + \nu\frac{\partial u_0}{\partial x}\right) \tag{3.155}$$

$$N'_{xy} = Gh\left(\frac{\partial u_0}{\partial y} + \frac{\partial v_0}{\partial x}\right) \tag{3.156}$$

In Figure 3.21 the force balance of a plate element is shown. The balance in the x-direction is

$$\begin{aligned}\left(N'_x + \frac{\partial N'_x}{\partial x}dx\right)dy - N'_x dy + \left(N'_{xy} + \frac{\partial N'_{xy}}{\partial y}dy\right)dx - N'_{xy}dx + F_x dxdy = \\ \left(\frac{\partial N'_x}{\partial x} + \frac{\partial N'_{xy}}{\partial y} + p_x\right)dxdy = 0\end{aligned} \tag{3.157}$$

Division by area $dxdy$ gives

$$\frac{\partial N'_x}{\partial x} + \frac{\partial N'_{xy}}{\partial y} + F''_x = 0 \tag{3.158}$$

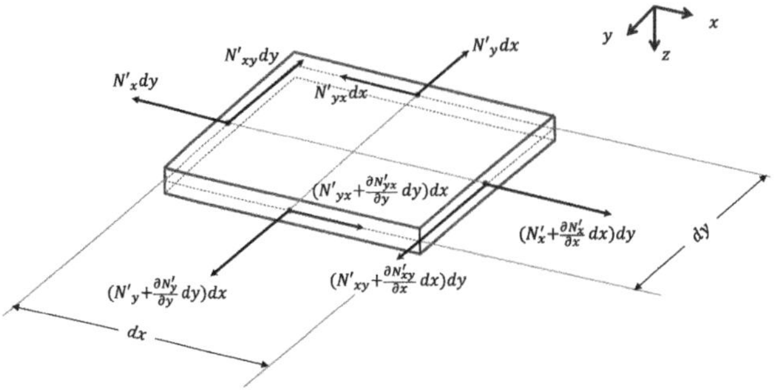

Figure 3.21 In-plane stress and shear force balance of cut-free plate element. *Source:* Alexander Peiffer.

and accordingly for the y-direction

$$\frac{\partial N'_y}{\partial y} + \frac{\partial N'_{xy}}{\partial x} + F''_y = 0 \tag{3.159}$$

With (3.154)–(3.156) we get finally, setting $u_0 = u$ and $v_0 = v$:

$$\frac{Eh}{1-\nu^2}\left(\frac{\partial^2 u}{\partial x^2} + \nu\frac{\partial^2 v}{\partial x \partial y}\right) + Gh\left(\frac{\partial^2 u}{\partial y^2} + \frac{\partial^2 v}{\partial x \partial y}\right) + F''_x =$$

$$\frac{Eh}{1-\nu^2}\frac{\partial^2 u}{\partial x^2} + Gh\frac{\partial^2 u}{\partial y^2} + \left(\frac{Eh\nu}{1-\nu^2} + Gh\right)\frac{\partial^2 v}{\partial x \partial y} + F''_x =$$

$$\frac{Eh}{1-\nu^2}\frac{\partial^2 u}{\partial x^2} + Gh\frac{\partial^2 u}{\partial y^2} + Gg\frac{1+\nu}{1-\nu}\frac{\partial^2 v}{\partial x \partial y} + F''_x = 0 \tag{3.160}$$

Equalising the force per area to the external pressure plus inertia forces and using $m'' = \rho_0 h$

$$F''_x = -\rho_0 h\frac{\partial^2 u}{\partial t^2} - F''_{x,\text{ext}} \tag{3.161}$$

leads to the wave equation for in-plane waves. Together with the following abbreviations

$$C = \frac{Eh}{(1-\nu^2)} \tag{3.162a}$$

$$S = Gh = \frac{Eh}{2(1+\nu)} \tag{3.162b}$$

we get

$$C\frac{\partial^2 u}{\partial x^2} + S\frac{\partial^2 u}{\partial y^2} + S\frac{1+\nu}{1-\nu}\frac{\partial^2 v}{\partial x \partial y} - \rho_0 h\frac{\partial^2 u}{\partial t^2} = F''_{x,\text{ext}} \tag{3.163}$$

and in a similar way for the y-direction:

$$C\frac{\partial^2 v}{\partial y^2} + S\frac{\partial^2 v}{\partial x^2} + S\frac{1+\nu}{1-\nu}\frac{\partial^2 u}{\partial x \partial y} - \rho_0 h\frac{\partial^2 v}{\partial t^2} = F''_{y,\text{ext}} \tag{3.164}$$

The motion in the in-plane direction is coupled: a displacement in u leads to motion in v and vice versa. The solution of both equations leads to two waves analogical to the longitudinal and shear waves of the solid material in section 3.4.

3.7.3 Longitudinal Waves

In longitudinal waves the wave motion is in the direction of propagation. Without loss of generality we choose the x-direction as propagation direction. Thus, there is no change of u over y and v is zero.

$$\frac{\partial u}{\partial y} = 0 \qquad v = 0 \tag{3.165}$$

Using these assumptions in (3.163) we get

$$\frac{\partial^2 u}{\partial x^2} - \frac{1}{c_{LP}}\frac{\partial^2 u}{\partial t^2} = 0 \quad \text{with} \quad c_{LP} = \sqrt{\frac{E}{(1-\nu^2)\rho_0}} \tag{3.166}$$

and the wavenumber yields

$$k_L = \frac{\omega}{c_{LP}} = \omega\sqrt{\frac{\rho_0 h}{C}} = \omega\sqrt{\frac{\rho_0(1-\nu^2)}{E}} \tag{3.167}$$

We see that this third longitudinal wave speed for plates is valued between the solution for the longitudinal waves in solids and beams.

$$c_L = \sqrt{\frac{E(1-\nu)}{\rho_0(1+\nu)(1-2\nu)}} > c_{LP} = \sqrt{\frac{E}{\rho_0(1-\nu^2)}} > c_{LB} = \sqrt{\frac{E}{\rho_0}}$$

The more constrained the bulk material is the stiffer it becomes due to the lateral contraction. The strain in thickness can be calculated from (3.32).

$$\varepsilon_{zz} = -\frac{\nu(1-\nu)}{E}\sigma_{xx} \tag{3.168}$$

Integrating the strain over the thickness delivers the shape of the wave motion.

3.7.4 Shear Waves

Shear waves have displacement perpendicular to their direction of propagation. For shear wave propagation in the x-direction we assume:

$$u = 0 \quad \frac{\partial v}{\partial y} = 0 \tag{3.169}$$

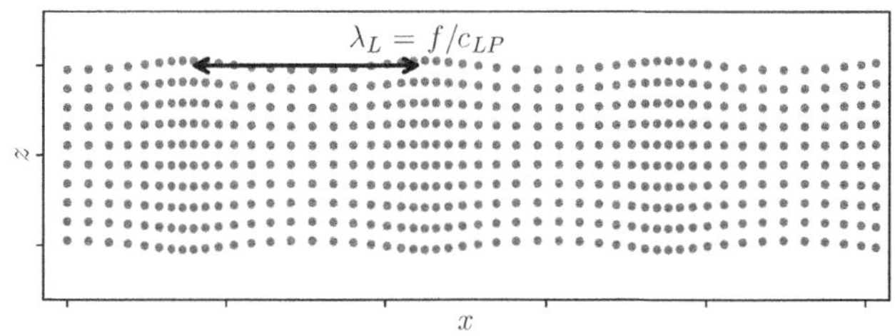

Figure 3.22 Shape of longitudinal wave in plate. *Source:* Alexander Peiffer.

Using (3.164) with these conditions gives the shear wave equation in the x-direction for plates:

$$\frac{\partial^2 v}{\partial x^2} - \frac{1}{c_S}\frac{\partial^2 v}{\partial t^2} = 0 \quad \text{with} \quad c_S = \sqrt{\frac{G}{\rho_0}} \tag{3.170}$$

$$k_S = \frac{\omega}{c_S} = \omega\sqrt{\frac{\rho_0 h}{S}} \tag{3.171}$$

The shear wave speed in plates or solid material is similar, because this wave type is free of dilatation due to condition 3.169 and thus not dependent on geometrical constraints.

$$c_S = \sqrt{\frac{G}{\rho_0}} \tag{3.172}$$

3.7.5 Combination of Longitudinal and Shear Waves

The combined dynamics of in-plane wave motion is found by entering the ansatz

$$\begin{Bmatrix} \boldsymbol{u} \\ \boldsymbol{v} \end{Bmatrix} = \begin{Bmatrix} \boldsymbol{u}_i \\ \boldsymbol{v}_i \end{Bmatrix} e^{-jk_x x - jk_y y + j\omega t} \tag{3.173}$$

with $i \in (S, L)$ being the index for both wave types. Entering this into (3.163) and (3.164) leads to the following system of equations

$$\begin{bmatrix} Ck_L^2 - Ck_x^2 - Sk_y^2 & -S\frac{1+\nu}{1-\nu}k_x k_y \\ -S\frac{1+\nu}{1-\nu}k_x k_y & Sk_S^2 - Ck_y^2 - Sk_x^2 \end{bmatrix} \begin{Bmatrix} \boldsymbol{u}_i \\ \boldsymbol{v}_i \end{Bmatrix} = \begin{Bmatrix} 0 \\ 0 \end{Bmatrix} \tag{3.174}$$

We used $\rho_0 h\omega^2 = Ck_L^2 = Sk_S^2$ following from (3.167) and (3.171). Equation (3.174) can be seen as a generalized eigenvalue problem with eigenvalue $\lambda = Ck_L^2 = Sk_S^2$. Dividing the above equation by S leads to a form that is more practical for the next steps

$$\begin{bmatrix} k_S^2 - \frac{C}{S}k_x^2 - k_y^2 & -\frac{1+\nu}{1-\nu}k_x k_y \\ -\frac{1+\nu}{1-\nu}k_x k_y & k_S^2 - \frac{C}{S}k_y^2 - k_x^2 \end{bmatrix} \begin{Bmatrix} \boldsymbol{u}_i \\ \boldsymbol{v}_i \end{Bmatrix} = \begin{Bmatrix} 0 \\ 0 \end{Bmatrix} \tag{3.175}$$

With the relationship

$$\frac{C}{S} = \frac{2}{1-\nu} = 1 + \frac{1+\nu}{1-\nu} \tag{3.176}$$

This can be further modified to

$$\begin{bmatrix} k_x^2 + k_y^2 - k_S^2 + \frac{1+\nu}{1-\nu}k_x^2 & \frac{1+\nu}{1-\nu}k_x k_y \\ \frac{1+\nu}{1-\nu}k_x k_y & k_x^2 + k_y^2 - k_S^2 + \frac{1+\nu}{1-\nu}k_y^2 \end{bmatrix} \begin{Bmatrix} \boldsymbol{u}_i \\ \boldsymbol{v}_i \end{Bmatrix} = \begin{Bmatrix} 0 \\ 0 \end{Bmatrix} \tag{3.177}$$

For non-trivial solutions of this matrix the determinant must vanish, and this leads to the equation

$$\left(k_x^2 + k_y^2 - k_S^2 + \frac{1+\nu}{1-\nu}k_x^2\right)\left(k_x^2 + k_y^2 - k_S^2 + \frac{1+\nu}{1-\nu}k_y^2\right) - \left(\frac{1+\nu}{1-\nu}k_x k_y\right)^2$$

This is of the form $(a+b)(a+c) - bc = a(a+b+c) = 0$. So the two solutions are

$$a = 0 \qquad k_x^2 + k_y^2 - k_S^2 = 0 \Rightarrow k_x^2 + k_y^2 = k_S^2 \tag{3.178}$$

$$a+b+c = 0 \qquad k_x^2 + k_y^2 - k_S^2 + \frac{1+\nu}{1-\nu}(k_x^2 + k_y^2) = 0 \Rightarrow k_x^2 + k_y^2 = k_L^2 \tag{3.179}$$

using relationship (3.176) again. Finally, we have derived the above wavenumbers for shear and longitudinal waves in a more formal way. Entering k_S^2 and k_L^2 for $k_x^2 + k_y^2$ into (3.177) provides the solution for the *shape* of the wave propagation. With $k_{xS}^2 + k_{yS}^2 = k_S^2$ we get:

$$\begin{bmatrix} k_{xS}^2 & k_{xS}k_{yS} \\ k_{xS}k_{yS} & k_{yS}^2 \end{bmatrix} \begin{Bmatrix} \boldsymbol{u}_S \\ \boldsymbol{v}_S \end{Bmatrix} = \begin{Bmatrix} 0 \\ 0 \end{Bmatrix} \tag{3.180}$$

So $k_{xS}\boldsymbol{u}_S = -k_{yS}\boldsymbol{v}_S$, and when we choose $\boldsymbol{u}_S = \boldsymbol{\Psi}_S k_{xS}$ we get $\boldsymbol{v}_S = -\boldsymbol{\Psi}_S k_{yS}$; finally the displacement due to the shear wave motion is given by

$$\begin{Bmatrix} \boldsymbol{u} \\ \boldsymbol{v} \end{Bmatrix}_S = \boldsymbol{\Psi}_S \begin{Bmatrix} k_{yS} \\ -k_{xS} \end{Bmatrix} e^{-jk_{xS}x - jk_{yS}y + j\omega t} \tag{3.181}$$

With $k_{xL}^2 + k_{yL}^2 = k_L^2$ the same equation reads as:

$$\begin{bmatrix} -k_{yL}^2 & k_{xL}k_{yL} \\ k_{xL}k_{yL} & -k_{xL}^2 \end{bmatrix} \begin{Bmatrix} \boldsymbol{u}_L \\ \boldsymbol{v}_L \end{Bmatrix} = \begin{Bmatrix} 0 \\ 0 \end{Bmatrix} \tag{3.182}$$

providing the longitudinal wave motion

$$\begin{Bmatrix} \boldsymbol{u} \\ \boldsymbol{v} \end{Bmatrix}_L = \boldsymbol{\Psi}_L \begin{Bmatrix} k_{xL} \\ k_{yL} \end{Bmatrix} e^{-jk_{xL}x - jk_{yL}y + j\omega t} \tag{3.183}$$

The in-plane wave propagation is given by the superposition of both waves with:

$$\begin{Bmatrix} \boldsymbol{u} \\ \boldsymbol{v} \end{Bmatrix} = \left(\boldsymbol{\Psi}_S \begin{Bmatrix} k_{yS} \\ -k_{xS} \end{Bmatrix} e^{-j\mathbf{k}_S\mathbf{x}} + \boldsymbol{\Psi}_L \begin{Bmatrix} k_{xL} \\ k_{yL} \end{Bmatrix} e^{-j\mathbf{k}_L\mathbf{x}} \right) e^{j\omega t} \tag{3.184}$$

The displacement of longitudinal waves is parallel to the propagation direction given by $\mathbf{k}_L = \{k_{xL}, k_{yL}\}^T$, and the shear wave displacement is perpendicular to the propagation direction orientation $\mathbf{k}_S$ as shown in Figure 3.23

Similar to solids the wavelengths of longitudinal and transversal waves are very high. Thus, in-plane waves can be neglected in many technical acoustics problems at audible frequency, but they are required for a full understanding of the physics when plates are coupled.

3.7.6 Bending Wave Equation

For the derivation of the bending wave dynamics we define the moments per length M' parallel to the edges and as a torsional moment.

$$\begin{Bmatrix} M'_x \\ M'_y \\ M'_{xy} \end{Bmatrix} = \int_{-h/2}^{h/2} \begin{Bmatrix} \sigma_{xx} \\ \sigma_{yy} \\ \sigma_{xy} \end{Bmatrix} z\, dz \tag{3.185}$$

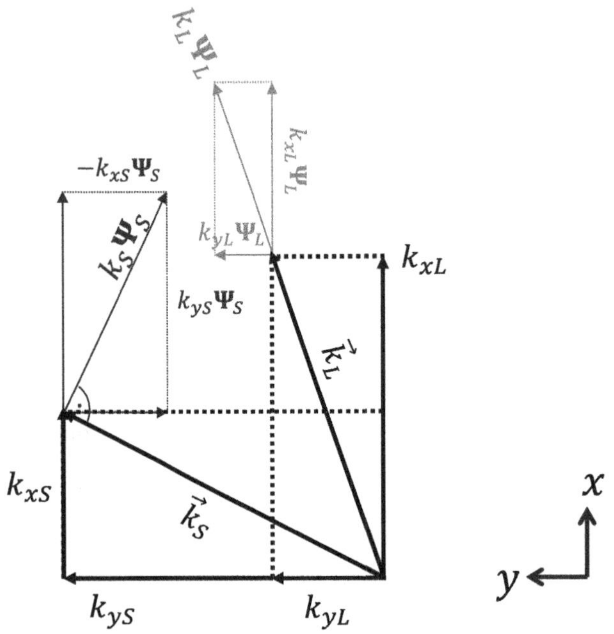

Figure 3.23 Propagation direction of in-plane waves and orientation of displacement. *Source:* Alexander Peiffer.

The thickness integration over the shear stresses gives the shear forces per length:

$$\begin{Bmatrix} Q'_x \\ Q'_y \end{Bmatrix} = \int_{-h/2}^{h/2} \begin{Bmatrix} \sigma_{xz} \\ \sigma_{yz} \end{Bmatrix} dz \tag{3.186}$$

With those definitions and according to Figure 3.24 we can now set up the relationships for the required equilibrium of the total forces in the z-direction

$$p\,dxdy + \frac{\partial Q'_x}{\partial x} dxdy + \frac{\partial Q'_y}{\partial y} dydx = 0 \tag{3.187}$$

$$p + \frac{\partial Q'_x}{\partial x} + \frac{\partial Q'_y}{\partial y} = 0 \tag{3.188}$$

and for the moments around the x-axis,

$$\frac{\partial M'_{xy}}{\partial x} dxdy + \frac{\partial M'_y}{\partial y} dydx - 2Q'_y dx \frac{dy}{2} - \frac{\partial Q'_y}{\partial y} dydx \frac{dy}{2} = 0 \tag{3.189}$$

and the y-axis accordingly:

$$\frac{\partial M'_{yx}}{\partial y} dydx + \frac{\partial M'_x}{\partial x} dxdy - 2Q'_x dy \frac{dx}{2} - \frac{\partial Q'_x}{\partial x} dxdy \frac{dx}{2} = 0 \tag{3.190}$$

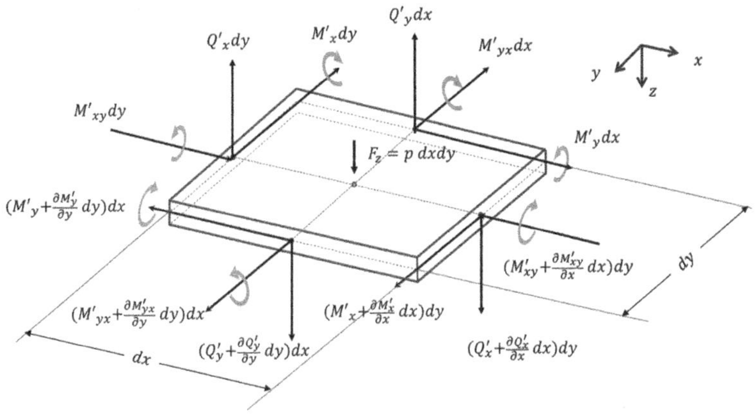

Figure 3.24 Moments and shear forces balance of cut-free plate element.
Source: Alexander Peiffer.

Neglecting terms of second order and with the moment equilibrium $M'_{xy} = M'_{yx}$ we get

$$\frac{\partial M'_{xy}}{\partial x} + \frac{\partial M'_y}{\partial y} - Q'_y = 0, \quad \frac{\partial M'_{xy}}{\partial y} + \frac{\partial M'_x}{\partial x} - Q'_x = 0 \tag{3.191}$$

Entering Q'_x and Q'_y from (3.188) in (3.191) gives

$$\frac{\partial^2 M'_x}{\partial x^2} + 2\frac{\partial^2 M'_{xy}}{\partial x \partial y} + \frac{\partial^2 M'_y}{\partial y^2} = -p \tag{3.192}$$

As mentioned above the bending motion is decoupled from the in-plane motion. So, we assume $u_0 = v_0 = 0$ and get with (3.145) for the displacement in the thin layer an approximately linear dependency from z

$$u(z) = -z\frac{\partial w}{\partial x} \quad v(v) = -z\frac{\partial w}{\partial y} \tag{3.193}$$

Entering (3.193) into (3.151)-(3.153) leads to the in-plane stresses as function of w and z

$$\sigma_{xx}(z) = -\frac{E\,z}{1-\nu^2}\left(\frac{\partial^2 w}{\partial x^2} + \nu\frac{\partial^2 w}{\partial y^2}\right) \tag{3.194}$$

$$\sigma_{yy}(z) = -\frac{E\,z}{1-\nu^2}\left(\frac{\partial^2 w}{\partial y^2} + \nu\frac{\partial^2 w}{\partial x^2}\right) \tag{3.195}$$

$$\sigma_{xy}(z) = -2G\,z\left(\frac{\partial^2 w}{\partial x \partial y}\right) \tag{3.196}$$

This is applied in the definitions (3.185) and reads for $w(x, y)$

$$M'_x = -B\left(\frac{\partial^2 w}{\partial x^2} + \nu\frac{\partial^2 w}{\partial y^2}\right) \tag{3.197}$$

$$M'_y = -B\left(\frac{\partial^2 w}{\partial y^2} + \nu\frac{\partial^2 w}{\partial x^2}\right) \tag{3.198}$$

$$M'_{xy} = -\frac{Gh^3}{6}\frac{\partial^2 w}{\partial x \partial y}. \tag{3.199}$$

with the bending stiffness B of the plate

$$B = \int_{-h/2}^{h/2} \frac{Ez^2}{1-\nu^2} dz = \frac{Eh^3}{12(1-\nu^2)} \tag{3.200}$$

With equations (3.191) and (3.197)–(3.199) the shear forces can also be expressed as functions of w

$$Q'_x = -B\frac{\partial}{\partial x}\left(\frac{\partial^2 w}{\partial x^2} + \frac{\partial^2 w}{\partial y^2}\right) = -B\frac{\partial}{\partial x}\Delta w \tag{3.201}$$

$$Q'_y = -B\frac{\partial}{\partial y}\left(\frac{\partial^2 w}{\partial x^2} + \frac{\partial^2 w}{\partial y^2}\right) = -B\frac{\partial}{\partial y}\Delta w \tag{3.202}$$

.

From (3.192) and with equations (3.197)–(3.199) we can derive the plate equation

$$B\left(\frac{\partial^4 w}{\partial x^4} + 2\frac{\partial^4 w}{\partial x^2 \partial y^2} + \frac{\partial^4 w}{\partial y^4}\right) = B\Delta\Delta w = p \tag{3.203}$$

In order to get the wave equation for plates we set the force per area equal to external pressure plus the inertia force

$$p = -m''\frac{\partial^2 w}{\partial t^2} + p_{ext} = -\rho_0 h\frac{\partial^2 w}{\partial t^2} + p_{\text{ext}} \tag{3.204}$$

Hence,

$$\Delta\Delta w + \frac{\rho_0 h}{B}\frac{\partial^2 w}{\partial t^2} = \frac{p_{\text{ext}}}{B} \tag{3.205}$$

Now we are prepared to determine the wave speed for bending waves in plates using the solution in frequency domain $w(x, y, t) = \boldsymbol{w}(x, y)e^{j\omega t}$

$$\Delta\Delta\boldsymbol{w} - \omega^2\frac{\rho_0 h}{B}\boldsymbol{w} = \frac{\boldsymbol{p}_{\text{ext}}}{B} \tag{3.206}$$

Considering waves $\boldsymbol{w} = \boldsymbol{w}_B e^{-jkx}$ in the x-direction without loss of generality the results for k_B are similar to the results of the beam bending waves

$$k_B^4 = \frac{\rho_0 h}{B}\omega^2 \tag{3.207}$$

and the wave speed

$$c_B = \sqrt[4]{\frac{B\omega^2}{\rho_0 h}} \tag{3.208}$$

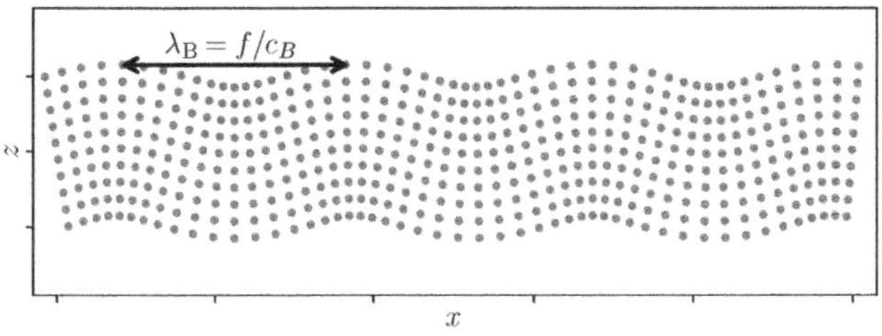

Figure 3.25 Shape of bending wave of a plate. *Source:* Alexander Peiffer.

Equation (3.206) can be written in Helmholtz form

$$(\Delta\Delta - k_B^4)\boldsymbol{w} = \frac{p_{ext}}{B} \tag{3.209}$$

We get again two real and two complex solutions:

$$k_{B1} = \pm j\sqrt[4]{\frac{\omega^2 \rho_0 h}{B}} \quad k_{B2} = \pm\sqrt[4]{\frac{\omega^2 \rho_0 h}{B}} \tag{3.210}$$

Thus the solution includes four propagating waves. Using only the positive absolute values per definition with

$$\boldsymbol{k}_{B1} = j\sqrt[4]{\frac{\omega^2 \rho_0 h}{B}} \quad k_{B2} = \sqrt[4]{\frac{\omega^2 \rho_0 h}{B}}$$

we can write for the four waves in the x-direction

$$\boldsymbol{w}(x,\omega) = \boldsymbol{w}_{1D} e^{jk_{B1}x} + \boldsymbol{w}_{2D} e^{-jk_{B1}x} + \boldsymbol{w}_2 e^{jk_{B2}x} + \boldsymbol{w}_2 e^{-jk_{B2}x} \tag{3.211}$$

The dispersion effects discussed in the beam section are also valid.

3.7.6.1 Cylindrical Solution of Bending Wave Equation

In order to derive the mechanical impedance for a plate we must agree on a certain set of boundary conditions and assumptions that allow for a solution:

- The solution is rotationally symmetric.
- The point of excitation is rotation free.
- The solution fulfils the Sommerfeld condition: i.e. that the wave decays in large distances.

Starting with (3.206), using the wavenumber result from (3.207) and rewriting it to

$$\Delta\Delta\boldsymbol{w} - k_B^4\boldsymbol{w} = (\Delta - k_B^2)(\Delta + k_B^2)\boldsymbol{w} = 0 \tag{3.212}$$

gives the following system of equations:

$$(\Delta + k_B^2)\boldsymbol{w} = 0 \tag{3.213a}$$

$$(\Delta - k_B^2)\boldsymbol{w} = 0 \tag{3.213b}$$

The first Equation (3.213a) is similar to the Helmholtz equation for fluids but for two dimensions. The solution is the Hankel function of second kind:

$$\boldsymbol{w}_1(r) = C_1 H_0^{(2)}(k_B r) \tag{3.214}$$

With r being the distance from the point of excitation. For the far field properties and the behavior at the excitation point we use the asymptotic behavior for small and large values of x

$$H_0^{(2)}(x) \approx \frac{2j}{\pi}\ln x \quad \text{for } |x| \ll 1 \quad H_0^{(2)}(x) \approx \sqrt{\frac{2}{\pi x}}e^{-j(x-\pi/4)} \quad \text{for } |x| \gg 1$$

The solution of the second Equation (3.213b) can be found by replacing $k_B r$ with $-jk_B r$, hence

$$\boldsymbol{w}_2(r) = C_2 H_0^{(2)}(-jk_B r). \tag{3.215}$$

For large kr we get

$$H_0^{(2)}(-jk_B r) \approx \sqrt{\frac{2}{-j\pi k_B r}}e^{+j\pi/4}e^{-k_B r} = j\sqrt{\frac{2}{\pi k_B r}}e^{-k_B r}$$

Thus, Equation (3.215) represents the exponentially decaying near field, as in (3.106). The total solution is

$$\boldsymbol{w}(r) = C_1 H_0^{(2)}(k_B r) + C_2 H_0^{(2)}(-jk_B r) \tag{3.216}$$

With the assumption that the rotation

$$\frac{\partial \boldsymbol{w}}{\partial r} = C_1 k_B\left(-j\frac{2}{\pi k_B r} + \dots\right) - C_2 k_B\left(\frac{2}{\pi k_B r} + \dots\right) = \frac{2j}{\pi r}(-C_1 - C_2) + \dots \tag{3.217}$$

must vanish at $r = 0$, so that

$$C_1 = -C_2$$

we finally get

$$\boldsymbol{w}(r) = C_1\left(H_0^{(2)}(k_B r) - H_0^{(2)}(-jk_B r)\right) \tag{3.218}$$

3.7.6.2 Power, Impedance, and Energy

The value at $r = 0$ follows from the asymptotic expression for small values:

$$\begin{aligned}\boldsymbol{w}(0) = \hat{\boldsymbol{w}}_0 &= C_1\left[-\frac{2j}{\pi}\ln\frac{k_B r}{2} + \dots + \frac{2j}{\pi}\left(\ln\frac{k_B r}{2} + \ln(-j)\right) + \dots\right] \\ &= C_1\frac{2j}{\pi}\ln(-j) = C_1\end{aligned} \tag{3.219}$$

So C_1 equals the displacement amplitude at the excitation point.

$$\boldsymbol{w} = \hat{\boldsymbol{w}}_0\left(H_0^{(2)}(k_B r) - H_0^{(2)}(-jk_B r)\right) \tag{3.220}$$

Some shapes at specific phases of the force are shown in Figure 3.26.

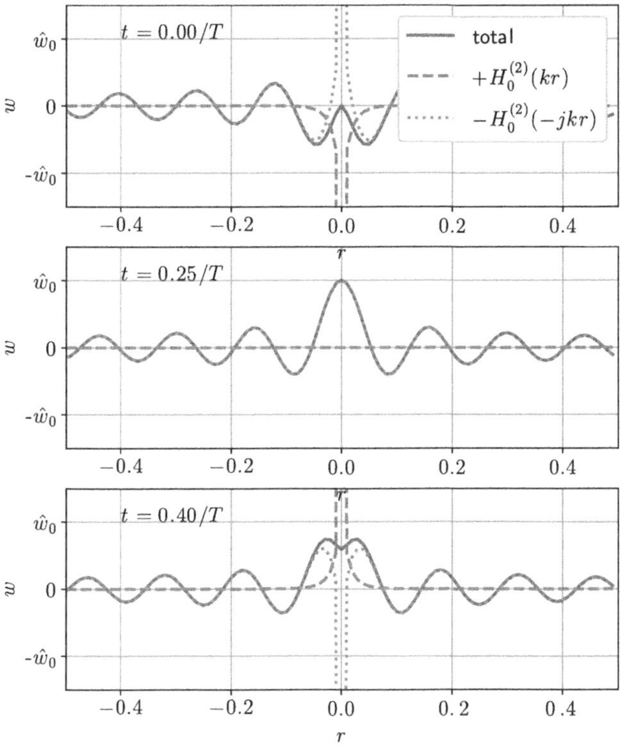

Figure 3.26 Bending wave forms of plate excited by a force $\boldsymbol{F}_z$ at $r = 0$.
Source: Alexander Peiffer.

In order to derive the displacement amplitude $\hat{\boldsymbol{w}}_0$ we consider a small circle around the the point of excitation and use equations (3.201)–(3.202) in cylindrical coordinates

$$\boldsymbol{Q}'_r = -B\frac{\partial \boldsymbol{w}}{\partial r} \tag{3.221}$$

Entering this into (3.218) leads to:

$$\boldsymbol{Q}'_r = Bk^3\hat{\boldsymbol{w}}_0\left(\frac{dH_0^{(2)}(kr)}{dkr} + \frac{dH_0^{(2)}(-jkr)}{dkr}\right) \tag{3.222}$$

Using the approximation of the Hankel function for small arguments gives

$$\boldsymbol{Q}'_{r0} = \frac{4jBk^2}{\pi r_0}\boldsymbol{w}_0 \tag{3.223}$$

Multiplication of (3.223) with the perimeter of a small circle finally provides the force

$$\boldsymbol{F}_z = 2\pi r_0 \boldsymbol{Q}_{r0} = 8jBk^2\boldsymbol{w}_0 \tag{3.224}$$

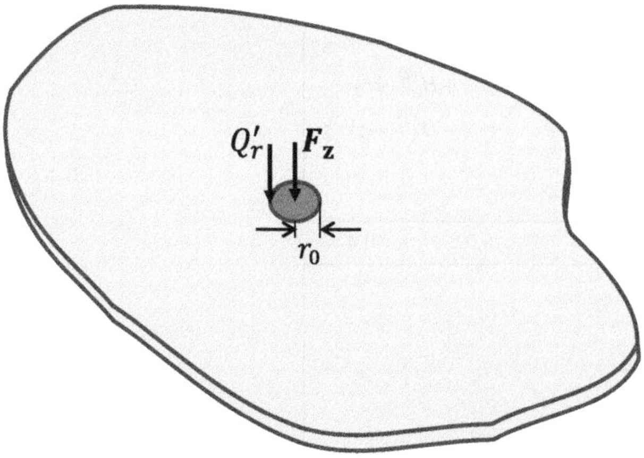

Figure 3.27 Force excitation at a small disc of radius r_0. *Source:* Alexander Peiffer.

Entering this into (3.220) we get the shape of the plate due to force excitation

$$\boldsymbol{w} = \frac{\boldsymbol{F}_z}{8jBk^2}\left(H_0^{(2)}(kr) - H_0^{(2)}(-jkr)\right) \tag{3.225}$$

With the velocity at the centre $\boldsymbol{v}_z(0) = j\omega\boldsymbol{w}_0$ we get a surprisingly simple expression for the mechanical impedance at $r = 0$:

$$\boldsymbol{Z}_0 = \frac{\boldsymbol{F}_z}{\boldsymbol{v}_{z0}} = \frac{8Bk^2}{\omega} = 8\sqrt{Bm''} = \frac{8\omega m''}{k_B^2} \tag{3.226}$$

Regarding the energy in the plate, we rely on the considerations of bending waves in beams in section 3.5. The property characterising bending for beams was the Young's modulus times the second moment of area, e.g. $B_y = EI_{yy}$. For plates this must be replaced by the bending stiffness that is defined per length as given in equation 3.200. Thus, after adjusting Equation (3.125) to this length reference we get area energy density $E'' = E/A$

$$\begin{aligned} E''_{\text{tot}} &= h(e_{\text{kin}} + e_{\text{pot}}) = \frac{\hat{w}^2}{2}(\rho_0 h\omega^2 \sin^2 k_B x + Bk_B^4 \cos^2 k_B x) \\ &= \frac{Bk_B^4\hat{w}^2}{2} = \frac{\rho_0 h\omega^2\hat{w}^2}{2} \end{aligned} \tag{3.227}$$

The equality of kinetic and potential energy, as far as the fact that the energy is not pulsating but constant over time, is also similar to beams. The expression for the power flow has to be adjusted to power per length or length intensity with

$$\Pi' = 2\frac{\omega}{k_B}E''_{tot} = 2c_B E''_{tot} = c_B\rho_0 h\omega^2\hat{w}_2^2 \tag{3.228}$$

Note that the energy is transported with twice the phase wave speed, which is not in contradiction to the energy conservation because of the fact that the energy is equally distributed over the phase cycle. For the point force excitation the introduced energy flows through the edge of a circle of radius r with the circumference $2\pi r$.

$$\Pi' = \frac{\Pi}{2\pi r} \tag{3.229}$$

The input power is known through the point impedance (3.226)

$$\Pi = \frac{1}{2}Re(\boldsymbol{F}^*\boldsymbol{v}) = \frac{1}{2}Re(\hat{F}\boldsymbol{Z}_0) = \frac{\omega\hat{F}^2}{16Bk_B^2} \tag{3.230}$$

Combining equations (3.228) to (3.229) we get an expression for the displacement amplitude derived from energy considerations:

$$\hat{w}^2(r) = \frac{\hat{F}}{32\pi B^2 k_B^5}\frac{1}{r} \tag{3.231}$$

The same expression can be derived from Equation (3.225) when using the asymptotic expression for the Hankel function for $kr \gg 1$.

3.7.6.3 Damping

The damping of wave propagation is related to energy or the transport of energy. This complicates the treatment for dispersive waves. To overcome this issue we use the damping loss as the complex ratio of stiffness related quantities from (1.62). For plates this is the complex Young's modulus $\boldsymbol{E}$:

$$\boldsymbol{E} = E(1 + j\eta) \tag{3.232}$$

impacting linearly the bending stiffness:

$$\boldsymbol{B} = B(1 + j\eta) \tag{3.233}$$

When we introduce this into Equation (3.207)

$$\boldsymbol{k}_B^4 = \frac{\rho_0 h}{B(1 + j\eta)}\omega^2 = k_B^4\frac{1 - j\eta}{1 + \eta^2} \approx k_B^4(1 - j\eta)$$

and

$$\boldsymbol{k}_B = k_B\sqrt[4]{1 - j\eta} \approx k_B(1 - \frac{j}{4}\eta) \tag{3.234}$$

In contrast to Equation (2.59) this is one-fourth of the wavenumber. The reason for this is that due to dispersion of bending waves the energy propagates twice as fast as the phase.

3.8 Propagation of Energy in Dispersive Waves

The bending wave considerations have shown that the phase velocity does not represent the speed for energy transportation. It is twice the phase wave speed as in Equation (3.126). An illustrative way to present the relationship between group and phase velocity is to investigate the interference of two one-dimensional waves.

$$\begin{aligned} w_{\text{ifr}}(x,t) &= \sin(\omega_1 t - k_1 x) + \sin(\omega_2 t - k_2 x) \\ &= 2\sin\left(\frac{\omega_1+\omega_2}{2}t - \frac{k_1+k_2}{2}x\right)\cos\left(\frac{\omega_1-\omega_2}{2}t - \frac{k_1-k_2}{2}x\right) \\ &= 2\cos(\omega_{\text{gr}} t - k_{\text{gr}} x)\sin(\omega_{\text{mean}} t - k_{\text{mean}} x) \end{aligned} \tag{3.235}$$

The interference of the two sine waves leads to a *carrier* wave with mean frequency $\omega_{\text{mean}} = (\omega_1 + \omega_2)/2$ and wave number $k_{\text{mean}} = (k_1 + k_2)/2$ whose amplitude is modulated by a much lower frequency $\omega_{gr} = (\omega_1 - \omega_2)/2$ and the much lower wavenumber $k_{\text{gr}} = (k_1 - k_2)/2$.

The energy in the wave is proportional to the squared amplitude of the wave. Thus, the energy is transported with the speed of the amplitude modulation:

$$c_{\text{gr}} = \frac{\omega_1 - \omega_2}{k_1 - k_2} = \frac{\Delta\omega}{\Delta k} \tag{3.236}$$

In the limit of $\Delta\omega \to 0$ the group velocity will be

$$c_{\text{gr}} = \frac{d\omega}{dk} \tag{3.237}$$

In Figure 3.28 an example for bending waves is depicted. We see how the envelope propagates with the group velocity c_{gr}. If we follow one phase value of the carrier we would notice that it propagates with a slower speed. This can also be recognized by inspecting the relative location of the carrier in the envelope. The maxima are located at different positions in the envelope.

We conclude that the speed of energy propagation in waves is the group velocity. If there is no frequency dependence phase and group velocity are the same. We have found this relationship already in the context of power flow for bending waves in beams in Equation (3.126).

The consequence of the fact that energy travels with group velocity is that the relationship between energy density and intensity (2.53) must consider the group velocity.

$$\mathbf{I} = \mathbf{n} c_{\text{gr}} e \tag{3.238}$$

3.9 Findings

With chapters 2 and 3 we have described several modes of wave propagation with different properties. Those modes are different in terms of sound speed, excitation impedance, and the relationship between energy density and the wave property. In Table 3.1 those properties are summarized.

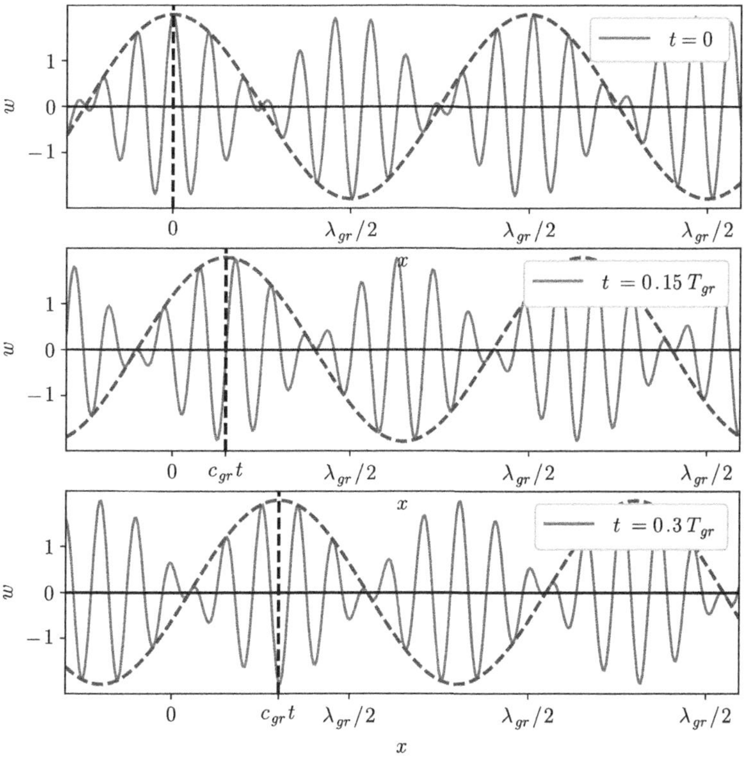

Figure 3.28 Group and phase velocity of two sine waves with similar frequency. *Source:* Alexander Peiffer.

Table 3.1 Wave speeds of structural waves.

Mode type	Group velocity	Phase velocity
Longitudinal waves in bars	$c_{gr} = c_{LB}$	$c_{LB} = \sqrt{E/\rho}$
Torsional waves in bars	$c_{gr} = c_T$	$c_T = \sqrt{GJ/\rho I_p}$
Bending waves	$c_{gr} = 2c_B$	$c_B = \sqrt{\omega}\sqrt[4]{\frac{B}{m}}$
Longitudinal waves in plates	$c_{gr} = c_{LP}$	$c_{LP} = \sqrt{\frac{E}{\rho_0(1-\nu^2)}}$
Shear waves	$c_{gr} = c_S$	$c_S = \sqrt{G/\rho}$
Waves in elastic solids	$c_{gr} = c_L$	$c_L = \sqrt{\frac{E(1-\nu)}{\rho_0(1+\nu)(1-2\nu)}}$

Bibliography

Klaus-Jürgen Bathe. *Finite Element Procedures in Engineering Analysis*. Prentice-Hall Civil Engineering and Engineering Mechanics Series. Prentice-Hall, Englewood Cliffs, N.J, 1982. ISBN 978-0-13-317305-5.

Lothar Cremer, Manfred Heckl, and Björn Petersson. *Structure-Borne Sound: Structural Vibrations and Sound Radiation at Audio Frequencies*. Springer Verlag, Berlin, Germany, 3rd edition edition, December 2005. ISBN 978-3-540-26514-6.

Karl F. Graff. *Wave Motion in Elastic Solids*. Dover Publications, New York, 1991. ISBN 978-0-486-66745-4.

Reinhard Lerch and H. Landes. Grundlagen der Technischen Akustik, September 2012.

I. Szabo. *Einführung in Die Technische Mechanik: Nach Vorlesungen*. Springer Verlag, Berlin, Germany, 2013a. ISBN 978-3-662-11624-1.

I. Szabo. *Höhere Technische Mechanik: Nach Vorlesungen*. Klassiker Der Technik. Springer Verlag, Berlin, Germany, 2013b. ISBN 978-3-642-56795-7.

Eduard Ventsel and Theodor Krauthammer. *Thin Plates and Shells: Theory: Analysis, and Applications*. CRC Press, August 2001. ISBN 978-0-8247-0575-6.

4

Fluid Systems

In Chapters 1–3, wave propagation is considered in unbound media, and boundaries occurred only in the context of scattering, reflection, or radiation. Beams or shells were characterized by their boundaries, but their effect was the creation of different forms of wave propagation. For example, the dimension (thickness) of a plate causes the different wave types, bending, in-plane shear, and dilation, but in principle there was always at least one infinite dimension. In Chapter 4 we will approach the system from Chapter 1, meaning that we consider bounded systems, for example a closed cavity with specific input and output positions.

4.1 One-dimensional Systems

A one-dimensional subsystem can be a small tube with radius R small enough that the wavelength is larger than the diameter $2R$ for the highest considered frequency (Pierce, 1991). The tube has rigid surfaces and is filled with a fluid, e.g. air with given density ρ_0 and speed of sound c_0.

The solution of the wave Equation (2.42) in the tube is given by the one-dimensional solution of the wave equation for pressure (2.34) and velocity (2.36)

$$\boldsymbol{p}(x) = \boldsymbol{A}e^{-jkx} + \boldsymbol{B}e^{jkx} \tag{4.1}$$

$$\boldsymbol{v}_x(x) = \frac{1}{z_0}(\boldsymbol{A}e^{-jkx} - \boldsymbol{B}e^{jkx}) \tag{4.2}$$

Interaction with the tube fluid takes place at the end and the beginning of the tube. So, we look for a representation similar to a 2DOF system (1.73) but with the velocities $\boldsymbol{v}_{x1} = \boldsymbol{v}_x(0)$ and $\boldsymbol{v}_{x2} = \boldsymbol{v}_x(L)$ instead of the displacement. We must clearly separate between state and external variables in order to avoid confusion. So, we separate between the state pressure $p_R(0)$, $p_L(L)$, denoting the pressure at the right side at $x = 0$ and at the left side at $z = L$. A positive pressure $\boldsymbol{p}_R(0)$ at $x = 0$ leads to a force in the negative x-direction that must be compensated by a positive pressure $\boldsymbol{p}_L(0)$ due to the face normal and x-coordinate both pointing in positive directions. At $x = L$ a positive pressure $p_L(L)$ leads to positive blocked force in the x-direction that must be compensated by a negative pressure $\boldsymbol{p}_R(L)$. We must keep this in mind when we introduce external forces and for coping with the transfer matrix method. Consequently the external force for the system description is replaced by the pressures $\boldsymbol{p}_1 = \boldsymbol{p}_L(0)$ and

Vibroacoustic Simulation: An Introduction to Statistical Energy Analysis and Hybrid Methods,
First Edition. Alexander Peiffer.

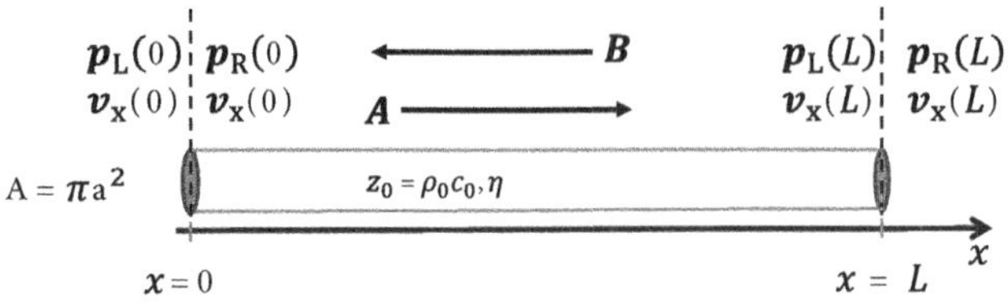

Figure 4.1 One-dimensional fluid system with two connections. *Source:* Alexander Peiffer.

$$\boldsymbol{p}_2 = \boldsymbol{p}_R(L) = -\boldsymbol{p}_L(L)$$

$$\begin{bmatrix} z_{11} & z_{12} \\ z_{21} & z_{22} \end{bmatrix} \begin{Bmatrix} \boldsymbol{v}_{x1} \\ \boldsymbol{v}_{x2} \end{Bmatrix} = \begin{Bmatrix} \boldsymbol{p}_R(0) \\ -\boldsymbol{p}_L(L) \end{Bmatrix} = \begin{Bmatrix} \boldsymbol{p}_1 \\ \boldsymbol{p}_2 \end{Bmatrix} \tag{4.3}$$

Using Equation (4.2) at the related positions of x we get for the velocities

$$z_0 \boldsymbol{v}_{x1} = \boldsymbol{A} - \boldsymbol{B} \tag{4.4}$$

$$z_0 \boldsymbol{v}_{x2} = \boldsymbol{A} e^{-jkL} - \boldsymbol{B} e^{+jkL} \tag{4.5}$$

The difference (4.4) – (4.5) $\cdot e^{\pm jkL}$ gives

$$\boldsymbol{A} = z_0 \frac{\boldsymbol{v}_{x1} e^{\,jkL} - \boldsymbol{v}_{x2}}{j2 \sin \boldsymbol{k}L} \tag{4.6}$$

$$\boldsymbol{B} = z_0 \frac{\boldsymbol{v}_{x1} e^{-jkL} - \boldsymbol{v}_{x2}}{j2 \sin \boldsymbol{k}L} \tag{4.7}$$

and entering this into (4.1) at $x = 0, L$ we get the following expressions for $\boldsymbol{p}_1$ and $\boldsymbol{p}_2$ depending on the velocities

$$\boldsymbol{p}_1 = \boldsymbol{A} + \boldsymbol{B} = z_0 \left(\frac{1}{j \tan(\boldsymbol{k}L)} \boldsymbol{v}_{x1} - \frac{1}{j \sin(\boldsymbol{k}L)} \boldsymbol{v}_{x2} \right) \tag{4.8}$$

$$\boldsymbol{p}_2 = -\boldsymbol{p}_L(L) = -(\boldsymbol{A} e^{-jkL} + \boldsymbol{B} e^{jkL}) = z_0 \left(-\frac{1}{j \sin(\boldsymbol{k}L)} \boldsymbol{v}_{x1} + \frac{1}{j \tan(\boldsymbol{k}L)} \boldsymbol{v}_{x2} \right) \tag{4.9}$$

This provides the following 2DOF system matrix in impedance form

$$\begin{bmatrix} \frac{z_0}{j \tan(\boldsymbol{k}L)} & \frac{-z_0}{j \sin(\boldsymbol{k}L)} \\ \frac{-z_0}{j \sin(\boldsymbol{k}L)} & \frac{z_0}{j \tan(\boldsymbol{k}L)} \end{bmatrix} \begin{Bmatrix} \boldsymbol{v}_{x1} \\ \boldsymbol{v}_{x2} \end{Bmatrix} = \begin{Bmatrix} \boldsymbol{p}_1 \\ \boldsymbol{p}_2 \end{Bmatrix} \tag{4.10}$$

This is the matrix description of the tube as a system with two (external) DOFs. If the velocities are given, the system response can be derived directly. For calculating the response to pressure excitation, the matrix must be inverted to give the mobility form

$$\frac{1}{z_0} \begin{bmatrix} \frac{1}{j \tan(\boldsymbol{k}L)} & \frac{1}{j \sin(\boldsymbol{k}L)} \\ \frac{1}{j \sin(\boldsymbol{k}L)} & \frac{1}{j \tan(\boldsymbol{k}L)} \end{bmatrix} \begin{Bmatrix} \boldsymbol{p}_1 \\ \boldsymbol{p}_2 \end{Bmatrix} = \begin{Bmatrix} \boldsymbol{v}_{x1} \\ \boldsymbol{v}_{x2} \end{Bmatrix} \tag{4.11}$$

The pressure and velocity field can be derived from (4.1) and (4.2) and the solutions for $\boldsymbol{A}$ and $\boldsymbol{B}$:

$$\boldsymbol{p}(x) = \frac{z_0}{j\sin(\boldsymbol{k}L)}\left[\boldsymbol{v}_{x1}\cos(\boldsymbol{k}(L-x)) - \boldsymbol{v}_{x2}\cos(\boldsymbol{k}x)\right] \tag{4.12}$$

$$\boldsymbol{v}_x(x) = \frac{1}{j\sin(\boldsymbol{k}L)}\left[\boldsymbol{v}_{x1}\sin(\boldsymbol{k}(L-x)) + \boldsymbol{v}_{x2}\sin(\boldsymbol{k}x)\right] \tag{4.13}$$

4.1.1 System Response

In order to interpret the above result we reduce our system to a 1DOF system by setting an impedance boundary condition at $x = L$. Thus,

$$p_L(L)/\boldsymbol{v}_{x2} = -p_R(L)/\boldsymbol{v}_{x2} = \boldsymbol{z}_2 \tag{4.14}$$

Using Equation (4.10) we get for a given velocity amplitude $\boldsymbol{v}_{x1}$ the pressure response at $x = 0$

$$\boldsymbol{p}_1 = z_0\frac{\boldsymbol{z}_2\cos(\boldsymbol{k}L) + jz_0\sin(\boldsymbol{k}L)}{z_0\cos(\boldsymbol{k}L) + j\boldsymbol{z}_2\sin(\boldsymbol{k}L)}\boldsymbol{v}_{x1} \tag{4.15}$$

In a first step we consider a rigid boundary condition with $|\boldsymbol{z}_2| \gg z_0$ leading to

$$\boldsymbol{p}_1 = \frac{z_0}{j\tan(\boldsymbol{k}L)}\boldsymbol{v}_{x1} \tag{4.16}$$

and from that the radiation impedance assuming a damping according to (2.59)

$$\boldsymbol{z}_{a,1} = \frac{z_0}{j\tan(\boldsymbol{k}L)} \quad \text{with } \boldsymbol{k} = k(1 - j\frac{\eta}{2}) \tag{4.17}$$

The tan function is zero at $kL = n\pi$ for $n = 0, 1, 2, \ldots$ when we assume a real $\boldsymbol{k}$. This is the case for $L = n\lambda/2$ when half wavelengths fit into the tube. The wave transmitted by the piston at $x = 0$ is reflected at $x = L$ and matches perfectly in phase at $x = 0$. So, the pressure would add up to infinity without damping, and this is called the resonance or first mode of the tube.

Solving expression (4.17) for $kL = n\pi$ gives for the amplitude of $\boldsymbol{z}_{a,1}$ at the resonances

$$\boldsymbol{z}_{a,1} = z_0\coth\left(\frac{n\pi\eta}{2}\right) \tag{4.18}$$

In Figure 4.2 the result is shown for values of kL. For low frequencies ($kL \approx 1$) we see a clear deterministic behavior. The impedance switches between the typical values. Going higher in frequency, the peaks are very close to each other. For high frequencies, $kl \gg 100$ there is no reflected wave because of the damping. Consequently, this is like radiation into half space and the impedance equals the characteristic impedance of air.

4.1.2 Power Input

The power introduced to the subsystem is according to (2.45) the acoustic intensity times the surface area. For this we link the wavenumber to the damping loss

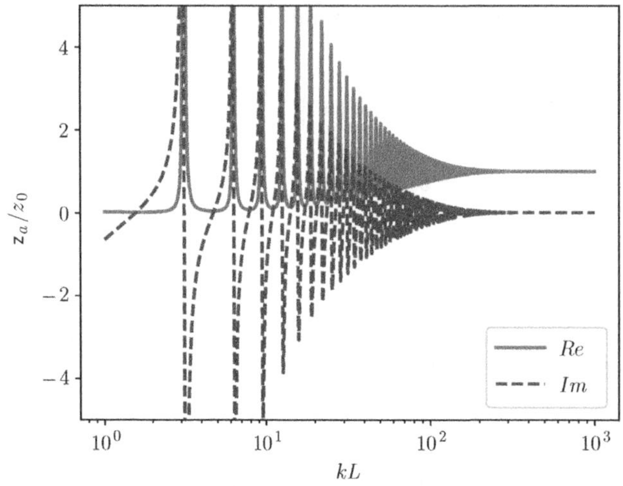

Figure 4.2 Relative radiation impedance at $x = 0$ for a one-dimensional fluid pipe for $\eta = 0.02$. *Source:* Alexander Peiffer.

$\boldsymbol{k} = k(1 - j\frac{\eta}{2})$ and separate the real from the imaginary part

$$\boldsymbol{z}_{a,1} = \frac{z_0}{j\tan(\boldsymbol{k}L)} = z_0\frac{\sinh(k\eta L) - j\sin(2kL)}{\cosh(k\eta L) - cos(2kL)} \tag{4.19}$$

Thus the introduced power is

$$\Pi = \frac{A}{2}Re(\boldsymbol{z}_{a,1}\boldsymbol{v}_{x1}\boldsymbol{v}_{x1}^*) = \frac{A}{2}\frac{z_0\sinh(k\eta L)}{\cosh(k\eta L) - cos(2kL)}|\boldsymbol{v}_{x1}|^2 \tag{4.20}$$

and for $k\eta \to \infty$ the input power becomes

$$\lim_{k\eta\to\infty}\Pi = \frac{A}{2}z_0|\boldsymbol{v}_{x1}|^2 \tag{4.21}$$

Due to the absence of reflection by damping the radiation into the tube equals radiation into the free field.

4.1.3 Pressure Field

In order to derive the interior pressure field inside the tube, we replace $\boldsymbol{v}_{x_2}$ in Equation (4.5) by

$$\boldsymbol{v}_{x2} = \frac{1}{\boldsymbol{z}_2}(\boldsymbol{A}e^{-j\boldsymbol{k}L} + \boldsymbol{B}e^{j\boldsymbol{k}L})$$

After some algebra we get for the two amplitudes

$$\boldsymbol{A} = z_0\boldsymbol{v}_{x1}\frac{(\boldsymbol{z}_2 + z_0)e^{j\boldsymbol{k}L}}{j2\boldsymbol{z}_2\sin(\boldsymbol{k}L) + 2z_0\cos(\boldsymbol{k}L)} \tag{4.22}$$

$$\boldsymbol{B} = z_0\boldsymbol{v}_{x1}\frac{(\boldsymbol{z}_2 - z_0)e^{-j\boldsymbol{k}L}}{j2\boldsymbol{z}_2\sin(\boldsymbol{k}L) + 2z_0\cos(\boldsymbol{k}L)} \tag{4.23}$$

Entering this into (4.1) and (4.2) we get

$$\boldsymbol{p}(x) = z_0 \boldsymbol{v}_{x1} \frac{\boldsymbol{z}_2 \cos(\boldsymbol{k}(L-x)) + jz_0 \sin(\boldsymbol{k}(L-x))}{z_0 \cos(\boldsymbol{k}L) + j\boldsymbol{z}_2 \sin(\boldsymbol{k}L)} \tag{4.24}$$

$$\boldsymbol{v}_x(x) = \boldsymbol{v}_{x1} \frac{j\boldsymbol{z}_2 \sin(\boldsymbol{k}(L-x)) + z_0 \cos(\boldsymbol{k}(L-x))}{z_0 \cos(\boldsymbol{k}L) + j\boldsymbol{z}_2 \sin(kL)} \tag{4.25}$$

and for a rigid end $\boldsymbol{z}_2 \gg z_0$ this yields

$$\boldsymbol{p}(x) = z_0 \boldsymbol{v}_{x1} \frac{\cos(\boldsymbol{k}(L-x))}{j \sin(\boldsymbol{k}L)} \tag{4.26}$$

$$\boldsymbol{v}_x(x) = \boldsymbol{v}_{x1} \frac{\sin(\boldsymbol{k}(L-x))}{j \sin(\boldsymbol{k}L)}. \tag{4.27}$$

In Figure 4.3 the pressure and velocity amplitudes along the tube axis are shown.

4.1.4 Modes

In Chapter 1 it was shown that every dynamic and closed system has natural states of vibration: the modes of the system. For the harmonic oscillator that was the resonance of the system, and for a system of connected springs and masses that was the modal solution (1.104). Continuous and bounded systems also have modes and natural shapes of vibration linked to a certain frequency.

4.1.4.1 Rigid Boundaries

Here, we are looking for the mode shapes of our system with given boundary conditions, eg. rigid conditions such as $\boldsymbol{v}_x(0) = \boldsymbol{v}_x(L) = 0$. So, the solution (4.2) must fulfil these conditions:

$$\boldsymbol{v}_x(0) = \frac{1}{z_0}(\boldsymbol{A} - \boldsymbol{B}) = 0 \tag{4.28}$$

$$\boldsymbol{v}_x(L) = \frac{1}{z_0}\left(\boldsymbol{A}e^{-jkL} - \boldsymbol{B}e^{jkL}\right) = 0 \tag{4.29}$$

The first equation gives $\boldsymbol{A} = \boldsymbol{B}$. Inserting this in the second equation leads to the following solutions for pressure and velocity

$$\boldsymbol{v}_{xn}(x) = \frac{j\hat{p}_n}{z_0} \sin(k_n x) \tag{4.30}$$

$$p_n(x) = \hat{p}_n \cos(k_n x) \qquad \text{for } k_n = \frac{n\pi}{L} \text{ with } n = 0, 1, ... \tag{4.31}$$

The choice $n = 0$ corresponds to a static pressure. These wave numbers correspond to modal frequencies and wave lengths

$$\omega_n = \frac{n\pi c_0}{L} \qquad \lambda_n = \frac{L}{2\pi n} \tag{4.32}$$

4.1.4.2 Modal Coordinates and Matrix Representation

Mode shapes are *natural* shapes of the system. That means the system tends to vibrate in this form at the specific frequency. We know that the mode shapes can be used for

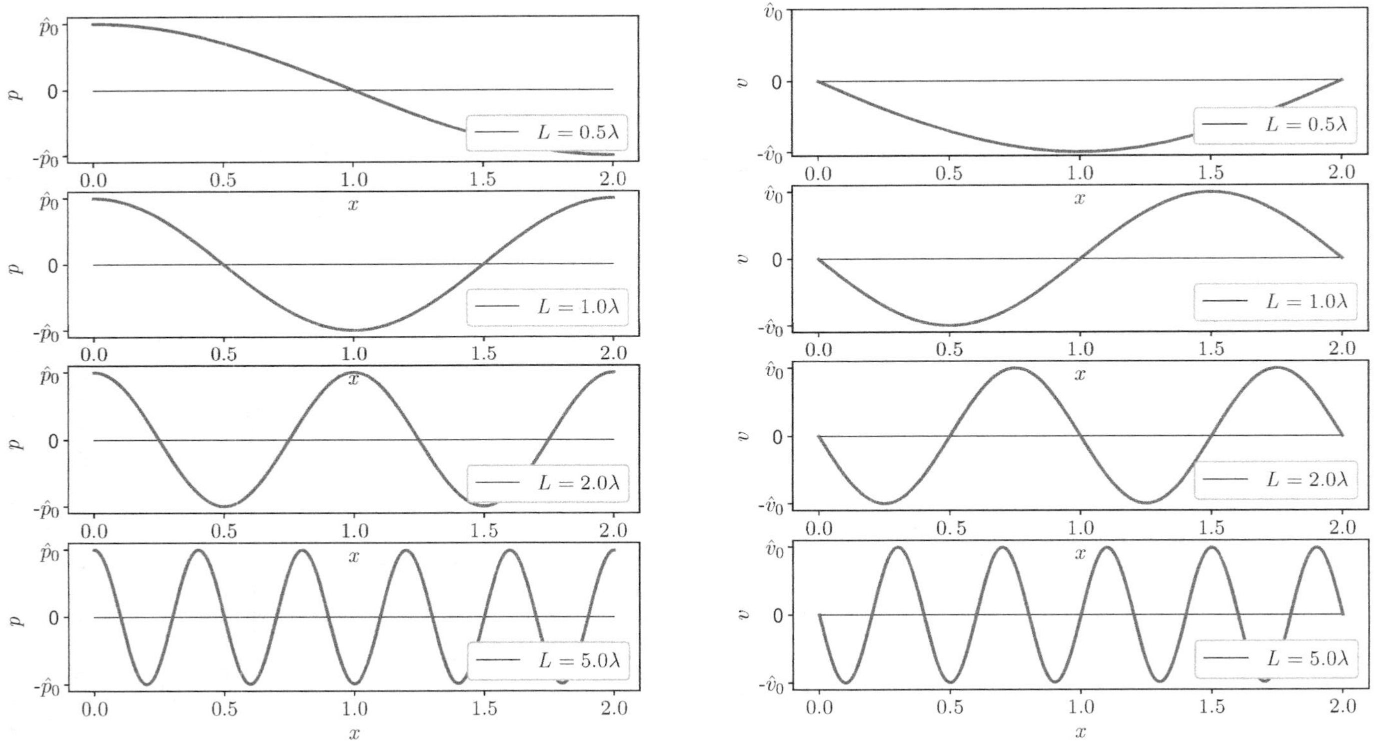

Figure 4.3 Pressure and velocity amplitude in the tube for multiples of half wavelengths. *Source:* Alexander Peiffer.

coordinate transformations such as when the state of the system is given in modal coordinates $\boldsymbol{p}'_n$

$$\boldsymbol{p}(x) = \sum_{n=0}^{\infty} \boldsymbol{p}'_n \Psi_{p,n}(x) \tag{4.33}$$

Entering this into the one-dimensional inhomogeneous Helmholtz Equation (2.123) gives

$$\sum_n (k_n^2 - k^2)\boldsymbol{p}'_n \Psi_{p,n}(x) = -\boldsymbol{f}_q(x) \tag{4.34}$$

For the derivation of the modal coordinates $\boldsymbol{p}'_n$ we need the orthogonality relations according to equations (1.111) and (1.112). As we don't have a discrete system the vector product is replaced by a scalar product of the continuous mode

$$\Psi^*_{p,m} \times \Psi_{p,n} = \int_0^L \Psi^*_{p,m}(x)\Psi_{p,n}(x)dx = \begin{cases} 0 & \text{for m} \neq \text{n} \\ \frac{\hat{p}_n^2 L}{2} & \text{for m} = \text{n} \end{cases} \tag{4.35}$$

The complex conjugate does not have any effect, as the mode shape is supposed to be real. But in later application we may have complex modes that require the scalar product in complex form. For practical reasons we normalize the modes with

$$\Phi_{p,n} = \sqrt{\frac{\Lambda_n}{L}} \cos(k_n x) \text{ with } \Lambda_0 = 1 \text{ and } \Lambda_n = 2\, \forall\, n > 0 \tag{4.36}$$

so that

$$\int_0^L \Phi^*_{p,m}(x)\Phi_{p,n}(x)dx = \delta_{mn} \tag{4.37}$$

We make use of the orthogonality by multiplying Equation (4.34) from the left with $\Phi^*_{p,m}$, and keeping in mind that the source term $\boldsymbol{f}_q$ contains the volume source strength density $\boldsymbol{q}_s$, so that we have to multiply with the tube cross-section A to get the dimensions right

$$A\int_0^L \Phi^*_{p,m}(x)\left(\sum_n (k_n^2 - k^2)p'_n\Phi_{p,n}(x)\right)dx = A(k_m^2 - k^2)p'_m$$
$$= -\int_0^L \Phi^*_{p,m}(x)\boldsymbol{f}_q(x)dx \tag{4.38}$$

providing an expression similar to (1.30) and (1.126)

$$p'_m = -\frac{\int_0^L \Phi^*_{p,m}(x)\boldsymbol{f}_q(x)dx}{A(k_m^2 - k^2)} = +j\omega\rho_0 c_0^2 \frac{\int_0^L \Phi^*_{p,m}(x)\boldsymbol{q}_s(x)}{A(\omega_m^2 - \omega^2)} \tag{4.39}$$

An equivalent relationship can be found with the free boundary modes.

4.1.4.3 Modal Density

When systems are excited by broadband signals, it is relevant how many modes are in the frequency range of interest. The more modes are available, the higher the probability for the exciting force to excite resonators. We define the number of modes until frequency ω as mode count

$$N = \frac{\omega}{\Delta\omega_n} = \frac{\omega L}{\pi c_0} \tag{4.40}$$

$\Delta\omega_n$ is the distance between each modal frequency as shown in Figure 4.4. The modal density is the number of modes per band given by

$$n_{1D}(\omega) = \frac{\Delta N}{\Delta\omega} \qquad \lim_{\Delta\omega\to 0} n_{1D} = \frac{dN}{d\omega} = \frac{L}{\pi c_0} \tag{4.41}$$

These quantities are indicators of the dynamic complexity. A high number of modes in a band may denote high dynamic complexity. In addition they are criteria for the probability to excite a mode when excited by signals with a specific bandwidth.

4.1.4.4 Damping

Damping is considered with Equation (2.59) by using a complex wavenumber

$$\boldsymbol{k}_n = k_n(1 - j\frac{\eta}{2}) \Rightarrow \boldsymbol{k}_n^2 = k_n(1 + j\eta - \eta^2/4) \approx k_n^2(1 + j\eta) \tag{4.42}$$

With this assumption we end up with a modal damping according to (1.63) and (1.127). So, the damping in the wave propagation is in line with the modal damping expressions from Chapter 1.

4.1.4.5 Modal Frequency Response

The first example with a piston excitation at $x = 0$ is taken as first application of the modal method. According to Equation (4.39) the response is

$$\boldsymbol{p}_n' = j\omega\rho_0 \frac{\int_0^L \Phi_{p,m}^*(x)\boldsymbol{q}_s(x)dx}{A(k_n^2 - k^2)} \tag{4.43}$$

We assume a piston vibrating with $\hat{v}_x$ at $x = 0$, so the volumetric source strength density would be $\boldsymbol{q}_s(x) = \hat{Q}\delta(x - x_0) = \delta(x - x_0)A\hat{v}_{x1}$. Using the sifting property of the delta function the result is

$$\boldsymbol{p}_n' = \sqrt{\frac{\Lambda_n}{L}} \frac{j\omega\rho_0\hat{Q}}{A(k_n^2 - k^2)} = \sqrt{\frac{\Lambda_n}{L}} \frac{j\omega\rho_0\boldsymbol{v}_x}{(k_n^2 - k^2)} \tag{4.44}$$

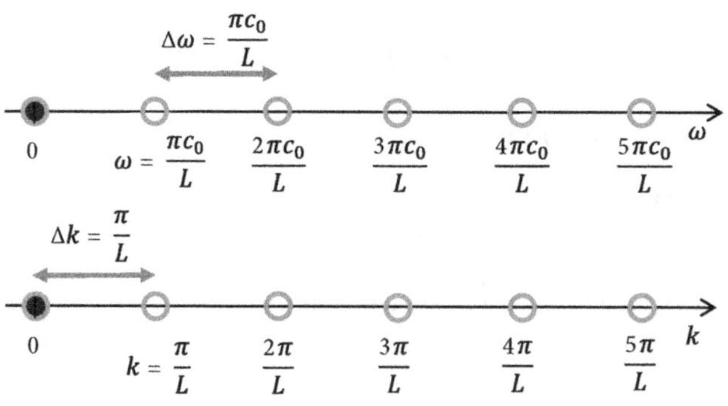

Figure 4.4 Modal frequencies along frequency and wavenumber axes. *Source:* Alexander Peiffer.

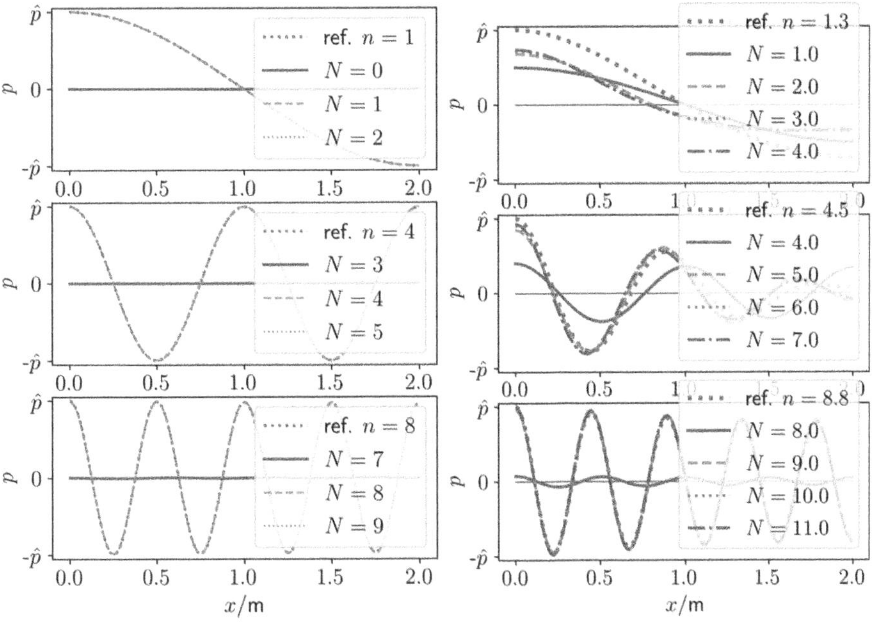

Figure 4.5 Modal reconstruction of system response to frequencies at resonance (LHS) and out of resonance (RHS). *Source:* Alexander Peiffer.

and transforming back into real space with (4.33) gives

$$\boldsymbol{p}(x) = \sum_{n=0}^{\infty} \frac{\Lambda_n}{L} \frac{j\omega\rho_0\boldsymbol{v}_x}{k_n^2 - k^2} \cos(k_n x) \tag{4.45}$$

Obviously, in numerical implementation the infinite series must be replaced by a certain number N of modes. As we now the exact solution from Equation (4.26), we can calculate the error

$$\delta p(x) = \sum_{n=0}^{N} \frac{2}{L} \frac{j\omega\boldsymbol{v}_{x1}}{k_n^2 - k^2} \cos(k_n x) - z_0\boldsymbol{v}_{x1} \frac{\cos(\boldsymbol{k}(L-x))}{j\sin(\boldsymbol{k}L)} \tag{4.46}$$

A modal solution may require a large number of modes to synthesize the exact solution similar to the Fourier transform of a rectangular signal. If the solution is similar to a mode, saying that k is chosen as $k = k_n$, the approximation is naturally exact if the damping is not too high. In Figure 4.5 six modes ($N = 6$) in the sum of Equation (4.45) are shown. On the left hand side the system is excited at resonance frequencies, the upper two are well represented, but the case $n = 8$ shows that six modes are not sufficient. The required shape to represent the solution is simply not available in the used mode set. When the system is excited out of resonance as shown on the right hand side, even for low frequencies the representation is not perfect, the different colors show increasing numbers of modes. Investigating the absolute value of the modal coordinates as shown in Figure 4.6 the above considerations are confirmed. The out-of-resonance excitation requires a reasonable number of modes, with decreasing contribution, the more the modal frequency is separated from the frequency of excitation.

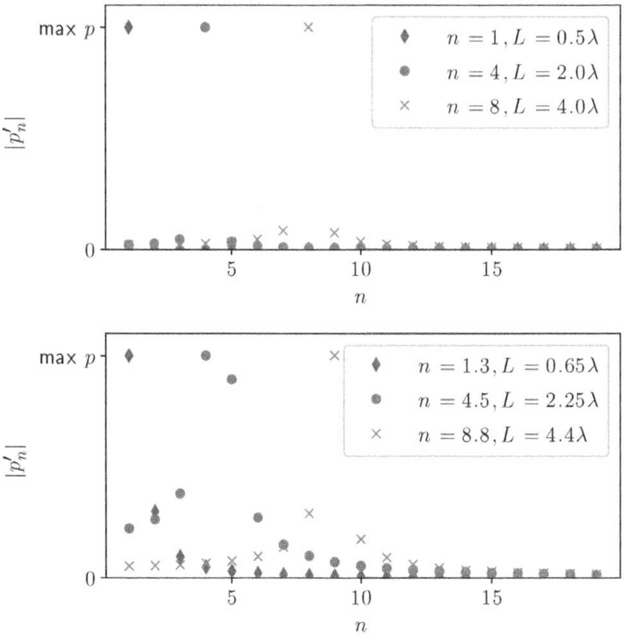

Figure 4.6 Modal coefficients for different wavenumbers. *Source:* Alexander Peiffer.

Choosing a high value of $n = 50$ we see that due to large damping even the result of resonance frequencies $N = 6$ leads to visible deviations, but for $N = 100$ both curves coincide. Consequently for out-of-resonance frequencies 100 modes are required for reasonable precision. In other words, for systems large compared to the wavelength and with damping, a larger number of modes is required to describe the dynamics correctly.

4.2 Three-dimensional Systems

In this section we deal with large three-dimensional cavities or rooms. We focus on simple geometries to describe the principle behavior over the full frequency range and for being able to work out the details on analytical solutions. Treatments on room acoustics can be found for example in Morse and Ingard (1968), Pierce (1991) and Jacobsen (2008).

4.2.1 Modes

We consider a room of rectangular shape as shown in Figure 4.8. We solve the Helmholtz Equation (2.42) in Cartesian coordinates with rigid boundary conditions $v_n = 0$. With (2.35) this yields

$$\frac{\partial \boldsymbol{p}}{\partial x} = 0 \text{ at } x = 0, L_x \tag{4.47}$$

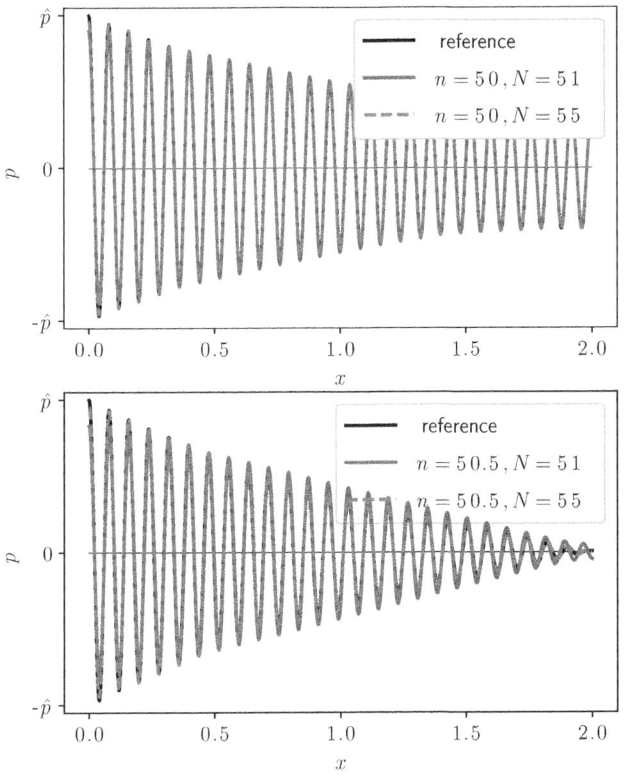

Figure 4.7 Exact and modal solutions for different total mode number, resonant and anti-resonant case, and high frequencies. *Source:* Alexander Peiffer.

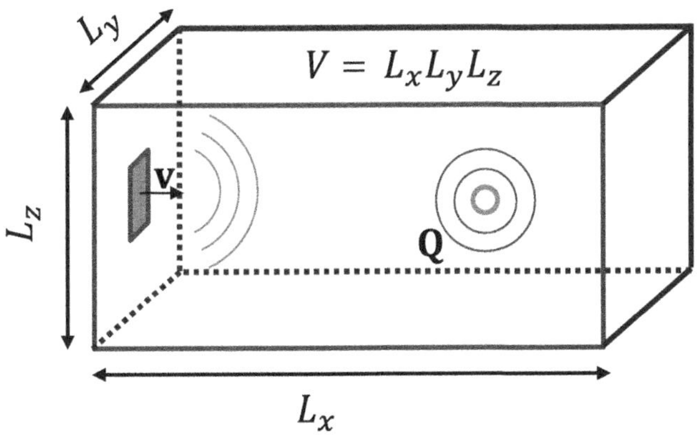

Figure 4.8 Rectangular room and dimensions. *Source:* Alexander Peiffer.

$$\frac{\partial \boldsymbol{p}}{\partial y} = 0 \text{ at } y = 0, L_y \tag{4.48}$$

$$\frac{\partial \boldsymbol{p}}{\partial z} = 0 \text{ at } z = 0, L_z \tag{4.49}$$

We assume that the solution can be factorized, meaning that the solution of p can be expressed as the product of three independent functions of x, y and z such as

$$\mathbf{p}(x,y,z) = \mathbf{p}_x(x)\mathbf{p}_y(y)\mathbf{p}_z(z) \tag{4.50}$$

Entering this into (2.42) gives

$$\frac{1}{\mathbf{p}_x(x)}\frac{d^2\mathbf{p}_x(x)}{dx^2} + \frac{1}{\mathbf{p}_y(y)}\frac{d^2\mathbf{p}_y(y)}{dy^2} + \frac{1}{\mathbf{p}_z(z)}\frac{d^2\mathbf{p}_z(z)}{dz^2} + k^2 = 0 \tag{4.51a}$$

All terms must be independent from the other space variables, e.g. the term with p_x is independent from y, z and equals k_x.

$$\frac{d^2\mathbf{p}_x(x)}{dx^2} + k_x^2\mathbf{p}(x) = 0 \tag{4.52a}$$

$$\frac{d^2\mathbf{p}_y(y)}{dy^2} + k_y^2\mathbf{p}(y) = 0 \tag{4.52b}$$

$$\frac{d^2\mathbf{p}_z(z)}{dz^2} + k_z^2\mathbf{p}(z) = 0 \tag{4.52c}$$

From Equation (4.51a) follows the constraint

$$k_x^2 + k_y^2 + k_z^2 = k^2 \tag{4.53}$$

The solutions for the equations can be adapted from the tube case

$$\mathbf{p}_x(x) = \mathbf{A}e^{-jk_xx} + \mathbf{B}e^{jk_xx} \tag{4.54a}$$

$$\mathbf{p}_y(y) = \mathbf{C}e^{-jk_yy} + \mathbf{D}e^{jk_yy} \tag{4.54b}$$

$$\mathbf{p}_z(z) = \mathbf{E}e^{-jk_zz} + \mathbf{F}e^{jk_zz} \tag{4.54c}$$

and the factorisation leads to

$$\mathbf{p}(x,y,z) = (\mathbf{A}e^{-jk_xx} + \mathbf{B}e^{jk_xx})(\mathbf{C}e^{-jk_yy} + \mathbf{D}e^{jk_yy})(\mathbf{E}e^{-jk_zz} + \mathbf{F}e^{jk_zz}) \tag{4.55}$$

The rigid boundary condition at $x = 0, y = 0$ and $z = 0$ implies that $\mathbf{A} = \mathbf{B}$ and $\mathbf{C} = \mathbf{D}$ and $\mathbf{E} = \mathbf{F}$ leading to a set of cosine functions

$$\mathbf{p}(x,y,z) = 2\mathbf{A}\cos(k_xx)2\mathbf{C}\cos(k_yy)2\mathbf{E}\cos(k_zz) \tag{4.56}$$

The boundary conditions at the opposite walls $x = L_x, y = L_y$ and $z = L_z$ require that a natural multiple of half-wavelength fit into the room dimensions

$$k_xL_x = n_x\pi, \qquad k_yL_y = n_y\pi, \qquad k_zL_z = n_z\pi \tag{4.57}$$

with $n_i = 0,1,2,3,\ldots$ and $i = x,y,z$. Finally, we denote every mode Ψ of the room with (n_x, n_y, n_z)

$$\Psi_{n_xn_yn_z}(x,y,z) = \Lambda_{n_xn_yn_z}\cos\left(\frac{n_x\pi x}{L_x}\right)\cos\left(\frac{n_y\pi y}{L_y}\right)\cos\left(\frac{n_z\pi z}{L_z}\right) \tag{4.58}$$

and the solution in modal coordinates $\mathbf{p}'_{n_xn_yn_z}$ is

$$\mathbf{p}(x,y,z) = \sum_{n_x=0}^{N_x}\sum_{n_y=0}^{N_y}\sum_{n_z=0}^{N_z}\mathbf{p}'_{n_xn_yn_z}\Psi_{n_xn_yn_z}(x,y,z) \tag{4.59}$$

Here, we have to include *static* modes in any direction. Thus Λ is defined as

$$\Lambda_{n_x n_y n_z} = \sqrt{\epsilon_{n_z}\epsilon_{n_z}\epsilon_{n_z}} \text{ with } \epsilon_0 = 1 \text{ and } \epsilon_n = 2\,\forall\, n > 0. \tag{4.60}$$

For simplicity the three indexes are combined into one index n

$$\boldsymbol{p}(x, y, z) = \sum_{n=0}^{N} \boldsymbol{p}'_n \Psi_n(x, y, z) \tag{4.61}$$

where n represents the three integers n_x, n_y and n_z. Using Equation (4.53) and (4.57) we get an expression for the natural frequencies of the room

$$\omega_n = k_n c_0 = c_0 \sqrt{\left(\frac{n_x \pi}{L_x}\right)^2 + \left(\frac{n_y \pi}{L_y}\right)^2 + \left(\frac{n_z \pi}{L_z}\right)^2} \tag{4.62}$$

The modes are orthogonal as will be shown by entering (4.58) into the Helmholtz Equation (2.42)

$$\Delta\Psi_m + k_m^2 \Psi_m = 0 \qquad \Delta\Psi_n + k_n^2 \Psi_n = 0 \tag{4.63}$$

Multiplying the first with Ψ_n and the second with Ψ_m leads to the difference

$$\Psi_n(\Delta + k_m^2)\Psi_m - \Psi_m(\Delta + k_n^2)\Psi_n = 0 \tag{4.64}$$

that can be rewritten as

$$\nabla(\Psi_n \nabla\Psi_m - \Psi_m \nabla\Psi_n) + (k_m^2 - k_n^2)\Psi_m \Psi_n = 0 \tag{4.65a}$$

In accordance with the one-dimensional system we integrate over the volume and apply Gauss's theorem on the first term

$$\int_{\partial V} (\Psi_n \nabla\Psi_m - \Psi_m \nabla\Psi_n) d\mathbf{S} + (k_m^2 - k_n^2) \int_V \Psi_m \Psi_n dV = 0 \tag{4.66}$$

with ∂V denoting the room surface. At rigid walls any normal component of the gradient is zero according to (4.47)–(4.49), so the functions are orthogonal regarding the three-dimensional functional

$$\int_V \Psi_m \Psi_n dV = 0 \text{ for all } m \neq n. \tag{4.67}$$

It can be shown as in (4.35) that the modes can be normalized with the following expression

$$\Phi_n = \frac{1}{\sqrt{V}} \Psi_n \text{ with } \int_V \Phi_m \Phi_n = \delta_{mn} \tag{4.68}$$

4.2.1.1 Modal Density

The number of modes in a three-dimensional system must be naturally higher than for one-dimensional systems. We can count the possible modes below a specific frequency using Equation (4.62). As the mode count becomes important when the system gets complex and thus many modes exist, we need a more appropriate way to estimate the number of modes that occur until frequency ω or $k = \omega/c_0$.

A geometrical approach is used to estimate the mode count with some corrections as proposed by Maa (1939). In Figure 4.9 the plane of modal wavenumbers for the tangential modes $n_z = 0$ is shown. Each oblique mode occupies a full rectangle of area $k_x k_y = \pi^2/(L_x L_y)$ in the quarter circle of radius $k = \omega/c_0$. The number of modes would be the quarter circle area divided by the rectangle area. However, the axial modes with additionally $n_x = 0$ or $n_y = 0$ are not correctly considered. Only half of their occupied area is taken into account by the quarter circle. This is corrected by adding half-rectangles of area $k\pi/(2L_x)$ and $k\pi/(2L_y)$. The static mode is covered by one quarter through the quarter circle, but two quarters are already compensated by rectangular area. So, only a correction by $\pi^2/(4L_x L_y)$ is necessary.

The total area is given by sum of

1. Quarter circle of area $\pi k^2/4$.
2. Two rectangles of size $\pi k/(2L_x)$ and $\pi k/(2L_y)$.
3. A quarter rectangle of size $\pi^2/(4L_x L_y)$.

The total area of all these components divided by the area of one oblique mode area provides an estimate for two-dimensional systems.

For the three-dimensional wavenumber grid in k-space a similar compensation is required. See Figure 4.10 for details. The octant of radius k is divided by the cube space any oblique mode is occupying, i.e. $\Delta V = \pi^3/(L_x L_y L_z)$ to estimate the mode count. The oblique modes with all $n_i \neq 0$ are considered correctly.

The cubes corresponding to the tangential modes are cut in half; this is compensated by three quarter slabs. Only the quarter of the axial modes is covered by the octant, but one half is already considered by the two adjacent quarter slabs. So, we compensate by the columns with length k and quarter area. To be fully consequent the static mode is considered by one-eighth from the octant and two times the contribution of the three slabs and columns. Finally, a missing one-eighth must be added to the full volume.

The summarized total mode count volume has the following components:

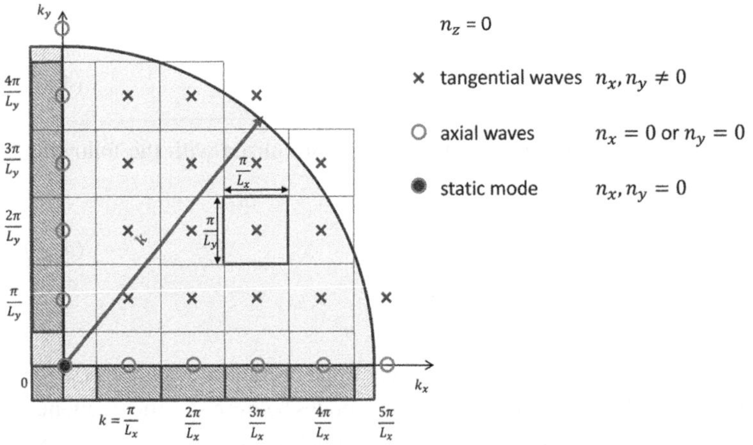

Figure 4.9 Wavenumber pattern for mode count determination in the k_x, k_y plane. *Source:* Alexander Peiffer.

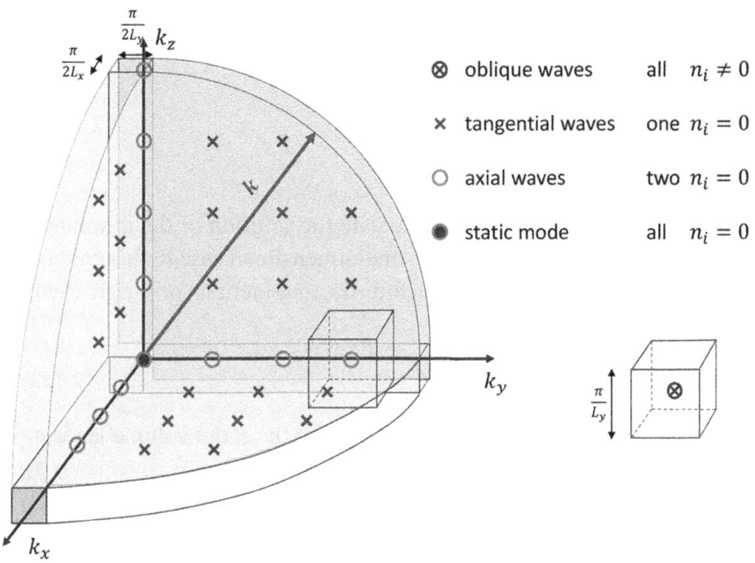

Figure 4.10 Wavenumber grid in three-dimensional space. *Source:* Alexander Peiffer.

1. An octant of a sphere of radius $k = \omega/c_0$ with $V_{\text{obl}} = \frac{\pi k^3}{6}$.
2. The sum of the volume of three quarter slabs of radius k and with thicknesses $\pi/(2L_x), \pi/(2L_y)$ and $\pi/(2L_z)$ respectively.
3. The sum of the volume of three rectangular columns of length k and cross-sectional area $\pi^2/(4L_xL_y)$, $\pi^2/(4L_yL_z)$ and $\pi^2/(4L_xL_z)$.
4. A volume $\pi^3/(8L_xL_yL_z)$ for the static mode.

Thus, the total volume divided by the volume on one mode $\pi^3/(L_xL_yL_z) = \pi^3/V$ is

$$N(\omega) \approx \frac{Vk^3}{6\pi^2} + \frac{\pi k^2}{8\pi^2}(L_yL_z + L_xL_z + L_xL_y) + \frac{k\pi^2}{4\pi^3}(L_z + L_x + L_y) + \frac{1}{8} \tag{4.69}$$

Introducing the quantities perimeter $P = 4(L_x + L_y + L_z)$ and surface $S = 2(L_xL_y + L_yL_z + L_xL_z)$ we get the approximate expression for the mode count formulated for basic geometric quantities V, S, and P

$$N_{3D}(\omega) \approx \frac{V\omega^3}{6\pi^2c_0^3} + \frac{S\omega^2}{16\pi c_0^2} + \frac{P\omega}{16\pi c_0} + \frac{1}{8} \tag{4.70}$$

and for the modal density

$$n_{3D}(\omega) = \frac{dnN(\omega)}{d\omega} = \frac{V\omega^2}{2\pi^2c_0^3} + \frac{S\omega}{8\pi c_0^2} + \frac{P}{16\pi c_0} \tag{4.71}$$

In Figure 4.11 the exact curves derived by counting all modes with $\omega_n \leq \omega$ and the estimated curves from function (4.70) are shown. At high mode count the estimation is precise enough. The more irregular the cavity the worse the estimation becomes. Irregular cavities mean having a large perimeter and surface compared to the volume, hence flat and column shaped cavities.

Figure 4.12 shows precise and estimated modal density for the same rooms. Here, the derivative leads to more sensitivity to small errors. Irregular cavities lead to inaccurate modal density estimations.

4.2.2 Modal Frequency Response

The mode shapes of the room can be used to calculate the solution of the inhomogeneous Helmholtz Equation (2.123) similar to the one-dimensional case in this chapter. For consideration of absorbing boundaries we include the treatment on non-rigid cavity surfaces. We start with

$$(k^2 + \Delta)\, \boldsymbol{p}(\mathbf{r}, \omega) = -j\omega\rho_0 \boldsymbol{q}_s + \nabla\mathbf{f}' = -\boldsymbol{f}_q(\mathbf{r}) \tag{2.123}$$

Multiplying from the left with $\Phi_n(\mathbf{r})$ and integrating over the room volume leads to

$$\int_V \Phi_n(\mathbf{r})(\boldsymbol{k}^2 + \Delta)\boldsymbol{p}(\mathbf{r})dV = -\int_V \Phi_n(\mathbf{r})\boldsymbol{f}_q(\mathbf{r})dV \tag{4.72}$$

With the Green identity, the first term on the left hand side can be converted and we get

$$\int_V \boldsymbol{p}(\mathbf{r})(\boldsymbol{k}^2 + \Delta)\Phi_n(\mathbf{r})dV + \int_{\partial V} \Phi_n(\mathbf{r})\nabla\boldsymbol{p}(\mathbf{r}) - \boldsymbol{p}(\mathbf{r})\nabla\Phi_n(\mathbf{r})dS$$
$$= -\int_V \Phi_n(\mathbf{r})\boldsymbol{f}_q(\mathbf{r})dV \tag{4.73}$$

The second term in the surface integral is zero because of the boundary conditions in the mode shape determination. The pressure gradient can be replaced with (2.35)

$$\nabla\boldsymbol{p}(\mathbf{r})_{\partial V} = \frac{\partial \boldsymbol{p}}{\partial n} = -j\omega\rho_0\boldsymbol{v}_n \tag{4.74}$$

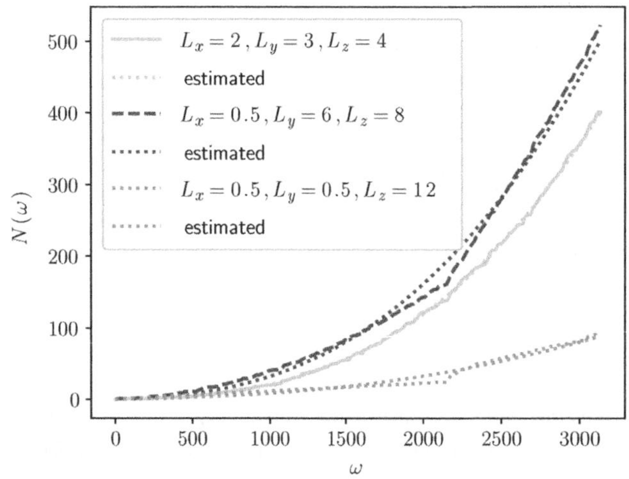

Figure 4.11 Mode count of different room configurations: cubic, flat and lengthy. *Source:* Alexander Peiffer.

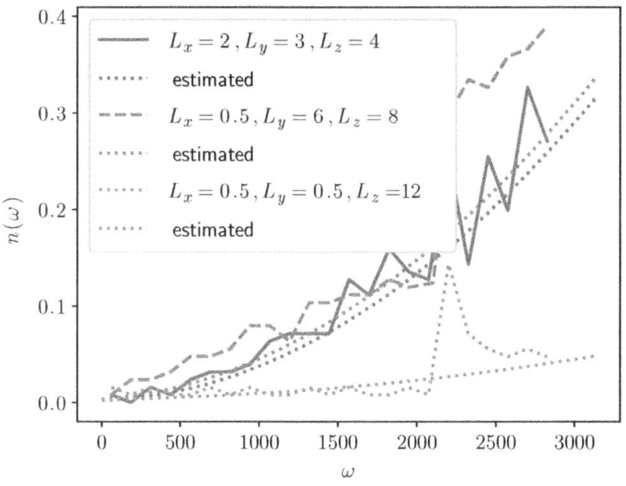

Figure 4.12 Modal density of different room configurations, $\Delta\omega = 40\pi$Hz.
Source: Alexander Peiffer.

and $\Delta\Phi_n(\mathbf{r}) = -k_n^2\Phi_n(\mathbf{r})$ This equation can be used for implementing boundary conditions, for example vibrating surfaces. Here we consider a locally reacting surface with

$$\frac{\partial \boldsymbol{p}}{\partial n} = -j\omega\rho_0\frac{\boldsymbol{p}}{\boldsymbol{z}} = -j\omega\boldsymbol{k}z_0\frac{\boldsymbol{p}}{\boldsymbol{z}} \tag{4.75}$$

Using this in (4.73) and the modal sum for the pressure

$$\boldsymbol{p}(\mathbf{r}) = \sum_{n=0}^{\infty} \boldsymbol{p}_n'\Phi_n(\mathbf{r}) \tag{4.76}$$

we get

$$\begin{aligned}&\int_V \sum_m^{\infty} \boldsymbol{p}_m'\Phi_m(\mathbf{r})(\boldsymbol{k}^2 - k^2)\Phi_n(\mathbf{r})dV\\ &- jk\int_{\partial V} \Phi_n(\mathbf{r})\frac{z_0}{\boldsymbol{z}(\mathbf{r})}\sum_m^{\infty} \boldsymbol{p}_m'\Phi_m(\mathbf{r})dS\\ &= -\int_V \Phi_n(\mathbf{r})\boldsymbol{f}_q(\mathbf{r})dV\end{aligned} \tag{4.77}$$

and by using the orthogonality relationship (4.68) this reads as

$$\boldsymbol{p}_n'(\boldsymbol{k}^2 - k_n^2) - jk\int_{\partial V}\sum_m^{\infty}\Phi_n\frac{z_0}{\boldsymbol{z}}\Phi_m dS = -\int_V \Phi_n(\mathbf{r})\boldsymbol{f}_q(\mathbf{r})dV \tag{4.78}$$

The surface integral leads to mode coupling, meaning that all modes contribute to the n-th modal coordinate. We define

$$B_{nm} = \int_{\partial V}\sum_m^{\infty}\Phi_n(\mathbf{r})\frac{z_0}{\boldsymbol{z}}\Phi_m(\mathbf{r})dS \tag{4.79}$$

In rooms with nearly rigid surfaces the off-diagonal components can be neglected and only B_{nn} is considered. Using this leads to

$$p_n'(k^2 - k_n^2 - jkB_{nn}) = -\int_V \Phi_n(\mathbf{r}) f_q(\mathbf{r}) dV \tag{4.80}$$

From the source term $f_q(\mathbf{r}) = -\delta(\mathbf{r} - \mathbf{r}_0)$ we would get the modal coordinates of the Green's function of rectangular rooms:

$$G_n' = -\frac{\Phi_n(\mathbf{r}_0)}{k^2 - k_n^2 - jkB_{nn}} \tag{4.81}$$

or the Green's function as a series

$$G(\mathbf{r}, \mathbf{r}_0) = -\sum_n \frac{\Phi_n(\mathbf{r})\Phi_n(\mathbf{r}_0)}{k^2 - k_n^2 - jkB_{nn}} \tag{4.82}$$

This is the generalized Green's function of the rectangular room, because it provides the point source response under given boundary conditions. Using acoustic point sources $f_q(\mathbf{r}) = j\omega\rho_0 Q\delta(\mathbf{r} - \mathbf{r}_0)$ and equations (4.80) and (4.76) we get the modal response of the rectangular room with point source of strength $Q(\mathbf{r}_0)$ located at $\mathbf{r}_0$

$$p(\mathbf{r}) = -j\omega\rho_0 Q \sum_n \frac{\Phi_n(\mathbf{r})\Phi_n(\mathbf{r}_0)}{k^2 - k_n^2 - jkB_{nn}} \tag{4.83}$$

The practical aspect of the above equation is that it considers damping due to surface absorption and propagation simultaneously. For rigid walls B_{nn} vanishes and the equation corresponds again to the usual modal response form.

4.2.3 System Responses

Whether the Green's function or the modal response form (4.83) is used we may calculate the response of the system. For the understanding of low- and high-frequency behavior we investigate the source characteristics and wave field of point sources. They can by located in the room volume, at the walls, edges, or corners. In our first considerations we neglect wall absorption and apply an overall modal damping.

$$p(\mathbf{r}) = -j\omega\rho_0 Q \sum_n \frac{\Phi_n^*(\mathbf{r})\Phi_n(\mathbf{r}_0)}{k^2 - k_n^2(1 + j\eta)} \tag{4.84}$$

4.2.3.1 Point Sources

A point source at the wall is represented by a small piston with $R \ll \lambda$ as shown in Figure 4.14. In this case the velocity at the wall leads to the following volume source

$$v(\mathbf{r}) = v \text{ for all } |\mathbf{r} - \mathbf{r}_0| \le R \Rightarrow Q = \pi R^2 v \tag{4.85}$$

Together with (4.84) the response can be calculated. This expression can also be derived by introducing the vibration as a boundary condition applying $\nabla p(\mathbf{r}_0) = j\omega\rho_0 v$. As we need the radiation impedance for later applications the mechanical impedance is derived by using the source area A.

$$\boldsymbol{Z} = \frac{\boldsymbol{F}}{\boldsymbol{v}} = \frac{\boldsymbol{p}A}{\boldsymbol{v}} = -j\omega\rho_0 A^2 \sum_m \frac{\Phi_m(\mathbf{r}_0)\Phi_m(\mathbf{r}_0)}{k^2 - k_m^2} \tag{4.86}$$

This impedance can be used when vibrating surfaces are described by discrete meshes. Every vibrating node is related to a specific area and thus experiences a mechanical radiation impedance from the fluid. In Figure 4.13 the mechanical impedance of a point source in the room wall is shown and compared to the impedance of the circular piston. In the range of validity for the point source, i.e. $\lambda \ll R$ or $kR < 1$ the real part of the room point impedance equals the piston result.

4.2.3.2 Radiation Impedance and Power

With the above defined impedance the radiated power is given by

$$\Pi = \frac{1}{2}\boldsymbol{F}\boldsymbol{v}^* = \frac{1}{2}\boldsymbol{Z}\boldsymbol{p}^2 = -\frac{1}{2}j\omega\rho_0 A\boldsymbol{v}^2 \sum_m \frac{\Phi_m(\mathbf{r})\Phi_m(\mathbf{r})}{k^2 - k_m^2} \tag{4.87}$$

From the considerations in section 2.4.1 we know that the radiation impedance is more appropriate to describe radiation of point sources. For the rectangular room this reads as

$$\boldsymbol{Z}_a(\mathbf{r}) = \frac{\boldsymbol{p}(\mathbf{r})}{\boldsymbol{Q}(\mathbf{r})} = -j\omega\rho_0 \sum_m \frac{\hat{\Phi}_m^2(\mathbf{r})}{k^2 - k_m^2(1 + j\eta)} \tag{4.88}$$

for the power radiated into the room we get

$$\Pi_{\text{rad}}(\mathbf{r}) = \frac{1}{2} Re\left(\boldsymbol{p}(\mathbf{r})\boldsymbol{Q}^*(\mathbf{r})\right) = -\frac{j\omega\rho_0}{2} \sum_m \frac{\Phi_m^2(\mathbf{r})\hat{Q}^2(\mathbf{r})}{k^2 - k_m^2(1 + j\eta)} \tag{4.89}$$

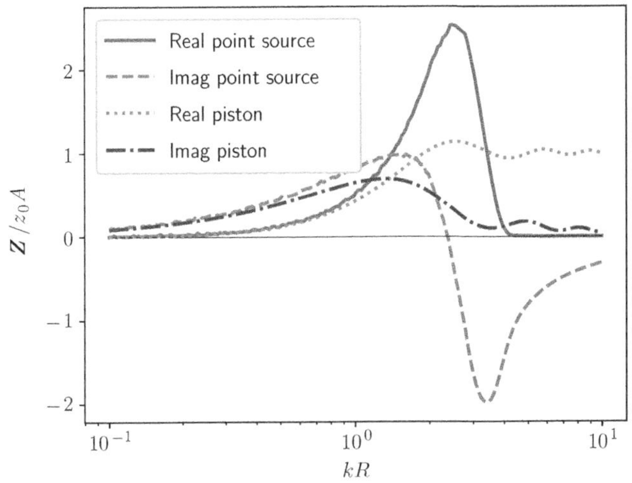

Figure 4.13 Mechanical impedance of point radiator in the room wall $\mathbf{r} = (0, 1, 0.7)$. *Source:* Alexander Peiffer.

4.2.3.3 Rectangular Piston in the Wall

The room itself is assumed to be source free, so $f(\mathbf{r}_0) = 0$. Therefore, there is no contribution from the volume term in the Kirchhoff integral (2.137). We assume a rectangular piston in the wall at $x = 0$; the dimensions are as shown in Figure 4.14.

$$\boldsymbol{p}(\mathbf{r}) = -j\omega\rho_0 \int_{\partial V} G(\mathbf{r}, \mathbf{r}_0)\boldsymbol{v}(\mathbf{r}_0)d\mathbf{r}_0 \tag{4.90}$$

The surface integral over the room surface is zero except at the wall at $x_0 = 0$ and the limit of the rectangular surface

$$\int_{\partial V} G(\mathbf{r}, \mathbf{r_0})\boldsymbol{v}(\mathbf{r}_0)d\mathbf{r}_0 = \boldsymbol{v} \int_{y_1}^{y_2} \int_{z_1}^{z_2} G(\mathbf{r}, \mathbf{r_0})dy_0 dz_0 \tag{4.91}$$

This integral can be analytically solved for the modal Green's function from Equation (4.82)

$$\begin{aligned} \boldsymbol{p}(\mathbf{r}) &= -j\omega\rho_0 \sum_m \frac{\boldsymbol{v}\Phi_m(\mathbf{r})}{k_m^2 - k^2} \int_{y_1}^{y_2} \int_{z_1}^{z_2} \Phi_m(\mathbf{r_0})dy_0 dz_0 \\ &= -j\omega\rho_0 \sum_m \frac{\boldsymbol{v}\Phi_m(\mathbf{r})}{k_m^2 - k^2} A_m \end{aligned} \tag{4.92}$$

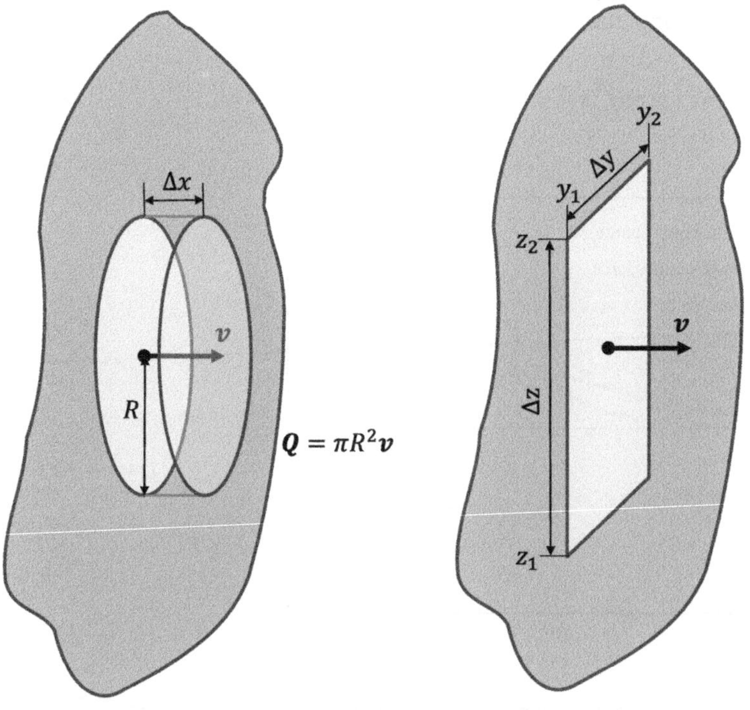

Figure 4.14 Point source and rectangular piston in room wall. *Source:* Alexander Peiffer.

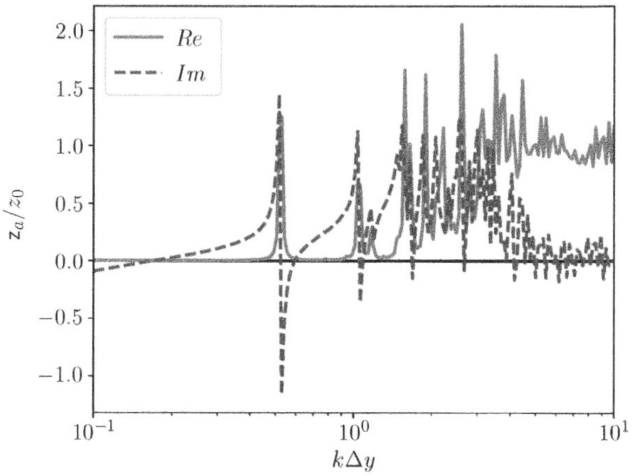

Figure 4.15 Acoustic radiation impedance of rectangular piston in room with dimensions $L_x = 3, L_y = 1.5, L_z = 2$ and piston dimensions $y_1 = 0.3, y_2 = 0.8, z_1 = 0.7, z_2 = 1.2$. *Source:* Alexander Peiffer.

Assuming $x_0 = 0$ the solution for $A_m = A_{m_x,m_y,m_z}$ is

$$A_m = \frac{\Lambda_m}{V}\frac{L_yL_z}{m_ym_z\pi^2}$$
$$\left[\sin\left(\frac{m_y\pi y_2}{L_y}\right) - \sin\left(\frac{m_y\pi y_1}{L_y}\right)\right]\left[\sin\left(\frac{m_z\pi z_2}{L_z}\right) - \sin\left(\frac{m_z\pi z_1}{L_z}\right)\right] \quad (4.93)$$

With Equation (4.92) the pressure field due to the vibrating piston can be derived. But, we are interested in the radiation impedance to the piston, in order to get the power radiated into the room.

For this we have to integrate the pressure over the piston as in Equation (2.146)

$$\begin{aligned} \mathbf{Z} &= -j\omega\rho_0 \sum_m \frac{1}{\mathbf{v}} \int_{y_1}^{y_2} \int_{z_1}^{z_2} p(\mathbf{r})dydz \qquad (4.94) \\ &= -j\omega\rho_0 \sum_m \sum_m \frac{A_m}{k_m^2 - k^2} \int_{y_1}^{y_2} \int_{z_1}^{z_2} \Phi_m(\mathbf{r})dydz \\ &= -j\omega\rho_0 \sum_m \frac{A_m^2}{k_m^2 - k^2} \end{aligned}$$

because we have to validate the same integral again. In Figure 4.15 the impedance is shown. Please note that the radiation impedance approaches $\mathbf{z}_a = \rho_0 c_0$ in the free field limit.

4.3 Numerical Solutions

For most cavity shapes and geometries an analytical expression of a closed form solution is not possible. This is the field of numerical solutions. Due to the homogenity of

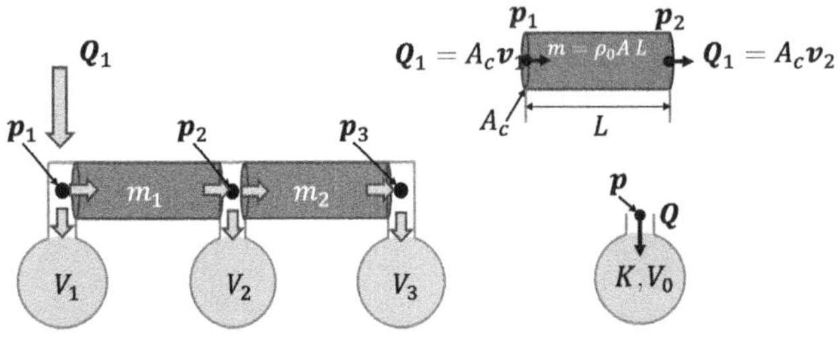

Figure 4.16 Lumped components of a fluid acoustic network. *Source:* Alexander Peiffer.

the fluid medium (or the existance of the Green's function) there are two options for numerical solutions:

FEM Finite Element Method. Here the fluid is discretized into small fluid elements, of for example tetrahedal, pentahedral, or hexahedral shape. Most solvers discretize the wave equation for pressure.

BEM Boundary Element Method. The second option is, to make use of Equation (2.137) and to integrate over elements of a surface mesh. Thus, the mesh is a surface mesh, which makes the meshing procedure mush easier. But, the resulting matrices are numerically complex.

The theories of FEM and BEM are not part of this book, but it makes sense to create model conceptions about the meaning of acoustic finite elements.

4.3.1 Acoustic Finite Element Methods

In Chapter 1 we introduced a discrete equation of motion that was created automatically by using discrete elements as springs, dampers and masses. The discrete acoustic wave equation of a fluid looks similar to equations (1.89) or (1.103) but with a different meaning of the matrices.

$$\left[\boldsymbol{D}_a\right]\{\boldsymbol{p}\} = \left[-\omega^2\left[M_a\right] + j\omega\left[C_a\right] + \left[K_a\right] + j\left[B_a\right]\right]\{\boldsymbol{p}\} = j\omega\{\boldsymbol{Q}\} \tag{4.95}$$

$\left[M_a\right]$ is the acoustic mass matrix, but has nothing to do with masses. It is called the mass matrix because of the ω^2 factor. The coefficients correspond to the inverse bulk modulus of the discrete fluid element. So, it is more a stiffness related quantity. The acoustic stiffness matrix $\left[K_a\right]$ corresponds to the inertia of the fluid element. Thus, in contrast to mechanical systems where the discrete equations can be visualized by a set of point masses and springs we have to find a different model.

This is because the pressure is the degree of freedom in the wave equation. When we would have chosen the displacements the discrete equation would be more alike to mechanical systems.

The sum of all three matrices is the acoustic dynamic stiffness matrix $\left[\boldsymbol{D}_a\right]$. It is frequency dependent and complex. However, it is helpful to develop physical models for the discrete presentation and the components of the discrete acoustic mass and stiffness matrix.

As shown in Figure 4.16 there are two elements to be considered. The fist element is a local compressible volume V_i. For the change in volume per pressure change we have

$$\frac{\partial V}{\partial P} = N_a = \frac{V}{K}. \tag{4.96}$$

Expressing the change in volume and pressure by complex amplitudes and using $\boldsymbol{V} = \boldsymbol{Q}/j\omega$ this reads as

$$\boldsymbol{Q}_i = j\omega N_a \boldsymbol{p}_i = j\omega \frac{V_i}{K} \boldsymbol{p}_i \tag{4.97}$$

and with the source term $j\omega\boldsymbol{Q}_i$ from (4.95) we get

$$-\omega^2 \frac{V_i}{K} \boldsymbol{p}_i = j\omega \boldsymbol{Q}_i \tag{4.98}$$

Thus, the components of the *mass* matrix in Equation (4.95) correspond to element volume V_i and the compressibility or inverse bulk modulus.

$$[M_a] = \begin{bmatrix} V_1/K & \dots & 0 \\ 0 & V_2/K & 0 \\ \dots & 0 & V_3/K \end{bmatrix} \tag{4.99}$$

In contrast to this, the second lumped element we must deal with is an acoustical mass. In the simplified approach of the lumped acoustic network, we assume an incompressible fluid of mass m_i. That means the wavelength in the fluid must be larger than the element length, to fulfill that condition. The relationship between the fluid velocity and pressures follow from Newton's law

$$\boldsymbol{F} = m\boldsymbol{a} = j\omega m\mathbf{v} \text{ with } \boldsymbol{F} = A(\boldsymbol{p}_1 - \boldsymbol{p}_2) \text{ and } m = \rho_0 LA \tag{4.100}$$

This reads as

$$\boldsymbol{p}_1 - \boldsymbol{p}_2 = j\omega \frac{\rho_0 L}{A} \boldsymbol{Q} \tag{4.101}$$

using $\boldsymbol{Q} = A\boldsymbol{v}$. When we consider the external flow into the volumes is considered as positive and thus $\boldsymbol{Q} = \boldsymbol{Q}_1 = -\boldsymbol{Q}_2$ the system of equations for the mass element is:

$$\frac{A}{\rho_0 L} \begin{bmatrix} 1 & -1 \\ -1 & 1 \end{bmatrix} \begin{Bmatrix} \boldsymbol{p}_1 \\ \boldsymbol{p}_2 \end{Bmatrix} = j\omega \begin{Bmatrix} \boldsymbol{Q}_1 \\ \boldsymbol{Q}_2 \end{Bmatrix} \tag{4.102}$$

The *stiffness* matrix of the acoustic system is linked to the mass of the connecting fluid pipes.

The exchange of physical meaning results from the exchange of state and excitation variables. The mechanical equation of motion was defined for displacement and the excitation by forces; in acoustics the state variable is pressure (better related to forces) and the excitation given by volume flow (better linked to displacement).

The sum of all volume flows at each node must equal the externally introduced volume flow $\boldsymbol{Q}_i$. The final set of equations for the system shown in Figure 4.16 is as follows

$$\left(\frac{A}{\rho_0 L} \begin{bmatrix} 1 & -1 & 0 \\ -1 & 2 & -1 \\ 0 & -1 & 1 \end{bmatrix} - \omega^2 \frac{1}{K} \begin{bmatrix} V_1 & & \\ & V_2 & \\ & & V_3 \end{bmatrix} \right) \begin{Bmatrix} \boldsymbol{p}_1 \\ \boldsymbol{p}_2 \\ \boldsymbol{p}_3 \end{Bmatrix} = j\omega \begin{Bmatrix} \boldsymbol{Q}_1 \\ \boldsymbol{Q}_2 \\ \boldsymbol{Q}_3 \end{Bmatrix} \tag{4.103}$$

4.3.2 Deterministic Acoustic Elements

When dealing with one-dimensional fluid systems such as air conditioning, mufflers, and hydraulic pipes the system can be described by frequency dependent system functions without discretisation. We learned above that the appropriate quantities for an acoustic network are pressure $\boldsymbol{p}$ and volume flow $\boldsymbol{Q}$. Thus Equation (4.11) must be slightly modified to get this relationship for one-dimensional fluid pipes by multiplying both lines by $j\omega A$.

$$\frac{A\omega}{z_0}\begin{bmatrix} \frac{1}{\tan(\boldsymbol{k}L)} & \frac{1}{\sin(\boldsymbol{k}L)} \\ \frac{1}{\sin(\boldsymbol{k}L)} & \frac{1}{\tan(\boldsymbol{k}L)} \end{bmatrix}\begin{Bmatrix} \boldsymbol{p}_1 \\ \boldsymbol{p}_2 \end{Bmatrix} = j\omega \begin{Bmatrix} \boldsymbol{Q}_1 \\ \boldsymbol{Q}_2 \end{Bmatrix} \tag{4.104}$$

The internal pressure and velocity field can be derived from (4.1) and (4.2) now solving $\boldsymbol{A}$ and $\boldsymbol{B}$ for the pressure, because this is the state variable in our network, or by simply multiplying (4.12) by z_0 with consideration of the differing propagation directions of both waves.

$$\boldsymbol{p}(x) = \frac{1}{j\sin(\boldsymbol{k}L)}\left[\boldsymbol{p}_1 \sin(\boldsymbol{k}(L-x)) + \boldsymbol{p}_2 \sin(\boldsymbol{k}x)\right] \tag{4.105}$$

$$\boldsymbol{v}_x(x) = \frac{1}{\sin(\boldsymbol{k}L)}\left[\boldsymbol{v}_{x1}(\boldsymbol{k}(L-x)) + \boldsymbol{v}_{x2} \sin(\boldsymbol{k}x)\right] \tag{4.106}$$

4.4 Reciprocity

When inspecting equations (4.83) and (2.137) one can see that the source and receiver locations can be exchanged. That means that in an acoustic system that relates the relationship of a source strength $Q(r)$ at location 1 to a pressure at location 2, source and receiver positions can be exchanged; hence

$$\frac{\boldsymbol{p}(\mathbf{r}_2)}{\boldsymbol{Q}(\mathbf{r}_1)} = \frac{\boldsymbol{p}(\mathbf{r}_1)}{\boldsymbol{Q}(\mathbf{r}_2)} \tag{4.107}$$

Globally the reciprocity is a general principle for linear systems. Thus, the only restriction to the variables is that they must be conjugate, meaning their product is related to power; thus, we use here Q and p. The same principle holds for all linear networks, structures, etc. One conclusion from this principle is that the stiffness matrix from Equation (4.95) must be symmetric to fulfill the reciprocity relationship.

Bibliography

Finn Jacobsen. THE SOUND FIELD IN A REVERBERATION ROOM. Technical Report Note no 31261, Danmarks Tekniske Universitet (DTU), Lyngby, Denmark, August 2008.

Dah-You Maa. Distribution of Eigentones in a Rectangular Chamber at Low Frequency Range. *The Journal of the Acoustical Society of America*, 10(3):235–238, January 1939. ISSN 0001-4966. doi: 10.1121/1.1915981.

P.M.C. Morse and K.U. Ingard. *Theoretical Acoustics*. International Series in Pure and Applied Physics. Princeton University Press, 1968. ISBN 978-0-691-02401-1.

Allan D. Pierce. *Acoustics - An Introduction to Its Physical Principles and Applications.* Acoustical Society of America (ASA), Woodbury, New York 11797,U.S.A., one thousand, nine hundred eighty-ninth edition, 1991. ISBN 0-88318-612-8.

5

Structure Systems

5.1 Introduction

The complex situation of wave propagation in structures becomes even worse for closed structural systems. When specific features like holes, rigs, beading or any complicated shape is given, this is definitely the world of numerical methods. Current finite element methods (FEM) in combination with pre and post processors can handle very complex and large systems as for example trains or aircraft. But, the modelling procedure, the creation of the mesh, and the population of the property and material database are time consuming. Analogous to fluid systems we stay with academic cases to work out the full frequency range.

Even though we are not dealing with the details of FEM we will rely on the discrete representation of systems as introduced in section 1.4. Some of the following treatments rely on the discrete matrix formulation of structures, that is the dynamic stiffness matrix. Thus, the principle definitions are given without description of the finite element method. Please rely for example on the textbook from Bathe (1982) for more details. In Chapter 3 the equations of motion are expressed in displacements u, v, w, the components of the stress tensor for the bulk material σ_{ij} and forces and moments F_i, M_i. Rotations β_i are approximated as the derivative of displacement components.

In finite element theory the rotational components are considered as dedicated degrees of freedom. Thus, in this chapter the natural coordinates will therefore be the displacements u, v, w, the rotations $\beta_x, \beta_y, \beta_z$ and the according forces F_x, F_y, F_z, and moments M_x, M_y, M_z.

Consequently all discrete structural subsystems are described using a dynamic stiffness matrix of the form:

$$\begin{bmatrix} \boldsymbol{D}_{11} & \boldsymbol{D}_{12} & \cdots & \boldsymbol{D}_{1N} \\ \boldsymbol{D}_{21} & \boldsymbol{D}_{22} & \cdots & \boldsymbol{D}_{2N} \\ \vdots & \vdots & \vdots & \vdots \\ \boldsymbol{D}_{N1} & \boldsymbol{D}_{N2} & \cdots & \boldsymbol{D}_{NN} \end{bmatrix} \begin{Bmatrix} \boldsymbol{q}_1 \\ \boldsymbol{q}_2 \\ \cdots \\ \boldsymbol{q}_N \end{Bmatrix} = \begin{Bmatrix} \boldsymbol{F}_1 \\ \boldsymbol{F}_2 \\ \vdots \\ \boldsymbol{F}_N \end{Bmatrix} = [\boldsymbol{D}]\{\boldsymbol{q}\} = \{\boldsymbol{F}\} \tag{5.1}$$

$\{\boldsymbol{q}\}$ is the vector of generalised displacement degrees of freedom including rotations. The vector is called *generalised* because the displacement can be represented in any

Vibroacoustic Simulation: An Introduction to Statistical Energy Analysis and Hybrid Methods,
First Edition. Alexander Peiffer.

base coordinate system, for example modal coordinates, wavelets or wavenumber. Consequently, $\{\boldsymbol{F}\}$ is the generalised force vector including forces and moments of the accompanying displacements.

Due to the fact that the displacement is the natural degree of freedom we switch from the impedance to the dynamic stiffness concept. In the complex notation we may easily change between velocity and displacement using the $j\omega$ factor for the derivative.

$$\boldsymbol{v}_x = j\omega\boldsymbol{u} \qquad\qquad \boldsymbol{D} = j\omega\boldsymbol{Z}$$

However, even if structure systems are very complicated it is necessary to elaborate some solutions for representative systems in order to be prepared for random methods, and the modal density is one of those quantities where analytical solutions yield results for the full frequency range that can later be used for a class of random systems.

5.2 One-dimensional Systems

5.2.1 Longitudinal Waves in Finite Beams

Rods are beams without bending capabilities or with exclusive treatment of dilatation of longitudinal waves. Thus, the solution is similar to the propagation of fluid waves in one-dimensional systems but with different physical quantities. We solve the equation

$$\frac{\partial^2 \boldsymbol{u}}{\partial x^2} + \frac{1}{k_{LB}}\boldsymbol{u} = 0 \quad \text{with} \quad k_{LB} = \sqrt{\frac{\omega^2 \rho_0}{E}} \tag{3.77}$$

for the given boundary conditions. Using the solution of the one dimensional wave Equation (3.78) and relationship (3.81)

$$\boldsymbol{u}(x) = \boldsymbol{A}e^{-jkx} + \boldsymbol{B}e^{+jkx} \tag{5.2}$$

$$\boldsymbol{\sigma}_1(x) = j\omega z_{LB}\left(-\boldsymbol{A}e^{-jkx} + \boldsymbol{B}e^{+jkx}\right) \tag{5.3}$$

With the following boundaries regarding the connections at both ends expressed in discrete degrees of freedom

$$\boldsymbol{u}(0) = \boldsymbol{u}_1 \qquad\qquad \boldsymbol{u}(L) = \boldsymbol{u}_2 \tag{5.4a}$$

$$\boldsymbol{\sigma}_1(0) = -\frac{\boldsymbol{F}_1}{A} \qquad\qquad \boldsymbol{\sigma}_1(L) = \frac{\boldsymbol{F}_2}{A} \tag{5.4b}$$

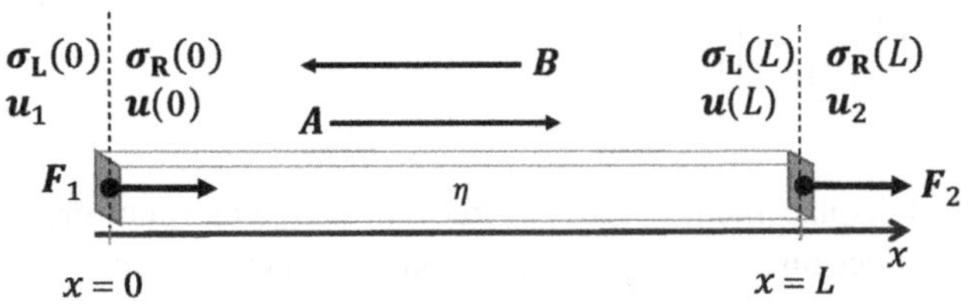

Figure 5.1 Finite beam with boundary conditions as two-port system.

Comparing the above Equations to (4.1) and (4.2) we see that all solutions are provided by section 4.1 when we exchange the following variables:

$$\boldsymbol{p} \Leftrightarrow \boldsymbol{u} \qquad \boldsymbol{v}_x \Leftrightarrow -\boldsymbol{\sigma}_1 \qquad 1/z_0 = j\omega z_{LB} \tag{5.5}$$

This leads to the dynamic stiffness matrix for the rod

$$j\omega A z_{LB} \begin{bmatrix} \frac{1}{j\tan(\boldsymbol{k}L)} & -\frac{1}{j\sin(\boldsymbol{k}L)} \\ -\frac{1}{j\sin(\boldsymbol{k}L)} & \frac{1}{j\tan(\boldsymbol{k}L)} \end{bmatrix} \begin{Bmatrix} \boldsymbol{u}_1 \\ \boldsymbol{u}_2 \end{Bmatrix} = AE\boldsymbol{k}_{LB}$$
$$\begin{bmatrix} \frac{1}{\tan(\boldsymbol{k}L)} & -\frac{1}{\sin(\boldsymbol{k}L)} \\ -\frac{1}{\sin(\boldsymbol{k}_)} & \frac{1}{\tan(\boldsymbol{k}L)} \end{bmatrix} \begin{Bmatrix} \boldsymbol{u}_1 \\ \boldsymbol{u}_2 \end{Bmatrix} = \begin{Bmatrix} \boldsymbol{F}_1 \\ \boldsymbol{F}_2 \end{Bmatrix} \tag{5.6}$$

It is worth mentioning that for $k \ll 1$ all denominators are approximately $\sin(\boldsymbol{k}L) \approx \boldsymbol{k}L$ and hence

$$\lim_{k\to 0} [\boldsymbol{D}]_{\text{bar}} = \frac{EA}{L} \begin{bmatrix} 1 & -1 \\ -1 & 1 \end{bmatrix} = \begin{bmatrix} k_s & -k_s \\ -k_s & k_s \end{bmatrix} \text{ with } k_s = EA/L \tag{5.7}$$

with k_s being the stiffness constant of the spring realised by a rod – it is the same as (1.95). The matrix can be used in a mechanical network representation. All expressions derived for power and impedance can also be used. For example the mechanical input impedance of a rod at port one and with a given mechanical impedance $\boldsymbol{Z} = \boldsymbol{F}_x/\boldsymbol{v}$ at port two after some lengthy exchange operations is

$$\boldsymbol{Z}_1 = \frac{\boldsymbol{F}_1}{j\omega \boldsymbol{u}_1} = Az_{LB} \frac{\boldsymbol{Z}_2 \cos(\boldsymbol{k}L) + jAz_{LB} \sin(\boldsymbol{k}L)}{Az_{LB} \cos(\boldsymbol{k}L) + j\boldsymbol{Z}_2 \sin(\boldsymbol{k}L)} \tag{5.8}$$

All considerations regarding system response, power input, etc. are also equivalent to section 4.1.

5.2.1.1 Modes

The expressions for the mode shapes can be derived using the same variable exchange from (5.5). The modes can be fixed or free. In technical systems, beams and plates are fixed at the ends, and an excitation is only possible for free DOFs. The normalized mode shapes for fixed boundaries ($u(0) = 0, u(L) = 0$) are

$$\Phi(x)_{u,n} = \sqrt{\frac{\epsilon_n}{L}} \sin(k_n x) \text{ with } \epsilon_n = \begin{cases} 1 & \text{for n} = 0 \\ 2 & \text{for n} > 0 \end{cases} \tag{5.9}$$

or for free boundaries ($\sigma(0) = 0, \sigma(L) = 0$)

$$\Phi(x)_{u,n} = \sqrt{\frac{\epsilon_n}{L}} \cos(k_n x) \text{ with } \epsilon_n = \begin{cases} 1 & \text{for n} = 0 \\ 2 & \text{for n} > 0 \end{cases} \tag{5.10}$$

both with the modal wavenumbers $k_n = \frac{n\pi}{L}$. The free modes allow for excitation on both ends, the fixed modes only at inner DOFs. Entering the modes into the harmonic inhomogeneous wave equation for longitudinal waves (3.77)

$$\rho_0 A\left(\frac{\partial^2 \boldsymbol{u}}{\partial x^2} + \boldsymbol{k}_{LB}\boldsymbol{u}\right) = F'(x) \tag{5.11}$$

and writing the force density per length expression with the use of the delta function

$$F(x_0) = \int_{-\infty}^{\infty} \boldsymbol{F}\delta(x - x_0)dx$$

yields the response shape

$$\boldsymbol{u}(x) = \sum_{n=0}^{\infty} \boldsymbol{u}'_n \Phi_{u,n} \tag{5.12}$$

The modal coordinates $\boldsymbol{u}'_n$ are derived by entering this into (5.11) and using the orthogonality of the shape functions providing

$$\boldsymbol{u}'_n = \frac{\Phi_{u,n}\boldsymbol{f}(x)}{\rho_0 A(k_n^2 - \boldsymbol{k}^2)} \tag{5.13}$$

Obviously, the modal density is equal to the pipe solution as it depends only on c_{LB} and L.

$$N = \frac{L\omega}{\pi c_{LB}} \qquad n(\omega) = \frac{L}{\pi c_{LB}}$$

5.2.2 Bending wave in Finite Beams

In beams, bending waves can propagate with elongation to the direction of the main axes. We skip the analytical closed solution for bending wave motion and stay with the modal shape description for the calculation of bending motion response.

5.2.2.1 Modes

We choose bending motion around the y-axis with displacement w in the z-direction. The boundary conditions are

$$w(0) = 0 \qquad w(L) = 0 \tag{5.14a}$$

$$M_y(0) = -B_y\frac{\partial^2 w}{\partial x^2}(0) = 0 \qquad M_y(L) = -B_y\frac{\partial^2 w}{\partial x^2}(L) = 0 \tag{5.14b}$$

When entering these conditions into the global solution (3.107)

$$\boldsymbol{w}(x,\omega) = \boldsymbol{w}_1 e^{jk_B x} + \boldsymbol{w}_2 e^{-jk_B x} + \boldsymbol{w}_{1D} e^{k_B x} + \boldsymbol{w}_{2D} e^{-k_B x}$$

this leads to $\boldsymbol{w}_1 = \boldsymbol{w}_2$ and $\boldsymbol{w}_{1D} = \boldsymbol{w}_{2D} = 0$. Thus, the mode shape function is

$$\Phi(x)_{w,n} = \sqrt{\frac{\epsilon_n}{L}}\sin(k_n x) \text{ with } \epsilon_n = \begin{cases} 1 & \text{for n} = 0 \\ 2 & \text{for n} > 0 \end{cases} \tag{5.15}$$

with $k_n = n\pi/L$. There are also free modes; the solution is a combination of cosine and hyperbolic cosine functions, however, because coefficients $\boldsymbol{w}_{1D}$ and $\boldsymbol{w}_{2D}$ are not vanishing in that case. The modal wavenumber is similar to the known one-dimensional formulas, but there is a difference for the modal frequency

$$\omega_n = \left(\frac{n\pi}{L}\right)^2 \sqrt{\frac{B_y}{m'}} \tag{5.16}$$

The distance between each modal frequency is not constant. So, it is convenient to use the constant wavenumber count of modes that are below k

$$N(k) = \frac{kL}{\pi} \tag{5.17}$$

With this equation we can derive the modal density by applying the chain rule

$$n_{1D}(\omega) = \frac{dN}{d\omega} = \frac{dN}{dk}\frac{dk}{d\omega} = \frac{L}{\pi c_{gr}} = \frac{L}{2\pi c_B} = \frac{L}{2\pi}\sqrt[4]{\frac{m'\omega^2}{B_y}} \tag{5.18}$$

This is the same formula as (4.41) but with the group velocity.

5.2.2.2 Modal Response

The displacement in the z-direction in modal coordinates is given by

$$\boldsymbol{w}(x) = \sum_{n=1}^{\infty} \boldsymbol{w}'_n \Phi(x)_{w,n} \tag{5.19}$$

Entering this into the inhomogeneous form of (3.104a)

$$-\omega^2 m' \boldsymbol{w} + B_y \frac{\partial^4 \boldsymbol{w}}{\partial x^4} = \boldsymbol{F}'_z \tag{5.20}$$

gives for the force distribution

$$\sum_{n=1}^{\infty} m'(\omega_n^2 - \omega^2)\boldsymbol{w}'_n \Phi_{w,n} = \boldsymbol{F}'_z(x) \tag{5.21}$$

Multiplication from the right with $\Phi^*_{w,m}$ and integration along the beam gives

$$\boldsymbol{w}'_m = \int_0^L \frac{\Phi^*_{w,m}\boldsymbol{F}'_z(x)}{m'(\omega^2 - \omega_m^2)} dx \tag{5.22}$$

A point force $\boldsymbol{F}_z$ located at x_0 is represented by

$$\boldsymbol{F}'_z(x) = \boldsymbol{F}_z \delta(x - x_0) \tag{5.23}$$

and thus

$$\boldsymbol{w}'_m = \frac{\Phi^*_{w,m}(x_0)\boldsymbol{F}_z}{m'(\omega^2 - \omega_m^2)} \tag{5.24}$$

From (3.110) we know that $\boldsymbol{F}_z = -\frac{\partial \boldsymbol{M}_y}{\partial x}$, so we can deal with moment excitation using the same modal base

$$\boldsymbol{w}'_m = -\int_0^L \frac{\Phi^*_{w,m}\frac{\partial \boldsymbol{M}'_y}{\partial x}(x)}{m'(\omega^2 - \omega_m^2)} = \int_0^L \frac{\frac{\partial \Phi^*_{w,m}}{\partial x}\boldsymbol{M}'_y(x)}{m'(\omega^2 - \omega_m^2)} \tag{5.25}$$

using the law of partial integration. With the moment $\boldsymbol{M}_y$ at x_0 described as

$$\boldsymbol{M}'_y(x) = \boldsymbol{M}_y \delta(x - x_0) \tag{5.26}$$

we get

$$\boldsymbol{w}'_m = \frac{\Phi'^*_{w,m}(x_0)\boldsymbol{F}_z}{m'(\omega^2 - \omega_m^2)} \text{ with } \Phi'^*_{w,m}(x) = \sqrt{\frac{\epsilon_n}{L}}\cos(k_n x) \tag{5.27}$$

As moments are linked to the rotations, the derivative of the mode shapes is used.

In Figure 5.2 the response of a beam excited at specific modal frequencies (top) and for a high frequency at different positions is shown.

The spectrum of the point impedance in Figure 5.3 approaches the value for infinite beams (3.136), but has many more peaks than the tube example for high frequencies. This results from the dispersion and the speed of sound that increases with frequency and therefore has fewer modes than rods or tubes.

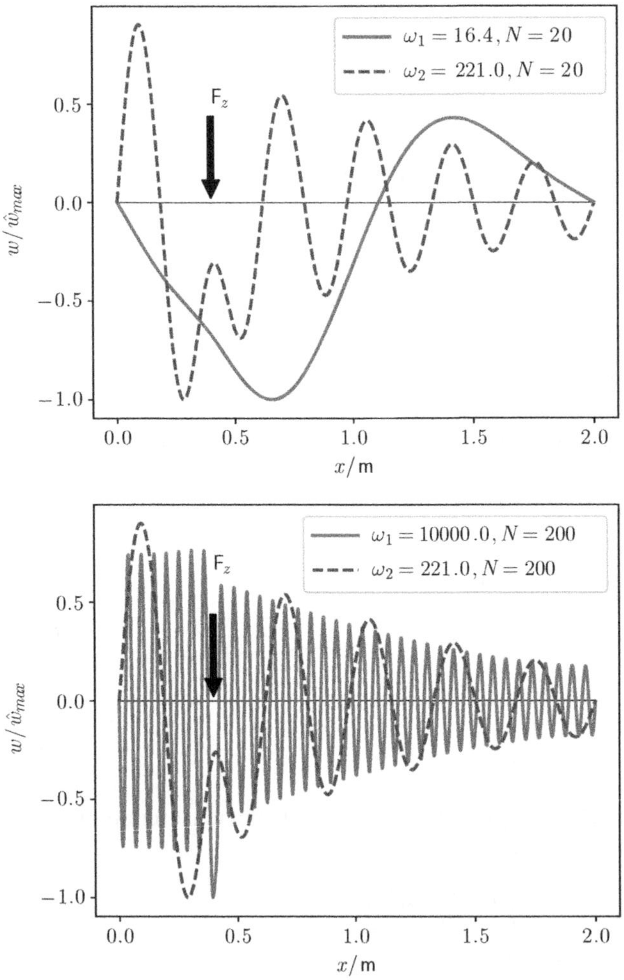

Figure 5.2 Beam bending wave response calculated with the modal method. *Source:* Alexander Peiffer.

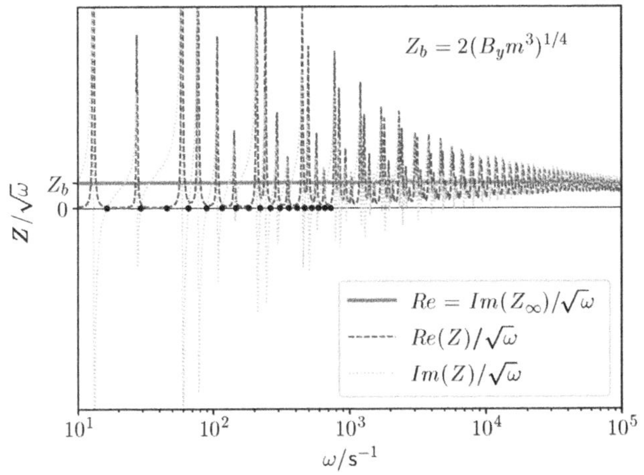

Figure 5.3 Scaled mechanical impedance of finite beam compared to infinite system. *Source:* Alexander Peiffer.

5.3 Two-dimensional Systems

We deal with plates as an example for a realistic and representative two-dimensional system. In section 3.7 the equation of motion was given for in-plane displacement (longitudinal and transversal) and out-of-plane displacement (bending). The in-plane displacement is characterized by dispersion-free and high-speed wave propagation. Thus, the practical relevance of such systems is not very high as wavelengths stay large in the audible frequency range. For example, a steel plate has longitudinal wave speed of $c_{LP} = 4600$ m/s, meaning that we have a wavelength of $\lambda = 4.6$ m at 1000 Hz. We would need very large systems to catch the first resonance of this wave type in the audible frequency range and in technical systems as for example a car. Thus, we stay with bending waves for the investigations on two-dimensional systems.

5.3.1 Bending Waves in Flat Plates

For the description of plate waves of finite systems we have to solve the homogenous form of (3.206)

$$\Delta\Delta \boldsymbol{w} - \omega_n^2 \frac{\rho_0 h}{B} \boldsymbol{w} = \frac{\boldsymbol{F}_z''(x,y)}{B}$$

with the following boundary conditions:

$$w(x,0) = 0 \qquad w(x,L_y) = 0 \tag{5.28a}$$

$$w(0,y) = 0 \qquad w(L_x,y) = 0 \tag{5.28b}$$

$$M_x(x,0) = -B\frac{\partial^2 w}{\partial y^2}(0) = 0 \qquad M_x(x,L_y) = -B\frac{\partial^2 w}{\partial y^2}(L_y) = 0 \tag{5.28c}$$

$$M_y(0,y) = -B\frac{\partial^2 w}{\partial x^2}(0) = 0 \qquad M_y(L_y,y) = -B\frac{\partial^2 w}{\partial x^2}(L_y) = 0 \tag{5.28d}$$

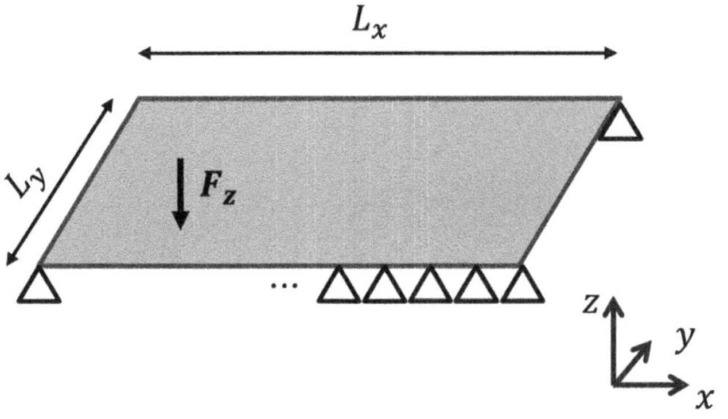

Figure 5.4 Rectangular plate with edge translations fixed. *Source:* Alexander Peiffer.

One can show using the same procedures as Sec. 5.2.2.1 that the function

$$\Psi_n(x,y) = \sin\left(\frac{n_x\pi x}{L_x}\right)\sin\left(\frac{n_y\pi y}{L_y}\right) \tag{5.29}$$

is a solution of Equation (3.206) with the modal frequencies

$$\omega_n = \sqrt{\frac{B}{\rho_0 h}}\left[\left(\frac{n_x\pi}{L_x}\right)^2 + \left(\frac{n_x\pi}{L_x}\right)^2\right] \tag{5.30}$$

and

$$k_n = \sqrt{\left(\frac{n_x\pi}{L_x}\right)^2 + \left(\frac{n_x\pi}{L_x}\right)^2} \tag{5.31}$$

where n is a double index $n = (n_x, n_y)$. These mode shapes are orthogonal and should be normalized for the product below the surface integral

$$\int_0^{L_x}\int_0^{L_y} \Psi_m^*(x,y)\Psi_n(x,y)dydy = \delta_{mn}\frac{L_x Ly}{4} \tag{5.32}$$

leading to the normalised shape functions

$$\Phi_n(x,y) = \frac{2}{\sqrt{L_x L_y}}\sin\left(\frac{n_x\pi x}{L_x}\right)\sin\left(\frac{n_y\pi y}{L_y}\right) \tag{5.33}$$

Some shapes are shown in Figure 5.6. The mode count is estimated by the area method but in the wavenumber domain due to dispersion as shown in Figure 5.5. Without tangential modes there are no additional correction areas required.

5.3.1.1 Modal Density

We estimate the number of modes that occur below the wavenumber k by comparing the areas of the quarter wavenumber circle $\pi k^2/4$ to the area covered by one

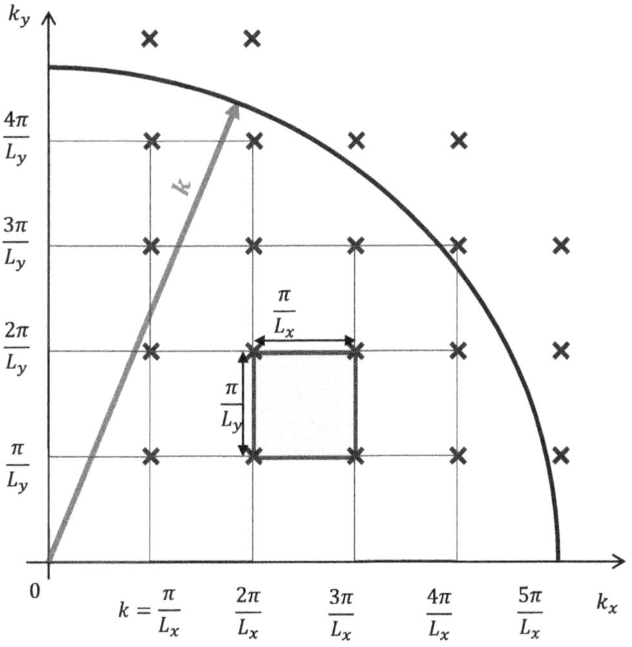

Figure 5.5 Wavenumber grid for plate waves. *Source:* Alexander Peiffer.

wavenumber rectangle $\pi^2/(L_x L_y)$

$$N_{2D}(k) \approx \frac{k^2 L_x L_y}{4\pi} \tag{5.34}$$

The modal density of the dispersive bending waves depends on the group velocity

$$n_{2D}(\omega) = \frac{dN}{d\omega} = \frac{dN}{dk}\frac{dk}{d\omega} \approx \frac{k L_x L_y}{2\pi c_{gr}} = \frac{k L_x L_y}{4\pi c_B} = \frac{L_x L_y}{4\pi}\sqrt{\frac{\rho_0 h}{B}} \tag{5.35}$$

The dispersion of bending waves leads to the surprising effect that the modal density is constant over frequency.

5.3.1.2 Modal Response

The response of an excited plate can be given by the double sum running over both modal indexes $n = (n_x, n_y)$.

$$\boldsymbol{w}(x, y) = \sum_{n=1}^{\infty} \boldsymbol{w}'_n \Phi_n(x, y) \tag{5.36}$$

Entering this into (3.206) we get the modal expression

$$\sum_{n=1}^{\infty} m''(\omega_n^2 - \omega^2)\boldsymbol{w}'_n \Phi_{w,n} = \boldsymbol{F}''_z(x, y) \tag{5.37}$$

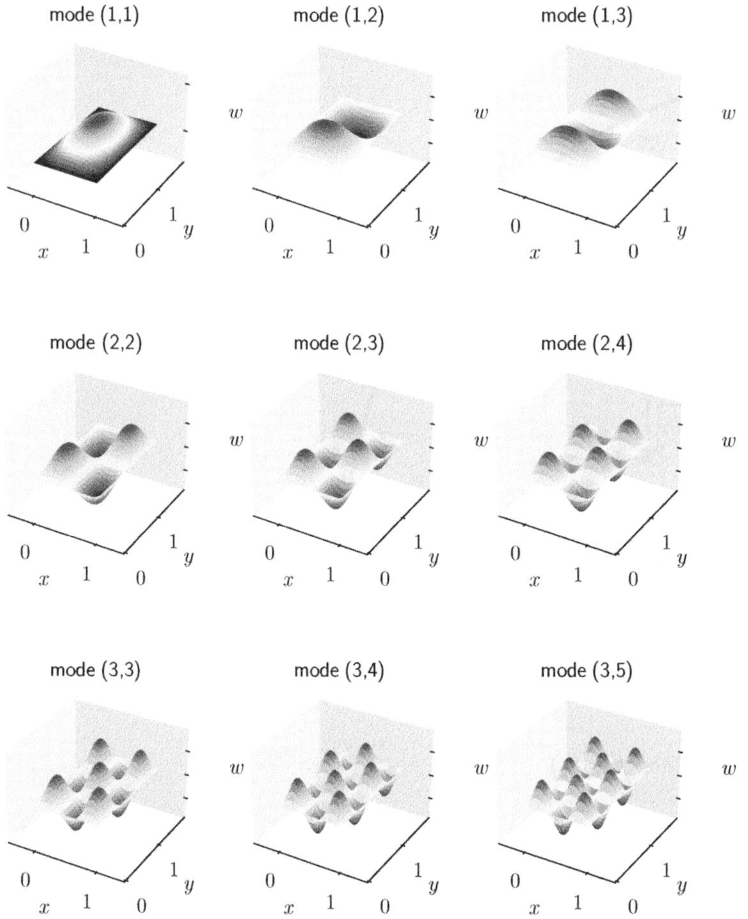

Figure 5.6 Some mode shapes of a flat rectangular plate. Dimensions $L_x = 1.0$ m, $L_y = 1.8$ m and $h = 2$ mm. *Source:* Alexander Peiffer.

and multiplication with mode shapes Φ_m^* and surface integration over the plate gives

$$\begin{aligned} \boldsymbol{w}'_m &= \frac{1}{m''(\omega_m^2 - \omega^2)} \int_0^{L_x} \int_0^{L_y} \Phi_m^*(x,y) \boldsymbol{F}''_z(x,y) dy dy \\ &= \frac{\boldsymbol{f}'_m}{m''(\omega_m^2 - \omega^2)} \end{aligned} \tag{5.38}$$

With modal forces

$$\boldsymbol{f}'_m = \int_0^{L_x} \int_0^{L_y} \Phi_m^*(x,y) \boldsymbol{F}''_z(x,y) dy dy \tag{5.39}$$

A point force at (x_0, y_0) is represented by a double delta function to create the required force per area function

$$\boldsymbol{F}''_z(x,y) = \delta(x - x_0)\delta(y - y_0)\boldsymbol{F}_z \tag{5.40}$$

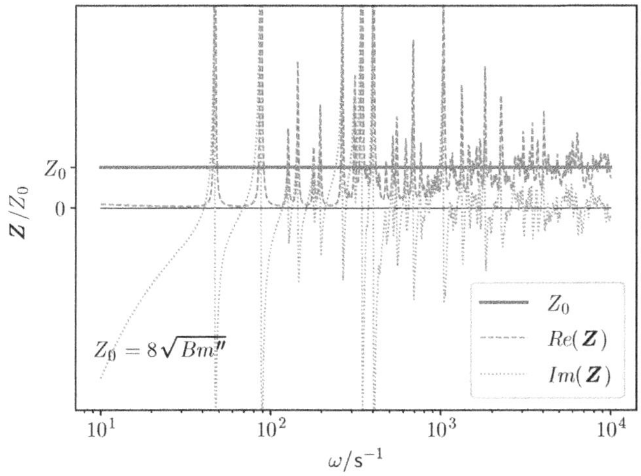

Figure 5.7 Point impedance of plate ($L_x = 1\,\text{m}, L_y = 1.8\,\text{m}, \eta = 0.02$) excited at $\mathbf{r}_0 = (0.7\,\text{m}, 0.36\,\text{m})$. *Source:* Alexander Peiffer.

leading to

$$\boldsymbol{w}(x,y) = \sum_{n=1}^{\infty} \frac{\Phi_n(x,y)\Phi_n^*(x_0,y_0)\boldsymbol{F}_z}{m''(\omega_n^2 - \omega^2)} \tag{5.41}$$

The mechanical point impedance is

$$\boldsymbol{Z}(x_0,y_0) = \frac{\boldsymbol{F}_z}{j\omega\boldsymbol{w}} = \frac{1}{j\omega}\left(\sum_{n=1}^{\infty} \frac{\Phi_n(x_0,y_0)\Phi_n^*(x_0,y_0)}{m''(\omega_n^2 - \omega^2)}\right)^{-1} \tag{5.42}$$

In Figure 5.7 an example for a point impedance is shown. The boundary impact can also be be neglected here for high frequencies and the impedance value for infinite values is reached. In the case of plates the shape of the impedance curve contains fewer peaks than the beam example. The higher dimensionality of the system provides more "space" for modes. Thus, the dynamic complexity of the system is reached earlier.

5.4 Reciprocity

Similar to the argument in Chapter 4, the reciprocity relationship also holds for structural equations. Inspecting (5.41) it becomes obvious that the quantities force at position 1 (and into direction 1) and the velocity at position 2 (and into direction 2) can be exchanged. Thus, reciprocity in structural dynamics reads as

$$\frac{\boldsymbol{v}(\mathbf{r}_2)}{\boldsymbol{F}(\mathbf{r}_1)} = \frac{\boldsymbol{v}(\mathbf{r}_1)}{\boldsymbol{F}(\mathbf{r}_2)} \tag{5.43}$$

The same relationship can be found for displacement and acceleration, even though these quantities are not conjugate. But, they are power related, and a $j\omega$ argument in Equation (5.41) does not change the above arguments. When proving the global principle, it can also be shown that this is also true for coupled systems. Thus, a volume source at position 1 generating a velocity at position 2 is equal to the force excitation at position 2 and the pressure response at position 1. This is of high practical use when the pressure signal of multiple force excitations at the engine mounts of a car are needed. A volume source with accelerometers at all mount positions replaces the experiment with one microphone at the head position and force excitation at every mount position. In complex geometries it may be hard to precisely excite the forces at positions that are not accessible. Placing accelerometers there is usually much easier.

5.5 Numerical Solutions

Due to the fact that we started with mechanical systems of point masses, dampers, and springs, the formulation of the physical interpretation of the mass and stiffness matrices of structural finite elements is clear. In section 5.1 of this chapter, the global dynamic equation in discrete form was formulated. However, even without a deeper understanding, some global properties of finite element models should be given in order to understand the properties and the limits of finite element formulation.

5.5.1 Normal Modes in Discrete Form

In the previous sections of this chapter we learned that modal condensation is a useful tool to calculate the response of mechanical system. The modal method is intensively used in numerical or finite element methods, too. For investigations in later chapters, we will use a discrete variant of the analytical mode shapes from Equation (5.33). The discrete degrees of freedom are $\{\boldsymbol{q}_i\}$ defined as shown in Figure 5.9 using a regular mesh. The node with index i is located in the $xy-$plane with nodal position $\mathbf{r}_i = \{x_i, y_i\}^T$ and the degree of freedom is the displacement in z, hence $\boldsymbol{q}_i = \boldsymbol{w}(x_i, y_i)$

Thus, the modal vectors are defined by the nodal position and Equation (5.33)

$$\{\Phi_n\} = A \begin{Bmatrix} \Phi_n(x_1, y_1) \\ \Phi_n(x_i, y_i) \\ \vdots \\ \Phi_n(x_N, y_N) \end{Bmatrix} \text{ with } N = N_x N_y \tag{5.44}$$

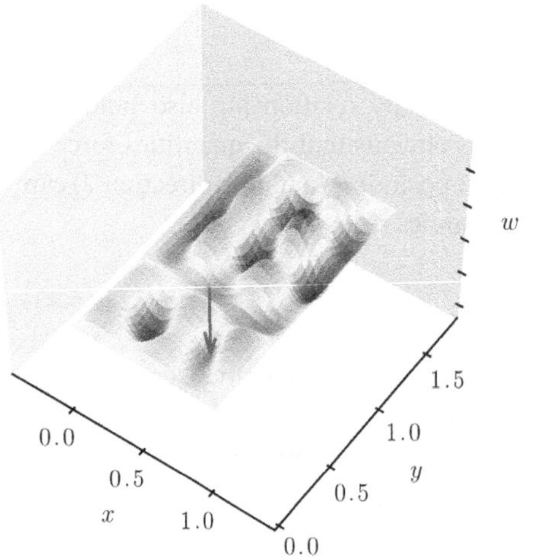

Figure 5.8 Plate reponse to point force at $\mathbf{r}_0 = (0.7\,\mathrm{m}, 0.36\,\mathrm{m}$ and frequency $\omega = 800$ Hz. *Source:* Alexander Peiffer.

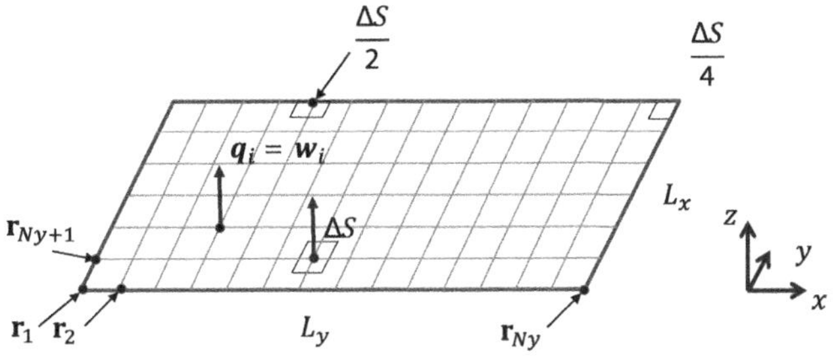

Figure 5.9 Mesh for discrete mode shapes. *Source:* Alexander Peiffer.

In order to ascertain mass normalization, the mass matrix is required. We assume that the mass of the discretized plate is given by the related lumped mass and thus derived from the specific mass m'' times the element surface element that belongs to the specific node. So, we get a diagonal matrix with mass entries $m_i = \Delta S m''$ for inner elements, $m_e = \Delta S m''/2$ for the edges, and $m_v = \Delta S m''/4$ for the vertex nodes.

$$[M] = \begin{bmatrix} m_v & & \cdots & & \cdots \\ & m_e & & & \\ \vdots & & \ddots & & \\ & & & m_i & \\ \vdots & & & & \ddots \end{bmatrix} \tag{5.45}$$

With this assumption the factor $A = \sqrt{\frac{L_x L_y}{M}}$ with M as total mass of the plate provides mass normalised modeshapes that can be used for modal frequency analysis or modal condensation

$$\Phi_n(x_i, y_i) = \frac{2}{\sqrt{M}} \sin\left(\frac{n_x \pi x_i}{L_x}\right) \sin\left(\frac{n_y \pi y_i}{L_y}\right) \text{ with } n = (n_x, n_y) \tag{5.46}$$

Due to the fact that the analytical solution is only sampled and the mesh doesn't have to be fine enough to allow a precise numerical solution, quite coarse meshes can be used, e.g. four elements per wavelength. For a finite element model of the same plate, at least six (linear) elements or nodes per wavelength would be necessary for precise results Bathe (1982).

Bibliography

Klaus-Jürgen Bathe. *Finite Element Procedures in Engineering Analysis.* Prentice-Hall Civil Engineering and Engineering Mechanics Series. Prentice-Hall, Englewood Cliffs, N.J, 1982. ISBN 978-0-13-317305-5.

6

Random Description of Systems

In the academic examples from Chapters 4 and 5 all parameters of the problem are given: Material constants, wave speed and damping, boundary conditions, and the excitation. However, even for those simple cases the resonant and deterministic response at low frequencies moves on to noisy but asymptotic signal history for high frequencies.

Obviously, real systems such as cars, buildings, or ships are much more complex. There are many subsystems and different materials, and there is complex geometry and a large combination of shapes. In addition, the environmental conditions lead to further changes. Thus, all real engineering systems are subject to uncertainty.

The test cases reveal that the small changes in the parameters don't change the dynamics of the system too much as long as the wavelength is just an order of magnitude below the length scale of the system. For high frequencies, the impact of small changes becomes high. The response of the dynamic system becomes very uncertain and varies strongly. Shorter and Langley (2005) introduced the term *dynamically complex system* for a system with fuzzy or uncertain parameters and small wavelengths compared to the system's length scale. Consequently, such a system is dealt with by statistical methods.

In statistical energy analysis (and statistical physics) we deal with this randomness by statistical parameters such as mean or root mean values. As this corresponds to energy quantities, there is the word *energy* in the name of the method. The statistics is performed over ensembles. That means we assume a *thought* representation of similar but varying systems (called an ensemble of systems) and investigate their average quantities.

Ensembles were introduced in the context of time signals in Chapter 1. There, we also thought about an ensemble of signals that was later replaced by an ensemble of time windows from one signal using the ergodic property.

In classical SEA the concept of ensemble average (Lyon and DeJong, 1995) is also used, but often the assumption of broadband excitation in combination with frequency averaging over frequency bands is made in addition. This leads to smooth responses in the frequency response functions but requires strict assumptions. In a pure SEA context the frequency averaging may be reasonable because the underlying assumption is that all systems are supposed to be random. But, there are strict requirements for the connections of the subsystems called *junctions*. The coupling must be weak and must not vary over frequency in each band. This cannot be guaranteed in many real systems. Imagine a rod as investigated in section 5.2.1 coupling two random systems. The

Vibroacoustic Simulation: An Introduction to Statistical Energy Analysis and Hybrid Methods,
First Edition. Alexander Peiffer.

coupling will be described by a stiffness matrix as given in (5.6) varying from stiff to soft connection depending on if the rod is operating in resonance or anti-resonance. In any case the coupling will not be constant in the frequency range.

This motivates the introduction of the ensemble averaging and especially the derivation of hybrid FEM/SEA theory that allows for narrow band simulation and deterministic components in the full model. In addition, the hybrid theory provides a very straightforward introduction to SEA.

When systems become random the modal or deterministic approach is not well suited for the description of such systems. They are much better described by the synthesis of non-coherent waves. Thus, before we start with randomized test cases the concept of the diffuse field is introduced. For the randomized and academic test cases it will be shown that the statistical approach is not only an approximation but also a necessity when dealing with random and dynamically complex systems.

6.1 Diffuse Wave Field

In dynamically complex systems the uncertainty and the multiple reflections of waves will constitute a wave field created by multiple waves with random phases. This initiates the concept of diffuse field where the wave field is described by a sum of multiple plane waves with randomly distributed phase, shape and orientation (Morse and Ingard, 1968; Pierce, 1991; Jacobsen, 1979). Because of the fact that this occurs in systems with several reflections the diffuse wave field is also called the reverberant field. Pierce (1991) proposed a simple and practical implementation of this model as a set of plane waves with uniform angular distribution depending on the space geometry (curves, surfaces, or volumes) of wave propagation. The amplitude is random with given expected value, and the phase is equally distributed over the phase interval $[-\pi, \pi)$. Thus, the plane wave diffuse field is represented by the following sum

$$\Psi(\mathbf{x}, t) = \sum_i \hat{\Psi}_i Re\left(e^{-j(\mathbf{k}_i\mathbf{x}+\phi_i)}\right) \text{ with } \mathbf{k}_i = k\mathbf{n}_i \tag{6.1}$$

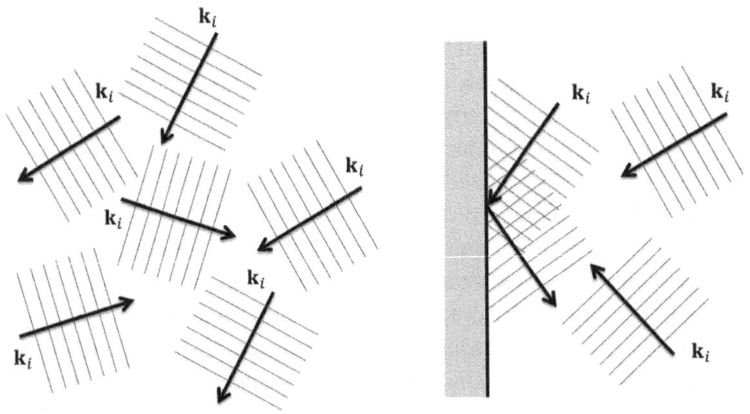

Figure 6.1 Multiple plane waves with random phase and amplitude but single frequency as representation of a diffuse field. *Source:* Alexander Peiffer.

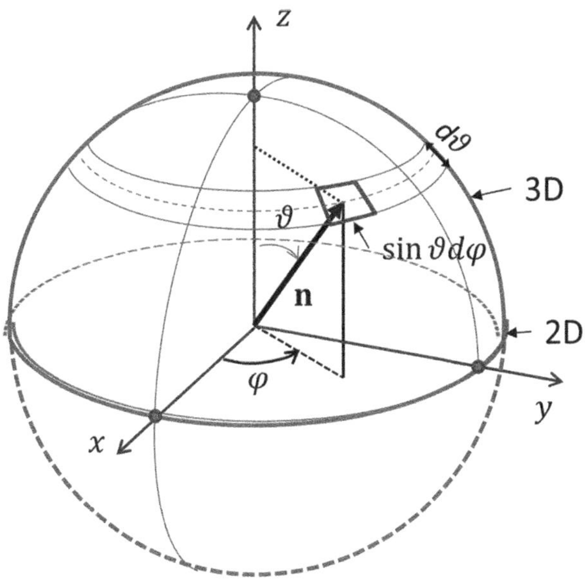

Figure 6.2 Spherical coordinate system for room angle integration of plane waves in the diffuse field. *Source:* Alexander Peiffer.

Here, $\mathbf{n}_i(\varphi_i, \vartheta_i)$ is the normal vector of directivity equally distributed over the (available) space represented by the solid angles φ_i and ϑ_i and is given by

$$\mathbf{n} = (n_x, n_y, n_z) \qquad n_x = \sin\vartheta\cos\varphi \qquad n_y = \sin\vartheta\sin\varphi \qquad n_z = \cos\vartheta \tag{6.2}$$

according to the spherical coordinate system (2.62).

The phase ϕ_i is randomly distributed over $\pm\pi$. The best way to characterize the random process of a diffuse field is to use statistical parameters. The wave amplitude is a complex random value with the following mean properties

$$\Psi^2_{\text{mean}} = \langle\hat{\Psi}_i\rangle_E = \lim_{M\to\infty}\frac{1}{M}\sum_{i=1}^{M}\hat{\Psi}_i \tag{6.3}$$

$$\Psi^2_{\text{ms}} = \left\langle\hat{\Psi}_i^{\,2}\right\rangle_E = \lim_{M\to\infty}\frac{1}{M}\sum_{i=1}^{M}\hat{\Psi}_i^2 \tag{6.4}$$

We introduce a probability density for each solid angle (ϑ_i, φ_i) and the random phase ϕ_i in order to get the expected value of the random process given by the above wave summation. The probability density for the phase is $p(\phi) = \frac{1}{2\pi}$. Obviously, the mean value vanishes

$$\langle\Psi(\mathbf{x})\rangle_E = \int_0^{2\pi} p(\phi)\hat{\Psi}Re(e^{-j(k\mathbf{nx}+\phi)}) = \int_0^{2\pi} p(\phi)\hat{\Psi}\cos(k\mathbf{nx}+\phi) = 0 \tag{6.5}$$

because the mean of $\cos(\phi)$ is zero. It is also clear that the mean square value exists. When we consider only the phase we get

$$\langle \Psi^2(\mathbf{x}) \rangle_E = \int_0^{2\pi} p(\phi)\hat{\Psi}^2 \cos^2(\phi) = \frac{\hat{\Psi}^2}{2} \tag{6.6}$$

This is independent from the dimension of space. For simplicity we define the root mean square value

$$\Psi_{\rm rms}^2 = \frac{1}{2}\hat{\Psi}^2 \tag{6.7}$$

For the average of a single position in space we don't need to consider the various orientations of the waves. We just take the average of all incoming and neglect their correlation. This is in contrast to the intensity or the correlation between two distinct points in space where the orientation must be taken into account. Consequently, the averaging procedure over all space directions depends on the dimensionality of our system. In order to consider the different dimensions, particular probability density functions for each dimension are defined.

3D The full solid angle according to the definition of spherical coordinates is used. The probability is given by the surface element of the unit sphere per total unit sphere surface

$$p(\vartheta, \varphi) = \frac{1}{4\pi}\sin(\vartheta) \qquad \text{for } \vartheta \in [0, \pi], \qquad \varphi \in [0, 2\pi) \tag{6.8}$$

2D In this case $\vartheta = \pi/2$ is constant and we get the modified probability density

$$p(\varphi) = \frac{1}{2\pi} \text{ for } \qquad \varphi \in [0, 2\pi) \tag{6.9}$$

with the reciprocal circumference of the unit circle. This covers all angles in the xy-plane.

1D The *angle* is reduced to two discrete directions. Thus, it is no longer a probability density but a probability value

$$p(\varphi) = \frac{1}{2} \text{ for } \varphi = 0, \pi \tag{6.10}$$

This means that there are effectively two orientations of $\mathbf{n} = (\pm 1, 0, 0)$.

Diffuse fields are an appropriate tool to describe random wave fields in terms of energy quantities. Therefore, we need to investigate the energy density and intensity especially at the boundaries of the systems where they interact with neighbor systems. In addition, this interaction with other systems requires the knowledge of the cross correlation in space and time. For the derivation of this correlation function we select one point $\mathbf{x}_1$ in the reverberant field without loss of generality and switch to the cosine function instead of using the real value of the exponential function

$$\Psi(\mathbf{x}_1, t) = \frac{1}{M}\sum_{i=1}^{M} \hat{\Psi}_i \cos(\omega t + \phi_i) \tag{6.11}$$

The sum of wave functions at a second location $\mathbf{x_2} = \mathbf{x_1} + \Delta\mathbf{x}$ and time $t + \Delta t$ is

$$\Psi(\mathbf{x}_2, t + \Delta t) = \frac{1}{M}\sum_{j=1}^{M} \hat{\Psi}_j \cos(\omega t - \mathbf{k}_j\Delta\mathbf{x} + \phi_j) \tag{6.12}$$

and the cross correlation of both expressions reads as

$$\langle\Psi(\mathbf{x}_1, t)\ \Psi^*(\mathbf{x}_2, t + \Delta t)\rangle_E$$
$$= E\left[\sum_i\sum_j \hat{\Psi}_i\hat{\Psi}_j \cos(\omega t + \phi_i)\cos(\omega(t + \Delta t) + \mathbf{k}_j\Delta\mathbf{x} + \phi_j)\right]$$

By definition there must be no correlation between waves of different index $i \neq j$, hence

$$= E\left[\sum_i \hat{\Psi}_i^2 \cos(\omega t + \phi_i)\cos(\omega(t + \Delta t) + \mathbf{k}_i\Delta\mathbf{x} + \phi_i)\right]$$
$$= \frac{1}{2}E\left[\sum_i \hat{\Psi}_i^2 \cos(\omega\Delta t + \mathbf{k}_i\Delta\mathbf{x})\right]$$
$$= \frac{1}{2}\int_\varphi\int_\vartheta p(\vartheta, \varphi)\hat{\Psi}^2 \cos(\omega\Delta t + k\mathbf{n}(\vartheta, \varphi)\Delta\mathbf{x}) \tag{6.13}$$

This expression has to be solved separately for each dimensionality.

6.1.1 Wave-Energy Relationships

In order to reconstruct the averaged wave amplitude from the total energy we must know how energy is related to the diffuse wave field amplitude. We keep the derivation here as general as possible. Thus, an arbitrary wave quantity Ψ was chosen. For the expression of energy density in terms of complex wave amplitudes we introduce the following relationship

$$e = \frac{B}{2}\hat{\Psi}^2 \tag{6.14}$$

For pressure waves ($\Psi = p$) this constant would be $B = 1/(\rho_0 c_0^2)$. In addition a link to the intensity is needed. According to Equation (3.238) this reads as

$$\mathbf{I} = \mathbf{n}c_{gr}e = \mathbf{n}\frac{c_{gr}B}{2}\hat{\Psi}^2 \tag{6.15}$$

because the group velocity defines the velocity of energy flow in contrast to the phase speed that defines the propagation of constant phase.

6.1.2 Diffuse Field Parameter of One-Dimensional Systems

In the context of one-dimensional waves the model of a diffuse reverberant field may seem artificial. Phase and direction are randomly but equally distributed over the phase angle 0 to 2π and the available *room* angle. The latter is just the positive and negative propagation direction for one-dimensional systems so $\mathbf{k}\Delta\mathbf{r} = \pm k\Delta x$. Thus the diffusivity of the one-dimensional field is given by the random phase and amplitude. The integration over the probability function for the expected value is replaced by a sum.

When we chose one point in the center of the one-dimensional system, the intensity is zero, because the flow of energy is equal in both directions. This changes when we look at the boundaries, for example at $x = 0$. In this case we have only one irradiating direction with probability 1/2:

$$I = \sum_{i=1}^{1} \frac{1}{2}\frac{c_{\text{gr}}B}{2}\hat{\Psi}^2 = \frac{c_{\text{gr}}B}{2}\Psi_{\text{rms}}^2 \tag{6.16}$$

Thus, the intensity is half the intensity of a free field wave with similar rms-amplitude. For the calculation of the cross correlation we also use the sum

$$\begin{aligned} S_{\Psi\Psi}(\Delta x, \Delta t) &= \frac{1}{2}\sum_{i=1}^{2} \hat{\Psi}_i^2 \cos(\pm k\Delta x + \omega\Delta t) \\ &= \frac{1}{2}\sum_{i=1}^{2} \hat{\Psi}_i^2 \cos(\pm k\Delta x)\cos(\omega\Delta t) \mp \sin(k\Delta x)\sin(\omega\Delta t) \\ &= \langle\hat{\Psi}^2\rangle_E \cos(k\Delta x)\cos(\omega\Delta t) \end{aligned} \tag{6.17}$$

6.1.3 Diffuse Field Parameter of Two-Dimensional Systems

The additional dimension creates more freedom in terms of available directions. For the integration we consider only the contributions that are normal to the edge. Choosing the edge parallel to the x-axis, the normal is $\mathbf{n} = \mathbf{e}_y$. The according integration angle is $\varphi \in [0, \pi]$ and the intensity normal to the edge is $I_y = I\sin(\varphi)$ (Figure 6.3)

$$\begin{aligned} I_y &= c_{\text{gr}}\frac{1}{2}B\int_0^{\pi} p(\varphi)\hat{\Psi}^2 \sin(\varphi)d\varphi \\ &= c_{\text{gr}}\frac{1}{2}B\int_0^{\pi} \frac{1}{2\pi}\hat{\Psi}^2 \sin(\varphi)d\varphi \\ &= c_{\text{gr}}\frac{B}{2\pi}\hat{\Psi}^2 = c_{\text{gr}}\frac{B}{\pi}\Psi_{\text{rms}}^2 \end{aligned} \tag{6.18}$$

Similar considerations are performed for the cross correlation and we choose without loss of generality $\Delta\mathbf{r} = \Delta r\,\mathbf{e}_y$. With $\mathbf{k}_i\Delta\mathbf{r} = k\Delta r\sin(\varphi_i)$

$$\begin{aligned} S_{\Psi\Psi}(\Delta x, \Delta t) &= \int_0^{2\pi} p(\varphi)\hat{\Psi}^2 \cos(k\Delta r\sin(\varphi) + \omega\Delta t)d\varphi \\ &= \langle\hat{\Psi}^2\rangle_E \cos(\omega\Delta t)2\int_0^{\pi} \frac{1}{\pi}\cos(k\Delta r\sin(\varphi))d\varphi \\ &= \langle\hat{\Psi}^2\rangle_E \cos(\omega\Delta t)J_0(k\Delta r) \end{aligned} \tag{6.19}$$

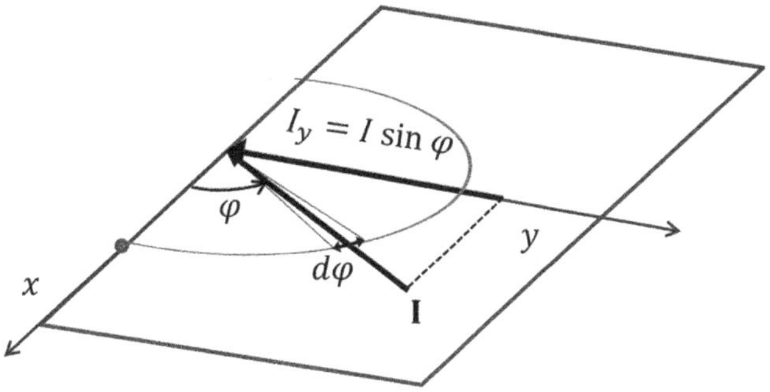

Figure 6.3 Area angle integration for edge intensity. *Source:* Alexander Peiffer.

This follows from the antisymmetry of the sine function and the integral representation of the Bessel function. The shape of this cross correlation is the same for location in the center of the diffuse field area because integration over the half circle leads to the same functional shape.

6.1.4 Diffuse Field Parameter of Three-Dimensional Systems

For this case, the normal vector $\mathbf{n} = \mathbf{e}_z$ of the xy-plane is chosen, and the intensity the integration runs from 0 to $\pi/2$ for ϑ. The normal intensity is $I_z = I\cos(\vartheta)$ and the integrand does not depend on φ

$$
\begin{aligned}
I_z &= \frac{c_{gr}B}{2}\int_0^{2\pi}\int_0^{\pi/2} p(\vartheta,\varphi)\hat{\Psi}^2\cos(\vartheta)d\vartheta d\varphi \\
&= \frac{c_{gr}B}{2}2\pi\int_0^{\pi/2}\frac{1}{4\pi}\hat{\Psi}^2\sin(\vartheta)\cos(\vartheta)d\vartheta \\
&= \frac{c_{gr}B}{8}\hat{\Psi}^2 = \frac{c_{gr}B}{4}\Psi_{\text{rms}}^2
\end{aligned}
\tag{6.20}
$$

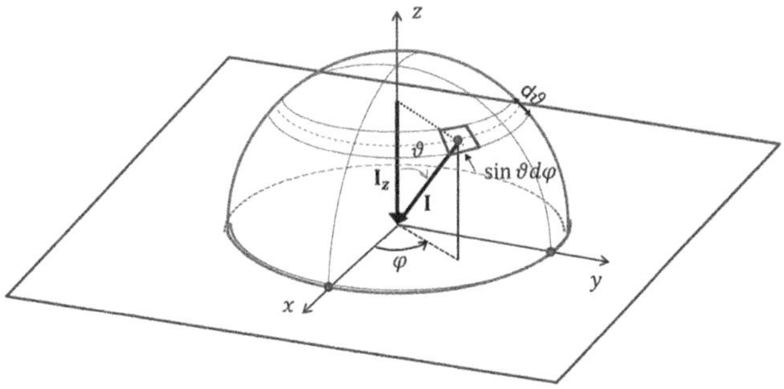

Figure 6.4 Half-sphere and room angle for intensity integration. *Source:* Alexander Peiffer.

We select $\mathbf{\Delta r} = \Delta r\,\mathbf{e}_z$ and the wave number distance vector product is given by $\mathbf{k\Delta r} = k\Delta r\sin(\vartheta)$.

$$
\begin{aligned}
S_{\Psi\Psi}(\Delta x, \Delta t) &= \int_0^{2\pi}\int_0^{\pi} p(\vartheta,\varphi)\hat{\Psi}^2 \cos(k\Delta r\cos(\vartheta) + \omega\Delta t)\sin(\vartheta)d\vartheta d\varphi \\
&= 2\pi\left\langle\hat{\Psi}^2\right\rangle_E \cos(\omega\Delta t)\int_0^{\pi}\frac{1}{4\pi}\cos(k\Delta r\cos(\vartheta))\sin(\vartheta)d\vartheta \\
&= \left\langle\hat{\Psi}^2\right\rangle_E \cos(\omega\Delta t)\frac{\sin(k\Delta r)}{(k\Delta r)}
\end{aligned}
\tag{6.21}
$$

The global form of the cross correlation is also kept for half-space integration. Two locations at the walls of a diffuse field have the above correlation function. This expression is used in experiments to check the quality of the diffuse field, see e.g. Jacobsen (1979) for details.

6.1.5 Topology Conclusions

With the different dimensions comes different behavior regarding correlation and surface intensity. The relationship between energy density and the normal surface intensity can be described by a proportionality factor β_D

$$
I_n = \beta_D c_{\text{gr}} e \tag{6.22}
$$

This factor scales from $\frac{1}{2}$ for one-dimensional systems to $\frac{1}{\pi}$ and $\frac{1}{4}$ for two-dimensional and three-dimensional systems, respectively, as summarized in Table 6.1.

In Figure 6.5 the different cross correlation functions are compared to each other.

6.1.5.1 Dissipation in the Reverberant Field

In the reverberant field there are two processes for sources of damping: first, the damping during wave propagation, for example described by the complex wave number $\boldsymbol{k} = k(1 - j\frac{\eta}{2})$, and second, the absorption due to the reflection at an absorbing surface at end sections, edges, or surfaces. In principle, all systems have a volume, a one-dimensional beam or tube has a cross section, and plates have thickness.

According to (1.67) the loss due to wave propagation is given by

$$
\Pi_{\text{diss,field}} = \omega\,\eta_{\text{field}} E \tag{6.23}
$$

Table 6.1 Intensity and cross correlation depending on system dimension.

Dimension	Intensity	$S_{\Psi\Psi}$
1D	$\frac{1}{2}c_{\text{gr}}\,e$	$\cos(k\Delta r)$
2D	$\frac{1}{\pi}c_{\text{gr}}e$	$J_0(k\Delta r)$
3D	$\frac{1}{4}c_{\text{gr}}e$	$\frac{\sin(k\Delta r)}{k\Delta r}$

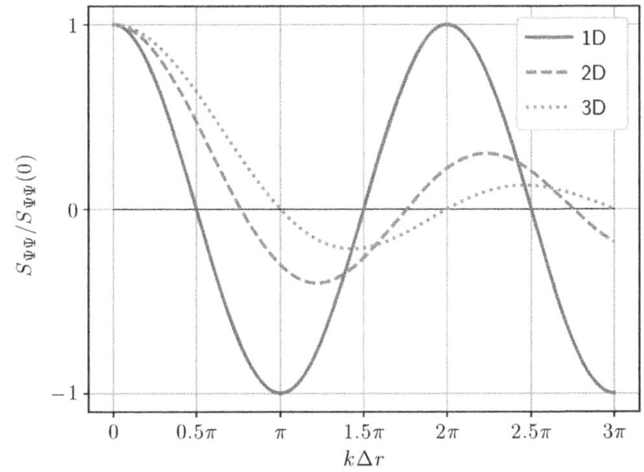

Figure 6.5 Normalized cross correlation functions of one- to three-dimensional diffuse reverberant fields. *Source:* Alexander Peiffer.

The dissipation due to surface absorption is given by the intensity irradiating all surfaces. The dissipated power due to boundary reflection in the diffuse field is

$$\Pi_{\text{diss},\alpha} = A_s I = A_s \beta_D c_{\text{gr}} e \tag{6.24}$$

A_s is the total absorption area and given by

$$A_s = \sum_i \alpha_{s,i} A_i = A_P \langle \alpha_s \rangle_S \tag{6.25}$$

A_P is the perimeter area or boundary. In order to get an expression that fits to the global damping loss factor, the dissipation at the perimeter is expressed by the total energy of the random field.

$$\Pi_{\text{diss},\alpha} = \frac{A_s}{V} \beta_D c_{\text{gr}} E \tag{6.26}$$

The sum of both dissipations provides the effective damping loss factor

$$\omega \eta E = \Pi_{\text{diss,field}} + \Pi_{\text{diss},\alpha} = \omega \left(\eta_{\text{field}} + \frac{A_P \langle \alpha_s \rangle_S \beta_D c_{\text{gr}}}{V\omega} \right) E \tag{6.27}$$

A random system with certain energy has a time decay $\eta\omega$ so that without any power flow into the system the energy will decay as

$$\langle E \rangle_E (t) = E_0 e^{-\eta\omega t} \tag{6.28}$$

6.1.6 Auto Correlation and Boundary Effects

The autocorrelation of the above given function follows from $\Delta r = 0$. Please note that this is

$$S_{\Psi\Psi}(0, \Delta t) = \frac{1}{2} \left\langle \hat{\Psi} \right\rangle_E \cos(\Delta t) \tag{6.29}$$

At the boundaries of such fields we have two effects. First, for the total sum of energy only half of the space is available. Thus, the mean square value of the amplitude is half of the amplitude if the full angle is available. Second the reflected wave amplitude is $\boldsymbol{R}\hat{\Psi}$. So, the boundary amplitude of every ensemble wave is

$$\Psi_i^{(\partial V)} = (1 + \boldsymbol{R})\hat{\Psi} \tag{6.30}$$

In case of hard walls and fixed edges this means a doubling of the pressure or stress, leading to four times higher mean square value of the amplitude. This is partly compensated by the fact that only waves from half space can reach the boundary

$$\left\langle \Psi^2_{(\partial V)} \right\rangle_E = \frac{(1 + \boldsymbol{R})^2}{2} \left\langle \hat{\Psi}^2 \right\rangle_E \overset{R\approx 1}{\approx} 2\left\langle \hat{\Psi}^2 \right\rangle_E \tag{6.31}$$

The amplitude of each noncorrelated wave doubles for full reflection and ($R = 1$), but for the summation of all waves only half of the room, area, or line is available. In other words, the double amplitude is partly compensated by the fact that only half the space is available

$$\Psi_{\text{rms,boundary}} = \sqrt{2}\Psi_{\text{rms,field}} \tag{6.32}$$

Note, that one coherent wave would give twice the amplitude. The factor of $\sqrt{2}$ instead of two in the reverberant field sometimes leads to confusion.

6.1.7 Sources in the Diffuse Acoustic Field – the Direct Field

Every system requires sources to get power input into the system. The waves that radiate from the sources or boundaries are obviously deterministic, because no reflection that may generate uncertainty has taken place. In the vicinity of the source around a radius of a few wavelengths there is a deterministic wave field with similar shape and behavior as in free field conditions as long as the wave amplitude reaches the diffuse field level. In room acoustics this distance is called the source radius. For correct measurement of the diffuse sound field level the distance between source and microphone must be larger than this radius. In the following examples, the direct field is included, except for the one-dimensional systems. The reason for this is that there is no geometrical attenuation for one-dimensional systems which is the case for two-dimensional ($\Psi(r) \sim r^{-1/2}$) and three-dimensional systems ($\Psi(r) \sim r^{-1}$).

6.1.8 Some Comments on the Diffuse Field Approach

The considerations in Sec. 6.1.7 regarding the diffuse field are very stringent. The system must be constructed and uncertain in such a way that the reverberant field assumptions are met. Thus, the properties of the system must allow a large number of reflections with uncertain phase so that there are enough waves with equally distributed angles and random phase.

This is hard to achieve when the system is excited, for example, only at one point. Consequently, the diffuse field approach (and also the statistical energy method that is based on this concept) is of limited precision when local and single excitation occurs. So, when this is the case we should consider using deterministic and numerical methods for systems that are locally excited.

However, when many excitations exist the diffuse wave field can be applied. When the frequency is high enough for geometrical acoustics, this is one further option to be applied to random wave fields but allowing energy distributions along the systems described by random wave fields.

As mentioned at the start of this chapter, one approach is to average in addition over frequency bands. However, with the ensemble concept the properties of the reverberant field can be kept, so the more critical question is how large the variation from the mean value is and under which conditions the mean value fulfills the reverberant field assumptions. Please note, that also frequency averaging does not solve the issues regarding the strict diffuse field assumptions. In the remainder of Chapter 6, we will deal with some examples of random systems in order to explain these phenomena.

6.2 Ensemble Averaging of Deterministic Systems

In Sections 6.3, 6.4, and 6.5, we will numerically simulate ensembles of representative systems. Therefore, we take the response of acoustical or structural systems with slight variations of input parameters. When the deterministic response of one ensemble representation i is given by $\mathbf{\Psi}_i(\mathbf{r})$, i will represent a specific parameter set $a_i, b_i, \ldots$ This could be the density $\rho_{0,i}$, the Young's modulus E_i, or the thickness of a plate h_i. In engineering systems this may also include local variations of those properties and the geometry.

From this we get the ensemble averages

$$E\left[\mathbf{\Psi}(\mathbf{r})\right] = \langle \mathbf{\Psi}_i(\mathbf{r})\rangle_E = \lim_{M\to\infty} \sum_{i=1}^{M} \mathbf{\Psi}_i(\mathbf{r}) \tag{6.33}$$

$$E\left[\mathbf{\Psi}^*(\mathbf{r}_1)\mathbf{\Psi}(\mathbf{r}_2)\right] = \left\langle \mathbf{\Psi}_i^*(\mathbf{r}_1)\mathbf{\Psi}_i(\mathbf{r})_2\right\rangle_E = \lim_{M\to\infty} \sum_{i=1}^{M} \mathbf{\Psi}_i^*(\mathbf{r}_1)\mathbf{\Psi}(\mathbf{r}_2) \tag{6.34}$$

For $\mathbf{r} = \mathbf{r}_1 = \mathbf{r}_2$ we get the autocorrelation and

$$\Psi_{\text{rms}}^2 = \frac{1}{2}E\left[\mathbf{\Psi}^*(\mathbf{r})\mathbf{\Psi}(\mathbf{r})\right] = \frac{1}{2}\left\langle \mathbf{\Psi}_i^*(\mathbf{r})\mathbf{\Psi}_i(\mathbf{r})\right\rangle_E = \frac{1}{2}\lim_{M\to\infty} \sum_{i=1}^{M} \mathbf{\Psi}_i^*(\mathbf{r})\mathbf{\Psi}(\mathbf{r}) \tag{6.35}$$

for the rms value. Even though the model conceptions of the reverberant field use uniform amplitude for all waves, the wavefield is not uniform for real systems. In order to capture this the space averages over the system extensions length, area, and volume are defined as follows

$$\langle X\rangle_L = \frac{1}{L}\int_L X dL \qquad \langle X\rangle_S = \frac{1}{S}\int_S X dS \qquad \langle X\rangle_V = \frac{1}{V}\int_V X dV \tag{6.36}$$

6.3 One-Dimensional Systems

The investigation starts with a system dimensionality that is rarely used in the context of reverberant fields, because it is hard to imagine that a reverberant field builds up from back and forth reflected waves. However, when the one-dimensional system is large (compared to the wavelength) and there is uncertainty in the ensemble in terms of wave speed, length, and boundary conditions, even one-dimensional systems may by diffuse.

The advantage of one-dimensional systems is that fast and analytical solutions are available, and we can easily investigate all aspects of an ensemble of systems.

6.3.1 Fluid Tubes

Let us imagine that we have an ensemble of tubes with different temperatures and consequently variations in the speed of sound or small imperfections in length. This would lead to propagating waves with random phase for both directions if the imperfections and variations in sound speed lead to differences in the propagation paths exceeding one wavelength.

Those imperfections will cause a diffuse sound field in an ensemble of one-dimensional systems. The wave quantity Ψ is the acoustic pressure p and the pressure field inside the tube is given by (4.24). Damping is introduced to ensure dissipation.

6.3.1.1 Energy

The reverberant pressure field is characterized be the mean square amplitude that is related to energy density of the wave and the total energy of the system. As the wave quantity is pressure, the constant in Equation (6.16) is $B = \frac{1}{\rho_0 c_0^2}$.

Thus, the random state is defined by the energy density e, the energy per length E', or the total energy E with a given length L

$$\langle e \rangle_E = \frac{\hat{p}_{\text{rms}}^2}{2\rho_0 c_0^2} \qquad \langle E' \rangle_E = A \langle e \rangle_E = A \frac{\hat{p}_{\text{rms}}^2}{2\rho_0 c_0^2} \qquad \langle E \rangle_E = AL \langle e \rangle_E = V \frac{\hat{p}_{\text{rms}}^2}{\rho_0 c_0^2} \tag{6.37}$$

In the diffuse field model the energy is equally distributed over the length. The irradiated intensity at both tube end cross-sections follows from (6.16)

$$I = \frac{c_0 \hat{p}}{4\rho_0 c_0^2} = \frac{1}{2} c_0 e \tag{6.38}$$

6.3.1.2 Power Input to the Reverberant Field

The power input into the field is defined by Equation (4.20). For large values $k\eta L \gg 1$ the radiation impedance of the system becomes a smooth function that approaches the impedance of a half infinite tube z_0, and the resonance peaks disappear.

We would now like to present an instructive explanation for this effect using the modal response.

We consider the half bandwidth of one modal peak that is given by $\Delta\omega_{3dB} = \omega\eta$. One can expect a smooth frequency response when the distance between the modal frequencies $\Delta\omega_n$ is equal or smaller than the half power bandwidth because the modes are *overlapping* in this case. This relationship is quantified by the model overlap factor

$$M = \frac{\Delta\omega_{3dB}}{\Delta\omega_n} = \frac{\eta\omega}{\Delta\omega/N} = \eta\omega n(\omega) \tag{6.39}$$

So, we have two parameters to control the smoothness of the response (independent from random effects): The damping loss η and the modal density $n(\omega)$. When the

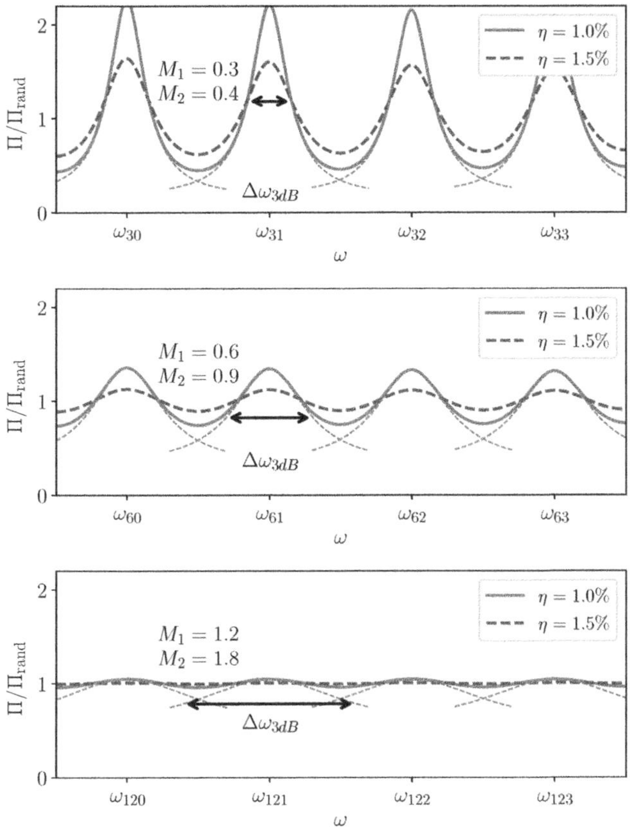

Figure 6.6 Power input for various modal overlap factors. *Source:* Alexander Peiffer.

model overlap factor is larger than 1 we can estimate the power input into the system by Equation (4.21)

$$\Pi_{in} = \lim_{k\eta\to\infty} \Pi = \frac{A}{2} z_0 |\boldsymbol{v}_{x1}|^2 = A z_0 \left\langle \boldsymbol{v}_{x1}^2 \right\rangle_T$$

This effect is depicted in Figure 6.6. When the modal overlap factor is less than one the gap between the modes is large enough so that single peaks can be identified; when M is approaching 1 there is still some waviness. For $M > 1$ the slope of the power input is nearly constant. Note that there is no ensemble averaging involved – the only condition is that all modes are equally distributed over the frequency band.

6.3.1.3 Dissipation

Surface dissipation occurs at the interaction with the tube end sections (neglecting the tube wall) with the total absorption area

$$A_s = A(\alpha_{s1} + \alpha_{s2}) \text{ or } = 2A \langle \alpha_s \rangle_S \tag{6.40}$$

With the tube volume $V = AL$ and $\beta_D = \frac{1}{2}$ the combined dissipated power is

$$\omega\eta E = \omega\left(\eta_{\text{field}} + \frac{\langle\alpha_s\rangle_S\, c_0}{L\omega}\right)E \tag{6.41}$$

providing the effective damping loss factor.

6.3.1.4 Power Balance

In a stable configuration the input power equals the dissipated power. This is the birth of our very first and simple SEA model

$$\Pi_{\text{in}} = \Pi_{\text{diss}} \Rightarrow \frac{A}{2} z_0 |\boldsymbol{v}_{x1}|^2 = \omega\eta\,\langle E\rangle_E \tag{6.42}$$

Looking at the above equation we see that the input power depends on the free field impedance of the pipe on one side; on the other side, the loss of energy is proportional to energy and frequency given by the factor η. With Equation (6.37), the rms pressure is then given by:

$$p_{SEA} = \left\langle p^2\right\rangle_E = \frac{z_0^2 c_0}{\eta\omega L}\left\langle \boldsymbol{v}_{x1}\right\rangle^2 \tag{6.43}$$

6.3.1.5 Energy Ray Tracing

For increasing frequencies, the wave attenuation (2.58) $\alpha = \eta k/2$ is getting larger, so the wave propagation is highly damped. Consequently, there are not enough reflections available to guarantee reverberance and an equal distribution of wave energy in the field. A description of random systems by diffuse wave fields is no longer valid, and methods that account for the damping along the propagation and system dimension are required.

The frequency limit of this effect can be estimated when we say that the energy damping along the system extension L must not exceed $\frac{1}{e}$, hence

$$e^{-2\alpha L} = e^{-\omega_{SEAmax}\eta L/c_0} < \frac{1}{e} \quad \Longrightarrow \omega_{SEAmax} = \frac{c_0}{\eta L} \tag{6.44}$$

Though SEA is often mentioned as a high frequency method, this important criteria for the application of SEA is neglected. If this criteria is not fulfilled, the error in the propagation of energy is not caused by an erroneous coupling of systems but the nonuniform energy distribution in the system. This issue can be overcome by geometrical acoustics. For very small wavelengths the wave propagation and scattering can

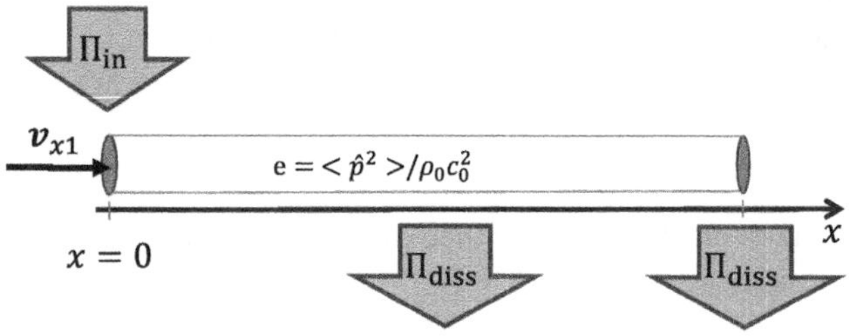

Figure 6.7 Power balance for the one-dimensional tube system. *Source:* Alexander Peiffer.

be described by rays (Pierce, 1991; Kuttruff, 2014). Each wave is traced similar to the ray model in optics, and the wave field is considered as the sum of these rays. For a one-dimensional system, this can be done analytically.

We consider each propagation in the positive and the negative direction as one ray. We still assume a vibrating piston with velocity $\boldsymbol{v}_{x1}$ on port 1. A *ray* in the positive direction will radiate into the tube the mean squared pressure as follows

$$p_{\text{rms}}^2(x) = p_{\text{rms}}^2(0)e^{-2\alpha x} \tag{6.45}$$

At the end of the tube, the wave is reflected with the reflection coefficient $\boldsymbol{R}_1$, so the reflected mean square pressure is

$$p_{\text{rms,1R}}^2(x) = p_{\text{rms}}^2(0)e^{-2\alpha L}(R_1)^2 e^{-2\alpha(L-x)} \tag{6.46}$$

with $R_1^2 = \boldsymbol{R}_1\boldsymbol{R}_1^*$; this wave is reflected at port 1 with the coefficient $\boldsymbol{R}_0$ and so on and on:

$$\begin{aligned} p_{\text{rms}}^2(x) &= p_{\text{rms}}^2(0)(e^{-2\alpha x} + R_1^2 e^{-2\alpha L} e^{-2\alpha(L-x)} + \\ &\qquad R_0^2 R_1^2 e^{-22\alpha L} e^{-2\alpha x} + R_1^2 R_0^2 R_1^2 e^{-32\alpha L} e^{-2\alpha(L-x)} \ldots) \\ &= p_{\text{rms}}^2(0)e^{-2\alpha x}[1 + (R_0R_1)^2 e^{-4\alpha L} + (R_0R_1)^4 e^{-8\alpha L} + \ldots] \\ &\qquad R_1^2 e^{-2\alpha(2L-x)}[1 + (R_0R_1)^2 e^{-4\alpha L} + (R_0R_1)^4 e^{-8\alpha L} + \ldots] \\ &= p_{\text{rms}}^2(0)(e^{-2\alpha x} + R_1^2 e^{-2\alpha(2L-x)}) \\ &\qquad [1 + (R_0R_1)^2 e^{-4\alpha L} + (R_0R_1)^4 e^{-8\alpha L} + \ldots] \end{aligned} \tag{6.47}$$

The last term in parenthesis is an infinite geometrical series and we get for the slope of the root mean square pressure

$$p_{\text{rms}}(x) = p_{\text{rms}}(0)\sqrt{\frac{e^{-2\alpha x} + R_1^2 e^{-2\alpha(2L-x)}}{1 - (R_0R_1)^2 e^{-4\alpha L}}} \tag{6.48}$$

6.3.1.6 Load at the Boundaries due to the Reverberant Field

The reverberant field will have an impact on the boundaries of the system. For deterministic systems the answer is given by Equation (4.24) or the pressure distribution itself. The field pressure at the boundaries creates a reaction force or pressure. The question now is what is the reaction force of a diffuse field that is only described by its energy or rms pressure average in the system?

The pressure in the diffuse field with energy E is:

$$\langle p^2 \rangle_E = \rho_0 c_0 \frac{2c_0}{LA} E = z_0 \frac{2c_0}{LA} E \tag{6.49}$$

The ratio $L/c_0 = \pi n(\omega)$ can be expressed by the modal density providing

$$\langle p^2 \rangle_E = z_0 \frac{2E}{\pi n(\omega)A} \tag{6.50}$$

According to Equation (6.31) the mean square pressure at rigid surface doubles

$$\langle p^2 \rangle_E = z_0 \frac{4E}{\pi n(\omega)A} \tag{6.51}$$

Finally, the auto spectrum of the blocked force that is generated by the diffuse field is given by

$$S_{ff} = A^2 \left\langle p^2 \right\rangle_E = A z_0 \frac{4E}{\pi n(\omega)} = Re(\mathbf{Z}) \frac{4E}{\pi n(\omega)} \tag{6.52}$$

Here $\mathbf{Z}$ is the mechanical free field impedance of the one-dimensional tube.

We have compiled all the information required to describe a random system by a diffuse field. This includes the interaction of diffuse field at the boundaries with external systems:

1. The input impedance of the reverberant field determining the power input.
2. The load generated by the reverberant field at the boundary.

6.3.1.7 Monte Carlo Experiment

The Monte Carlo method applies uniform random variation to parameters of a system that is solved numerically. By this method a large ensemble of systems and its response can be simulated. The pressure field for a deterministic system is known from Equation (4.24) and therefore an excellent case for fast calculation of a large set of results that can be averaged. In order to work out the limits of the diffuse reverberant field we deal with two realizations of damping.

1. Damping is dominated by dissipation in the wave propagation given by the complex wave number $\mathbf{k} = k(1 - j\frac{\eta}{2})$ with $\eta = 0.01$ and rigid ends.
2. Damping is dominated by absorption at the boundary by applying an impedance $\mathbf{z} = 19.1z_0$ leading to a reflection factor of $R = 0.9$. Damping loss for propagation is $\eta = 0.001$.

On the LHS of Figure 6.8 it can be seen that the system damping loss for case 2 starts with high values but is decreasing. The damping loss is defined per cycle, and for higher frequencies the wave passes many cycles before the reflection dissipates energy from the system. Thus, this case reaches a modal overlap of $M = 1$ at $kL > 300$ and case 2 is not in the investigated frequency range.

Those values are selected to investigate the statistical aspect and to reduce the effect of high modal overlap that would smooth even the deterministic response.

For ensemble realization, we randomize the speed of sound by $\Delta c_0 = \pm 2\% c_0$ and $\Delta L = \pm 1\% L$. We generate $N = 1000$ representations of this ensemble, with pressure

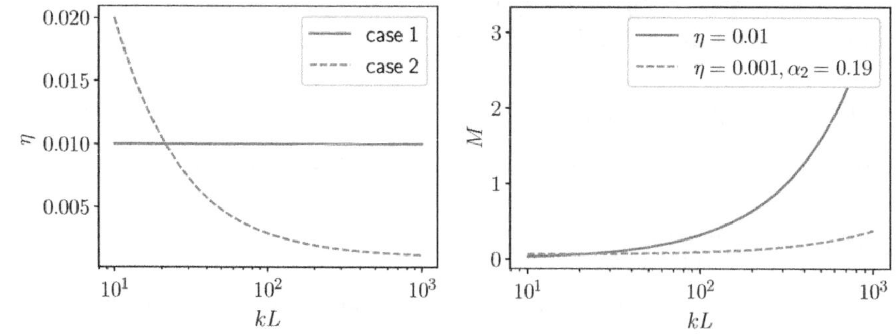

Figure 6.8 Damping loss and modal overlap of both cases. *Source:* Alexander Peiffer.

$p_n(x)$ and calculate the following ensemble averages:

$$p_{\text{mean}}(x) = \langle p(x) \rangle_E = \frac{1}{N} \sum_{n=1}^{N} p_n(x) \tag{6.53}$$

$$p_{\text{rms}}(x) = \sqrt{\langle p^2(x) \rangle_E} = \sqrt{\frac{1}{N} \sum_{n=1}^{N} p_n^2(x)} \tag{6.54}$$

The first average shall show that the mean value vanishes in an ensemble average; the rms value should verify the reverberant field assumption for high frequencies. The spatial average over length will also be presented in the following numerical experiments to check if a single value can be representative for the total system.

$$p_{\text{rms,L}} = \sqrt{\langle \langle p^2(x) \rangle_E \rangle_L} \tag{6.55}$$

6.3.1.8 Input Power and Impedance

From section 4.1.1 and the modal overlap discussions, it was shown that the radiation impedance approaches the free field impedance. This effect should occur even earlier in frequency due to incoherent reflection according to the diffuse field concept. In Figures 6.9 and 6.10 this effect is clearly visible for both cases. The light grey background represents the ensemble of different realizations.

For case 1 with dominating wave absorption the ensemble mean reaches very early ($kL \approx 50$) the free field limit with higher peaks in the ensemble variations. In case 2 the ensemble peaks are still quite sharp, because of the low modal overlap even for higher frequencies.

The same effect can be seen in Figure 6.10. The second case's mean value oscillates more, but the ensemble variation is smaller.

Doing the same investigation with the input power, this is further confirmed. In Figure 6.11, the free field limit is reached much earlier than for the deterministic case shown.

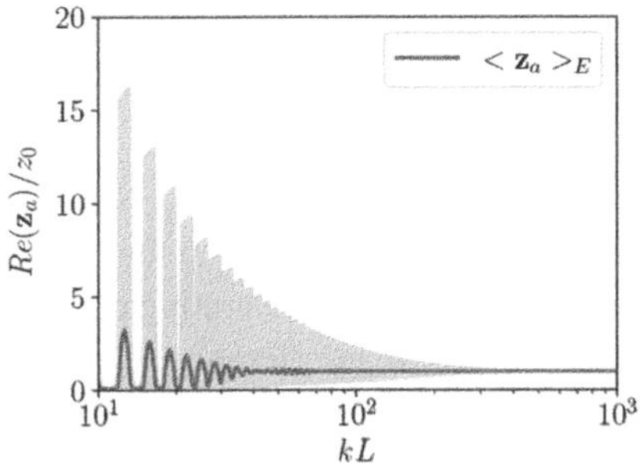

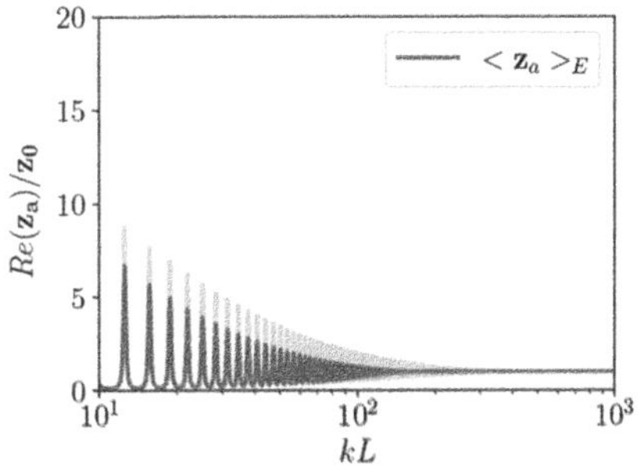

Figure 6.9 Relative radiation resistance for an ensemble of tubes with 2% variation in c_0 and 1% in length, LHS $\eta = 0.01$ rigid ends, RHS $\eta = 0.001$ at port 2 with absorbing boundary conditions. *Source:* Alexander Peiffer.

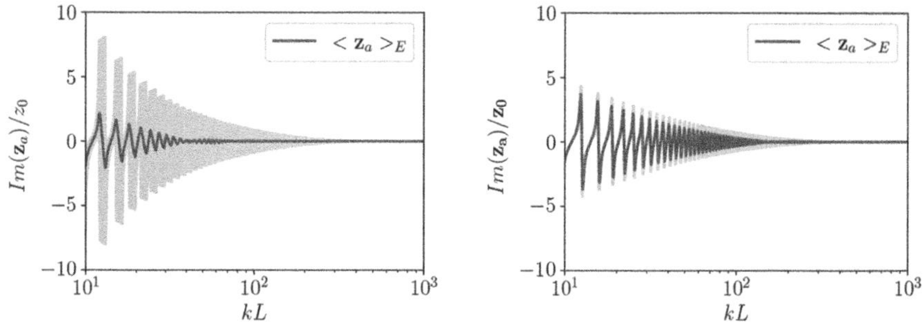

Figure 6.10 Relative radiation reactance for an ensemble of tubes with 2% variation in c_0 and 1% in length, LHS $\eta = 0.01$ rigid ends, RHS $\eta = 0.001$ at port 2 with absorbing boundary conditions. *Source:* Alexander Peiffer.

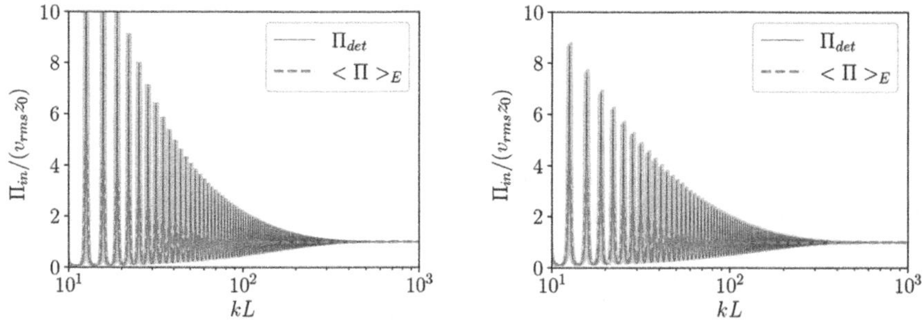

Figure 6.11 Power input to an ensemble of one-dimensional acoustic tube systems LHS case 1 RHS case 2. *Source:* Alexander Peiffer.

6.3.1.9 Random Fields

We investigate the rms and mean pressure along the tube. The results are shown in Figures 6.12 and 6.13 for cases 1 and 2, respectively. Several modal frequencies and overlap factors are shown to investigate the validity and limits of the random field assumptions in both cases. For the first frequency, the main impact of the uncertainty is a large variation in amplitude, but each wave form remains stable in phase. So, mean and rms averages result in the similar pressure distribution along the pipe. This is still true to frequencies around the 10th mode as shown in the upper right axes. When we compare the rms pressure averaged over length to the result of the SEA model the agreement is quite good.

Reaching higher frequencies as shown in the lower left axis, both rms values don't coincide, because damping along the tube prevents a constant pressure along the pipe. The mean pressure becomes zero for $x = L$ but not for $x = 0$ near the vibrating piston side. Here, the coherence of the near field keeps the constructive interference near the source; the wave field has not yet gathered enough uncertainty to establish a diffuse sound field. This is the direct field component which can be added to the reverberant background to improve the precision of the sound field description.

For the high frequency results in the lower right axis it can be seen that rms pressure shows a strong decay along x. The wave propagation is strongly affected by damping,

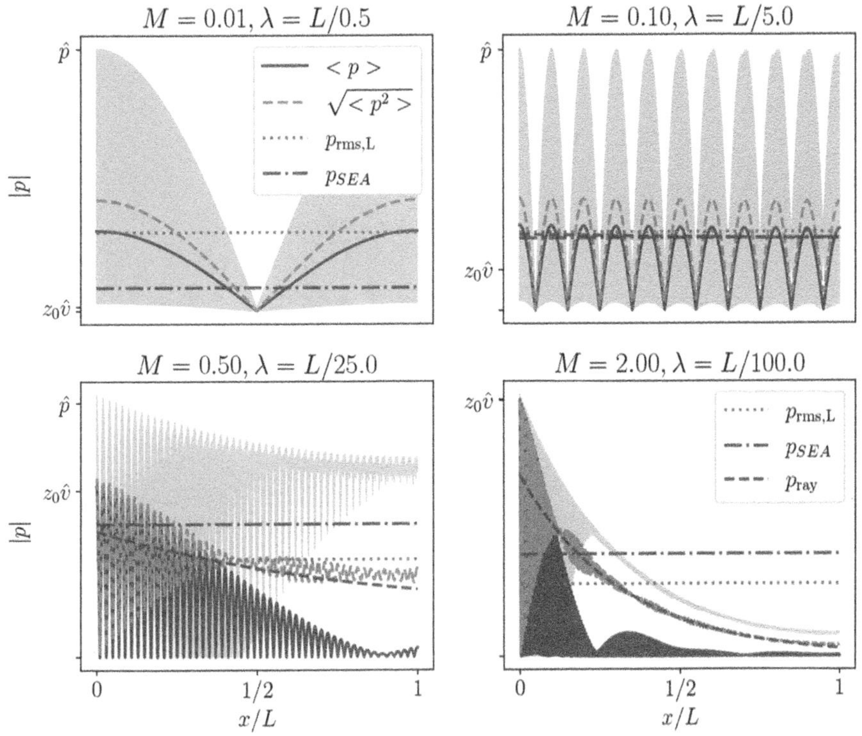

Figure 6.12 Monte Carlo simulation of an ensemble of tubes with $\pm 2\%$ speed of sound and $\pm 1\%$ length variation, case 1, $\eta = 0.01$, rigid ends. *Source:* Alexander Peiffer.

and the diffuse field assumption is not met any more. It is the assumption of equivalent amplitude for both directions that is explicitly violated. The system is dynamically complex, but the damping is too high to establish a reverberant field. In this case ray-tracing or geometrical acoustics shall be applied, which agrees very well with the average result.

However, there are many options to achieve a reverberant field. Our case is very stringent, because there is only one source at one end of a one-dimensional field. If there are more sources, a reverberant field can be realized even for damped waves.

So, even with such a *simple* system the effects of deterministic and random regimes are visible. At low (deterministic) frequencies, there is the typical wave dynamics with maxima and minima regarding the related wave shape. The wave motion, the local minima and maxima, can be clearly determined. In the random regime the wave behavior is smeared out. In the statistical energy description, neglecting the phase is sufficient to describe the dynamics of a complex system. For higher frequencies when damping along the system dimensions becomes important, the damping must be considered, for example, by energy ray tracing methods. This high frequency limit is often neglected in the literature. Lyon and DeJong (1995) and Le Bot (2015) derived the same equation as (6.44) but expressing an upper limit of system size

$$L < \frac{c_{\text{gr}}}{\omega \eta} \tag{6.56}$$

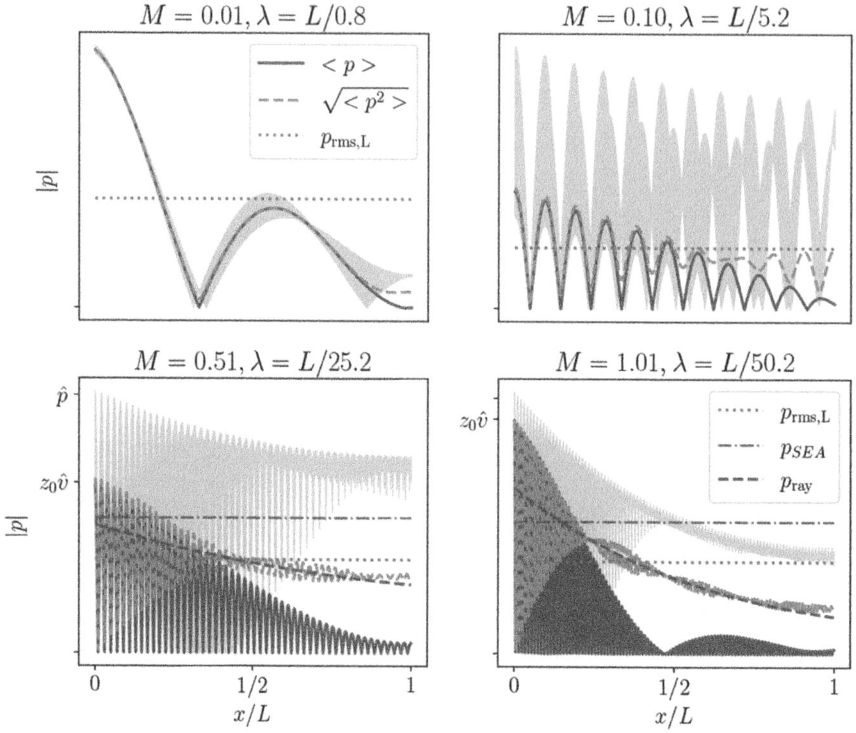

Figure 6.13 Monte Carlo simulation of an ensemble of tubes with $\pm 2\%$ speed of sound and $\pm 1\%$ length variation, case 2, $\eta = 0.001$, absorption at port 2. *Source:* Alexander Peiffer.

For the power input the local dynamics of the area near the excitation must be known precisely, meaning that the free field impedance is required. Thus, even when we later deal with fully random models the dynamic properties of the system in the vicinity of junctions must be precisely known and is thus considered as deterministic.

6.4 Two-Dimensional Systems

The rectangular plates is taken as a test case for two-dimensional (2D) systems. The bending wave is dispersive and this impacts the diffuse field assumptions and the link between out-of-plane motion and the energy density.

6.4.1 Plates

We apply a point force excitation somewhere at the plate and focus on bending waves, because the wave speed for the other propagation modes is too high to create large modal overlap at audible frequencies even for large plates. Similar to the acoustic tube, the high modal overlap causes the mechanical impedance to reach the free field radiation value (3.226) of the infinite plate. The modal approximation is used to calculate the system response for the Monte Carlo analysis. Plates are two-dimensional systems;

so, all equations for the two-dimensional wave field apply, and the field quantity is the displacement $\boldsymbol{w}$ in the z-direction.

6.4.1.1 Energy

In section 3.7.6.1, the energy per area is given by

$$E''_{tot} = \frac{\rho_0 h\omega^2\hat{w}^2}{2} \tag{3.227}$$

providing energy density and energy by

$$e = \frac{\rho_0\omega^2\hat{w}^2}{2} \qquad E = \frac{m''A\omega^2\hat{w}^2}{2} \tag{6.57}$$

Thus, the constant B is

$$B = \rho_0\omega^2 \tag{6.58}$$

and for the two-dimensional expression of diffuse field intensity at the edges (6.18) we get

$$I = \frac{1}{\pi}c_{\text{gr}}\, e \tag{6.59}$$

Please note that the intensity expression is per edge area – for intensity per length this expression must be multiplied with the plate thickness h. In addition, the group velocity is different to the phase velocity and depends on the frequency

$$I' = \frac{1}{\pi}c_{\text{gr}}\, eh \tag{6.60}$$

6.4.1.2 Power Input to the Reverberant Field

For systems and frequencies of high modal overlap the mechanical impedance converges to the free field impedance of an infinite plate. Thus, the input power can be calculated using the assumption (3.226)

$$\Pi_{\text{in}} = \frac{1}{2}Re\left(\boldsymbol{F}^*\boldsymbol{V}\right) = \frac{1}{2}Re\left(\frac{\boldsymbol{F}^*\boldsymbol{F}}{\boldsymbol{Z}_0}\right) = \frac{1}{16\sqrt{Bm''}}\boldsymbol{F}^*\boldsymbol{F} = \frac{1}{16\sqrt{Bm''}}\hat{F}^2 \tag{6.61}$$

However, the free field assumption is not fulfilled for low frequencies so we will use Equation (5.42) in addition to correct the power input.

6.4.1.3 Dissipation and Power Balance in the Reverberant Field

The same dissipation assumptions as for one-dimensional systems apply and the inert damping due to wave propagation is the same

$$\Pi_{\text{diss,field}} = \eta\omega E$$

For two-dimensional systems, the edge intensity is determined from our considerations in section 6.1.1, the volume is $V = Ah$, and the dimensional constant is $\beta_D = 1/\pi$.

The power balance provides the energy of the subsystem with given input power

$$E = \frac{\Pi_{\text{in}}}{\omega\eta} \tag{6.62}$$

with $E = \frac{m'' A\omega^2 \hat{w}^2}{2}$ the average displacement result is

$$\hat{w}^2 = \frac{\hat{F}}{8A\sqrt{Bm''^3}\omega^3\eta} \tag{6.63}$$

6.4.1.4 Direct Field Correction

For the direct field part in the vicinity of the point force, we add the results from Equation (3.231)

$$\hat{\boldsymbol{w}}^2(r) = \frac{\hat{F}}{32\pi B^2 k_B^5}\frac{1}{r}$$

to the reverberant field.

6.4.2 Monte Carlo Simulation

The dimensions of the rectangular plate test case are $L_x = 1.0$ m, $L_y = 1.8$ m and $h = 2$ mm. The ensemble is realized by varying the thickness h by $\pm 2\%$ and the area mass m'' by $\pm = 2\%$, and the system average response is presented by the rms displacement averaged over the plate area

$$w_{\text{rms,S}} = \sqrt{\langle\langle w^2(x)\rangle_E\rangle_S} \tag{6.64}$$

6.4.2.1 Input Power and Impedance

In Figure 6.14 the mean value and the ensemble impedances are shown together with the modal overlap. Due to the properties of the plate, the modal density is constant over frequency (5.35); the consequence for the modal overlap is, that it increases

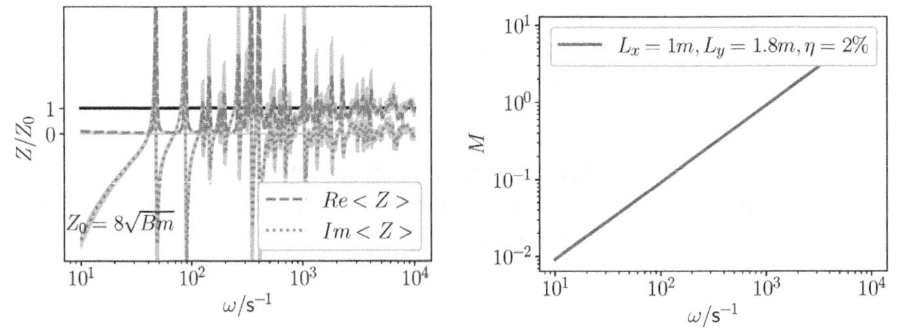

Figure 6.14 Point impedance at **x** = (0.7 m, 0.38 m) with 2% thickness variation and 1% density variation and modal overlap over frequency. *Source:* Alexander Peiffer.

linearly with frequency

$$M = \omega\eta n_{plate}(\omega) \tag{6.65}$$

The modal overlap is above 1 around a frequency of $\omega > 1000\ \mathrm{s}^{-1}$. Even though there is an ensemble average the input impedance of the free field is hardly approached. There is a remaining variability even at high frequencies. The reason for this is that the point excitation does not excite all modes equivalently. This is different to the tube case where we excited at the $x = 0$ and thus made sure all the modes are excited. This is not the case for the plate. A mode that has zero amplitude $\hat{w} = 0$ at the point of excitation will not be excited and will not contribute to the statistics of the assumed random field. This is the reason why we also use the real input power calculated from (5.42) to correct the power input for the final random field comparison. It is worth mentioning that commercial software codes implement the point force in the same way leading to similar error. Thus, if the response of plate structures to point forces is targeted, the input power should be derived by alternative methods, e.g. experiments or modal methods as suggested above.

6.4.2.2 Random Displacement Field

When we do the random response analysis based on the power balance according to Equation (6.63) and the Monte Carlo analysis with the various random quantities, we get the result in Figures 6.15 and 6.17 for $\omega = 2000\ \mathrm{s}^{-1}$. The rms wave field shows still some variation even in the ensemble average. However, when we compare the displacement averaged over the surface, the order of magnitude is met, and the corrected power input leads to an improvement in the prediction quality.

The assumptions of the diffuse field are better fulfilled at frequencies around $\omega = 10\,000\ \mathrm{s}^{-1}$ as shown in Figures 6.16 and 6.18. The oscillation of the displacement rms still exists. But the SEA result and the surface average coincide very well. Here, the SEA random result plus the direct field component shows a much better agreement with the Monte Carlo result.

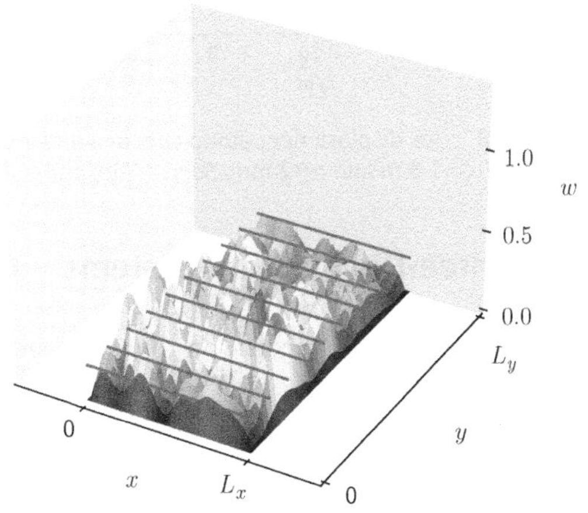

Figure 6.15 Ensemble rms reponse of plate at $\omega = 2000\ \mathrm{s}^{-1}$, dimensions $L_x = 1.0$ m, $L_y = 1.8$ m and $h = 2$ mm. *Source:* Alexander Peiffer.

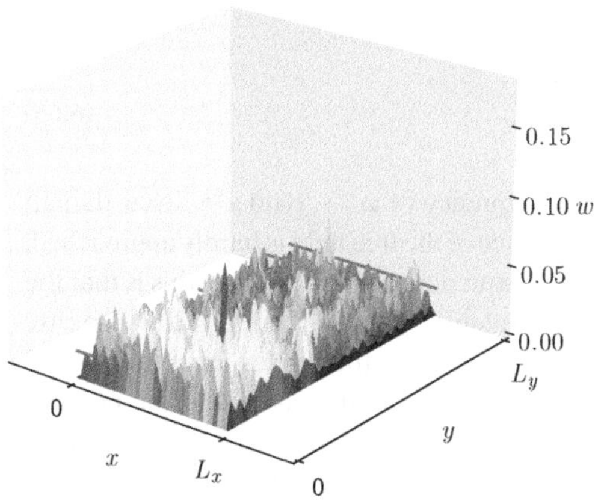

Figure 6.16 Ensemble rms reponse of plate at $\omega = 10000\ \text{s}^{-1}$, dimensions $L_x = 1.0$ m, $L_y = 1.8$ m and $h = 2$ mm. *Source:* Alexander Peiffer.

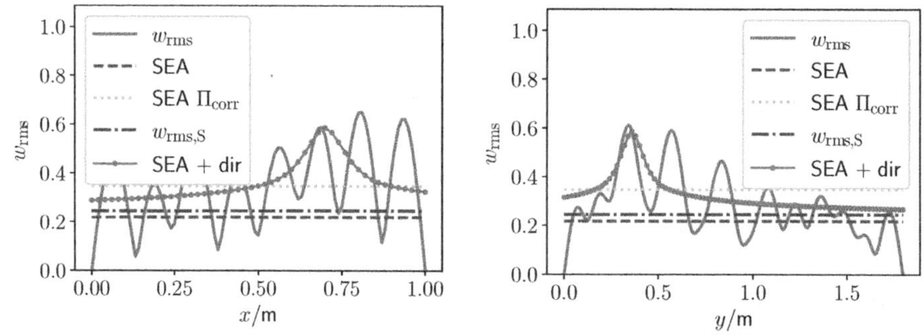

Figure 6.17 rms displacement along cuts in x and y-directions for $\omega = 2000\ \text{s}^{-1}$, dimensions $L_x = 1.0$ m, $L_y = 1.8$ m and $h = 2$ mm. *Source:* Alexander Peiffer.

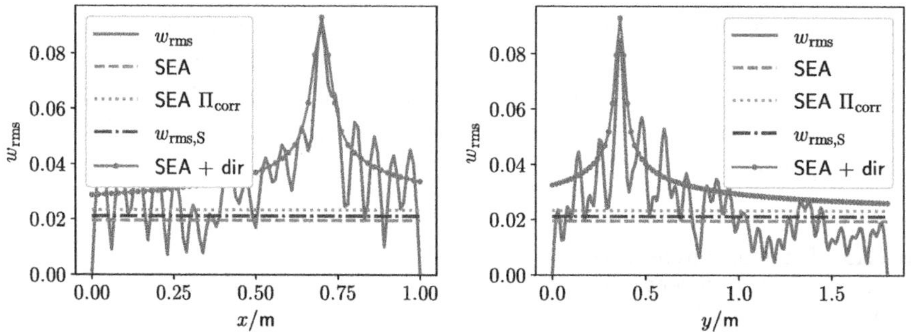

Figure 6.18 rms displacement along cuts in x and y-directions for $\omega = 10\,000\ \text{s}^{-1}$, dimensions $L_x = 1.0$ m, $L_y = 1.8$ m and $h = 2$ mm. *Source:* Alexander Peiffer.

6.5 Three-Dimensional Systems – Cavities

The first historical treatments on diffuse reverberant fields were in the context of acoustic fields in large spaces shown, for example, in Morse and Ingard (1968), Pierce (1991), or Jacobsen (1979). Diffuse acoustic fields are experienced in daily life – reverberant rooms can be found in many industrial or office spaces. The reverberant field in churches can be so disturbing that speech recognition is not given, or music must be adapted to the special conditions.

Large rooms easily reach high modal overlap factors and modal densities at relatively low frequencies. This has led to many test set-ups and norms that apply reverberation chambers in order to experimentally create a standardized diffuse field excitation. Systems of solid materials rarely occur in transportation or even building constructions, and they can hardly be described by diffuse random fields due to the high wave speed of longitudinal and even transversal waves; those systems are simulated by deterministic methods. Thus, the only system we will deal with in the three-dimensional (3D) context are fluid cavities. Cavities meet the conditions of the diffuse field assumptions easily, but the upper SEA limit (6.56) is also often reached. So, ray tracing or optical methods can be used as in the tube example, because the ratio between wavelength and room dimensions are on the order of magnitude that allow for geometrical acoustics. For example, concert hall acoustics is an excellent case for the application of ray-tracing methods as described in Kuttruff (2014).

6.5.1 Energy and Intensity

The link between pressure field and system energy is given by

$$\langle e \rangle_E = \frac{\hat{p}^2}{2\rho_0 c_0^2} \qquad \langle E \rangle_E = V\frac{\hat{p}^2}{2\rho_0 c_0^2} = V\frac{p_{\text{rms}}^2}{\rho_0 c_0^2} \tag{6.66}$$

The intensity at the boundary is

$$I = \frac{1}{4}c_0 \langle e \rangle_E = \frac{\hat{p}^2}{8\rho_0 c_0} = \frac{p_{\text{rms}}^2}{4\rho_0 c_0} \tag{6.67}$$

The above expression is known as the irradiated intensity in a twin chamber arrangement where the average pressure is measured in the room. The power irradiating a specimen of surface A at the room walls is

$$\Pi_{\text{in}} = \frac{A p_{\text{rms}}^2}{4\rho_0 c_0} \tag{6.68}$$

6.5.2 Power Input to the Reverberant Field

From section 4.2 and Equations (2.95) and (2.96) the radiated power is determined by:

$$\Pi_{in,\infty} = \frac{k^2 \rho_0 c_0}{4\pi} \boldsymbol{Q}^2 \tag{6.69}$$

The power input based on modal room theory is given by Equation (4.89)

$$\Pi_{\text{in}}(\mathbf{r}) = -\frac{j\omega\rho_0}{2} \sum_m \frac{\hat{\Phi}_m^2(\mathbf{r})\hat{Q}^2(\mathbf{r})}{k^2 - k_m^2(1 + j\eta)}$$

6.5.3 Dissipation

The dissipation in room cavities depends on propagation and boundary absorption (6.41). For the surface absorption we must consider the angular dependency of the absorption according to Equation (2.103) related to the normal component of the intensity.

$$\langle\alpha_s\rangle_E = \frac{\int_0^{\pi/2} \alpha_s(\vartheta)\cos(\vartheta)\sin(\vartheta)d\vartheta}{\int_0^{\pi/2} \cos(\vartheta)\sin(\vartheta)d\vartheta} = 2\int_0^{\pi/2} \alpha_s(\vartheta)\cos(\vartheta)\sin(\vartheta)d\vartheta \tag{6.70}$$

with

$$\alpha_s(\vartheta) = (1 - |\boldsymbol{R}(\vartheta)|^2) \text{ and } \boldsymbol{R}(\vartheta) = \frac{\boldsymbol{z}_s\cos\vartheta - z_0}{\boldsymbol{z}_s\cos\vartheta + z_0} \tag{6.71}$$

Usually, the angular integration is performed numerically. Assuming that the room is surrounded by surfaces A_i and with absorption $\alpha_{s,i}$ the total absorption area A_s follows from Equation (6.25), and the final dissipation formula (6.27) reads in case of cavities as

$$\omega\eta E = \Pi_{\text{diss,field}} + \Pi_{\text{diss},\alpha}$$
$$\eta = \left(\eta_{\text{field}} + \frac{A_s c_0}{4V\omega}\right) \tag{6.72}$$

6.5.4 Power Balance

The pressure in the room follows from the balance of energy flow $\Pi_{\text{in}} = \Pi_{\text{diss}}$

$$\hat{p}^2 = \frac{kz_0^2}{2\pi\eta V}\hat{Q}^2 \tag{6.73}$$

Similar to the considerations for the plate, the direct pressure field amplitude of the point source (2.91) is added to the random field component.

$$\hat{p}^2 = \frac{k^2 z_0^2 \hat{Q}^2}{16\pi^2 r^2}$$

6.5.5 Monte Carlo Simulation

The test case applies the modal room acoustics from section 4.2 considering two different cases.

room 1 Damping is dominated by dissipation in the wave propagation with $\eta_{\text{field}} = 0.01$ and rigid wall surfaces.

room 2 Damping is dominated by absorption at the boundary by applying an impedance $\boldsymbol{z} = 10z_0$ at floor $z = 0$ and ceiling $z = L_z$ leading to a diffuse field absorption coefficient $\alpha_s = 0.165$ and an absorption area $A_s = 1.5\ \text{m}^2$. Wave damping is $\eta_{\text{field}} = 0.001$.

The ensemble variation of the room lengths $L_{x,y,z}$ is $\pm 2\%$ and for the speed of sound $\pm 1\%$.

6.5.5.1 Room Absorption

In Equation (4.79), a complex wavenumber was introduced to determine the impact of absorbing boundaries on the modal coefficients. For nearly rigid walls, the off diagonal terms can be neglected and we get

$$B_{nn} = \int_{\partial V} \frac{z_0}{\boldsymbol{z}} \Phi_n^2(\mathbf{r}) dS \tag{6.74}$$

Thus, we select $\boldsymbol{Z} = 20z_0$ and get specifically for the floor area the following expression

$$B_{nn} = \frac{1}{V} \int_0^{L_x} \int_0^{L_y} \frac{z_0}{\boldsymbol{z}} \cos^2(\frac{n_x \Pi x}{L_x}) \cos^2(\frac{n_y \Pi y}{L_y}) dx dy = \frac{L_x L_y z_0}{V a \boldsymbol{z}}$$

$$\text{with } a = \begin{cases} 4 & \text{for } \mathrm{n_x, n_y} > 0 \\ 2 & \text{for } \mathrm{n_x} = 0, \mathrm{n_y} > 0 \\ 2 & \text{for } \mathrm{n_x} > 0, \mathrm{n_y} = 0 \\ 1 & \text{for } \mathrm{n_x} = 0, \mathrm{n_y} = 0 \end{cases} \tag{6.75}$$

When we evaluate expression (6.70) for the floor area $A = L_x L_y$ we get for the average absorption coefficient for the floor $\langle \alpha_s \rangle_E = 15.6\,\%$ leading to a total absorption area of $A_s = 1.5\ \mathrm{m}^2$.

6.5.5.2 Input Power and Impedance

Studying the ensemble average radiation impedance, the results from the sections before are further verified. The modal overlap becomes larger than one (Figure 6.19) above $\omega = 2000\ \mathrm{s}^{-1}$.

Both real ensemble impedances (Figures 6.20 and 6.21) show resonance variation at low frequencies but approach the free field value. The modal peaks of room 2 are lower at low frequencies because of the larger modal overlap from the beginning, whereas the variations between each ensemble representation are higher for room 2 because of the lower damping contributions from the fluid.

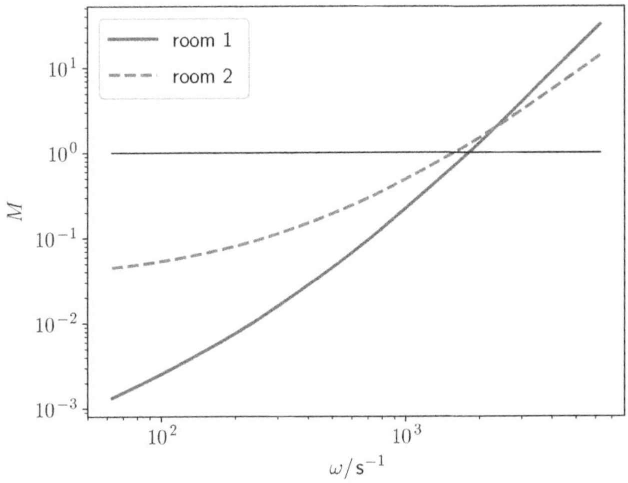

Figure 6.19 Modal overlap for both rooms. *Source:* Alexander Peiffer.

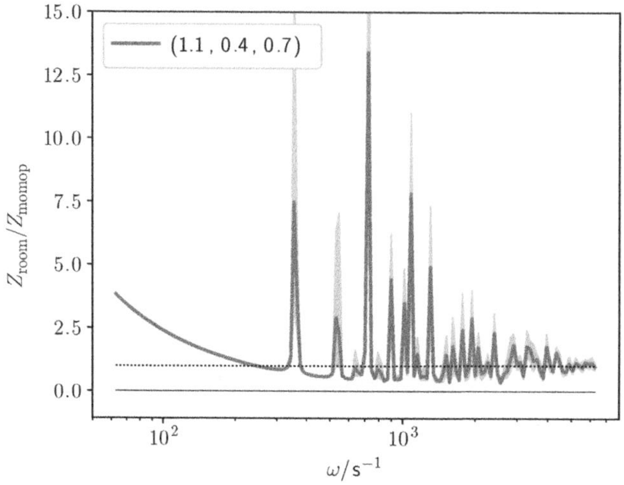

Figure 6.20 Ensemble acoustic radiation impedance of room 1 and monopole ($L_x = 3$ m, $L_y = 1.5$ m, $L_z = 2$ m). *Source:* Alexander Peiffer.

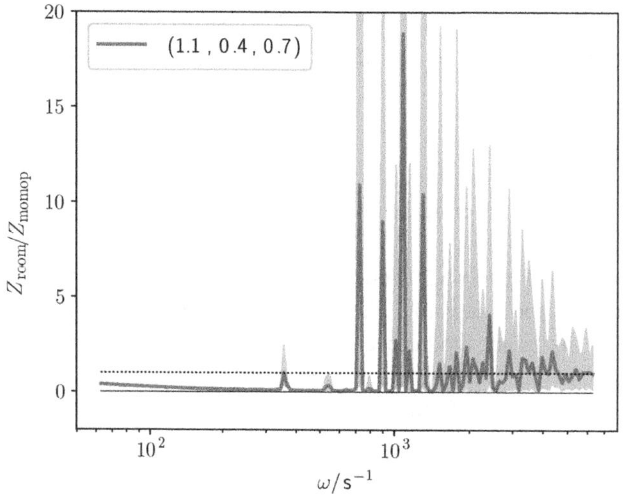

Figure 6.21 Ensemble acoustic radiation impedance of room 2 and monopole ($L_x = 3$ m, $L_y = 1.5$ m, $L_z = 2$ m). *Source:* Alexander Peiffer.

6.5.5.3 Random Pressure Field – Room 1

The pressure field at lines passing through the source is shown in the Figures 6.22 and 6.23 for $\omega = 5000\ \text{s}^{-1}$. The need for near field consideration is further confirmed, but with this simple direct field consideration the random field pressure corresponds surprisingly well to the Monte Carlo result. At the boundaries the pressure is roughly $\sqrt{2}$ higher than the field pressure as predicted by Equation (6.31). Due to the high expense of the calculation time, only 100 averages were performed leading to limited statistics.

6.5.5.4 Random Pressure Field – Room2

At 5000 s^{-1}, the second room with damping via absorption shows a similar agreement between random and the Monte Carlo results (Figures 6.24 and 6.25). The suitability of three-dimensional systems for reverberant field modelling is further verified.

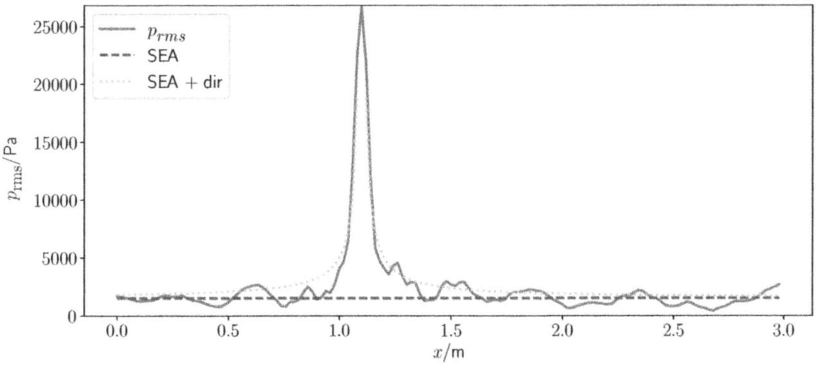

Figure 6.22 Ensemble pressure field in rectangular room 1 ($\eta_{\text{fluid}} = 0.01, A_s = 0$ m^2) along x-line through source location (x_0, y_0, z_0). *Source:* Alexander Peiffer.

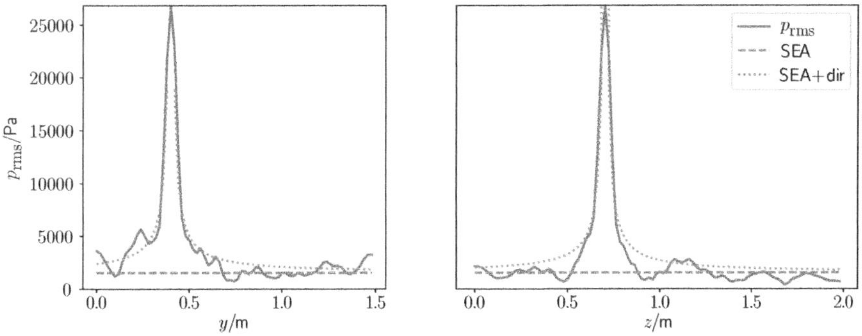

Figure 6.23 Ensemble pressure field in rectangular room 1 ($\eta_{\text{fluid}} = 0.01, A_s = 0$ m^2) along y and z-lines through source location (x_0, y_0, z_0). *Source:* Alexander Peiffer.

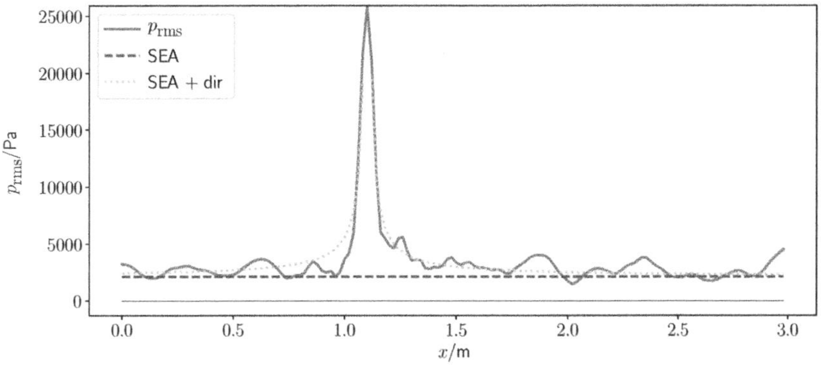

Figure 6.24 Ensemble pressure field in rectangular room 2 ($\eta_{\text{fluid}} = 0.001, A_s = 1.5$ m^2) along x-line through source location (x_0, y_0, z_0). *Source:* Alexander Peiffer.

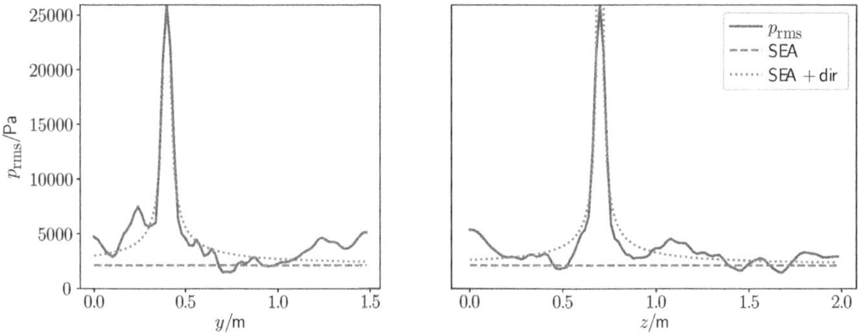

Figure 6.25 Ensemble pressure field in rectangular room2 ($\eta_{\text{fluid}} = 0.001, A_s = 1.5\ \text{m}^2$) along y and z-lines through source location (x_0, y_0, z_0). *Source:* Alexander Peiffer.

6.6 Surface Load of Diffuse Acoustic Fields

The reflection of the diffuse wave field at the room boundaries creates random load at the boundary surfaces. The amplitude doubles due to Equation (6.30) for each room angle (ϑ, ϕ). The cross correlation spectrum at the surface is derived by the integration over half the room angle. Modifying (6.21) accordingly reads

$$\begin{aligned} S_{pp}^{\partial V}(\Delta r) &= \int_0^{2\pi}\int_0^{\pi/2} p(\vartheta,\varphi)(1+\boldsymbol{R})^2 \hat{\boldsymbol{p}}^2 \cos(k\Delta r\cos(\vartheta))\sin(\vartheta)d\vartheta d\varphi \\ &= 2\pi(1+\boldsymbol{R})^2 \left\langle \hat{p}^2 \right\rangle_E \int_0^{\pi/2} \frac{1}{4\pi}\cos(k\Delta r\cos(\vartheta))\sin(\vartheta)d\vartheta \\ &= \frac{(1+\boldsymbol{R})^2}{2}\left\langle \hat{p}^2 \right\rangle_E \frac{\sin(k\Delta r)}{k\Delta r} \end{aligned} \tag{6.76}$$

This is in line with the discussion in section 6.1.6. The pressure at the boundary is called the blocked pressure (force) when the walls are rigid and we have total reflection ($\boldsymbol{R} = 1$). In the random description of subsystems global quantities such as volume, energy, and the modal density are given. Expressing the field pressure $\left\langle \hat{p}^2 \right\rangle_E$ in these terms gives

$$\left\langle \hat{p}^2 \right\rangle_E = 2\rho_0 c_0^2 \frac{E}{V} \tag{6.77}$$

With the modal density of rooms and when perimeter and surface effects can be neglected $n(\omega) = \frac{V\omega^2}{2\pi^2 c_0^3}$ we get

$$\left\langle \hat{p}^2 \right\rangle_E = \frac{\rho_0 \omega^2}{\pi^2 c_0 n(\omega)} \tag{6.78}$$

Entering this into (6.76) we get for the cross correlation of the diffuse acoustic field with energy E and modal density $n(\omega)$

$$S_{pp}^{\partial V}(\Delta r) = \frac{2\rho_0\omega^2 E}{\pi^2 c_0 n(\omega)} \frac{\sin(k\Delta r)}{k\Delta r} \tag{6.79}$$

When we discretize the surface by a regular mesh with element area ΔA, the cross correlation of the blocked force load is

$$S_{ff}^{\partial V}(\Delta r) = \Delta A^2 \frac{2\rho_0\omega^2 E}{\pi^2 c_0 n(\omega)} \frac{\sin(k\Delta r)}{k\Delta r} \tag{6.80}$$

Note that the grid length in this mesh must be much smaller than the wavelength. The sinc function in the above description can be replaced by a more generic expression. The transfer impedance of one vibrating element at $\mathbf{r}$ to another element located at $\mathbf{r} + \Delta\mathbf{r}$ is given by the Rayleigh integral of the semi infinite half space with rigid wall (2.141) and constant velocity over the small element area.

$$\boldsymbol{p}(\mathbf{r} + \Delta\mathbf{r}) = \frac{j\omega\rho_0}{2\pi\Delta r} e^{-jk\Delta r} \Delta A \boldsymbol{v}(\mathbf{r}) \text{ with } \Delta r = |\mathbf{r} - \mathbf{r}_0| \tag{6.81}$$

and the transfer impedance for the mesh is

$$\boldsymbol{Z}_a(\Delta r) = \frac{\boldsymbol{p}(\mathbf{r} + \Delta r)}{\boldsymbol{v}(\mathbf{r})} = \frac{j\omega\rho_0}{2\pi\Delta r} e^{-jk\Delta r} \Delta A \tag{6.82}$$

providing a mechanical impedance when we multiply the acoustic impedance by ΔA. The real part of this provides the wanted sinc function

$$Re(\boldsymbol{Z}(\Delta r)) = \Delta A^2 \frac{\omega^2\rho_0}{2\pi c_0} \frac{\sin(k\Delta r)}{k\Delta r} \tag{6.83}$$

Entering this into (6.80) we get the diffuse field reciprocity relationship for acoustic wave fields:

$$S_{ff}^{\partial V}(\Delta r) = \frac{4E}{\pi n(\omega)} Re(\boldsymbol{Z}(\Delta r)) \tag{6.84}$$

This is an important and useful expression. We can determine the cross spectral load function of diffuse fields only from the total energy, the modal density, and the real part of the radiation impedance. Thus, the response of every deterministic system described by a transfer matrix $[\boldsymbol{H}(\omega)]$ and that is excited by a diffuse acoustic field can be calculated using the matrix multiplication from section 1.7.2.

This expression was proven by Shorter et al. (2005) for general diffuse wave fields. So, the response of any deterministic system excited by a reverberant field can be calculated using the above expression.

6.7 Mode Wave Duality

For the investigation of the diffuse field concept applied to ensembles of random systems, modal solutions of two- or three-dimensional systems are used. It seems that the modal description in combination with high modal density is a further appropriate model for reverberant wave fields, but we have not established a clear theoretical relationship between the modal and wave concepts. Lyon and DeJong (1995) investigated a system described by modal response with uncertainty in order to prove that the result from the diffuse wave field model can also be shown using the modal approach.

6.7.1 Diffuse Field Energy

In the diffuse field model with equal power from all directions, the homogeneous energy distribution is evident. For systems described by modes, it is not that obvious that modes (with a specific uncertainty) lead to an equally distributed mean square amplitude in the wave field.

6.7.2 Free Field Power Input

In the discussions of the modal overlap factor of section 6.3.1.2 we have established the criteria that allows treating the point impedances of resonant and deterministic systems as free field impedances. In that case the damping is so high that no reflection is received at the point of excitation (independent from the fact of whether the reflection is coherent or not).

We will now show that the free field assumption, which is easy to explain in the diffuse wave field model, can also be proven when we assume a Gaussian orthogonal ensemble of modes (see details in Langley (2016)). In this approach, the uncertainty in an ensemble of systems leads to small variations in the modal frequencies as shown in Figure 6.26 for one ensemble representation. When dealing with structural systems, the input power into the system excited by a harmonic source at a specific location and frequency ω is according to (1.49)

$$\Pi = \frac{1}{2}Re\left(\frac{\boldsymbol{F}\boldsymbol{F}^*}{\boldsymbol{Z}}\right) = \frac{1}{2}Re\left(F^2\boldsymbol{Y}\right) = \frac{1}{2}F^2G \text{ with } \boldsymbol{Y} = G + jB \tag{6.85}$$

The real part is called conductance, the imaginary part susceptance. For the derivation of the conductance G of modal systems we take Equation (5.42) as a typical example.

$$\boldsymbol{Y}(\mathbf{x},\mathbf{x}_0) = j\omega\sum_{n=1}^{\infty}\frac{\Phi_n(\mathbf{x})\Phi_n^*(\mathbf{x}_0)}{m''[\omega_n^2(1+j\eta)-\omega^2]} \tag{6.86}$$

In principle, every closed dynamic system can be described by the above equation. Independent from the fact of whether a discrete or homogeneous aproach is selected.

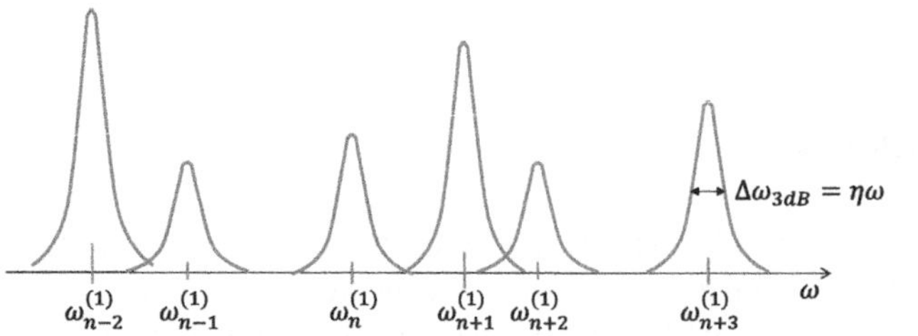

Figure 6.26 Distribution of modal frequencies near ω_n and the receptance function along the frequency axis for one ensemble representation. *Source:* Alexander Peiffer.

In order to assess the real and imaginary parts of the mobility we rationalize the denominator.

$$\boldsymbol{Y}(\mathbf{x}, \mathbf{x}_0) = \sum_{n=1}^{\infty} \frac{\Phi_n(\mathbf{x})\Phi_n^*(\mathbf{x}_0)}{m''} \frac{\eta\omega\omega_n^2 + j\omega(\omega_n^2 - \omega^2)}{(\omega_n^2 - \omega^2)^2 + \eta^2\omega_n^4} \tag{6.87}$$

and the conductance reads as

$$\begin{aligned} G(\mathbf{x}, \mathbf{x}_0, \omega) &= \sum_n \frac{\Phi_n(\mathbf{x})\Phi_n^*(\mathbf{x}_0)}{m''\eta\omega} \frac{\eta^2\omega^2\omega_n^2}{(\omega_n^2 - \omega^2)^2 + \eta^2\omega_n^4} \\ &= \sum_n a_n(\omega) g_n(\omega) \end{aligned} \tag{6.88}$$

with $a_n = \frac{\Phi_n(\mathbf{x})\Phi_n^*(\mathbf{x}_0)}{m''\eta\omega}$, and $g_n(\omega)$ that can be further simplified assuming that the major contribution is for frequencies near the resonances $\omega \approx \omega_n$.

$$\begin{aligned} g_n(\omega) &= \frac{\eta^2\omega^2\omega_n^2}{(\omega_n^2 - \omega^2)^2 + \eta^2\omega_n^4} \\ &= \frac{\eta^2\omega^2\omega_n^2}{[(\omega_n - \omega)(\omega_n + \omega)]^2 + \eta^2\omega_n^4} \\ &\approx \frac{\eta^2\omega^2\omega_n^2}{[(\omega_n - \omega)2\omega]^2 + \eta^2\omega_n^4} = \frac{\eta^2\omega_n^2}{[2(\omega_n - \omega)]^2 + \eta^2\omega_n^2} \\ &= \frac{1}{\xi^2 + 1} \qquad \text{with} \qquad \xi = \frac{2(\omega - \omega_n)}{\eta\omega_n} \end{aligned} \tag{6.89}$$

We stay first with one mode n and investigate the ensemble average of g_n of modal frequency ω_n. When the modal frequencies are equally distributed over the frequency interval $\Delta\omega$ the probability density of ω_n in this interval is $p(\omega_n) = 1/\Delta\omega$ as shown in Figure 6.27. Thus, the expected value of $g_n(\omega)$ is

$$E\left[g_n(\omega)\right] = \langle g_n(\omega)\rangle_E = \frac{1}{\Delta\omega}\int_{-\infty}^{\infty} g_n(\omega)d\omega = \frac{\eta\omega_n}{2\Delta\omega}\int_{-\infty}^{\infty}\frac{1}{\xi^2+1}d\xi = \frac{\eta\omega_n\pi}{2\Delta\omega} \tag{6.90}$$

This interval $\Delta\omega$ must be larger than the modal bandwidth $\eta\omega$ for getting a smooth average, or in other words, the variation in modal frequency must be so large that the peak in frequency is shifted such that every value of $g_n(\omega)$ is possible.

The sum over all $g_n(\omega)$ is approximated by the number of modes N times $\langle g_n(\omega)\rangle_E$. N is simply the product of the frequency interval $\Delta\omega$ and the modal density. Thus, the

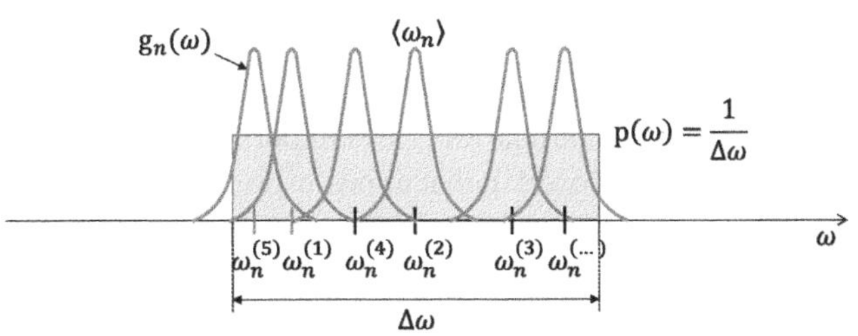

Figure 6.27 Ensemble of modal frequency distributions for one specific mode of frequency ω_n. *Source:* Alexander Peiffer.

expected value of (6.88) is

$$E\left[G(\omega)\right] = \left\langle \sum_n a_n g_n(\omega) \right\rangle_E = n(\omega)\Delta\omega \left\langle a_n g_n(\omega) \right\rangle_E$$

$$= n(\omega)\Delta\omega \left\langle \frac{\Phi_n(\mathbf{x})\Phi_n^*(\mathbf{x}_0)}{m''\eta\omega} \right\rangle_E \frac{\eta\omega\pi}{2\Delta\omega} = \left\langle \frac{\Phi_n(\mathbf{x})\Phi_n^*(\mathbf{x}_0)}{m''} \right\rangle_E \frac{n(\omega)\pi}{2} \tag{6.91}$$

6.7.2.1 Point Conductance

For the point conductance, excitation and response points are the same $\mathbf{x} = \mathbf{x}_0$, and the ensemble average over the sum of all modes is given as

$$\left\langle \frac{\Phi_n^2(\mathbf{x}_0)}{m''} \right\rangle_E \tag{6.92}$$

The mode shape function from (5.33) is

$$\Phi_n^2(x,y) = \frac{4}{L_x L_y} \sin^2\left(\frac{n_x \pi x}{L_x}\right) \sin^2\left(\frac{n_y \pi x}{L_y}\right)$$

Similar to the diffuse wave field, we assume that the mode shapes are not correlated. So, arguments in the sine functions in an ensemble of mode shapes have uncorrelated phase, and we take the expected value of each squared sine function $E\left[\sin^2(\varphi)\right] = 1/2$. Thus,

$$\left\langle \frac{\Phi_n^2(\mathbf{x}_0)}{m''} \right\rangle_E = \frac{4}{L_x L_y m''} \frac{1}{4} = \frac{1}{M} \tag{6.93}$$

and finally

$$E\left[G(\omega)\right] = \frac{\pi n(\omega)}{2M} \tag{6.94}$$

When we use the modal density for a flat plate (5.35), the result is

$$E\left[G(\omega)\right] = \frac{\pi}{2M} \frac{L_x L_y}{4\pi} \sqrt{\frac{\rho_0 h}{B}} = \frac{1}{8\sqrt{Bm''}} \tag{6.95}$$

which corresponds exactly to the point impedance of infinite plates (3.226). We can conclude that for an ensemble of modal systems with modal frequency uncertainty with stable modal density and frequency variation $\Delta\omega \gg \eta\omega$, the point conductance and also the impedance are equal to the free field expression.

6.8 SEA System Description

Random subsystems are described and modelled by a few properties that characterize the random field. These are geometrical entities such as volume, area, and length. In addition, the ratio of these quantities to the boundaries is also important. This determines the modal density, therefore the modal overlap that allows estimating when the wave field can be assumed as a diffuse random wave field. The power input into the systems is determined by the free field impedance.

The power loss due to damping of the systems may be caused by dissipation in the wave propagation or absorption at the system boundaries. From the power balance, the energy state of the system can be calculated. Once the energy is computed, the wave field property can be recovered from the link between energy density and each appropriate engineering unit, e.g. pressure, displacement, and velocity.

Thus, a complicated and detailed system is reduced to a small set of global parameters that allow simple determination of the random system response. This is the reason why SEA is very practical when dealing with large systems of high dynamic complexity and uncertainty.

However, the diffuse field assumption has its limitations, especially when it comes to large propagation damping. In this case, methods similar to optics may be applied that consider effects of damping and diffraction or shadowing of wave fields.

6.8.1 Power Balance in Diffuse Fields

In the introduction to Section 6.8, the model of the diffuse acoustic field was applied to describe the state of a dynamically complex system, as shown in Figure 6.28. For the single system, the equation is simple, and the energy of the system that determines the amplitude of pressure, displacement, or other wave amplitudes is given by

$$\Pi_{\text{in}} = \Pi_{\text{diss}} = \eta\omega E \tag{6.96}$$

Engineering systems consist of several components and systems. Therefore there are usually many diffuse acoustic fields (encapsulated in specific systems) that are exchanging energy. This transmitted energy is related to the system surface intensity as given by Equations (6.16), (6.18), and (6.20) as far as the connecting area and the transmission of radiated waves into the other connected subsystems. The details of this coupling will be discussed in Chapter 7. In the following we use the upper index (n) for the nth subsystem for quantities that are already using a descriptive index as for example Π_{in}, Π_{diss}.

In Figure 6.29, two connected random systems are depicted. In order to cope with the damping loss η_m, we use a similar expression for the loss due to transmission into the connected subsystem. Thus, the power transmitted by subsystem (1) into system (2) reads as

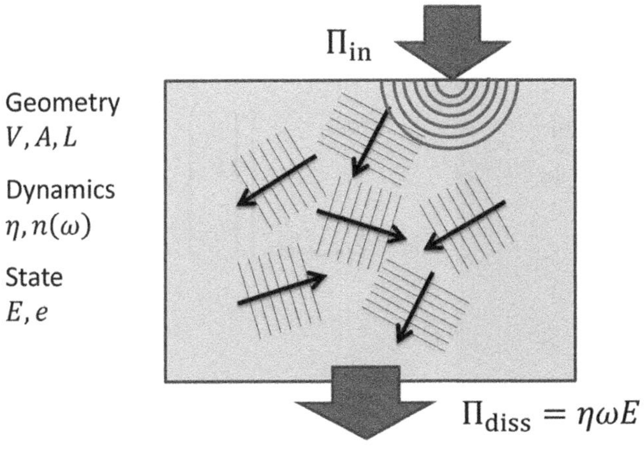

Figure 6.28 Single SEA system power flow. *Source:* Alexander Peiffer.

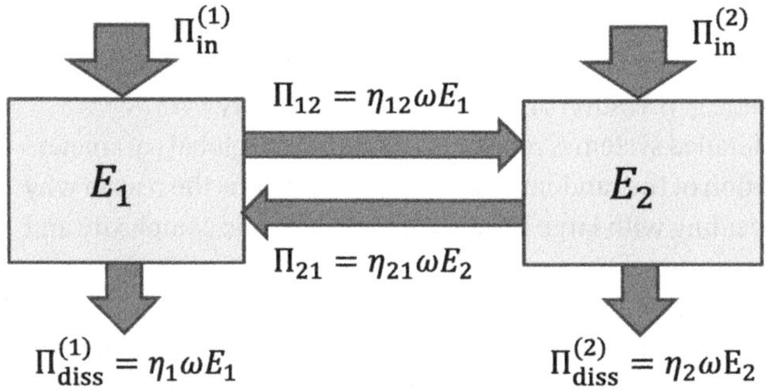

Figure 6.29 Two connected SEA systems power flow. *Source:* Alexander Peiffer.

$$\Pi_{12} = \eta_{12}\omega E_1 \tag{6.97}$$

The expression for the other direction is

$$\Pi_{21} = \eta_{21}\omega E_2 \tag{6.98}$$

The coefficient η_{mn} is called *coupling loss factor* in accordance with the definition of the loss factor. The total power balance for subsystem (1) is

$$\Pi_{in}^{(1)} = \eta_{12}\omega E_1 + \eta_1\omega E_1 - \eta_{21}\omega E_2 \tag{6.99}$$

and for subsystem (2)

$$\Pi_{in}^{(2)} = -\eta_{12}\omega E_1 + \eta_{21}\omega E_2 + \eta_2\omega E_2 \tag{6.100}$$

When we consider a number N of subsystems the power balance of the mth subsystem would be given by

$$\Pi_{in}^{(m)} = \eta_m\omega E_m + \omega \sum_{n=1,n\neq m}^{N} (\eta_{mn}E_m - \eta_{nm}E_n) \tag{6.101}$$

and this can be written in full matrix form as

$$\omega \begin{bmatrix} \eta_1 + \sum_{n\neq 1}^N \eta_{1n} & -\eta_{21} & \cdots & -\eta_{N1} \\ -\eta_{12} & \eta_2 + \sum_{n\neq 2}^N \eta_{2n} & & \vdots \\ \vdots & & \ddots & \\ -\eta_{1N} & \cdots & & \eta_N + \sum_{n\neq N}^N \eta_{Nn} \end{bmatrix} \begin{Bmatrix} E_1 \\ E_2 \\ \vdots \\ E_N \end{Bmatrix} = \begin{Bmatrix} \Pi_{in}^{(1)} \\ \Pi_{in}^{(2)} \\ \vdots \\ \Pi_{in}^{(N)} \end{Bmatrix} \tag{6.102}$$

or in short form

$$\omega [L]\{E\} = \{\Pi_{in}\} \tag{6.103}$$

$[L]$ is called the power-flow matrix or the SEA matrix.

6.8.2 Reciprocity Relationships

Reciprocity is a general principle that relates the conjugate input variables of a two port system. Applied to the mechanical systems of Chapter 5, that means that the velocity at position 1 generated by a force at position 2 is similar to a velocity at position 2 generated by the same force at position 1. This general principle applies to all systems that are composed of linear, passive, and bilateral elements.

The same principle must be fulfilled for random systems. Thus, if we apply a force at system 1, the velocity response at system 2 at one point and therefore also at the average must be similar to the reciprocal configuration – that is, the force excitation at system 2 and the velocity response at 1.

$$\frac{\boldsymbol{F}_1}{\boldsymbol{v}_2} = \frac{\boldsymbol{F}_2}{\boldsymbol{v}_1} \tag{6.104}$$

Now imagine a two system SEA configuration as given by Equation (6.99). When the force is exciting exclusively system 1 ($\Pi_2 = 0$), the solution of this system with regards to the energy of system 2 is

$$\frac{\Pi_1}{E_2} = \omega\frac{\eta_1\eta_2 + \eta_1\eta_{21} + \eta_2\eta_{12}}{\eta_{12}} \tag{6.105}$$

and a similar solution is found for the excitation at subsystem 2

$$\frac{\Pi_2}{E_1} = \omega\frac{\eta_1\eta_2 + \eta_1\eta_{21} + \eta_2\eta_{12}}{\eta_{21}} \tag{6.106}$$

The reciprocity relationship must now be converted into energy and power expressions. The energy of the subsystem is given by the total mass m_i and the average velocity of the subsystem

$$E_i = \frac{1}{2}m_i\langle v_i\rangle_S \tag{6.107}$$

and the power input is determined by the conductance

$$\Pi_i = \frac{1}{2}G_iF_i^2 \tag{6.108}$$

When expressing power and energy by the above expressions and squaring the reciprocity relationship (6.104) we get

$$\frac{\Pi_1}{E_2} = \omega\frac{\eta_1\eta_2 + \eta_1\eta_{21} + \eta_2\eta_{12}}{\eta_{12}}\frac{m_2}{G_1} = \omega\frac{\eta_1\eta_2 + \eta_1\eta_{21} + \eta_2\eta_{12}}{\eta_{21}}\frac{m_1}{G_2} \tag{6.109}$$

This leads to an expression that relates the SEA matrix coefficients using conductance and mass quantities.

$$\frac{\eta_{12}G_1}{m_2} = \frac{\eta_{21}G_2}{m_1} \tag{6.110}$$

In section 6.7.2 a relationship for the input conductance that links the input power created by force excitation and the modal density was found

$$G_i = \frac{\pi n_i(\omega)}{2M_i} \tag{6.111}$$

leading to

$$\eta_{12} n_1(\omega) = \eta_{21} n_2(\omega) \tag{6.112}$$

This is the reciprocity relationship of SEA and it leads to a symmetric form of the equation when dividing each energy by the modal density of the system. When we define the double index $\eta_{nn} = \eta_n$ for the damping loss, we may write

$$\omega \begin{bmatrix} n_1(\omega) \sum_{n=1}^{N} \eta_{1n} & -n_2(\omega)\eta_{21} & \cdots & -n_N(\omega)\eta_{N1} \\ -n_1(\omega)\eta_{12} & n_2(\omega) \sum_{n=1}^{N} \eta_{2n} & & \vdots \\ \vdots & & \ddots & \\ -n_1(\omega)\eta_{1N} & \cdots & & n_N(\omega) \sum_{n=1}^{N} \eta_{Nn} \end{bmatrix} \begin{Bmatrix} \frac{E_1}{n_1(\omega)} \\ \frac{E_2}{n_2(\omega)} \\ \vdots \\ \frac{E_N}{n_N(\omega)} \end{Bmatrix} = \begin{Bmatrix} \Pi_{\text{in}}^{(1)} \\ \Pi_{\text{in}}^{(2)} \\ \vdots \\ \Pi_{\text{in}}^{(N)} \end{Bmatrix} \tag{6.113}$$

This matrix is symmetric and better suited for computationally efficient solvers. In addition, only half of the coupling coefficients must be determined. We denote this symmetric matrix with $[L']$

$$\omega [L'] \{E/n\} = \{\Pi\} \tag{6.114}$$

6.8.3 Fluid Analogy

The principle of energy per modal density motivates the fluid analogy of SEA shown in Figure 6.30. The amount of water corresponds to the total energy E_m in each subsystem, the width to the modal density (linearly dependent on the number of modes in band), and the height of the fluid in the tank is the energy per modal density. The power input and losses are the amount of fluid per time that are entering or leaving the systems. The exchange of fluid depends on the energy difference in the systems.

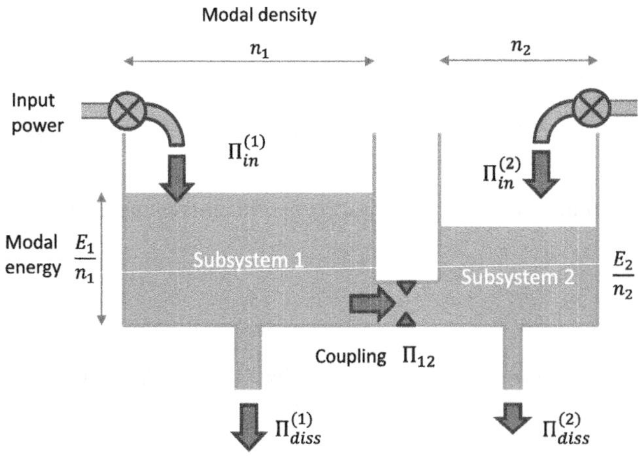

Figure 6.30 Fluid analogy of SEA. *Source:* Alexander Peiffer.

6.8.4 Power Input

The input power is determined from the source properties and the convenient property describing the reaction to this excitation. When we studied the different random systems we showed that in the ensemble average the dynamics of the system is equal to the free field radiation dynamics of the involved wave physics. In other words: sources connected to a random system get the dynamic reaction of the free field. This fact is quite obvious in the model construct of the diffuse wave field model, because it is one main assumption of this model that there is no coherent reflection from the boundaries of the diffuse wave field system. But, even in modal space this assumption can be proven given the right statistics and uncertainty for the involved modes. When damping is high enough, this fact can even by shown without any ensemble averaging. In this case, the reflecting waves are simply damped out.

Thus, the important message is that random systems have the same dynamic characteristics as the according free field. Assuming a surface impedance $\boldsymbol{Z}_{\text{rnd}}$ of a random system, the ensemble average of the free field radiation impedance is supposed to be

$$\langle \boldsymbol{Z}_{\text{rnd}} \rangle_E = \boldsymbol{Z}_\infty \tag{6.115}$$

6.8.5 Engineering Units

The random description of systems involves the energy of systems but not directly the required quantity. In general we must separate between the potential and the kinetic energy of the waves, but as the model of the diffuse field is linked to the propagation of plane waves, we can use the statements from the plane wave energy descriptions: the potential and kinetic energy is equally distributed, and we chose the typical wave quantity usually taken to describe the system.

For fluid systems such as pipes and cavities, this quantity is the acoustic pressure (6.37):

$$p^2_{m,rms} = \rho_0 c^2_{\text{gr}} \frac{E_m}{V} \tag{6.116}$$

and for plates it is the displacement (6.57):

$$w^2_{m,rms} = \frac{e_m}{\rho_0 \omega^2} = \frac{E_m}{Ah\rho_0 \omega^2} \tag{6.117}$$

For practical reasons and better scaling over the frequency range, the results are often expressed as velocity or acceleration

$$v^2_{m,rms} = \frac{E_m}{Ah\rho_0} \qquad a^2_{m,rms} = \frac{E_m \omega^2}{Ah\rho_0} \tag{6.118}$$

The same can be done for one-dimensional system such as beams and pipes. For pipes the first equation applies; but, concerning beams and rods, we will not enter deeper into the details of random modelling, because the high bending stiffness in engineering systems (this is what their task is in typical constructions – to be stiff and to carry loads) doesn't make random modelling of such systems realistic. In most practical cases, those systems must be considered as deterministic.

In the introduction to Chapter 6, we mentioned the ratio between wavelength and the dimensions of the subsystems as possible criteria to seperate between random and

Table 6.2 Modal densities for one-, two-, and three-dimensional subsystems.

Dimension type	Modal density $n(\omega)$
Bars, rods, beams, pipes	$n_{1D}(\omega) = \frac{L}{\pi c_{gr}}$
Plates, shells, thin cavities	$n_{2D}(\omega) = \frac{kA}{2\pi c_{gr}}$
Cavities	$n_{3D}(\omega) = \frac{V\omega^2}{2\pi^2 c_0^3} + \frac{S\omega}{8\pi c_0^2} + \frac{P}{16\pi c_0}$

deterministic systems. Either the modal overlap $M = \eta\omega n(\omega)$ must be greater than one, or the ensemble variation in modal frequency must be large enough to make the modal responses overlap to allow for a random description of the system. In all cases the major criteria is the modal density in combination with the damping as a factor in the first case or the frequency variation in the second case. To establish the link between the Helmholtz number $He = \omega L/c_0$, it is helpful to present the modal density in terms of the Helmholtz number. Usually λ is linked to the phase velocity, but we set $c_{gr} = c_0$ to allow for simple expressions that are used here as fuzzy criteria to separate between random or deterministic systems.

$$n_{1D}(\omega) \approx \frac{He}{\omega\pi} \tag{6.119}$$

$$n_{2D}(\omega) \approx \frac{He^2}{2\pi\omega} \tag{6.120}$$

$$n_{3D}(\omega) \approx \frac{He^3}{2\pi^2\omega} \tag{6.121}$$

For the different exponents of the Helmholtz number, we assumed that $A = L^2$ and $V = L^3$. With this view it becomes clear why two- and especially three-dimensional systems become random much earlier in frequencies than one-dimensional systems.

6.8.6 Multiple Wave Fields

In plates, there occur different wave types: the longitudinal, the shear, and the bending wave. In beams, there are even more types of wave propagation that are not treated in detail here. Those waves have different wavelengths, so there can be simultaneous waves that fulfill the requirements for a reverberant field and that do not. From the energy perspective, we may use a combination of wave types to describe a random system. It is good practice to consider the one wave type field as one SEA subsystem and not all wave fields in one system. We will see later that the calculation of the coupling loss factors will motivate a particular wave field selection. For fluids, the situation is simple, because there is only one mode of wave propagation.

Bibliography

Finn Jacobsen. The diffuse sound field. Technical Report 27, Danmarks Tekniske Universitet (DTU), Lyngby, Denmark, 1979.

Heinrich Kuttruff. *Room Acoustics, Fifth Edition*. 2014. ISBN 978-1-4822-6645-0 978-0-203-87637-4.

R.S. Langley. On the statistical properties of random causal frequency response functions. *Journal of Sound and Vibration*, 361:159–175, January 2016.

Alain Le Bot. Statistical Energy Analysis, July 2015.

R.H. Lyon and R.G. DeJong. *Theory and Application of Statistical Energy Analysis, Second Edition*. Butterworth Heinemann, second edition, 1995. ISBN 0-7506-9111-5.

P.M.C. Morse and K.U. Ingard. *Theoretical Acoustics*. International Series in Pure and Applied Physics. Princeton University Press, 1968. ISBN 978-0-691-02401-1.

Allan D. Pierce. *Acoustics - An Introduction to Its Physical Principles and Applications*. Acoustical Society of America (ASA), Woodbury, New York 11797,U.S.A., one thousand, nine hundred eighty-ninth edition, 1991. ISBN 0-88318-612-8.

P. J. Shorter and R. S. Langley. Vibro-acoustic analysis of complex systems. *Journal of Sound and Vibration*, 288(3):669–699, 2005.

P. J. Shorter, Y. Gooroochurn, and B. Rodewald. Advanced vibro-acoustic models of welded junctions. In *Proceedings Internoise 2005*, Rio, Brasil, August 2005.

7

Coupled Systems

When systems of similar physics are connected, the situation is relatively simple. The degrees of freedom of the coupling zone are shared, and the state variables, for example pressure and displacement, are equal. The situation changes for coupled deterministic systems of different wave types and physics. Here, conditions must be defined that link, for example, the pressure of one subsystem to the displacements of the other subsystem. The zone where the coupling dynamics takes place is called a junction.

However, simulating sound and vibration propagation in realistic systems means coupling systems of different types of wave propagation and different topology. This can be, for example, a fluid cavity that is connected to several plates surrounding it or a beam that is connected to other beams or plates. Hence, simulating the coupling dynamics can become a complicated task even for pure deterministic subsystems. However, the dynamics can be finally described by a large set of equations of motion with different degrees of freedom.

When random subsystems are connected, the description becomes more complicated. How can the coupling physics between random subsystems be described? If the state of random systems is given by energy, the power flow between the systems must be determined, given by the so-called *coupling loss factor*.

As shown in section 6.1, the boundary load of random systems is defined by a cross spectral density function that acts on the junction. In the junction itself, the dynamics is deterministic in the sense that the physics of the wave transmission through the connection is described in detail by deterministic equations of wave transmission. Thus, a junction between random systems must also be deterministic in order to catch the physics of wave transmission correctly.

A practical example for this can be two acoustic spaces or rooms connected by a thin wall or a window. In order to derive the energy exchange between the two spaces we must know precisely the wave behavior of this wall. In classical SEA literature, these phenomena are explained by wave transmission processes.

Here, we will apply the hybrid theory from Shorter and Langley (2005b) and derive the link to the classical approaches. The advantage of Shorter's hybrid approach is that it allows for systematic and logical derivation of SEA theory and coupling loss factors.

Hereinafter, deterministic and random systems are called FEM-systems and SEA-systems, respectively. The acronym FEM is selected because deterministic systems are usually modelled by finite element methods. In this chapter, we will deal with the connection of all these different systems.

Vibroacoustic Simulation: An Introduction to Statistical Energy Analysis and Hybrid Methods,
First Edition. Alexander Peiffer.

7.1 Deterministic Subsystems and their Degrees of Freedom

In previous sections, we used continuous coordinates for the description of continuous systems. They are necessary, because quantities such as wave number, speed of sound etc. are derived based on continuous equations. In practical vibroacoustics, deterministic systems are rarely described by continuous coordinates but discrete coordinates. Nearly every geometry, set-up, and material configuration can be approximated by a discrete mesh. There exist a variety of tools and methods to simulate the dynamics of discrete systems. The maturity of commercial FEM software allows for the simulation of large and complex structures with several millions of degrees of freedom. In addition, other disciplines in the design process of vibroacoustic systems such as crash performance, statics, or dynamics also require finite element (FE) models. Hence, FE-models are usually automatically available early in the design process.

Therefore, it is reasonable to rely on FE-models for the deterministic part. We use the generalized degrees of freedom for the displacement denoted by $\boldsymbol{q}$ and introduced in section 5.1. However, when dealing with the coupling physics, different choices of degrees of freedom can be useful for each specific problem. So, we will use the most appropriate description for all following dynamics of coupled systems. For example, for infinite systems the wave number is the best degree of freedom to describe the problem.

7.2 Coupling Deterministic Systems

When deterministic systems of similar natures are coupled, the joint degrees of freedom of both subsystems are the same. For example, a mechanical subsystem (1) as shown in Figure 7.1 can be described by the dynamic stiffness matrix $[\boldsymbol{D}]$. The displacement vector can be sorted in such a way by gathering the surface degrees of freedom $\boldsymbol{q}_s$ that can potentially be shared with other subsystems, junction degrees of freedom $\boldsymbol{q}_{j12}$ that are actually shared, and inner degrees of freedom $\boldsymbol{q}_i$ that cannot connect to other systems because they are internal

$$\begin{bmatrix} \boldsymbol{D}_{ii} & \boldsymbol{D}_{is} & \boldsymbol{D}_{ij12} \\ \boldsymbol{D}_{si} & \boldsymbol{D}_{ss} & \boldsymbol{D}_{sj12} \\ \boldsymbol{D}_{j12i} & \boldsymbol{D}_{j12s} & \boldsymbol{D}_{j12j} \end{bmatrix} \begin{Bmatrix} \boldsymbol{q}_i \\ \boldsymbol{q}_s \\ \boldsymbol{q}_{j12} \end{Bmatrix} = \begin{Bmatrix} \boldsymbol{F}_i = 0 \\ \boldsymbol{F}_s \\ \boldsymbol{F}_{j12} = 0 \end{Bmatrix} \tag{7.1}$$

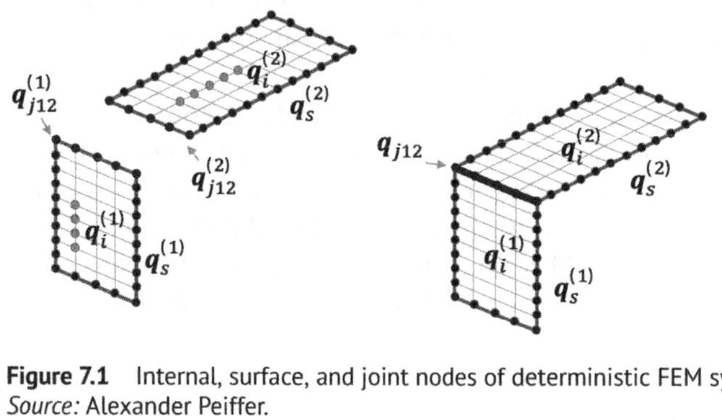

Figure 7.1 Internal, surface, and joint nodes of deterministic FEM systems.
Source: Alexander Peiffer.

The expression *external* is motivated by the fact that these are the degrees of freedom that can be potentially excited by external forces. When creating a simulation strategy for a vibroacoustic system, it is helpful to consider the potential subsystem configuration from the very beginning and to identify which degrees of freedom are inner, external, or junction degrees of freedom. When doing so we always have the deterministic equations in the appropriate form.

When dealing with such system matrices it may be useful to get rid of the internal degrees of freedom and focus on the external or junction degrees of freedom. For outlining this procedure, junction and external degrees of freedom are denoted as outer degrees of freedom by a common subscript $o \equiv s + j$. The inner degrees of freedom can be removed from the equations of motion. We start from the following simplified block matrix system:

$$\begin{bmatrix} \boldsymbol{D}_{ii} & \boldsymbol{D}_{io} \\ \boldsymbol{D}_{oi} & \boldsymbol{D}_{oo} \end{bmatrix} \begin{Bmatrix} \boldsymbol{q}_i \\ \boldsymbol{q}_o \end{Bmatrix} = \begin{Bmatrix} \boldsymbol{F}_i = 0 \\ \boldsymbol{F}_o \end{Bmatrix} \tag{7.2}$$

Due to the fact that no external forces $\boldsymbol{F}_i = 0$ are applied to the inner degrees of freedom, the upper block matrix of Equation (7.2) can be inverted

$$\{\boldsymbol{q}_i\} = \left[-\boldsymbol{D}_{ii}^{-1}\boldsymbol{D}_{io}\right]\{\boldsymbol{q}_o\} \tag{7.3}$$

Entering this in the lower part gives the stiffness matrix exclusively for the outer degrees of freedom.

$$\left[\boldsymbol{D}_{oo} - \boldsymbol{D}_{oi}\boldsymbol{D}_{ii}^{-1}\boldsymbol{D}_{io}\right]\{\boldsymbol{q}_o\} = [\boldsymbol{D}_o]\{\boldsymbol{q}_o\} = \{\boldsymbol{F}_o\} \tag{7.4}$$

This process is called condensation. It is of general importance to clearly define and handle the required degrees of freedom when dealing with vibroacoustic problems. The inner and outer degrees of freedom shall always be clearly differentiated.

A convention is required to sort these multiple definitions of degrees of freedom. According to the convention in the sections before, we use a subsystem superscript $\cdot^{(m)}$ for the m-th subsystem. The surface and inner degrees of freedom of subsystem m are denoted by $\boldsymbol{q}_s^{(m)}$ and $\boldsymbol{q}_i^{(m)}$, respectively. For the junction degrees of freedom, we use the subscript $_j$ in combination with the connected subsystems, for example $\boldsymbol{q}_{j12}$ for junction degrees of freedom connecting subsystems 1 and 2 or $\boldsymbol{q}_{j123}$ connecting subsystems 1, 2, and 3.

Now, let us assume that we have two subsystems as shown in Figure 7.1 on the right hand side. With these conventions, the stiffness matrix of the full deterministic system can be arranged in such a way that we have block matrices in the total matrix.

Using the reduced matrix from Equation (7.4) in order to get rid of the inner degree of freedoms we can write for subsystem 1 and 2:

$$\begin{bmatrix} \boldsymbol{D}_{ss}^{(1)} & \boldsymbol{D}_{sj12}^{(1)} \\ \boldsymbol{D}_{j12s}^{(1)} & \boldsymbol{D}_{j1212}^{(1)} \end{bmatrix} \begin{Bmatrix} \boldsymbol{q}_s^{(1)} \\ \boldsymbol{q}_{j12}^{(1)} \end{Bmatrix} = \begin{Bmatrix} \boldsymbol{F}_s^{(1)} \\ \boldsymbol{F}_{j12}^{(1)} \end{Bmatrix} \tag{7.5}$$

$$\begin{bmatrix} \boldsymbol{D}_{ss}^{(2)} & \boldsymbol{D}_{sj12}^{(2)} \\ \boldsymbol{D}_{j12s}^{(2)} & \boldsymbol{D}_{j1212}^{(2)} \end{bmatrix} \begin{Bmatrix} \boldsymbol{q}_s^{(2)} \\ \boldsymbol{q}_{j12}^{(2)} \end{Bmatrix} = \begin{Bmatrix} \boldsymbol{F}_s^{(2)} \\ \boldsymbol{F}_{j12}^{(2)} \end{Bmatrix} \tag{7.6}$$

Supposing the joint degrees of freedom are equal $\boldsymbol{u}_{j12}^{(1)} = \boldsymbol{u}_{j12}^{(2)} = \boldsymbol{u}_{j12}$, the above matrices would look in the total system as follows:

$$\begin{bmatrix} \boldsymbol{D}_{ss}^{(1)} & 0 & \boldsymbol{D}_{sj12}^{(1)} \\ 0 & 0 & 0 \\ \boldsymbol{D}_{j12s}^{(1)} & 0 & \boldsymbol{D}_{j1212}^{(1)} \end{bmatrix} \begin{Bmatrix} \boldsymbol{q}_s^{(1)} \\ \boldsymbol{q}_s^{(2)} \\ \boldsymbol{q}_{j12} \end{Bmatrix} = \begin{Bmatrix} \boldsymbol{F}_s^{(1)} \\ \boldsymbol{F}_s^{(2)} \\ \boldsymbol{F}_{j12} \end{Bmatrix} \tag{7.7}$$

$$\begin{bmatrix} 0 & 0 & 0 \\ 0 & \boldsymbol{D}_{ss}^{(2)} & \boldsymbol{D}_{sj12}^{(2)} \\ 0 & \boldsymbol{D}_{j12s}^{(2)} & \boldsymbol{D}_{j1212}^{(2)} \end{bmatrix} \begin{Bmatrix} \boldsymbol{q}_s^{(1)} \\ \boldsymbol{q}_s^{(2)} \\ \boldsymbol{q}_{j12} \end{Bmatrix} = \begin{Bmatrix} \boldsymbol{F}_s^{(1)} \\ \boldsymbol{F}_s^{(2)} \\ \boldsymbol{F}_{j12} \end{Bmatrix} \tag{7.8}$$

Summing both matrices and integrating them into one global matrix reads:

$$\begin{bmatrix} \boldsymbol{D}_{ss}^{(1)} & 0 & \boldsymbol{D}_{sj12}^{(1)} \\ 0 & \boldsymbol{D}_{ss}^{(2)} & \boldsymbol{D}_{sj12}^{(2)} \\ \boldsymbol{D}_{j12s}^{(1)} & \boldsymbol{D}_{j12s}^{(2)} & \boldsymbol{D}_{jj12}^{(1)} + \boldsymbol{D}_{jj12}^{(2)} \end{bmatrix} \begin{Bmatrix} \boldsymbol{q}_s^{(1)} \\ \boldsymbol{q}_s^{(2)} \\ \boldsymbol{q}_{j12} \end{Bmatrix} = \begin{Bmatrix} \boldsymbol{F}_s^{(1)} \\ \boldsymbol{F}_s^{(2)} \\ \boldsymbol{F}_{j12} \end{Bmatrix} \tag{7.9}$$

The above exercise can be performed for any number of subsystem connections. This process might look somehow academical, but it has a very practical background. As discussed above, large engineering systems consist of many different subsystems. When putting subsystems together for a large system set-up the stiffness matrices must be added under consideration of the correct degrees of freedom.

As mentioned above the degrees of freedom of connections or junctions are always deterministic. Thus, the shared degrees of freedom denote the connection of both SEA and FEM-systems. They are the glue for all the subsystems independent from the fact of whether they are random or deterministic.

7.2.1 Fluid Subsystems

All the derived expressions from the section introduction can be transferred to cavities by exchanging $\boldsymbol{p}$ instead of $\boldsymbol{q}$ for the generalized degrees of freedom and $j\omega\boldsymbol{Q}$ instead of $\boldsymbol{F}$ for the generalized force, hence:

$$\begin{bmatrix} \boldsymbol{D}_{a,ii} & \boldsymbol{D}_{a,is} \\ \boldsymbol{D}_{a,si} & \boldsymbol{D}_{a,ss} \end{bmatrix} \begin{Bmatrix} \boldsymbol{p}_i \\ \boldsymbol{p}_s \end{Bmatrix} = j\omega \begin{Bmatrix} \boldsymbol{Q}_i \\ \boldsymbol{Q}_s \end{Bmatrix} \tag{7.10}$$

Acoustic or fluid matrices are denoted by the extra subscript $_a$.

7.2.2 Fluid Structure Coupling

Whereas the coupling of pure structural or acoustic systems is well defined by shared degrees of freedom the situation is different for the coupling of fluid and structure. This is not a general problem. It comes from the fact that different quantities are used as degrees of freedom, the displacement for structures and the pressure (or velocity potential) for fluid waves. There are implementations that use the displacement as degree of freedom in fluid waves, which makes the coupling formulation easier but leads to other drawbacks. However, we have already defined the coupling and boundary condition when we derived the pressure–velocity relationship for acoustic waves with equation (2.35). The link to the displacement follows directly from $j\omega \boldsymbol{u} = \boldsymbol{v}_x$, thus

$$\mathbf{u} = \frac{1}{\omega^2 \rho_0} \nabla \boldsymbol{p} \tag{7.11}$$

At structure surfaces this concerns only the normal vector. From this coupling condition, the finite element equations of combined fluid and structure systems are given by

$$\left[\begin{bmatrix} K & \boldsymbol{A} \\ 0 & K_a \end{bmatrix} + j\omega \begin{bmatrix} B & 0 \\ 0 & B_a \end{bmatrix} - \omega^2 \begin{bmatrix} M & 0 \\ \boldsymbol{A}^H & M_a \end{bmatrix} \right] \begin{Bmatrix} \boldsymbol{u} \\ \boldsymbol{p} \end{Bmatrix} = \begin{Bmatrix} \boldsymbol{F} \\ j\omega \boldsymbol{Q} \end{Bmatrix} \tag{7.12}$$

as described in detail, for example, by Davidsson (2004). $\boldsymbol{A}$ is the coupling matrix and K, B and M or K_a, B_a and M_a the structural or acoustic system matrices. Note that the coupling structural and fluid meshes are usually not coincident and hence interpolation between both is required. To conclude, even the full deterministic coupling of fluid and structure domains is a complex task when assembling the discrete equations of motion.

7.2.3 Deterministic Systems Coupled to the Free Field

Technical systems are usually embodied in an exterior fluid. Thus, the vibrating systems that are connected to the surroundings radiate waves to the exterior. Usually, the simulation of radiation to the exterior is performed by boundary element analysis, but there are simpler and faster approximations that will be given in section 8.2.3. Technically, this means defining a radiation stiffness matrix that considers only the outer degrees of freedom. The physical interpretation of this is that every locally radiating element surface excites an acoustic pressure field. This pressure field leads to a force acting on the element itself and all others. Adding up all elements leads to the free field radiation stiffness.

In addition we have shown that the surface dynamics of random systems is given by free field dynamics when the modal overlap factor is high enough. For an ensemble average of uncertain systems this occurs even earlier when the system is dynamically complex. Thus, for the theory of coupling random systems, it is useful to elaborate the effect of semi infinite or finite systems on deterministic systems.

Imagine a deterministic system as shown in Figure 7.2. The FEM system in the middle is system (3). The free field can be described by the free field radiation stiffness that affects the degrees of freedom that are connected to the deterministic subsystem. The connected degrees of freedom are denoted by $\boldsymbol{q}_{jm3}$ for the m-connected free field. Theoretically this can be any free field, acoustic space, a line of points radiating into an infinite plate, or a point radiating into a one-dimensional free field.

The free field stiffness is defined exclusively for the connected degrees of freedom. Consider a deterministic system connected to a free field as shown in Figure 7.2. The free field stiffnesses $\left[\boldsymbol{D}_{\rm dir}^{(m)}\right]$ with shared degrees of freedom $\boldsymbol{q}_{jm}$ must be added to the dynamic stiffness matrix of the deterministic system as shown in Equation (7.9). Assuming that we have already condensed the dynamic stiffness matrix to the shared

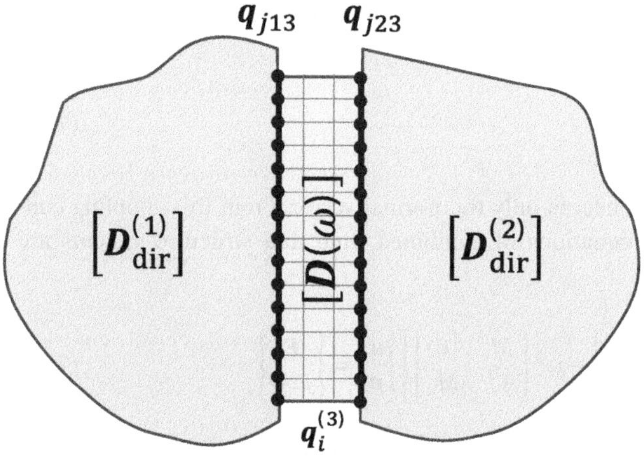

Figure 7.2 Deterministic system connected to two semi infinite systems.
Source: Alexander Peiffer.

degrees of freedom, the total dynamic stiffness reads as

$$\left[\boldsymbol{D}^{(3)}_{\mathrm{tot}}\right] = \begin{bmatrix} \boldsymbol{D}^{(3)}_{j1313} & \boldsymbol{D}^{(3)}_{j1323} \\ \boldsymbol{D}^{(3)}_{j2313} & \boldsymbol{D}^{(3)}_{j2323} \end{bmatrix} + \begin{bmatrix} \boldsymbol{D}^{(1)}_{\mathrm{dir},j13} & \\ & \end{bmatrix} + \begin{bmatrix} & \\ & \boldsymbol{D}^{(2)}_{\mathrm{dir},j23} \end{bmatrix} \tag{7.13}$$

In order to keep the equations readable, we assume that the correct degrees of freedom are selected automatically. With this convention, the above definition of the total matrix reads

$$\left[\boldsymbol{D}^{(3)}_{\mathrm{tot}}\right] = \left[\boldsymbol{D}^{(3)}\right] + \left[\boldsymbol{D}^{(1)}_{\mathrm{dir}}\right] + \left[\boldsymbol{D}^{(2)}_{\mathrm{dir}}\right] = \left[\boldsymbol{D}^{(3)}\right] + \sum_{m=1,2} \left[\boldsymbol{D}^{(m)}_{\mathrm{dir}}\right] \tag{7.14}$$

Here m are the indexes of all connected free fields. To conclude, the dynamic behaviour of a deterministic system emerged into semi infinite systems described by several free field radiation stiffness matrices is given by:

$$\left[\boldsymbol{D}_{\mathrm{tot}}\right]\{\boldsymbol{q}\} = \{\boldsymbol{F}\} \tag{7.15}$$

This total stiffness combines the dynamics of the deterministic system and connected SEA systems. In the following, we will make use of this fact first in the context of coupling loss factors. For this derivation, we neglect excitation at the deterministic system except the reverberant load from the diffuse wave fields. Finally, the general form of Equation (7.15) for the total stiffness reads

$$\left[\boldsymbol{D}_{\mathrm{tot}}\right] = \left[\boldsymbol{D}_s\right] + \sum_m \left[\boldsymbol{D}^{(m)}_{\mathrm{dir}}\right] \tag{7.16}$$

7.3 Coupling Random Systems

The usual approach in classical SEA literature derives the coupling between random systems by application of wave scattering at the junctions or considerations based on reciprocity and excitation exactly at the junction. Here, we would like to apply the systematic approach by applying the diffuse field reciprocity relationship and the fact that the dynamics of every junction is determined by the total stiffness matrix.

In Chapter 6 we have shown that the expected value of stiffness of an ensemble of random systems is equal to the free field radiation stiffness because of the high dynamical complexity that does not allow for coherent reflection. Hence, the dynamic stiffness matrix of a random subsystem can be replaced by the free field direct radiation stiffness for the ensemble of random systems.

$$\langle [\boldsymbol{D}_{j1}] \rangle_E = \left[\boldsymbol{D}_{\text{dir}}^{(1)}\right] \tag{7.17}$$

The impact of the random systems connected to the deterministic one is that each SEA system adds its free field radiation stiffness to the connected degrees of freedom of the deterministic system. Thus, the ensemble average dynamics of a deterministic system connected to several random systems can be described by the total stiffness matrix.

$$\left\langle \left[\boldsymbol{D}^{(3)}\right] + [\boldsymbol{D}_{j1}] + [\boldsymbol{D}_{j2}] \right\rangle_E = \left[\boldsymbol{D}^{(3)}\right] + \left[\boldsymbol{D}_{\text{dir}}^{(1)}\right] + \left[\boldsymbol{D}_{\text{dir}}^{(3)}\right] = [\boldsymbol{D}_{\text{tot}}] \tag{7.18}$$

Finally, we have defined the impact of random subsystems on the deterministic system. Note that the ensemble mean works as a filter that extracts the deterministic part of the equations of motion.

The next step is the determination of the load generated by reverberant fields irradiating the junctions. The cross spectral density of this load is given by the reciprocity relationship between the direct field radiation and the reverberant loading. Shorter and Langley (2005a) have proven this diffuse field reciprocity relationship for arbitrary diffuse fields. We will not go into the details of this quite complex proof and present the final result for stiffness and impedance

$$\left[S_{ff}\right] = \frac{4E_m}{\pi\omega n(\omega)} Im\left[\boldsymbol{D}_{\text{dir}}^{(m)}\right] \qquad \left[S_{ff}\right] = \frac{4E_m}{\pi n(\omega)} Re\left[\boldsymbol{Z}_{\text{dir}}^{(m)}\right] \tag{7.19}$$

What is the use of those relationships? They allow for the determination of random systems load at the boundaries just from energy, modal density and the free field radiation stiffness. Thus, for the simulation of the energy exchange between random systems they are an excellent foundation. The dynamics of a junction is determined by the free field radiation properties of the connected SEA systems and the load coming from the reverberant fields. With the presence of a deterministic system in the junction the dynamic stiffness of the FEM system must be added.

In Figure 7.3 the configuration is shown with direct and reverberant fields. Due to the connection to the random systems, the total stiffness is the sum of the dynamic stiffness of the deterministic system plus the two free field radiation stiffness matrices as in Equation (7.18).

On the right hand side of the Figure 7.3, there is no deterministic system in the connection of both random fields. In this case the junction boundary is considered as deterministic, and the related stiffness matrix is only the sum of the free field radiation stiffness of the connected subsystems.

$$[\boldsymbol{D}_{\text{tot}}] = \sum_m \left[\boldsymbol{D}_{\text{dir}}^{(m)}\right] \tag{7.20}$$

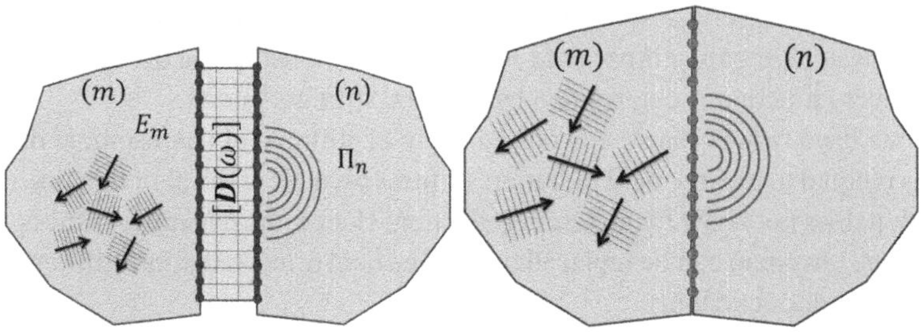

Figure 7.3 Random subsystem exciting the deterministic connection in between (LHS) or deterministic junction only (RHS). *Source:* Alexander Peiffer.

7.3.1 Power Input to System (m) from the nth Reverberant Field

Random load excitation is given by the cross spectral density function. When we replace the system response matrix by the inverse of the total stiffness matrix $[\boldsymbol{H}] = \left[\boldsymbol{D}_{\text{tot}}^{-1}\right]$ in Equation (1.208) we get

$$\left[\boldsymbol{S}_{qq}^{*(n)}\right] = [\boldsymbol{D}_{\text{tot}}]^{-1}\left[\boldsymbol{S}_{ff}^{*(m)}\right][\boldsymbol{D}_{\text{tot}}]^{-H} \tag{7.21}$$

The cross spectral density of a stochastic load is hermitian; in case of diffuse fields it is a real matrix, and for symmetric radiation stiffness also symmetric. The cross spectral density matrix of the reverberant field can be replaced using the diffuse field reciprocity relationship (7.19)

$$[S_{qq}] = [\boldsymbol{D}_{\text{tot}}]^{-1}\left(\frac{4E_m}{\pi\omega n_m} Im\left[\boldsymbol{D}_{\text{dir}}^{(m)}\right]\right)[\boldsymbol{D}_{\text{tot}}]^{-H} \tag{7.22}$$

This is the cross spectral density response of the FEM system of the junction degrees of freedom to the reverberant field with energy E_m in the mth connected subsystem.

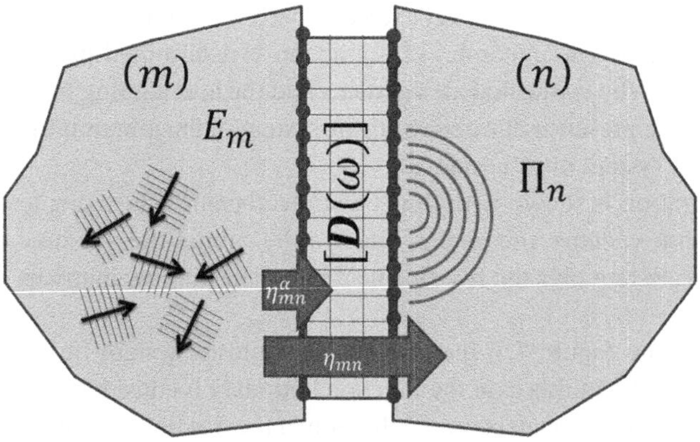

Figure 7.4 Power flow from reverberant field (m) into system (n). *Source:* Alexander Peiffer.

7.3.1.1 Power Radiated from Random Displacement

For the calculation of the power that is radiated into the nth system we consider those degrees of freedom that are connected to the nth system in combination with the radiation stiffness. According to Equation (1.99), and considering in addition that we need the expected value of the force-displacement product, the power into the n-system is given by

$$\left\langle \Pi_{\text{in}}^{(n)} \right\rangle_E = \frac{\omega}{2} \sum_i Im \left\langle \boldsymbol{F}_i \boldsymbol{q}_i^* \right\rangle_E$$

With the force $\boldsymbol{F}_i$ given by reaction forces due to the displacements, we get

$$\begin{aligned} &= \frac{\omega}{2} \sum_i Im \left\langle \left(\sum_j \boldsymbol{D}_{\text{dir},ij}^{(n)} \boldsymbol{q}_j \right) \boldsymbol{q}_i^* \right\rangle_E \\ &= \frac{\omega}{2} \sum_{i,j} Im \left\{ \boldsymbol{D}_{\text{dir},ij}^{(n)} \left\langle \boldsymbol{q}_j \boldsymbol{q}_i^* \right\rangle_E \right\} \end{aligned} \tag{7.23}$$

When the stiffness matrix coordinates are a real based system the matrix is symmetric (Langley, 2007). In this case and only then, the hermitian cross spectral matrix can be factored out. This form is given in Shorter et al. (2005) under the assumption of symmetric stiffness matrices. In matrix notation we get

$$\left\langle \Pi_{\text{in}}^{(n)} \right\rangle_E = \frac{\omega}{2} \sum_{i,j} Im \left\{ \boldsymbol{D}_{\text{dir},ij} \left[\boldsymbol{S}_{qq} \right]_{ij} \right\} = \underbrace{\frac{\omega}{2} \sum_{i,j} Im \boldsymbol{D}_{\text{dir},ij} \left[\boldsymbol{S}_{qq} \right]_{ij}}_{\boldsymbol{D}_{\text{dir}} \text{symmetric}} \tag{7.24}$$

with the cross spectral matrix (7.22), and by compiling all the information from the above considerations, we get the coupling loss factor in the hybrid formulation. From the energy power relationship (6.97) and the radiated power

$$\begin{aligned} \Pi_{m \to n} &= \frac{\omega}{2} \sum_{i,j} Im \boldsymbol{D}_{\text{dir},ij}^{(n)} \left(\left[\boldsymbol{D}_{\text{tot}} \right]^{-1} \left(\frac{4E_m}{\pi \omega n_m} Im \left[\boldsymbol{D}_{\text{dir}}^{(m)} \right] \right) \left[\boldsymbol{D}_{\text{tot}} \right]^{-H} \right)_{ij} \\ &= \frac{2E_m}{\pi n_m} \sum_{i,j} Im \boldsymbol{D}_{\text{dir},ij}^{(n)} \left(\left[\boldsymbol{D}_{\text{tot}} \right]^{-1} Im \left[\boldsymbol{D}_{\text{dir}}^{(m)} \right] \left[\boldsymbol{D}_{\text{tot}} \right]^{-H} \right)_{ij} \end{aligned} \tag{7.25}$$

follows the coupling loss factor η_{mn}. To conclude, only the following three quantities are required to calculate the coupling loss factor:

1. The free field radiation stiffness of the connected subsystems.
2. The modal density of the source system.
3. The dynamic stiffness matrix of the deterministic component of the junction.

In case of direct connection of random systems, the last item is neglected. When we now assume that there are several subsystems m, we have to sum up the contributions from all $(N-1)$ inputs to system n and include the external power input from sources in and at the boundary of the system $\Pi_{\text{in,ext}}$:

$$\Pi_{\text{in}}^{(n)} = \Pi_{\text{in,ext}}^{(n)} + \sum_m \omega\eta_{mn} E_m \tag{7.26}$$

with

$$\eta_{mn} = \frac{2}{\pi n_m \omega} \sum_{i,j} Im \boldsymbol{D}_{\text{dir},ij}^{(n)} \left(\left[\boldsymbol{D}_{\text{tot}}\right]^{-1} Im\left[\boldsymbol{D}_{\text{dir}}^{(m)}\right] \left[\boldsymbol{D}_{\text{tot}}\right]^{-H} \right)_{ij} \tag{7.27}$$

7.3.2 Power Leaving the (m)th Subsystem

The power loss in this section must not be confused with the internal power losses due to dissipation in wave propagation or absorption at the (random) boundaries. In previous sections and especially in Equation (7.27), we dealt with the power input from all neighbor systems. In contrast to this we deal here with the losses due to random load excitation at the junction and the dissipation in the deterministic part of the junctions plus the radiation to the connected systems. Thus, we replace the radiation stiffness in (7.24) by the total stiffness

$$\begin{aligned} \Pi_{\text{out}}^{(m)} &= \frac{\omega}{2} \sum_{jk} Im\{[\boldsymbol{D}_{\text{tot}}]_{ij}\} \left[\boldsymbol{S}_{qq}^{(m)}\right]_{ij} \\ &= \frac{2E_m}{\pi n_m} \sum_{jk} Im\{[\boldsymbol{D}_{\text{tot}}]_{ij}\} \left(\left[\boldsymbol{D}_{\text{tot}}\right]^{-1} Im\left[\boldsymbol{D}_{\text{dir}}^{(m)}\right] \left[\boldsymbol{D}_{\text{tot}}\right]^{-H} \right)_{ij} \end{aligned} \tag{7.28}$$

Introducing the detailed expression of the imaginary part of the total stiffness matrix

$$Im\{[\boldsymbol{D}_{\text{tot}}]_{ij}\} = Im\{[\boldsymbol{D}]_{ij}\} + \sum_n Im\{\left[\boldsymbol{D}_{\text{dir}}^{(n)}\right]_{ij}\} \tag{7.29}$$

and entering it into (7.28) provides

$$\begin{aligned} \Pi_{\text{out}}^{(m)} &= \frac{2E_m}{\pi n_m} \sum_{jk} Im\{[\boldsymbol{D}]_{ij}\} \left(\left[\boldsymbol{D}_{\text{tot}}\right]^{-1} Im\left[\boldsymbol{D}_{\text{dir}}^{(m)}\right] \left[\boldsymbol{D}_{\text{tot}}\right]^{-H} \right)_{ij} \\ &+ \frac{2E_m}{\pi n_m} \sum_n \sum_{jk} Im\{\left[\boldsymbol{D}_{\text{dir}}^{(n)}\right]_{ij}\} \left(\left[\boldsymbol{D}_{\text{tot}}\right]^{-1} Im\left[\boldsymbol{D}_{\text{dir}}^{(m)}\right] \left[\boldsymbol{D}_{\text{tot}}\right]^{-H} \right)_{ij} \\ &= \omega E_m \left(\eta_{mm}^{\alpha} + \sum_n \eta_{mn} \right) \end{aligned} \tag{7.30}$$

The right hand side of the above equation is the sum of all power leaving the mth system including itself. This corresponds to the sum in the diagonal terms of the SEA matrix in Equation (6.113). Compared to (6.113) there is an additional special damping loss factor in (7.30):

$$\eta_{mm}^{\alpha} = \frac{2}{\pi\omega n_m(\omega)} \sum_{i,j} Im\{\boldsymbol{D}_{ij}\} \left(\left[\boldsymbol{D}_{\text{tot}}\right]^{-1} Im\left[\boldsymbol{D}_{\text{dir}}^{(m)}\right] \left[\boldsymbol{D}_{\text{tot}}\right]^{-H} \right)_{ij} \tag{7.31}$$

For a non-dispersive deterministic system the imaginary part of the dynamic stiffness matrix is zero and thus $\eta_{mm}^{\alpha} = 0$. This term is a difference to classical SEA because it considers damping in the junction or deterministic systems that is not given in classical SEA theory (Lyon and DeJong, 1995).

However, the dissipation due to field and (random) boundary absorption is still given by

$$\Pi_{\text{diss}}^{(m)} = \eta_{mm}\omega E_m$$

and we compile all expressions to[1]

$$\Pi_{\text{out}}^{(m)} = \omega\eta_m E_m = \omega\left(\eta_{mm}^{\alpha} + \sum_n \eta_{mn}\right) = \omega\left(\eta_{mm}^{\alpha} + \sum_n \eta_{mn}\right) \tag{7.32}$$

With the above equations we can calculate all required coefficients for the SEA-matrix (6.113).

7.3.2.1 Assembling the Hybrid SEA Matrix

In Figure 7.5 the power components of the mth SEA subsystem are depicted:

$\Pi_{\text{in,ext}}^{(m)}$ power input from external sources, e.g. forces, volume sources radiated into the system via the direct field.

$\Pi_{\text{in}}^{(m)}$ power input from surrounding SEA subsystems.

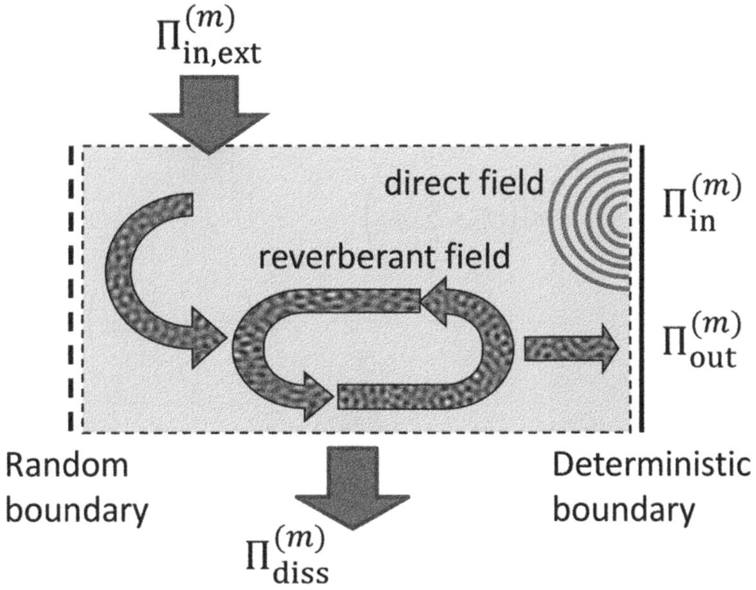

Figure 7.5 Power balance of random subsystem connected to deterministic boundary and system. *Source:* Alexander Peiffer.

1 The dynamics of an absorbing surface can be integrated via a dynamic stiffness matrix of this surface leading to the same surface absorption.

$\Pi_{\text{out}}^{(m)}$ power loss into the surrounding SEA subsystems and dissipation in the FEM subsystem due to interaction of the reverberant field with the junctions.

$\Pi_{\text{diss}}^{(m)}$ power loss due to field dissipation and absorption at random boundaries.

Setting up the power balance

$$\Pi_{\text{out}}^{(m)} + \Pi_{\text{diss}}^{(m)} = \Pi_{\text{in,dir}}^{(m)} + \Pi_{\text{in,ext}}^{(m)} \tag{7.33}$$

and assembling all contributions leads to the same form as in Equation (6.113)

$$\begin{bmatrix} n_1\left(\eta_{11}^{\alpha} + \sum\limits_{n=1}^{N} \eta_{1n}\right) & -n_2\eta_{21} & \cdots & -n_N\eta_{N1} \\ -n_1\eta_{12} & n_2\left(\eta_{22}^{\alpha} + \sum\limits_{n=1}^{N} \eta_{2n}\right) & & \vdots \\ \vdots & & \ddots & \\ -n_1\eta_{1N} & \cdots & & n_N\left(\eta_{NN}^{\alpha} + \sum\limits_{n=1}^{N} \eta_{Nn}\right) \end{bmatrix}$$

$$\ldots \omega \begin{Bmatrix} \frac{E_1}{n_1} \\ \frac{E_2}{n_2} \\ \vdots \\ \frac{E_N}{n_N} \end{Bmatrix} = \begin{Bmatrix} \Pi_{\text{in}}^{(1)} \\ \Pi_{\text{in}}^{(2)} \\ \vdots \\ \Pi_{\text{in}}^{(N)} \end{Bmatrix} \tag{7.34}$$

or

$$[L'] = \begin{bmatrix} n_1\left(\eta_{11}^{\alpha} + \sum\limits_{n=1}^{N} \eta_{1n}\right) & -n_2\eta_{21} & \cdots & -n_N\eta_{N1} \\ -n_1\eta_{12} & n_2\left(\eta_{22}^{\alpha} + \sum\limits_{n=1}^{N} \eta_{2n}\right) & & \vdots \\ \vdots & & \ddots & \\ -n_1\eta_{1N} & \cdots & & n_N\left(\eta_{NN}^{\alpha} + \sum\limits_{n=1}^{N} \eta_{Nn}\right) \end{bmatrix} \tag{7.35}$$

7.3.3 Some Remarks on SEA Modelling

Before we get into detailed SEA models it is time to make some remarks. In the preceding sections we discussed the diffuse wave field model to describe random systems. In this chapter we presented the numerical toolbox to determine the exchange of power between diffuse reverberant fields. So, we are totally aware of the practical simplicity but also the restrictions of the diffuse wave field model. Imagine a system with many random subsystems – each of them must fulfill these restrictions. Imagine now in addition that we model the power flow through a cascade of such random systems because we have to model a system of connected plates. In every step of power flow from one system to the other, we will make an error, so consequently when applying SEA to this configuration the system must be really complex.

So, how can this dilemma be overcome? First, try to avoid small subsystems and cascades of subsystems in major paths of power flow. In other words try to model these parts by deterministic methods.

Second, in many cases a subsystem does not only transport acoustic energy via reverberant fields, but very often there is a forced motion of the subsystem. In the modal approach this corresponds to excitation of modes out of their resonant frequency. In the SEA language this path of indirect transfer is called the indirect or non-resonant path. We will later use the hybrid junction description in modal coordinates to clarify this phenomena.

And last but not least, we can correct the reverberant field assumption by methods of geometrical acoustics. This is standard in room acoustics due to the historical application of ray tracing methods in concert hall design or the prediction of large acoustic environments such as work shops or sporting event areas.

7.4 Hybrid FEM/SEA Method

Until now we have seen deterministic and random approaches to describe subsystems. When we dealt with single systems, we derived the criteria for how to separate the FEM from the SEA world normally at one specific frequency determined by Helmholtz number, dimension, and modal overlap. Real engineering systems consist of many such subsystems, and this specific frequency is different for each subsystem. Thus, it may happen that there are random and deterministic subsystems in one global system. This dilemma is called the mid-frequency problem and motivated the creation of the hybrid FEM/SEA method that combines both approaches in one set-up. In the preceding sections and the chapters before we have dealt with all means that are required to set-up the hybrid method, especially by including FEM components into the calculation of coupling loss factors. The FEM part was considered as part of a junction and an illustrative derivation of the coupling loss factor.

However, we did not include the FEM part in the global system context and did not include excitation of FEM degrees of freedom. Therefore, the hybrid FEM/SEA theory is introduced here again in a more general way.

In Figure 7.6 a generic set-up of of potentially three subsystems is shown. When defining a model of this configuration for the full audible frequency range from $f \approx 20 - 20\,000$ Hz we need to cover three orders of magnitude for frequency or wavenumber.

At low frequencies (case 1) the full set-up is modelled by FEM denoted by a mesh presentation of all subsystems. With increasing frequency the first subsystem becomes random (case 2), and now we must couple the reduced FEM system to the new SEA subsystem. Further increase in frequency (case 3) makes an additional component switch to random behavior, and the middle system must be considered in the junction formulation as defined by Equation (7.27). For even higher frequencies a full SEA (case 4) configuration may be required involving only pure SEA junctions, meaning that the total stiffness of (7.27) contains only the radiation stiffness terms.

The consequence is that we must apply both modelling principles in one model, with full flexibility if we would like to avoid creating several different models for the full frequency range as shown, for example, in Peiffer et al. (2013) and in section 12.2.3.

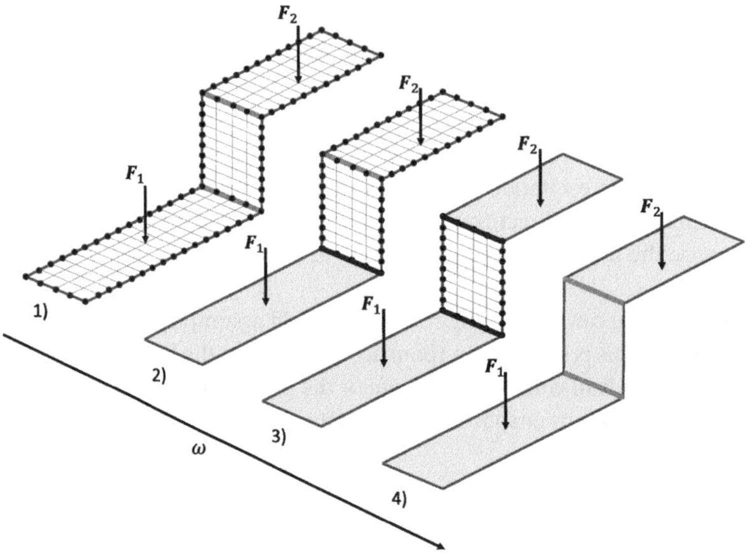

Figure 7.6 Subsystem configuration over frequency. *Source:* Alexander Peiffer.

However, Shorter and Langley (2005b) have formulated this principle. We have already derived the components of this theory to explain the ideas of hybrid modelling, and we will extend this by deterministic and random excitation of the FEM subsystems.

7.4.1 Combining SEA and FEM Subsystems

In Figure 7.7 a generic FEM subsystem that is connected to several SEA systems is shown. The dynamics of the FEM part is defined by a matrix equation as shown for example in (5.1) and given by $[\boldsymbol{D}_s]$. The SEA impact on the FEM subsystem is described by the dynamic radiation stiffness $\left[\boldsymbol{D}_{\text{SEA}}^{(m)}\right]$ of the surface degrees of freedom $\boldsymbol{q}_s$ that are

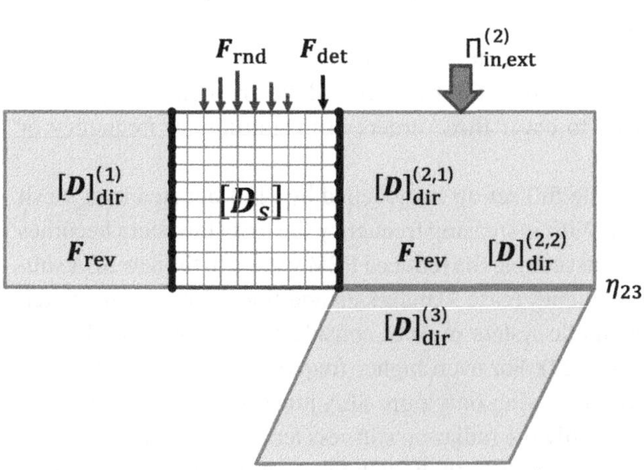

Figure 7.7 Separation of a hybrid FEM/SEA system into FEM and SEA parts. *Source:* Alexander Peiffer.

connected to FEM subsystem. We learned in section 7.3 that the total stiffness matrix is the matrix of all FEM subsystems plus the stiffness matrices of all connected SEA systems

$$[\boldsymbol{D}_s] + \sum_m \left[\boldsymbol{D}_{\text{SEA}}^{(m)}\right] \tag{7.36}$$

here still considered as a deterministic matrix.

There can be excitation by external forces (deterministic and random) and reverberant fields. The sum of all external forces

$$\{\boldsymbol{F}\} = \{\boldsymbol{F}\}_{\text{ext}} + \{\boldsymbol{F}\}_{\text{rev}} = \{\boldsymbol{F}\}_{\text{det}} + \{\boldsymbol{F}\}_{\text{rnd}} + \{\boldsymbol{F}\}_{\text{rev}} \tag{7.37}$$

is separated for practical reasons into the following components:

$\{\boldsymbol{F}\}_{\text{ext}} = \{\boldsymbol{F}\}_{\text{det}} + \{\boldsymbol{F}\}_{\text{rnd}}$	external forces.
$\{\boldsymbol{F}\}_{\text{det}}$	deterministic, coherent external forces.
$\{\boldsymbol{F}\}_{\text{rnd}}$	random force due to other random excitations, e.g. TBL, rain on the roof, jet noise.
$\{\boldsymbol{F}\}_{\text{rev}}$	random force due to the reverberant field of SEA systems.

The ensemble average of the two forces with random character is zero

$$\left\langle \{\boldsymbol{F}\}_{\text{rnd}} \right\rangle_E = \left\langle \{\boldsymbol{F}\}_{\text{rev}} \right\rangle_E = 0$$

Putting the above loads and the stiffness matrices together gives:

$$([\boldsymbol{D}_s] + [\boldsymbol{D}_{\text{SEA}}])\{\boldsymbol{q}\} = \{\boldsymbol{F}\} = \{\boldsymbol{F}_{ext}\} + \{\boldsymbol{F}_{rev}\} \tag{7.38}$$

This combination of matrices and excitation vectors must be reduced and separated into the FEM and the SEA part. We think of an ensemble of subsystems that is averaged to filter both components. In Figure 7.8 this filtering procedure is depicted. The mean value extracts the FEM components of the system; the expected value of the cross spectral density transforms the random systems into SEA components and keeps the deterministic system (for the coupling).

7.4.1.1 Ensemble Average of Linear State Variable

We start with the ensemble average of the equation of motion (7.38). On the left hand side the SEA matrix is replaced by the free field radiation stiffness because the ensemble average of random systems stiffness is equal to the free field radiation stiffness

$$\left\langle [\boldsymbol{D}_s] + \sum_m \left[\boldsymbol{D}_{\text{SEA}}^{(m)}\right] \right\rangle_E = [\boldsymbol{D}_s] + \sum_m \left[\boldsymbol{D}_{\text{dir}}^{(m)}\right] = [\boldsymbol{D}_{\text{tot}}] \tag{7.39}$$

and the ensemble average of the full equation reads, using the zero average of the random forces:

$$[\boldsymbol{D}_{\text{tot}}]\{\boldsymbol{u}\} = \{\boldsymbol{F}_{\text{det}}\} \tag{7.40}$$

Thus, for the deterministic calculation the only impact of the SEA systems is the contact impedance, given by the free field radiation stiffness as illustrated on the left hand side of Figure 7.9. This reponse will lead to a power input into the reverberant

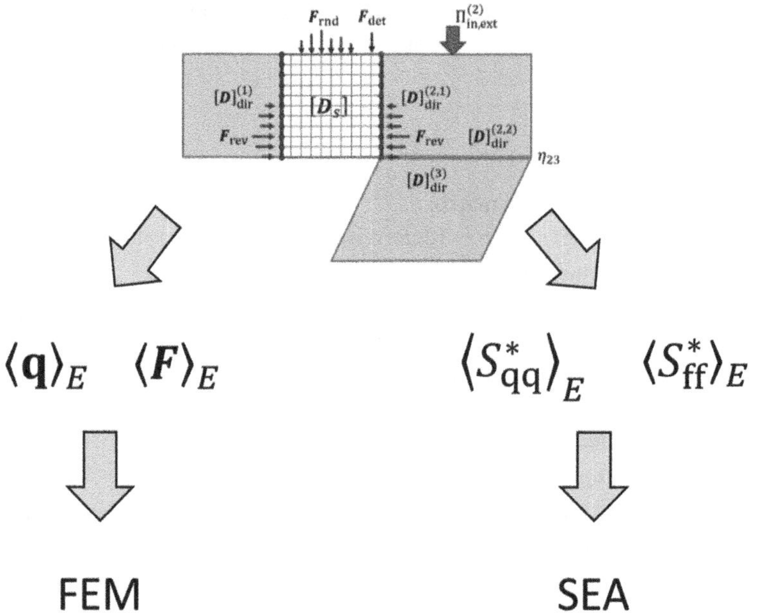

Figure 7.8 Application of ensemble averaging to a complex system. *Source:* Alexander Peiffer.

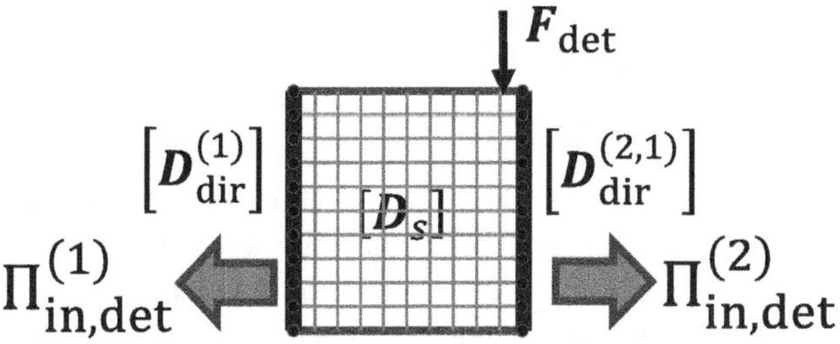

Figure 7.9 Remaining expressions after ensemble averaging of all systems.

Source: Alexander Peiffer.

fields of the SEA subsystems, but this is not considered in the deterministic response because of its non-coherent character and zero mean response.

7.4.1.2 Ensemble Average of Cross Spectral Density

The cross spectral density is targeting the random part. From section 1.7.2 we know that the response to random loads requires the application of cross spectral density. Starting from the inverted (7.38) and neglecting the deterministic part $\{\boldsymbol{F}_{\text{det}}\}$

$$\{\boldsymbol{q}\} = [\boldsymbol{D}_{\text{tot}}]^{-1} (\{\boldsymbol{F}_{rev}\} + \{\boldsymbol{F}_{rnd}\}) \tag{7.41}$$

and using the expected value expression for the cross spectral density 1.205 we get

$$[\boldsymbol{S}_{qq}] = \langle \boldsymbol{q}\boldsymbol{q}^H \rangle_E = [\boldsymbol{D}]_{\text{tot}}^{-1} [S_{ff}] [\boldsymbol{D}]_{\text{tot}}^{-H} \tag{7.42}$$

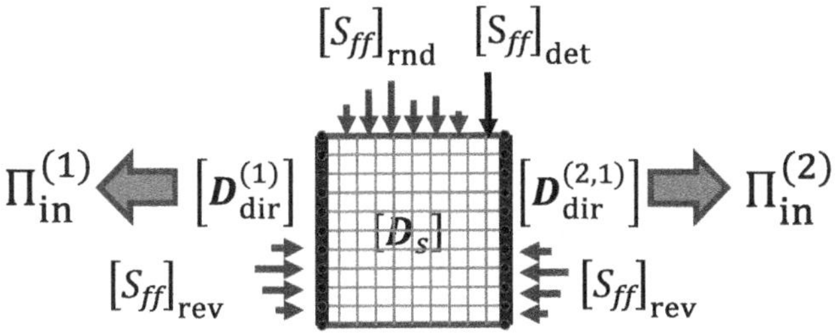

Figure 7.10 Remaining expressions after ensemble averaging of cross spectral density. *Source:* Alexander Peiffer.

with $[S_{ff}]$ being the cross spectral density of $\{F\}$. Separating the forces into the different components yields

$$[S_{qq}] = [D]_{\mathrm{tot}}^{-1}\left([S_{ff}]_{\mathrm{ext}} + [S_{ff}]_{\mathrm{rev}}\right)[D]_{\mathrm{tot}}^{-H}$$

where the reverberant force term can be replaced by the diffuse field reciprocity expression

$$= [D]_{\mathrm{tot}}^{-1}\left([S_{ff}]_{\mathrm{ext}} + \frac{4}{\pi\omega}\sum_m \frac{E_m}{n_m(\omega)} Im[D]_{\mathrm{dir}}^{(m)}\right)[D]_{\mathrm{tot}}^{-H} \tag{7.43}$$

Equation (7.43) allows the calculation of an FEM subsystem response connected to several SEA subsystems and excited by random loads. According to our nomenclature we have to keep in mind that each subsystem is connected to specific degrees of freedom when we apply (7.43).

For practical reasons we make some further definitions. Usually, the FEM system is connected to several SEA systems.

In the discussion of section 7.3, the derivation was applied to FEM systems that are in the junction context. Here, Equation (7.13) is applied to the full FEM subsystem meaning that every SEA subsystem adds a free field radiation stiffness to the remaining dynamic stiffness.

$$[D_{\mathrm{tot}}]_i = [D]_i + \sum_{m(i)} \left[D_{\mathrm{dir}}^{(m)}\right] \tag{7.44}$$

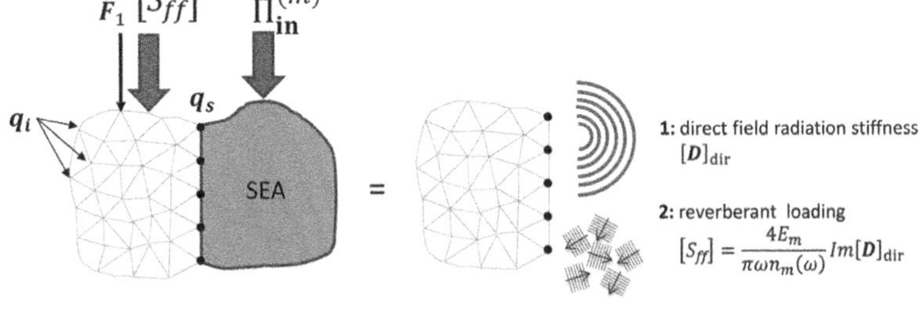

Figure 7.11 The two effects of a SEA system on a connected FEM subsystem. *Source:* Alexander Peiffer.

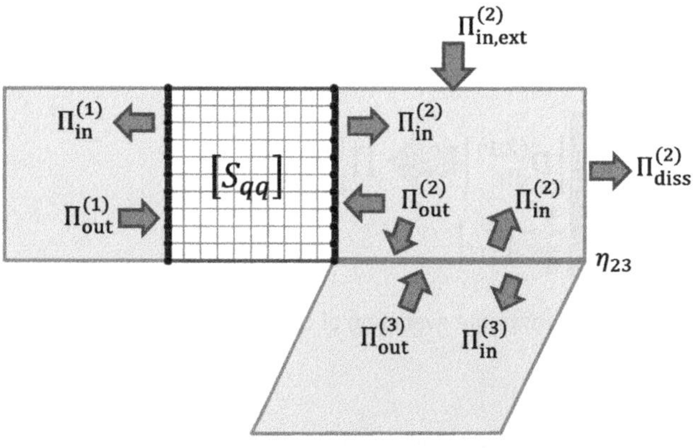

Figure 7.12 Remaining expressions for SEA power balance in the hybrid system configuration. *Source:* Alexander Peiffer.

Note that we must keep in mind the related degrees of freedom and the fact that the matrix $[\boldsymbol{D}]_i$ is different for every configuration (i) in Figure 7.6. For example case 2 $(i = 2)$ and case 3 $(i = 3)$ in Figure 7.6 require a different set $m(i)$ of SEA subsystems. The FEM part also depends on every configuration in any case (i).

7.4.1.3 Integration of FEM into the SEA Power Flow

This step was already dealt with in section 7.3 by using the deterministic formulation of the junction dynamics. Now, the (remaining) FEM system is given by all junctions and the deterministic systems. The SEA subsystems are modelled by the power flow Equation (7.34). The properties of the FEM subsystem are included in the total stiffness matrix $\boldsymbol{D}_{\text{tot}}$ that is used to determine the coupling loss factors.

7.4.2 Work Flow of Hybrid Simulation

The above derivation includes all steps of a workflow to calculate the dynamic response of deterministic and random subsystem over the full frequency range. Roughly speaking the workflow follows the above subsections 7.4.1.1 to 7.4.1.3. For reasons of simplicity the workflow is presented for one frequency domain with one subsystem configuration. That means we have already determined which subsystem is modelled

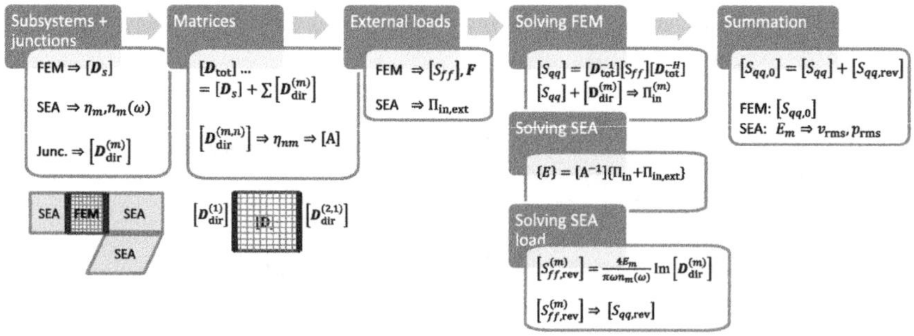

Figure 7.13 Workflow of hybrid simulation. *Source:* Alexander Peiffer.

as FEM and which one is considered as SEA. In later applications this decision has to be made for every frequency.

7.4.2.1 Setting up the System Configuration

The first step in hybrid modelling is the determination of the global configuration of FEM and SEA subsystems and the identification of junction areas or the specific degrees of freedom of the junction areas.

FEM1 Set up the combined discrete equations of motion of FEM subsystems according to Equation (7.12).

SEA1 Set up the properties of all SEA subsystems (modal density, damping, group wave speed) for random system description.

JUN1 Determine the related free field radiation stiffness of all junctions into the connected SEA subsystems.

7.4.2.2 Setting up the System Matrices and Coupling Loss Factors

In hybrid models there are two system matrices. Both need the total stiffness matrices and the dynamic stiffness matrix that leads to the combined total stiffness matrix. By the help of the total stiffness matrix, the coupling loss factors can be determined that finally provide the SEA matrix.

FEM2 Create the total stiffness matrix according to Equation (7.13).

SEA2 Create $[L']$ from the total stiffness and the (connected) radiation stiffnesses for each junction to determine the coupling loss factors according to Equations (7.27) and (7.31).

7.4.2.3 Apply External Loads

Usually external loads exist for all frequencies of our system. They must be considered in a different way depending on the fact of whether they excite a SEA system or a FEM subsystem.

FEM3.1 Define the generalized nodal forces for deterministic loads.

FEM3.2 Define the cross spectrum of generalized nodal forces for random loads.

SEA3 Determine the input power into the SEA system. This requires the free field radiation stiffness into the SEA subsystem at the point of excitation. Similar to the FEM excitation specific algorithms are required depending on the nature of the load.

7.4.2.4 Solving the Linear System of Equations

When the FEM systems are purely excited by deterministic loads, it is easier to use the classical equations of motion and solve for the deterministic response of $\{\boldsymbol{q}_s\}$. From this result the power radiated into the SEA subsystems is calculated with Equation (1.102) using generalized coordinates. However, when one external excitation is random, it makes sense to consider Equation (7.24) for both loads.

FEM4.1 Solve Equation (7.43) excluding the reverberant loads from the SEA subsystems.

FEM4.2 Determine the power from direct field radiation according to Equation (7.24).
SEA4.1 Solve the SEA matrix using the radiated power from FEM systems and power sources directly exciting the SEA systems.
SEA4.2 Solve the FEM equation using the reverberant load according to (7.19).

7.4.2.5 Combining Both Results

Finally, all results must be combined. Due to the fact that the reverberant fields from the SEA systems are not correlated to the results from the external excitation, the sum of both cross spectral matrices is taken.

FEM5 Add FEM results from external and reverberant loads.
SEA5 Calculate the physical units from energy results of SEA subsystems.

7.5 Hybrid Modelling in Modal Coordinates

Deterministic systems are characterized by the fact that few wavelengths fit into the system. One consequence of this is that there are only a reasonable number of normal modes required to calculate the dynamic response with modal methods. Consequently, we express the generalized displacement in modal coordinates by

$$\{\boldsymbol{q}\} = \sum_k q'_n \{\Phi_n\} \text{ or } \{\boldsymbol{q}\} = [\Phi]\{\boldsymbol{q}'\} \text{ with } [\Phi] = [\Phi_1 \Phi_2 \dots \Phi_N] \tag{7.45}$$

and the generalized forces by

$$[\Phi]^T \{\boldsymbol{F}\} = \{\boldsymbol{F}\}' \tag{7.46}$$

Entering this into the dynamic equation for the deterministic part of the hybrid theory (7.40) and multiplying from the right with $[\Phi]^H$, the result is

$$\begin{aligned} [\Phi]^H [\boldsymbol{D}_{\text{tot}}][\Phi]\{\boldsymbol{q}\}' &= \\ [\Phi]^H [\boldsymbol{D}_s][\Phi] + \sum_m [\Phi]^H \left[\boldsymbol{D}_{\text{dir}}^{(m)}\right][\Phi] &= [\Phi]^H \{\boldsymbol{F}_{\text{det}}\} \\ = [\boldsymbol{D}_s]' + \sum_m \left[\boldsymbol{D}_{\text{dir}}^{(m)}\right]' &= \{\boldsymbol{F}_{\text{det}}\}' \end{aligned} \tag{7.47}$$

$$[\boldsymbol{D}_{\text{tot}}]'\{\boldsymbol{q}\}' = \{\boldsymbol{F}_{\text{det}}\}' \tag{7.48}$$

Table 7.1 FEM and SEA system description.

FEM	SEA
$\left[\boldsymbol{D}_s^{(m)}\right]\left\{\boldsymbol{q}_s^{(m)}\right\} = \left\{\boldsymbol{F}_s^{(m)}\right\}$	$\omega[L]E_m = \Pi_{\text{in}}$

The first matrix is the diagonal dynamic stiffness matrix

$$[\boldsymbol{D}'_s] = [\Phi]^H [\boldsymbol{D}_s][\Phi] = \begin{bmatrix} \omega_1^2(1+j\eta)-\omega^2 & & & \\ & \ddots & & \\ & & \omega_n^2(1+j\eta)-\omega^2 & \\ & & & \omega_N^2(1+j\eta)-\omega^2 \end{bmatrix} \tag{7.49}$$

but the direct field radiation stiffness matrices in the sum are not. So, we keep a simplified condensed equation of motion for the FEM part, but the connection to the SEA systems described by the direct field radiation stiffness leads to fully populated but hermitian matrices. As a consequence the number of modes must be kept small to keep the advantage of the modal method.

The SEA part can be converted accordingly leading to exactly the same formulas but using matrices in modal space. Hence,

$$\eta_{mn} = \frac{2}{\pi n_m \omega} \sum_{i,j} Im\{\boldsymbol{D}'^{(n)}_{\text{dir},ij}\} \left([\boldsymbol{D}_{\text{tot}}]'^{-1} Im\left[\boldsymbol{D}^{(m)}_{\text{dir}}\right]' [\boldsymbol{D}_{\text{tot}}]'^{-H} \right)_{ij} \tag{7.50}$$

and

$$\eta^{\alpha}_{mm} = \frac{2}{\pi \omega n_m(\omega)} \sum_{i,j} Im\{[\boldsymbol{D}_{ij}]'\} \left([\boldsymbol{D}_{\text{tot}}]'^{-1} Im\left[\boldsymbol{D}^{(m)}_{\text{dir}}\right]' [\boldsymbol{D}_{\text{tot}}]'^{-H} \right)_{ij} \tag{7.51}$$

With (7.49) the second equation simplifies to

$$\eta^{\alpha}_{mm} = \frac{2}{\pi \omega n_m(\omega)} \sum_{i} \eta \omega_i^2 \left([\boldsymbol{D}_{\text{tot}}]'^{-1} Im\left[\boldsymbol{D}^{(m)}_{\text{dir}}\right]' [\boldsymbol{D}_{\text{tot}}]'^{-H} \right)_{ii} \tag{7.52}$$

The modal coordinates are one example for generalized coordinates.

Bibliography

Peter Davidsson. *Structure-Acoustic Analysis; Finite Element Modelling and Reduction Methods.* PhD thesis, Lund University, Lund, Sweden, August 2004.

R. S. Langley. On the diffuse field reciprocity relationship and vibrational energy variance in a random subsystem at high frequencies. *The Journal of the Acoustical Society of America*, 121(2):913–921, 2007.

R.H. Lyon and R.G. DeJong. *Theory and Application of Statistical Energy Analysis, Second Edition.* Butterworth Heinemann, second edition, 1995. ISBN 0-7506-9111-5.

Alexander Peiffer, Clemens Moeser, and Arno Röder. Transmission loss modelling of double wall structures using hybrid simulation. In *Fortschritte Der Akustik*, pages 1161–1162, Merano, March 2013.

P. J. Shorter and R. S. Langley. On the reciprocity relationship between direct field radiation and diffuse reverberant loading. *The Journal of the Acoustical Society of America*, 117(1):85–95, 2005a.

P. J. Shorter and R. S. Langley. Vibro-acoustic analysis of complex systems. *Journal of Sound and Vibration*, 288(3):669–699, 2005b. ISSN 0022-460X.

P. J. Shorter, Y. Gooroochurn, and B. Rodewald. Advanced vibro-acoustic models of welded junctions. In *Proceedings Internoise 2005*, Rio, Brasil, August 2005.

Bibliography

8

Coupling Loss Factors

The first step in SEA modelling is to find a realistic description of the subsystems by a diffuse wave field. Wavenumber, group wave speed, and geometry provide the modal density. From the physics of wave motion the relationship of the wave physical quantities (pressure, displacement) to physical units is derived. The damping properties are measured or estimated.

Once the properties of the subsystems are fixed, the next step is the derivation of the coupling loss factors using equation (7.27). So, why do we spend a full chapter with coupling loss factor derivation?

First, it is a complicated task that it is often neglected in textbooks. Starting from expression (7.27) there is a clear and straightforward rule, but we need a certain set of matrices and at least the radiation stiffness matrices for two coupled subsystems. Depending on the choice of degrees of freedom, the expressions for the radiation stiffness and the total stiffness can be very complex.

Second, the understanding of the coupling phenomena is very important for noise control in general. When we thoroughly understand how waves are transmitted through junctions, we have the key in our hand to reduce the amount of transmitted acoustic or vibrational power.

Third, the value of the coupling loss factor determines if it makes sense to perform the SEA approach for the chosen subsystem configuration. Lyon and DeJong (1995) quoted that for coupling loss factors higher than the damping loss factor of the systems, the assumption that the boundary is not impacting the diffuse field of the random systems is violated, and the SEA approach or the subsystems configuration must be questioned in general.

In the following sections several different versions of degrees of freedom will be applied. In some cases modal degrees of freedom are excellent; in other cases plane waves defined in wavenumber space are more useful. Some of the examples for coupling loss factors are useful for practical purposes, because the coupling loss factor and later the transmission coefficient are measures for the efficiency of acoustic isolation.

This book and Chapter 8 cannot provide the complete set of coupling loss factors, but some useful and important examples are shown in order to work out the principle procedures of the coupling loss factor derivation.

Vibroacoustic Simulation: An Introduction to Statistical Energy Analysis and Hybrid Methods,
First Edition. Alexander Peiffer.

8.1 Transmission Coefficients and Coupling Loss Factors

In the wave based approach of classical SEA literature, the derivation of coupling loss factors is based on the transmission of plane waves. Because of the diffuse wave field assumption, the wave transmission considers arbitrary angles of impact, depending on the geometry of the subsystems. It is quite instructive to use this wave transmission approach also here, especially when we use the projected wavenumber (that depends on the angle of incidence) as degree of freedom. Using wavenumbers as degrees of freedom provides a direct relationship between the coupling loss factor and the transmission coefficient.

The coupling loss factor relates energy and power

$$\eta_{12} = \frac{\Pi_2}{\omega E_1} \tag{8.1}$$

In contrast to this the transmission coefficient is defined as the ratio of transmitted and irradiated power (2.118).

$$\langle \tau \rangle = \frac{\Pi_{trans}}{\Pi_{in}}$$

Average of the transmission coefficient is needed, because it is usually defined for a specific angle. In order to establish a link between both quantities, we need the relationship between irradiated power and the energy in the diffuse field. This relationship depends on the dimension of the system or reverberant field. So, the link between transmission coefficient and coupling loss factor must be dealt with separately for each dimensionality.

8.1.1 τ-η Relationship from Diffuse Field Assumptions

We use the energy density and intensity relationships from section 6.1 that link the irradiated power per area and energy depending on the dimension of the diffuse field. The results are summarized in Table 6.1. The subsystem energy is related to the energy density e by each specific volume, independent from the fact of whether we have one-, two-, or three-dimensional systems:

$$E = Ve = A\,h\,e = L_x A_c\, e \tag{8.2}$$

This relationship is sufficient to establish the link between transmission coefficient and coupling loss factor when we assume a junction area S_j for all cases. This junction area can be expressed differently depending on the junction topology.

$$S_j = L_j h = A_c \tag{8.3}$$

With these definitions the irradiated power using (6.22) reads as

$$\Pi_{in} = S_j I_n = S_j \beta_D c_{\text{gr}} e = \frac{\beta_D c_{\text{gr}} E_1 S_j}{V} \tag{8.4}$$

leading to the transmitted power from $\Pi_{\text{trans}} = \tau \Pi_{\text{in}}$ according to (2.118)

$$\Pi_{trans} = \frac{\beta_D c_{\text{gr}} E_1 S_j \langle \tau \rangle_E}{V} \tag{8.5}$$

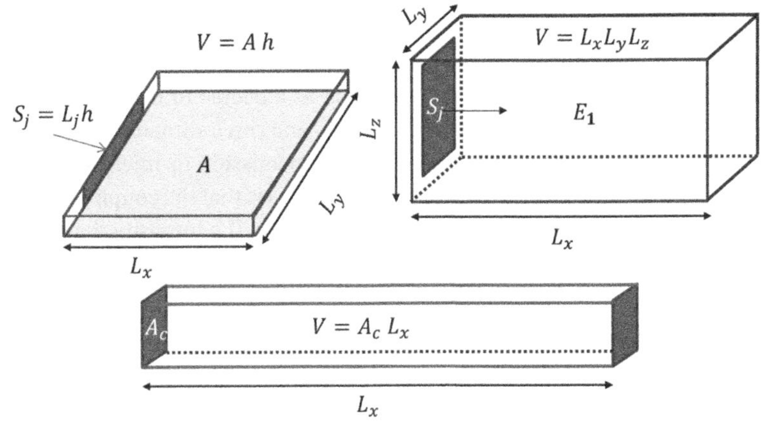

Figure 8.1 Relationships between junction area and system dimensions.
Source: Alexander Peiffer.

With (8.1) we get

$$\eta_{12} = \frac{\beta_D c_{\text{gr}} S_j \langle \tau_{12} \rangle_E}{V\omega} \tag{8.6}$$

With the constant β_D and the expression for subsystem volume V for each dimensionality, we get the relationship between τ_{12} and η_{12}.

$$\eta_{12} = \begin{cases} \frac{c_{\text{gr}}\langle \tau_{12} \rangle_E}{2L\omega} & \text{0D with} \quad V = A_c L \quad S_j = A_c \\ \frac{c_{\text{gr}} L_j \langle \tau_{12} \rangle_E}{\pi A\omega} & \text{1D with} \quad V = A\,h \quad S_j = L_c h \\ \frac{c_{\text{gr}} S_j \langle \tau_{12} \rangle_E}{4V\omega} & \text{2D} \end{cases} \tag{8.7}$$

By deriving the intensity–energy relationship, we have established a link between both quantities based on the assumptions of the diffuse wave field. When we express the ratio between the group velocity and the system relevant geometry quantity, the above expressions yields

$$\eta_{12} = \frac{\langle \tau_{12} \rangle_E}{2\pi n_{1D}(\omega)\omega} \qquad \langle \tau_{12} \rangle_E = 2\pi n_{1D}(\omega)\omega\eta_{12} \qquad \text{0D} \tag{8.8a}$$

$$\eta_{12} = \frac{kL_j \langle \tau_{12} \rangle_E}{2\pi^2 n_{2D}(\omega)\omega} \qquad \langle \tau_{12} \rangle_E = \frac{2\pi^2 n_{2D}(\omega)\omega}{kL_j}\eta_{12} \qquad \text{1D} \tag{8.8b}$$

$$\eta_{12} = \frac{k^2 S_j \langle \tau_{12} \rangle_E}{8\pi^2 n_{3D}(\omega)\omega} \qquad \langle \tau_{12} \rangle_E = \frac{8\pi^2 n_{3D}(\omega)\omega}{k^2 S_j}\eta_{12} \qquad \text{2D} \tag{8.8c}$$

Please note that the above expressions are based on the average transmission coefficients. The averaging procedure is generated by the assumption of wave field intensity of the diffuse field normal to the junction surface, and it is assumed that the transmission coefficient follows the same procedure. However, in order to develop a better understanding of the hybrid theory, we will describe the detailed relationship between the transmission coefficient depending on specific angle and also on degrees of freedom.

8.1.2 Angular Averaging

When the coupling loss factors are derived from plane wave transmission at the junctions, the incident angle of the plane wave is considered as a degree of freedom. It is useful to present the general averaging procedures for two- and three-dimensional subsystems and to derive the averaging procedures for the transmission or intensity and the coupling loss factor. The difference between both quantities is that the coupling loss factor is scalar whereas the transmission coefficient is related to the intensity that is a vector quantity and must be projected to the junction normal.

8.1.2.1 Two-Dimensional Systems

The semi infinite half plane is shown in Figure 6.3. The angular integration for the coupling loss factor is

$$\langle \eta_{12} \rangle_E = \frac{\int_0^{\pi} \eta_{12}(\varphi) d\varphi}{\int_0^{\pi} d\varphi} = \frac{1}{\pi} \int_0^{\pi} \eta_{12}(\varphi) d\varphi \tag{8.9}$$

For the transmission coefficient we must consider the normal intensity by the factor $\sin\varphi$:

$$\langle \tau \rangle_E = \frac{\int_0^{\pi} \tau(\varphi) \sin(\varphi) d\varphi}{\int_0^{\pi} \sin(\varphi) d\varphi} = \frac{1}{2} \int_0^{\pi} \tau(\varphi) \sin(\varphi) d\varphi \tag{8.10}$$

8.1.2.2 Three-Dimensional Systems

The half space integration sphere is shown in Figure 6.4. The coupling loss factor is averaged via integration over half the room angle divided by the room angle integral

$$\langle \eta_{12} \rangle_E = \frac{\int_0^{2\pi} \int_0^{\pi/2} \eta_{12}(\vartheta, \varphi) \sin\vartheta d\vartheta d\varphi}{\int_0^{2\pi} \int_0^{\pi/2} \sin\vartheta d\vartheta d\varphi} = \frac{1}{2\pi} \int_0^{2\pi} \int_0^{\pi/2} \eta_{12}(\vartheta, \varphi) \sin\vartheta d\vartheta d\varphi \tag{8.11}$$

In many cases the coupling loss factor is independent from φ, and the equation further simplifies:

$$\langle \eta_{12} \rangle_E = \int_0^{\pi/2} \eta_{12}(\vartheta) \sin\vartheta d\vartheta \tag{8.12}$$

According to the two-dimensional case the half sphere integration is modified by an additional $\sin\vartheta$ factor for the transmission coefficient

$$\begin{aligned}\langle \tau \rangle_E &= \frac{\int_0^{2\pi} \int_0^{\pi/2} \tau(\vartheta, \varphi) \cos(\vartheta) \sin(\vartheta) d\vartheta d\varphi}{\int_0^{2\pi} \int_0^{\pi/2} \cos(\vartheta) \sin(\vartheta) d\vartheta d\varphi} \\ &= \frac{1}{\pi} \int_0^{2\pi} \int_0^{\pi/2} \tau(\vartheta, \varphi) \cos(\vartheta) \sin(\vartheta) d\vartheta d\varphi \end{aligned} \tag{8.13}$$

With τ being independent from ϕ we can also further simplify.

$$\langle\tau\rangle_E = 2\int_0^{\pi/2} \tau(\vartheta)\cos(\vartheta)\sin(\vartheta)d\vartheta \tag{8.14}$$

8.2 Radiation Stiffness and Coupling Loss Factors

The values of these key properties for the coupling of random systems are shown in the next sections for point, line, and area junctions. The geometry is defined by the lowest dimension of all connected subsystems. For example, when a beam is connected to a plate, the coupling is defined by a point junction.

The coming sections adhere to the following two-step strategy for every topology:

- Determination of the direct field radiation stiffness.
- Calulation of the coupling loss factor based on the radiation stiffness.

8.2.1 Point Radiation Stiffness

When dealing with point connections, the stiffness matrices of equation (7.27) become a complex value if only one degree of freedom is concerned.

8.2.1.1 Bars and Tubes

Imagine two bars or tubes that are large and uncertain enough to carry a diffuse wave field. The free field stiffness of both systems involves longitudinal waves, and the stiffness is according to equations (2.39) and (3.84)

$$\boldsymbol{D}_i = j\omega A_z z_i \tag{8.15}$$

In both cases z_i is the characteristic impedance $z = \rho_0 c_0$, with c_0 given for fluids or $c_{LB} = \sqrt{\frac{E}{\rho_0}}$ for solid bars.

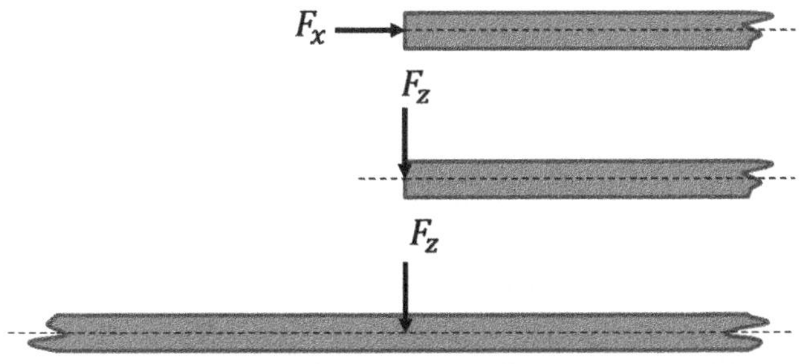

Figure 8.2 Different point stiffness of beam junctions. *Source:* Alexander Peiffer.

8.2.1.2 Beams

When bending waves are considered, the structure is called a beam. In section 3.5.4 with the equation for the half beam (3.132) we get:

$$\boldsymbol{D}_z = \frac{Bk_B^3}{2}(j-1) \tag{8.16}$$

or for the full beam excited in the middle

$$\boldsymbol{D}_z = 2Bk_B^3(j-1) \tag{8.17}$$

8.2.1.3 Plates

The point stiffness of plates is well determined by the treatments from section 3.7.6.2 and equation (3.226).

$$\boldsymbol{D}_{\text{plate}} = j\omega 8\sqrt{Bm''} \tag{8.18}$$

The point stiffness at the edges of plates is not derived here but given in Cremer et al. (2005).

$$\boldsymbol{D}_{\text{plate}} = j\omega 3.5\sqrt{Bm''} \tag{8.19}$$

8.2.1.4 Cavities

An acoustic pipe with cross section small compared to the wavelength connected to or radiating into a cavity requires the half space radiation stiffness of small pistons into cavities. An appropriate approximation for this is (2.154) with $S_j = \pi a^2$ thus $a = \sqrt{S_j/\pi}$, and finally multiplying with S_j we get:

$$\boldsymbol{F}_z = j\omega z_0 S_j^2 \left(\frac{1}{2}(ka)^2 + j\frac{8}{3\pi}ka\right)\boldsymbol{w}(\mathbf{r}_i) \tag{8.20}$$

or

$$\boldsymbol{D} = -\omega^2 \rho_0 S_j^2 \left(\frac{8\sqrt{S_j}}{3\,\pi^{3/2}} - j\frac{\omega S_j}{2c_0\pi}\right) \tag{8.21}$$

8.2.2 Point Junctions

Chapters 4 and 5 revealed that one-dimensional systems become random at relatively high frequencies. Thus, point junctions are not often used and applied to describe a full system by SEA means. Langley and Shorter (2003) gave a comprising overview of most required structural point junctions.

However, even when the point junctions are not frequently used in SEA models because of the lack of random one-dimensional systems, they are important in the context when engines or machines are connected to structures by bushes or mounts that isolate the structure from the source. These connections should be as small as possible in order to decouple best and can be considered as *point* junctions.

On the left hand side of Figure 8.3, the radiation stiffness of the first one-dimensional system is $\boldsymbol{D}^{(1)} = \boldsymbol{D}_1$, and the stiffness of the second system is $\boldsymbol{D}^{(2)} = \boldsymbol{D}_2$. We consider a direct connection, so dynamic stiffness of the junction is $\boldsymbol{D}_s = 0$. Thus, the total

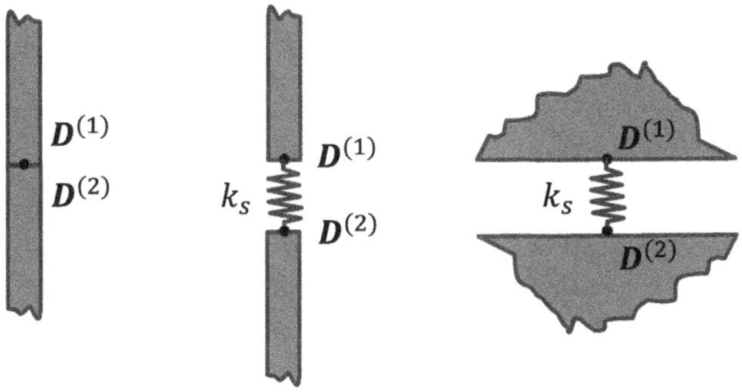

Figure 8.3 Point junctions with direct or indirect coupling. *Source:* Alexander Peiffer.

stiffness of the junction is $\boldsymbol{D}_{\text{tot}} = \boldsymbol{D}_1 + \boldsymbol{D}_2$. The inverse of the total stiffness is simply the reciprocal. With (7.27) the coupling loss factor of the one-dimensional connection is

$$\eta_{12} = \frac{2}{\pi n_1(\omega)\omega} \frac{Im\boldsymbol{D}_1 Im\boldsymbol{D}_2}{|\boldsymbol{D}_1 + \boldsymbol{D}_2|^2} \tag{8.22}$$

With equation (8.8a) the result for the transmission coefficient reads

$$\tau_{12} = \frac{4Im\boldsymbol{D}_1 Im\boldsymbol{D}_2}{|\boldsymbol{D}_1 + \boldsymbol{D}_2|^2} \tag{8.23}$$

If there is more than one degree of freedom per node a stiffness matrix must be used, meaning that we have to invert the matrices to get the coupling loss factor. This is also the case when points are connected via a spring of stiffness k_s. The stiffness matrix of system one, system two, and the spring are

$$\left[\boldsymbol{D}^{(1)}\right] = \begin{bmatrix} \boldsymbol{D}_1 & 0 \\ 0 & 0 \end{bmatrix} \qquad \left[\boldsymbol{D}^{(2)}\right] = \begin{bmatrix} 0 & 0 \\ 0 & \boldsymbol{D}_2 \end{bmatrix} \qquad [\boldsymbol{D}] = \begin{bmatrix} -k_s & k_s \\ k_s & -k_s \end{bmatrix} \tag{8.24}$$

and the total stiffness matrix and its inverse are

$$[\boldsymbol{D}_{\text{tot}}] = \begin{bmatrix} \boldsymbol{D}_1 - k_s & k_s \\ k_s & \boldsymbol{D}_2 - k_s \end{bmatrix} \tag{8.25}$$

$$[\boldsymbol{D}_{\text{tot}}]^{-1} = \frac{1}{\boldsymbol{D}_1\boldsymbol{D}_2 - (\boldsymbol{D}_1 + \boldsymbol{D}_2)k_s} \begin{bmatrix} \boldsymbol{D}_2 - k_s & -k_s \\ -k_s & \boldsymbol{D}_1 - k_s \end{bmatrix} \tag{8.26}$$

Entering this into equation (7.27), we get for the right parenthesis

$$\left([\boldsymbol{D}_{\text{tot}}]^{-1} Im\left[\boldsymbol{D}^{(1)}_{\text{dir}}\right][\boldsymbol{D}_{\text{tot}}]^{-H}\right) = \frac{Im\boldsymbol{D}_1}{|\boldsymbol{D}_1\boldsymbol{D}_2 - k_s(\boldsymbol{D}_1 + \boldsymbol{D}_2)|^2} \begin{bmatrix} |\boldsymbol{D}_2 - k_s|^2 & -k_s^*(\boldsymbol{D}_2 - k_s) \\ -k_s(\boldsymbol{D}_2 - k_s)^* & |k_s|^2 \end{bmatrix} \tag{8.27}$$

We sum over all matrix coefficients in (7.27), leaving only $i = j = 2$ because all other coefficients of $\left[\boldsymbol{D}^{(2)}\right]$ are zero, and get

$$\eta_{12} = \frac{2}{\pi n_1 \omega} \frac{Im\boldsymbol{D}_1 |k_s|^2 Im\boldsymbol{D}_2}{|\boldsymbol{D}_1\boldsymbol{D}_2 - k_s(\boldsymbol{D}_1 + \boldsymbol{D}_2)|^2} \tag{8.28}$$

or again as transmission coefficient

$$\tau_{12} = \frac{4ImD_1|k_s|^2 ImD_2}{|D_1D_2 - k_s(D_1 + D_2)|^2} \tag{8.29}$$

For large stiffness $k_s \gg 1$, (8.28) becomes (8.22). The use of this equation is the design of spring mounts for high frequency isolation of structureborne sound. The denominator contains the absolute value of the stiffnesses, independent from the fact of whether the stiffness is mass or stiffness controlled or resistive. So, a high absolute stiffness value allows for high isolation.

In the numerator we have the product of spring stiffness and both imaginary parts of the radiation stiffness. Here, a low spring stiffness and a low resistive or imaginary part leads to better decoupling and a low transmission coefficient.

So, when an isolating bush is aimed at structure borne noise reduction, stiff bush mounts and soft bushes are recommended for high isolation. When there is no imaginary part (radiation), this is even better.

8.2.2.1 Impedance Tube Example

To illustrate the effect of deterministic systems connected to the junction we start with a simple case: a mass point. For tubes this mass can be realized by a limb membrane that separates both fluids. The dynamic stiffness of a free mass is given $D_s = -\omega^2 M$. Thus, the total stiffness of the junction is given by

$$D_{\text{tot}} = D_1 + D_2 - \omega^2 M + = j\omega A_c z_1 + \omega A_c z_2 - \omega^2 A_c m'' \tag{8.30}$$

Using this in the above derivation, we get for the transmission coefficient

$$\tau_{12} = \frac{4z_1 z_2}{(z_1 + z_2)^2 + (\omega m'')^2} \tag{8.31}$$

Assuming only one fluid $z_0 = z_1 = z_2$, we get

$$\tau_{12} = \frac{4z_0^2}{4z_0^2 + \omega^2 m''^2} = \frac{1}{1 + (\omega m''/2z_0)^2} \tag{8.32}$$

For $\omega m'' \gg z_0$ this simplifies approximately to

$$\tau_{12} \approx \left(\frac{2z_0}{\omega m''}\right)^2 \tag{8.33}$$

The above relation is known as the oblique mass law of fluid waves.

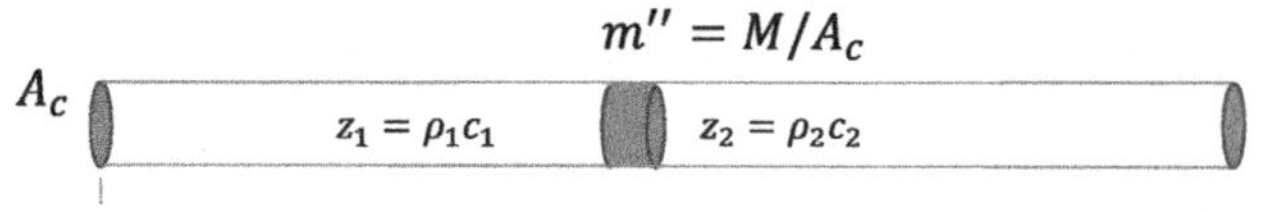

Figure 8.4 SEA pipe system connected by a mass connection. *Source:* Alexander Peiffer.

8.2.2.2 Connected Bars

As a first example we take the middle configuration from Figure 8.3. The radiation stiffnesses are purely imaginary according to (8.15).

$$\boldsymbol{D}_1 = j\omega A_1\sqrt{E_1\rho_1} \qquad \boldsymbol{D}_2 = j\omega A_2\sqrt{E_2\rho_2} \tag{8.34}$$

With (8.29) and assuming the same material for both bars we get

$$\tau = \frac{4\omega^2 A_1 A_2 E\rho k_s^2}{(\omega^2 A_1 A_2 E\rho)^2 + [\omega\sqrt{E\rho}(A_1 + A_2)k_s]^2} \tag{8.35}$$

As a numeric example we chose a spring stiffness corresponding to a bar of length $L = 1$ m with factor $\gamma = 0.01, 0.1, 1$, and chose the $A_1 = A_2 = A$ thus:

$$k_s = \gamma EA$$

In Figure 8.5 we see for a stiffness a tenth or hundredth of the bar stiffness leads to high transmission losses. When the stiffness is the same as the stiffness of both connected bars ($\gamma = 1$), the transmission is one except for high frequencies where the transmission becomes less than one. This results from the fact that the center beam is modelled as a massless spring and not as a bar.

8.2.2.3 Connected Beams

When beams are connected via a rectangular connection, two different wave types occur. In Figure 8.6 the configurations are shown. On the left hand we see a direct connection. When we focus on the diplacement in the x-direction, we must investigate the bending wave in system 1 and the longitudinal wave in system 2.

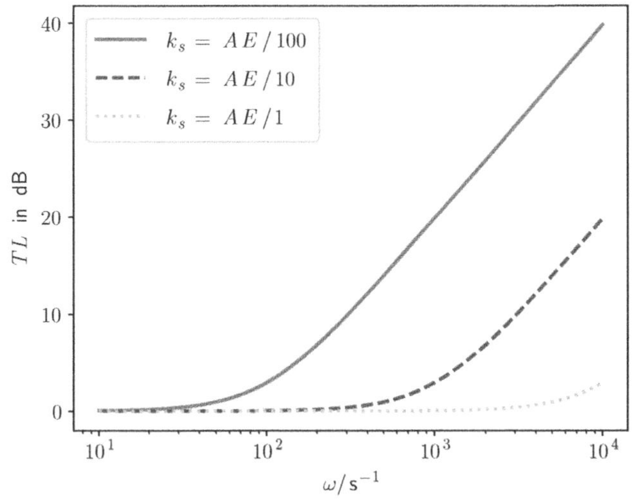

Figure 8.5 Transmission loss of different springs in the bar connection. *Source:* Alexander Peiffer.

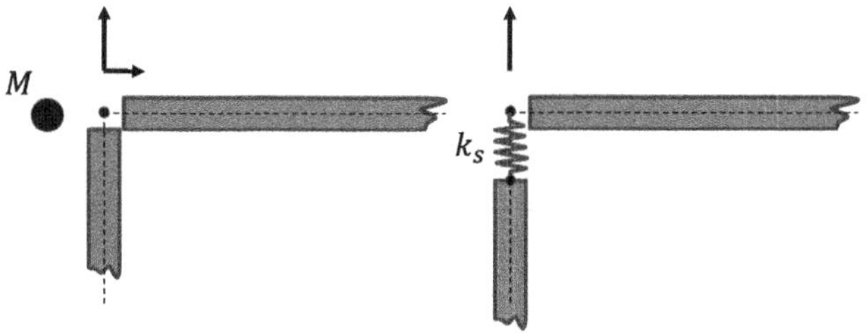

Figure 8.6 Two beams connected perpendicularly, with and without spring. *Source:* Alexander Peiffer.

This leads to the following point stiffnesses,

$$\boldsymbol{D}_1 = \frac{B_1 k_{B1}^2}{2}(j-1) \tag{8.36}$$

and

$$\boldsymbol{D}_2 = j\omega A_2 \sqrt{E_2 \rho_2} \tag{8.37}$$

With (8.23) we get after some math an expression that is independent from the frequency.

$$\tau = \frac{2\sqrt{\rho_1 A_1 B_1} A_2 \sqrt{E_2 \rho_2}}{\left(\frac{\sqrt{\rho_1 A_1 B_1}}{2} + A_2\sqrt{E_2\rho_2}\right)^2 + \left(\frac{\sqrt{\rho_1 A_1 B_1}}{2}\right)^2} \tag{8.38}$$

We consider an example for two similar rectangular aluminium beams with dimensions $L_x = 2$ mm, $L_y = 5$ mm. In order to determine the stiffness, we plot the imaginary part of D_1 and D_2 in Figure 8.7. The ratio of both stiffnesses is about 1000 so that the decoupling is supposed to be high even without extra means.

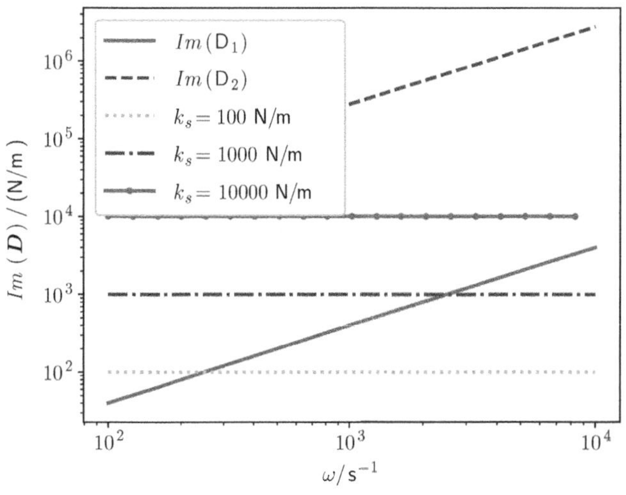

Figure 8.7 The radiation stiffness of bending and longitudinal waves in the beams $L_x = 2$ mm, $L_y = 10$ mm. *Source:* Alexander Peiffer.

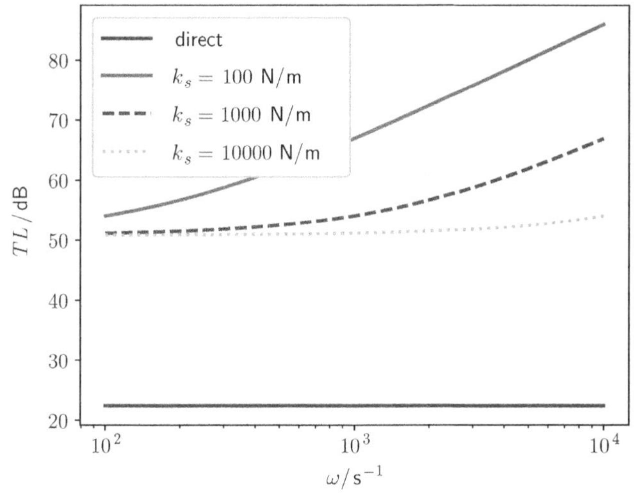

Figure 8.8 Transmission loss of a rectangular two-beam junction, without and with spring coupling. *Source:* Alexander Peiffer.

In addition we couple the beams via a spring as shown in Figure 8.6 on the right hand side. A decoupling effect occurs only when the spring stiffness is lower than the radiation stiffnesses. Thus, we chose the spring stiffness in a range from $k_s = 100$–10 000 N/m.

In Figure 8.8 the transmission loss for the directly and spring coupled situation is shown. The transmission coefficient is constant for the direct case and constanly higher for all springs. This comes from the fact that the spring is always lower than D_2. An additional effect can be seen when the spring is also lower than D_1.

A further option for decoupling are masses. Masses are useful if the connection is needed to be stable and stiff to withstand a certain load. For the consideration of masses the total stiffness must be modified by the additional *stiffness* similar to (8.30). For further arguments the dynamic stiffness of the masses and beams are shown in Figure 8.9.

In Figure 8.10 we see that masses are less efficient and work mainly at high frequencies because of the ω^2 scaling of the dynamic stiffness. The mass becomes efficient when the dynamic stiffness becomes larger than the larger of both radiation stiffnesses. However, keep in mind that the rotational inertia is neglected here, which leads to further decoupling in less academic cases.

8.2.2.4 Plates Connected via Springs

In technical acoustics one of the best noise control features are double leaves or walls. A second plate is mounted at a certain distance to the fist one, and absorptive material is placed in the middle (Beranek, 1988). The acoustics of such double leaf configurations will be dealt with in Chapter 9. In any case such plates must be mounted by mechanical connections that define the exact position of the second plate but without creating a structure borne path that reduces the effect of the double leaf. This is sometime done by soft mounts or springs, for example small elastomer bushes called shock mounts in aerospace applications. We consider two flat plates that are connected by springs

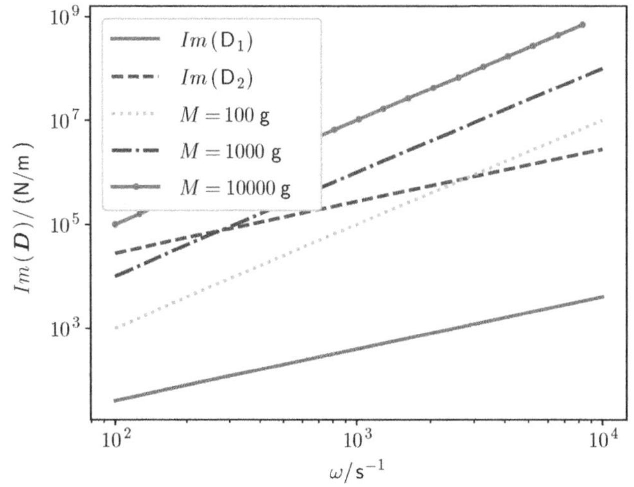

Figure 8.9 Radiation stiffness compared to the dynamic stiffness of the stop mass. *Source:* Alexander Peiffer.

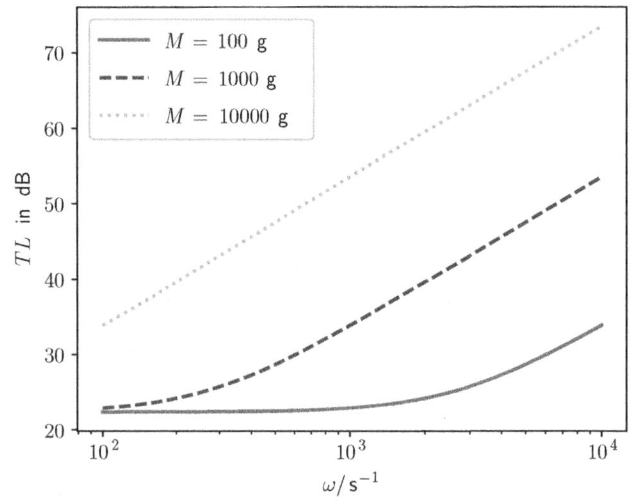

Figure 8.10 Transmission loss of beam junction with masses. *Source:* Alexander Peiffer.

as shown in Figure 8.11. We would need the stiffness for plate corners for the correct radiation stiffness. However, the nearest formula we have is (8.19).

The plates in Figure 8.11 are assumed to be an aluminium and a polystyrene plate of $h_1 = 2$ mm and $h_2 = 3$ mm thicknesses, respectively. In Figure 8.12 the dynamic stiffness of both edges is shown in relation to the stiffness of three different springs.

With this input and using equations (8.28) and (5.35), the coupling loss factor of one spring connection can by calculated as shown in Figure 8.13. The coupling loss factor is below 1% so the task of dynamically disconnecting the two plates is achieved for all springs.

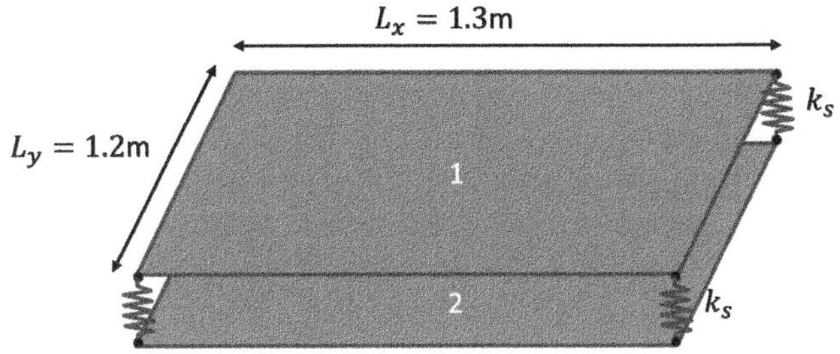

Figure 8.11 Plates connected via corners. *Source:* Alexander Peiffer.

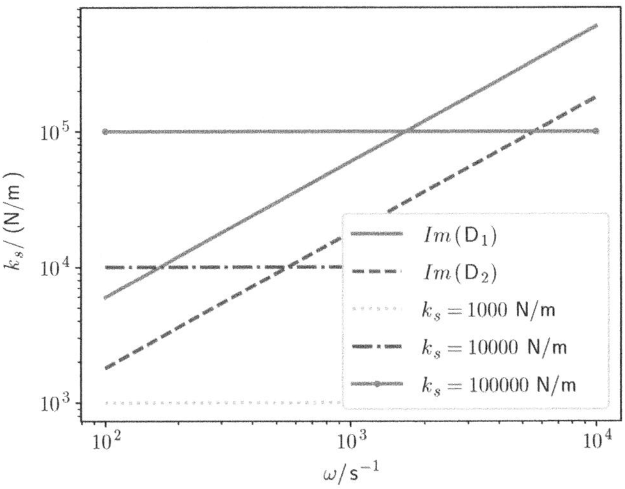

Figure 8.12 Plate stiffness related to spring stiffness. *Source:* Alexander Peiffer.

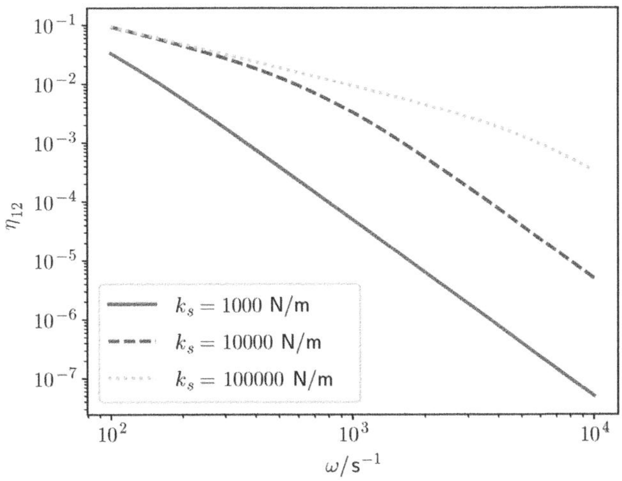

Figure 8.13 Coupling loss factor η_{12} of plates connected by springs. *Source:* Alexander Peiffer.

8.2.3 Area Radiation Stiffness

Plates that are connected to cavities constitute a two-dimensional junction that requires the determination of the area radiation stiffness. With the Rayleigh integral there is an expression available that defines the free field radiation stiffness of a cavity. This integral is used in two versions: in the space domain allowing discrete radiation stiffness approximation, and in the Fourier or wavenumber space representation.

A pressure field creates solely normal forces at a connected plate. Thus, the bending wave solutions of plates are appropriate for the area radiation stiffness of plates.

8.2.3.1 Cavities - Discrete Space Domain

The Rayleigh integral with surface receiver positions $z = 0$ reads as

$$\mathbf{p}(x,y,0) = \int_{-\infty}^{\infty} \frac{j\omega\rho_0}{2\pi l} e^{-jkl} \mathbf{v}_z(x_0,y_0) dx_0 dy_0$$
$$= \int_{-\infty}^{\infty} -\frac{\omega^2\rho_0}{2\pi l} e^{-jkl} \mathbf{w}(x_0,y_0) dx_0 dy_0 \text{ with } l = |\mathbf{r} - \mathbf{r}_0| \tag{8.39}$$

When we discretize the surface area into a regular mesh with nodal positions $\mathbf{r}_i = (x_0, y_0)$, $\mathbf{r}_j = (x, y)$ and area $\Delta S = \Delta x \Delta y$, the integral can be approximated by

$$\mathbf{p}(\mathbf{r}_j) = \sum_{i=1, i\neq j}^{N} -\frac{\omega^2\rho_0}{2\pi|\mathbf{r}_i - \mathbf{r}_j|} e^{-jk|\mathbf{r}_i - \mathbf{r}_j|} \mathbf{w}(\mathbf{r}_i)\Delta S \tag{8.40}$$

excluding the singularity for $i = j$ that can be overcome when using the approximation of the radiation impedance of a piston source in the wall (2.154) with $\Delta S = \pi R^2$ thus $R = \sqrt{\Delta S/\pi}$

$$\mathbf{p}(\mathbf{r}_j) = j\omega z_0 \Delta S_i \left(\frac{1}{2}(kR)^2 + j\frac{8}{3\pi}(kR)^2\right) \mathbf{w}(\mathbf{r}_j) \tag{8.41}$$

Multiplication with nodal element area ΔS_j, and writing the above sum in matrix form, we get the reaction forces of the fluid onto the structure

$$\left[\mathbf{D}_{ij}\right]_{\text{dir}} \{\mathbf{w}_i\} = \{\mathbf{p}(\mathbf{r}_j)\Delta S_j\} = \{\mathbf{F}_{zj}\} \tag{8.42}$$

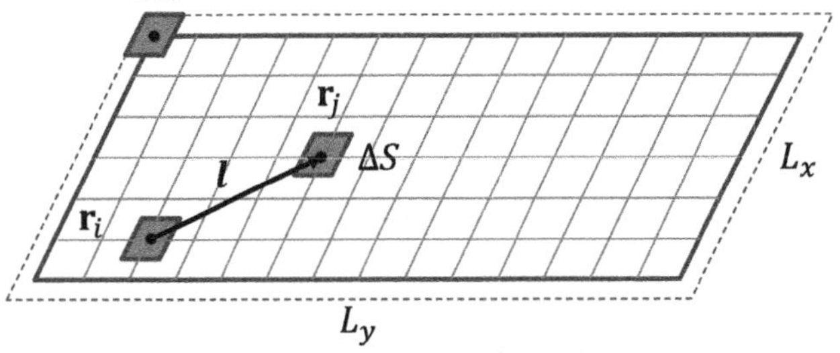

Figure 8.14 Regular mesh for sampling a surface of maximal extension L_x and L_y. This mesh is supposed to be regular for practical reasons. Note, the additional edge area due to the fact that every node is supposed to have a radiating area of ΔS. *Source:* Alexander Peiffer.

with

$$\boldsymbol{D}_{ij} = -\frac{\Delta S_i \Delta S_j \omega^2 \rho_0}{2\pi|\mathbf{r}_i - \mathbf{r}_j|} e^{-jk|\mathbf{r}_i-\mathbf{r}_j|} \qquad i \neq j$$

$$\boldsymbol{D}_{ii} = j\omega^2 \rho_0 \Delta S_i^2 \left(\frac{\omega}{2\pi c_0} + j\frac{8}{3\pi^2}\sqrt{\frac{\pi}{\Delta S_i}} \right)$$

This is the free field radiation stiffness of rigid semi infinite baffles. For regular meshes this means $\Delta S = \Delta S_i = \Delta S_j$. Langley (2007) developed a similar formula based on wavelets theory, being more precise for the mass load effect of the fluid.

8.2.3.2 Cavities – Wavenumber Domain

From the diffuse field theory the transmission coefficient is often derived by averaging over certain angles of incidence. This leads to a two-dimensional wavenumber as degree of freedom in equation (7.27) describing the surface pressure of the incoming plane wave.

Imagine a plane wave with wavenumber $\mathbf{k} = (k_x, k_y, k_z)$ radiating a surface at $z = 0$. Using spherical coordinates or angles of the incoming wave ϑ, ϕ leads to the Cartesian components of the wavenumber

$$k_x = -k \sin\vartheta \cos\phi \qquad k_y = -k \sin\vartheta \sin\phi \qquad k_z = -k \cos\vartheta \tag{8.43}$$

To derive the coupling loss factor or transmission coefficient the free field radiation stiffness of the fluid in half space is expressed by k_x and k_y in wavenumber domain. This was already derived from the two-dimensional Fourier transform of the Rayleigh integral (2.159), but the same expression can be developed from plane wave assumptions. We start with the radiation impedance and the reaction of a fluid to the surface velocity of infinite half space

$$\boldsymbol{v}(x, y) = \boldsymbol{v}_0 e^{-j(k_x x + k_y y)} \tag{8.44}$$

The wavenumber vector $\{k_x, k_y\}$ is called the trace wavenumber. We propose the same solution for the pressure in the fluid in the positive half space, so

$$\boldsymbol{p}(x, y, z) = \boldsymbol{p}_0 e^{-j(k_x x + k_y y + k z_z)} \tag{8.45}$$

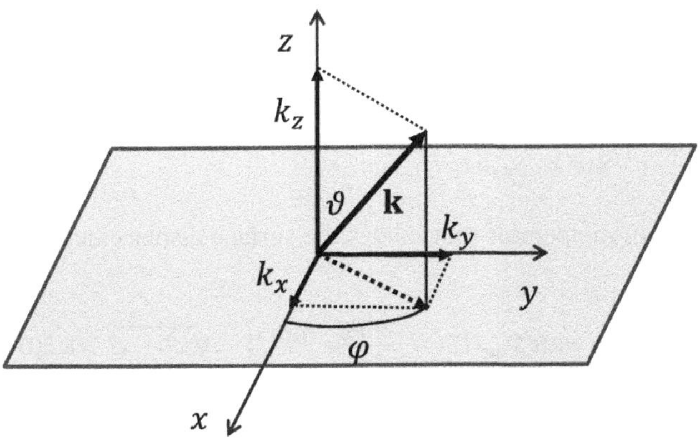

Figure 8.15 Wavenumber projection onto junction area. *Source:* Alexander Peiffer.

For fulfilling the Helmholtz equation, the wavenumber must satisfy

$$k_z^2 = k_a^2 - k_x^2 - k_y^2 \tag{8.46}$$

with (2.35) the velocity according to the pressure at $z = 0$ must be

$$\boldsymbol{v}_0 e^{-j(k_x x + k_y y)} = \frac{\boldsymbol{p}_0 k_z}{\omega \rho_0} e^{-j(k_x x + k_y y)} \tag{8.47}$$

providing finally

$$\boldsymbol{z}(k_x, k_y) = \frac{\omega \rho_0}{\sqrt{k_a^2 - k_x^2 - k_y^2}} \tag{8.48}$$

This is equal to the factor in the wavenumber space version of the Rayleigh integral (2.159). The radiation stiffness in wavenumber domain is then

$$\boldsymbol{D}''_{\text{dir}}(k_x, k_y) = -j \frac{\omega^2 \rho_0}{\sqrt{k_a^2 - k_x^2 - k_y^2}} \tag{8.49}$$

The double prime denotes a stiffness per area because the stiffness here is the ratio of pressure and displacement. This expression has three cases for the denominator with specific phyiscal meaning. For simplicity we assume $k_y = 0$ so $k = k_x$.

$k = k_a$ The stiffness becomes infinite. When the acoustic wavenumber coincides with the wavenumber of a radiating plate, this effect is called coincidence.

$k < k_a$ The radiation stiffness is purely imaginary. Thus, radiation into the half space is possible and efficient.

$k > k_a$ The radiation stiffness becomes real, $D_1 = -\frac{\omega \rho_0^2}{\sqrt{k^2 - k_a^2}}$. The semi infinite fluid generates a mass load on the radiating surface, because the fluid motion does not lead to radiation but to local movements near the surface.

Note that this results from the infinite half space. Finite areas will not lead to infinite expressions but cannot be presented in such an easy way.

8.2.3.3 Cavities – the Concept of Shape Stiffness

The effect of finite panel size can be dealt with by assuming a velocity profile like the modes of plates. Thus,

$$\begin{aligned} \boldsymbol{p}(x, y, 0) &= \int_{-\infty}^{\infty} \frac{j\omega \rho_0}{2\pi l} e^{-jkl} \boldsymbol{v}_z(x_0, y_0) dx_0 dy_0 \\ &= \int_{-\infty}^{\infty} -\frac{\omega^2 \rho_0}{2\pi l} e^{-jkl} \boldsymbol{w}(x_0, y_0) dx_0 dy_0 \end{aligned} \tag{8.39}$$

can be interpreted as a convolution operator that relates the surface displacement to the surface pressure.

$$\boldsymbol{p}(x, y) = \boldsymbol{D}_{\text{dir}}(x, y) \star \boldsymbol{w}(x, y) \text{ with } \boldsymbol{D}_{\text{dir}}(l) = -\frac{\omega^2 \rho_0}{2\pi l} e^{-jkl} \quad l = \sqrt{x^2 + y^2} \tag{8.50}$$

Each vibrating part of the surface excites a pressure field. Therefore, we need the convolution in contrast to the multiplication in the wavenumber space. The continuous convolution becomes a matrix multiplication in discrete coordinates.

For the calculation of radiated power a second surface integration is required as shown in section 2.7.4.

$$\begin{aligned}\Pi &= \frac{1}{2} Re\{\boldsymbol{v}_z^*(x,y)\boldsymbol{p}(x,y)\} \\ &= \frac{\omega}{2}\int_S Im\{\boldsymbol{w}^*(x,y)\int_S \mathbf{D}_{\text{dir}}[\sqrt{(x-x')^2+(y-y')^2}]\boldsymbol{w}(x,y)dS'\}dS \end{aligned} \tag{8.51}$$

This double integral is somehow unhandy, and we approximate the expression by discrete approximations and matrices. Using the stiffness matrix from equation (8.42) the above expression can be expressed in matrix form. Assuming a surface shape sampled at nodes $\mathbf{r}_i$ such as $\boldsymbol{w}_i = \boldsymbol{w}(\mathbf{r}_i)$, such that the vector $\{\boldsymbol{w}\}$ is a discrete approximation of the shape, we get with equation (8.42):

$$\{\boldsymbol{F}_z\} = [\boldsymbol{D}]_{\text{dir}}\{\boldsymbol{w}\} \tag{8.52}$$

leading to the following expression for the radiated power.

$$\Pi = \frac{\omega}{2} Im\left\{\{\boldsymbol{w}\}^H\{\boldsymbol{F}_z\}\right\} = \frac{\omega}{2} Im\{\{\boldsymbol{w}\}^H[\boldsymbol{D}]_{\text{dir}}\{\boldsymbol{w}\}\} \tag{8.53}$$

The term in the parentheses can be considered as the matrix transformation when we take a set of mode shapes for $\boldsymbol{w}$. By the use of mode shapes Φ_n, the radiation stiffness is converted into the modal space by coordinate transformation

$$[\boldsymbol{D}_{\text{dir}}]' = [\Phi]^H[\boldsymbol{D}_{\text{dir}}][\Phi] \text{ or } \boldsymbol{D}'_{\text{dir},mn} = \Phi_m^H[\boldsymbol{D}_{\text{dir}}]\Phi_n \tag{8.54}$$

The diagonal of the modal radiation stiffness matrix is equal to equation (8.53).

$$\{\Pi\} = \frac{\omega}{2} Im\left\{diag\left([\boldsymbol{D}_{\text{dir}}]'\right)\right\} \text{ or } \Pi_{nn} = \frac{\omega}{2} Im\{\boldsymbol{D}'_{\text{dir},nn}\} \tag{8.55}$$

Coming back to the definition of the radiation efficiency this quantity can be easily derived from the modal radiation stiffness. For the radiated power Π_{nn} and the related radiation stiffness

$$\sigma_{rad,nn} = \frac{\Pi_{nn}}{\Pi_0} = \frac{\Pi_{nn}}{\frac{S}{2}\rho_0 c_0 \omega^2 \langle \Phi_n^2 \rangle_S} \tag{8.56}$$

we get for regular meshes with nodal area ΔS the approximated expression

$$\sigma_{rad,nn} = \frac{Im\{\Phi_n^H[\boldsymbol{D}_{\text{dir}}]\Phi_n\}}{\Delta S \rho_0 c_0 \omega \Phi_n^H \Phi_n} = \frac{Im\boldsymbol{D}'_{\text{dir},nn}}{\Delta S z_0 \omega \Phi_n^H \Phi_n} \tag{8.57}$$

The modal stiffness matrix allows determining the radiation efficiency of mode shapes. On the other hand, the radiation efficiency can replace the imaginary part of the radiation stiffness to calculate the coupling loss factor.

$$Im\boldsymbol{D}'_{\text{dir},nn} = \Delta S z_0 \omega \Phi_n^H \Phi_n \sigma_{rad,nn} \tag{8.58}$$

8.2.3.4 Cavities – Radiation Efficiency of Finite, Rectangular Shapes

We are interested in the coupling loss factor of the plate diffuse wave field to the cavity. The infinite plate model leads to simple expressions regarding the fluid plate coupling, but further simplifications are required to consider the finite size of the panel and to avoid the singularity at coincidence.

Unfortunately, even for simple sinusoidal mode shapes, there is no closed form solution available. Leppington et al. (1982) derived an approximative expression for the radiation efficiency that will be outlined in the following. Simply supported rectangular plates have mode shapes of the form (5.33)

$$w_{n_x,n_y}(x,y) = \sin(k_x x)\sin(k_y x) \text{ with } k_x = \frac{n_x\pi}{L_x} \quad k_y = \frac{n_y\pi}{L_y} \tag{8.59}$$

The task is to find an expression for equation (8.51) using the above shape function. Leppington took the dimensions of a radiating rectangular plate with length L_x, L_y and introduced the following dimensionless wavenumbers

$$\alpha = \frac{k_x}{k_a} \qquad \beta = \frac{k_y}{k_a} \qquad \mu = \sqrt{\alpha^2 + \beta^2} \tag{8.60}$$

$\mu = 1$ represent the quarter circle of coincidence. The estimated radiation efficiency inside of the circle ($\mu < 1$) is the above coincidence $\omega > \omega_c$ regime with efficient radiation, and outside of the circle ($\mu > 1$) is for small acoustic wavenumber $k_a \gg 1$ below coincidence and low radiation. The dimensionless wavenumbers α and β span the first quadrant of a coordinate system as shown in Figure 8.16.

This extensive paper gives approximative but very precise solutions depending on the area in the wavenumber plane outside the unit cycle. Each specific area has detailed approximations and expression. When we keep in mind that we are mainly interested in the radiation efficiency of the diffuse plate wave field, those detailed equations are

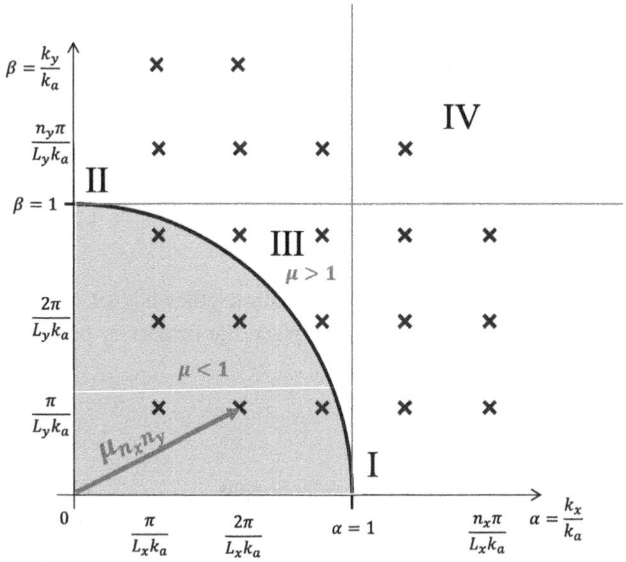

Figure 8.16 Regions of normalized structural wavenumbers of a rectangular plate. *Source:* Alexander Peiffer.

not required. We use Leppington's simplified versions for the following areas:

$$\sigma \approx (1-\mu^2)^{-\frac{1}{2}} \qquad \mu < 1 \tag{8.61}$$

$$\sigma \approx \frac{L_x + L_x}{\pi\mu k_a L_x L_y \sqrt{\mu^2-1}}\{\ln\left(\frac{\mu+1}{\mu-1}\right) + \frac{2\mu}{\mu^2-1}\} \qquad \mu > 1 \tag{8.62}$$

The approximation for $\mu \approx 1$ is a complicated expression that is presented here without further comments

$$\sigma \approx \mp\frac{1}{\sqrt{\pi|\epsilon|}} Im\left[e^{\mp j\pi/4}\left\{\left(2-\frac{j}{N_1}\right)F(N_1^{\frac{1}{2}}) + jF(N_1^{-\frac{1}{2}})e^{jN_1}\right\}\right.$$
$$\left. + Te^{\mp j\pi/4}\left\{\frac{3}{2}N_1^{-\frac{3}{2}}e^{jN_1} - \left(\frac{j}{N_1} - \frac{3}{2N_1^2}\right)F(N_1^{\frac{1}{2}})\right\}\right] \tag{8.63}$$

with

$$\epsilon = 2(\mu - 1) \tag{8.64}$$

and

$$N_1 = \frac{1}{2}(k_a L_x/\alpha)|\epsilon|, \quad T = \frac{L_x\beta}{L_y\alpha} \quad \text{if} \quad L_x\beta < L_y\alpha \tag{8.65}$$

$$N_2 = \frac{1}{2}(k_a L_y/\beta)|\epsilon|, \quad T = \frac{L_y\alpha}{L_x\beta} \quad \text{if} \quad L_x\beta > L_y\alpha \tag{8.66}$$

The upper and lower signs in formula (8.63) refer to $\epsilon > 0$ and $\epsilon < 0$. Equation (8.63) is analytical for $\mu = 1$

$$\sigma \approx \frac{2}{15}\sqrt{\frac{k_a L_x}{\pi\alpha}}\left(5 - \frac{L_x\beta}{L_y\alpha}\right) \quad \text{at } \mu = 1 \tag{8.67}$$

L_x, α must be interchanged with L_y, β in case of $L_x\beta < L_y\alpha$. For the transition between the three regimes, Leppington proposes to consider a ring of width

$$\epsilon_1 = \frac{\pi}{k_a\sqrt{L_x L_y}} \tag{8.68}$$

near the coincidence circle defined by $|\mu - 1| < \epsilon_1$.

We take a plate of dimensions $L_x = 0.8$ m and $L_y = 0.5$ m and investigate several modes as shown in Figure 8.17. The results of the Leppington approximation equations (8.61)–(8.63) are compared to the numerical approach using (8.57). Here, each shape is kept constant over frequency. We notice a good agreement between both methods except some oscillations for low frequencies even though it is not perfect. For precise results the more accurate formulas from Leppington et al. (1982) have to be used.

8.2.3.5 Plates – Wavenumber Domain

The radiation stiffness of infinite plates can be expressed in wavenumber coordinates, too. We start with equation (3.206), considering an excitation of the form

$$p(x, y) = \mathbf{p}(k_x, k_y)e^{-j(k_x x + k_y y)} \tag{8.69}$$

Here, only the bending wave is considered, because we focus on fluid–structure coupling. Only the out-of-plane motion of the bending leads to a velocity contribution

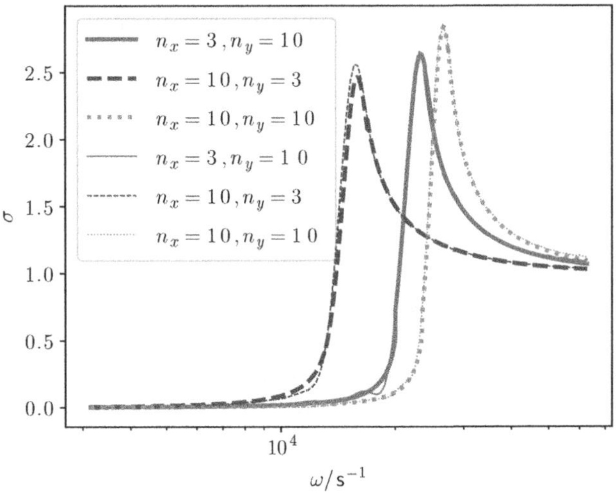

Figure 8.17 Radiation efficiency of plate of size $L_x = 0.8$ m and $L_y = 0.5$ m. Bold line: Leppington theory, thin line: Piston. *Source:* Alexander Peiffer.

perpendicular to the boundaries that couple the plate to the fluid. For infinite plates, and only then the displacement must have the same wavenumber,

$$w(x,y) = \boldsymbol{w}(k_x,k_y)e^{-j(k_x x + k_y y)} \tag{8.70}$$

Entering this into the bending wave equation of the plate (3.206) leads to:

$$B\left[\left(k_x x^2 + k_y y^2\right)^2 - k_B^4\right]\boldsymbol{w}(k_x,k_y) = \boldsymbol{p}(k_x,k_y) \tag{8.71}$$

leaving the radiation stiffness in the plate in the wavenumber domain as

$$\boldsymbol{D}''^{(plate)}_{\text{dir}}(k_x,k_y) = B\left[\left(k_x^2+k_y^2\right)^2 - k_B^4\right] = m''\omega^2\left[\frac{\left(k_x^2+k_y^2\right)^2}{k_B^4} - 1\right] \tag{8.72}$$

Without damping this expression has no imaginary component and becomes zero when the wavenumber equals the wavenumber of the bending wave, except for plates with dissipation. We introduce damping by the complex bending wavenumber from equation (3.234)

$$\boldsymbol{D}''^{(plate)}_{\text{dir}}(k_x,k_y) = m''\omega^2\left[\frac{\left(k_x^2+k_y^2\right)^2}{\boldsymbol{k}_B^4} - 1\right] = m''\omega^2\left[\frac{\left(k_x^2+k_y^2\right)^2}{k_B^4(1-j\eta)} - 1\right]$$

$$\approx m''\omega^2\left[\frac{\left(k_x^2+k_y^2\right)^2}{k_B^4}(1+j\eta) - 1\right] \tag{8.73}$$

The damping loss η of the plate determines the imaginary part of the radiation stiffness

$$Im\left[\boldsymbol{D}''^{(plate)}_{\text{dir}}(k_x,k_y)\right] =\approx m''\omega^2\eta\frac{\left(k_x^2+k_y^2\right)^2}{k_B^4} \tag{8.74}$$

8.2.3.6 Plates – Discrete Space Domain

The discrete approximation for plates is given by sampling the point force response (3.225) at mesh positions $\mathbf{r}_i = \{x_i, y_i\}^T$. With

$$\boldsymbol{w} = \frac{\boldsymbol{F}_{z0}}{8jBk^2}\left[H_0^{(2)}(k_B r) - H_0^{(2)}(-jk_B r)\right] \text{ with } r = |\mathbf{r}_i - \mathbf{r}_j|$$

and the point receptance of a plate (3.224) the matrix form is given by:

$$\{\boldsymbol{w}_i\} = \left[\boldsymbol{R}_{ij}\right]_{\text{dir}}\left\{\boldsymbol{F}_{zj}\right\} \tag{8.75}$$

with coefficients

$$\boldsymbol{R}_{ij} = \frac{1}{8jBk_B^2}\left[H_0^{(2)}(k_B r) - H_0^{(2)}(-jk_B r)\right] \qquad i \neq j$$

$$\boldsymbol{R}_{ii} = \frac{1}{8jBk_B^2}$$

The discrete stiffness matrix follows from the inverse of the receptance matrix

$$\left[\boldsymbol{D}_{\text{dir}}^{(plate)}\right] = \left[\boldsymbol{R}_{\text{dir}}\right]^{-1} \tag{8.76}$$

8.2.4 Area Junctions

From the previous sections we got two options for degrees of freedom to calculate the properties of area junctions from radiation stiffness: the wavenumber or the discrete space domain. The wavenumber version is useful for the understanding of principle properties of sound transmission and radiation. The discrete version is excellent for precise calculation of junction properties. However, area junctions are important because they describe the airborne noise transmission through and radiation from walls, plates or encapsulations.

8.2.4.1 Fluid–Fluid Coupling

We apply the hybrid coupling loss factor formalism to present a different approach for deriving the transmission coefficient at the interface between two fluids of Chapter 2.

When a plane wave approaches the two-dimensional interface plane, this defines the wavenumber coordinates for the radiation stiffness of the infinite surface. For $\phi = 0$ we get $k_x^2 + k_y^2 = k_{an}^2 \sin^2 \vartheta$ as

$$\boldsymbol{D}_{\text{dir}}^{\prime\prime(n)}(k, \vartheta) = -j\frac{\omega^2 \rho_0}{\sqrt{k_{an}^2 - k_{an}^2 \sin^2 \vartheta}} = -j\omega\frac{z_n}{\cos\vartheta} \tag{8.77}$$

with k_{an} for the acoustic wavenumber of the nth fluid. When we take fluid 1 with properties c_1, ρ_1 irradiating the interface to fluid 2 with properties c_2, ρ_2 the radiation stiffnesses are as follows

$$D''^{(1)}(\vartheta_1) = -j\frac{\omega^2\rho_1}{\sqrt{k_1^2 - k_1^2\sin^2\vartheta_1}} = -j\omega\frac{z_1}{\cos\vartheta_1} \tag{8.78}$$

$$D''^{(2)}(\vartheta_1) = -j\frac{\omega^2\rho_2}{\sqrt{k_2^2 - k_2^2\sin^2\vartheta_1}} = -j\omega\frac{z_2}{\cos\vartheta_2} \tag{8.79}$$

with $\cos\vartheta_2 = c_2/c_1\cos\vartheta_1$ from Snell's law (2.112). For infinite planes a single angle or surface wavenumber represents the complete area and is considered as a zero-dimensional (0D) junction. Thus, for the relationship between transmission and coupling loss factor the rule of a 0D junction (8.8a) applies

$$\tau(\vartheta_1) = \frac{4 Im D''^{(1)} Im D''^{(2)}}{|D''^{(1)} + D''^{(2)}|^2} = \frac{4\frac{z_1}{\cos\vartheta_1}\frac{z_2}{\cos\vartheta_2}}{\left(\frac{z_1}{\cos\vartheta_1} + \frac{z_2}{\cos\vartheta_2}\right)^2} \tag{8.80}$$

leading to the same result as in equation (2.120).

8.2.4.2 Diffuse Field Transmission from Wavenumber Space

In order to apply this method to the diffuse wave field and real two-dimensional junctions, we use the normal intensity and power flow averaged over the half space angle. Expressing the energy power relationship (8.1) and the transmission coefficient with the angle as argument gives

$$\eta_{12}(\vartheta) = \frac{\Pi_2}{\omega E_1} \qquad \Pi_{\text{trans}} = \tau_{12}(\vartheta)\Pi_{\text{in}} \tag{8.81}$$

Let us consider an area of S_j and a plane wave arriving under angle ϑ so that the projected power relates to the energy by

$$\Pi_{\text{in}} = I_{\text{in}}\cos(\vartheta)S_j \tag{8.82}$$

The energy density at the surface results from the incoming and reflected intensity

$$e_1 = \frac{I_{\text{in}} + I_{\text{refl}}}{c_{\text{gr}}} \tag{8.83}$$

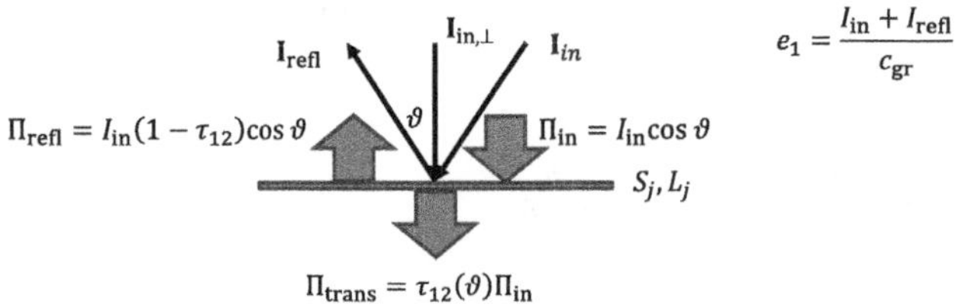

Figure 8.18 Conventions for reconstruction of incoming power and field energy of a wave field at boundaries. *Source:* Alexander Peiffer.

We know that the reflected wave intensity is $I_{\text{refl}} = I_{\text{in}}(1-\tau_{12})$ when dissipation in the junction is neglected, and with the volume we get the link between subsystem energy and incoming power using (8.82)

$$E_1(\vartheta) = Ve_1 = V\frac{I_{\text{in}}(2-\tau_{12})}{c_{\text{gr}}} = V\frac{\Pi_{\text{in}}(2-\tau_{12})}{\cos\vartheta S_j c_{\text{gr}}} \tag{8.84}$$

Using this in (8.81) provides the two-dimensional relationship between angular coupling loss factor and transmission coefficient. When we update this

$$\eta_{12}(\vartheta) = \frac{c_{\text{gr}}S_j}{\omega V}\frac{\tau_{12}(\vartheta)}{2-\tau_{12}(\vartheta)} \tag{8.85}$$

As we are interested in the diffuse field coupling loss factor, we apply the angular average (8.12) and get

$$\langle\eta_{12}\rangle_E = \frac{c_{\text{gr}}S_j}{\omega V}\int_0^{\pi/2}\frac{\tau_{12}(\vartheta)\cos\vartheta\sin\vartheta}{2-\tau_{12}(\vartheta)}d\vartheta \tag{8.86}$$

or by replacing $\frac{c_{\text{gr}}}{V}$ with the modal density expression for three-dimensional systems

$$\langle\eta_{12}\rangle_E = \frac{k_1^2 S_j}{4\pi^2 n_{3D}(\omega)\omega}2\int_0^{\pi/2}\frac{\tau_{12}(\vartheta)\cos\vartheta\sin\vartheta}{2-\tau_{12}(\vartheta)}d\vartheta$$

When we assume τ to be small ($\tau \ll 1$) so that the denominator is approximately 2 we get with (8.14)

$$\langle\eta_{12}\rangle_E \approx \frac{k_1^2 S_j}{8\pi^2 n_{3D}(\omega)\omega}\langle\tau_{12}\rangle_E \tag{8.87}$$

This equation is equal to (8.7)

$$\langle\eta_{12}\rangle_E = \frac{k^2 S_j\langle\tau_{12}\rangle_E}{8\pi^2 n_{3D}(\omega)\omega} = \frac{c_{\text{gr}}S_j}{4\omega V}\langle\tau_{12}\rangle_E$$

8.2.4.3 Plate-Fluid Connection

When the plate is modelled as a random system, it is represented by a diffuse wave field of plane bending waves with wavenumber k_B propagating in the angle range of 2π. When k_B equals k_a the coincidence will lead to singularities in the radiation stiffness of the semi infinite fluid. In addition, the radiation stiffness of the plate in wavenumber domain (8.73)

$$D_{\text{dir}}^{''(plate)}(k_x,k_y) = m''\omega^2\left(\frac{\left(k_x^2+k_y^2\right)^2}{k_B^4}(1+j\eta)-1\right) \tag{8.73}$$

is zero in absence of damping. Thus, the wavenumber space is not appropriate for deriving the coupling loss factor between plates and cavities. A different approach is required using the radiation efficiency for finite plates and radiation area.

Independent from the numerical consequences, the coincidence frequency is an important parameter in noise control, because it is a weak spot for the plate isolation performance. The design goal of an acoustic engineer is to keep the coincidence frequency as high as possible, meaning that the bending stiffness has to be low.[1] From $k_B = k_a$ we get the following expressions for the coincidence frequency:

$$\omega_c = c_0^2 \sqrt{\frac{m''}{B}} \qquad f_c = \frac{c_0^2}{2\pi} \sqrt{\frac{m''}{B}} \tag{8.88}$$

8.2.4.4 Approximation from Radiation Efficiency

The radiation efficiency relates average vibration and radiated power. Thus, it is an ideal quantity to determine the coupling loss factor – the vibration is linked to the plate energy, and the radiation defines the power flow into the cavity. Equation (2.165) relates the average velocity level of the diffuse acoustic field with the radiated power into the semi infinite half space.

$$\Pi = S\sigma_{rad}\rho_0 c_0 \left\langle v_{z,rms}^2 \right\rangle_S \tag{8.89}$$

When we compare this to the definition of the coupling loss factor

$$\Pi_{12} = \omega\eta_{12}E_1 = \omega\eta_{12}Am''v_{\text{rms}}^2 \tag{8.90}$$

the coupling loss factor follows directly from this:

$$\eta_{12} = \frac{\rho_0 c_0 A v_{rms}^2 \sigma}{\omega m'' A v_{rms}^2} = \frac{\rho_0 c_0 \sigma}{\omega m''} \tag{8.91}$$

The reverberant field in the plate is constituted by plane waves, so the wavenumber vector is $\{k_x, k_y\}^T = k(\omega)\{\cos\varphi, \sin\varphi\}^T$ with $\varphi \in [0, 2\pi)$. For homogenous flat plates of isotropic material there is only one specific wavenumber independent from the angle $k(\varphi) = k$. The Leppingtion approximation from section 8.2.3.4 provides the radiation efficiency for one pair for k_x and k_y in the first quadrant of wavenumbers and the frequency ω. As the mode shape function used there consists of waves travelling in both directions, one may argue that the radiation efficiency is symmetric to the x and y-axes, and we deal only with angles $\varphi \in [0; \pi/2)$. In the shape radiation formulation of Leppington, k_x, k_y and ω are independent variables.

In Figure 8.19 the modal wavenumber pattern is shown, denoted by the crosses. Each node corresponds to a modal frequency of the plate. The quarter circle represents the wavenumber of the reverberant field of flat plates that does not depend on the angle. Hence, the bending wave follows from frequency $k = k_B(\omega)$ so $\sigma(k_x, k_y, \omega) = \sigma(k_B(\omega), \varphi, \omega)$.

Even though the theory is derived for a discrete pair of k_x and k_y, it works for continuous values. Note that especially for rectangular plates ($L_x \neq L_y$), the radiation efficiency at constant absolute value of wavenumber depends on the angle in the Leppington

1 It is especially this fact that makes noise engineers unpopular in typical engineering teams, because this requirement is very often in contradiction to all other design targets of structural design.

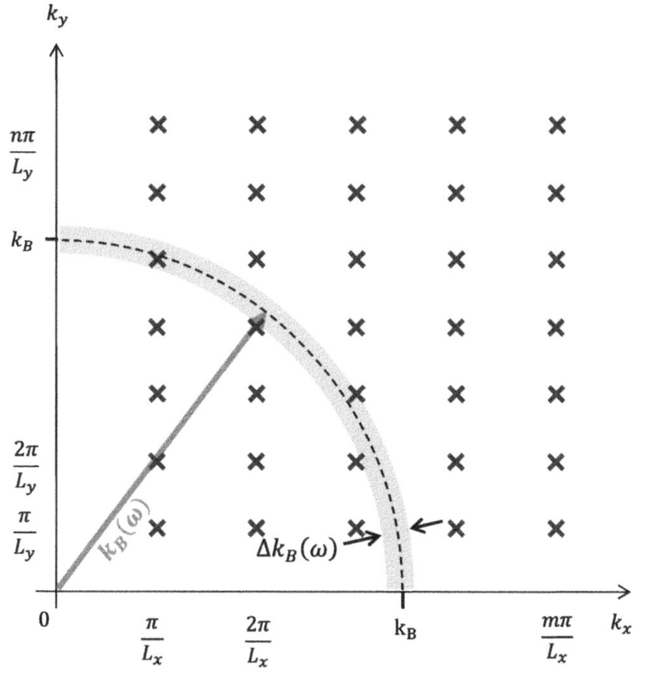

Figure 8.19 Averaging the radiation efficiency of one plate wavenumber using the discrete Leppington model.
Source: Alexander Peiffer.

theory. Thus, we must average the values over the quarter angle.

$$\langle\sigma(\omega)\rangle_S = \frac{\int_0^{\pi/2} \Delta k_B(\omega,\varphi)\sigma(k_B(\omega,\varphi),\varphi,\omega)d\varphi}{\int_0^{\pi/2} \Delta k_B(\omega,\varphi)d\varphi} \stackrel{k_B(\varphi)=k_B}{=} \frac{\int_0^{\pi/2} \sigma(k_B,\varphi,\omega)d\varphi}{\pi/2} \tag{8.92}$$

With this formulation the wavenumber singularity of (8.73) can be overcome for plates of rectangular shapes. In Figure 8.20 the result is shown for an aluminium plate of thickness $h = 4$ mm.

8.2.4.5 Fluid–Plate–Fluid Connection

We ignore the issues of the wavenumber domain of this junction. The plate is considered as a deterministic system in the middle. For simplicity we choose both fluids to be similar; we assume also that the plate is isotropic, so we choose $k_x^2 + k_y^2 = k_a^2 \sin^2\vartheta$ and $k_y = 0$, leading to equation(8.77) for the fluid radiation stiffness:

$$\boldsymbol{D}''_{\text{dir},1/2}(k_a,\vartheta) = -j\omega\frac{\rho_0 c_0}{\cos\vartheta} \tag{8.93}$$

and the stiffness of an infinite plate in wavenumber domain (8.72). This gives the total radiation stiffness:

$$D''_{\text{tot}}(k_a,\vartheta) = m''\omega^2\left[\frac{k_a^4\sin^4\vartheta}{k_B^4} - 1\right] - j\frac{2\rho_0 c_0}{\cos\vartheta} \tag{8.94}$$

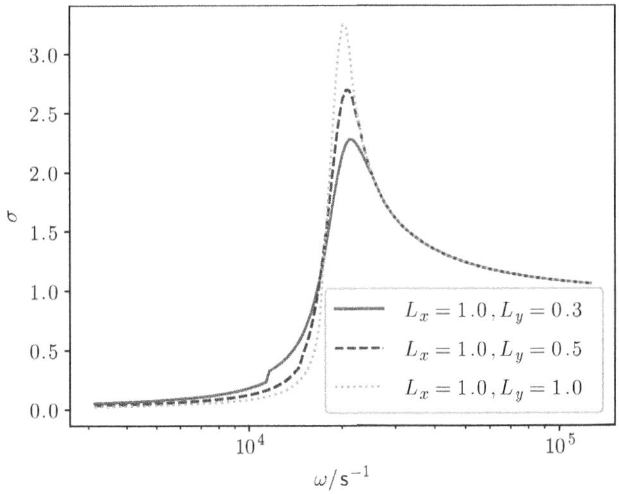

Figure 8.20 Radiation efficiency for aluminium plate of $h = 4$ mm, $L_x = 1$ m and $L_y = 0.7$ m from angular averaging. *Source:* Alexander Peiffer.

With (8.8a) the transmission coefficient reads

$$\tau = \frac{\frac{4\omega^2(\rho_0 c_0)^2}{\cos^2\vartheta}}{\left| m''\omega^2\left(\frac{k_a^4 \sin^4\vartheta}{k_B^4} - 1\right) + j\omega\frac{2\rho_0 c_0}{\cos\vartheta}\right|^2}$$

$$= \frac{\frac{4\omega^2(\rho_0 c_0)^2}{\cos^2\vartheta}}{m''^2\omega^4\left(\frac{k_a^4 \sin^4\vartheta}{k_B^4} - 1\right)^2 + \frac{4\omega^2(\rho_0 c_0)^2}{\cos^2\vartheta}}$$

$$= \left[\left(\frac{m''\omega}{2\rho_0 c_0}\right)^2 \left(\frac{k_a^4 \sin^4\vartheta}{k_B^4} - 1\right)^2 \cos^2\vartheta + 1\right]^{-1} \tag{8.95}$$

This formula was, for example, also derived by Cremer Cremer et al. (2005) based on the solution of plane waves. We see that the coincidence occurs in the formulation – when the acoustic trace wavenumber $k_a \sin\vartheta$ equals the bending wavelength k_b, the transmission coefficient equals unity $\tau = 1$, and the acoustic isolation is zero for infinite plates. However, when the frequency is far below coincidence, thus $k_a << k_B$, the bending term can be neglected, and we get:

$$\tau(\vartheta) = \left[1 + \left(\frac{m''\omega}{2\rho_0 c_0}\right)^2 \cos^2\vartheta\right]^{-1} \tag{8.96}$$

or

$$TL(\vartheta) = 10\log\tau(\vartheta)^{-1} = 10\log\left[1 + \left(\frac{m''\omega}{2\rho_0 c_0}\right)^2 \cos^2\vartheta\right] \tag{8.97}$$

This equation is called the angular dependent mass-law of airborne sound transmission. This is the upper limit of single plate transmission loss. No single leaf can provide better isolation as given by the mass law if not fixed at the edges. For perpendicular wave irradiation $\vartheta = 0$ this reads

$$\tau(0) = \left[1 + \left(\frac{m''\omega}{2\rho_0 c_0}\right)^2\right]^{-1} \tag{8.98}$$

or

$$TL(0) = R_0 = 10\log\left[1 + \left(\frac{m''\omega}{2\rho_0 c_0}\right)^2\right] \tag{8.99}$$

For the diffuse sound field transmission coefficient we use (8.14) and numerical integration. In Figure 8.21 the results for specific angles and the diffuser field average are shown.

In order to provide first results for plate transmission and to verify the validity of the mass law, we apply the discrete version of the area junction using the common form of the coupling loss factor (7.27) together with the discrete area radiation stiffness of the cavities (8.42) and the radiation stiffness of the infinite flat plate (8.76). The transmission coefficient reads with equation (8.8c)

$$\langle\tau\rangle_E = \frac{16\pi}{k_1^2 S_j}\sum_{i,j} Im\boldsymbol{D}^{(n)}_{\text{dir},ij}\left(\left[\boldsymbol{D}_{\text{tot}}\right]^{-1} Im\left[\boldsymbol{D}^{(m)}_{\text{dir}}\right]\left[\boldsymbol{D}_{\text{tot}}\right]^{-H}\right)_{ij} \tag{8.100}$$

Using the radiation stiffness (8.76) means that, though we used a finite mesh, the system representing this mesh is still infinite.

In case of limb plates, we are in the lucky situation that there is a simple discrete model available. For frequencies far below the coincidence frequency, the dynamic stiffness matrix of the plate can be considered as a pure mass matrix, and with a regular

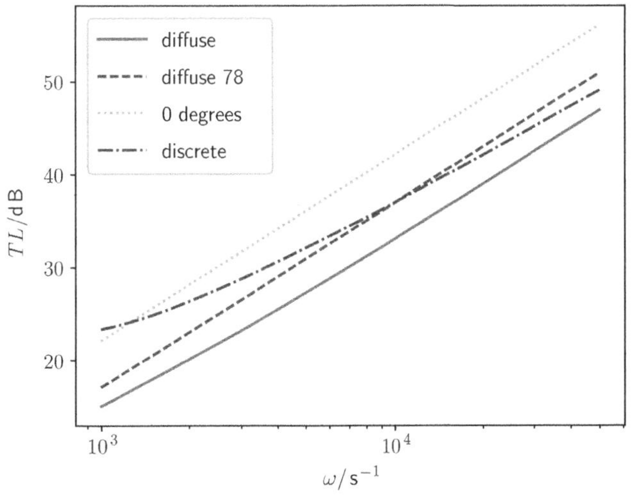

Figure 8.21 Transmssion loss at certain angles and diffuse field calculated by wavenumber and discrete methods. Both fluids are air; the plate for the discrete version has plate dimension $L_x = 0.8$ m,$L_y = 0.5$ m. *Source:* Alexander Peiffer.

mesh as shown in Figure 8.14, the dynamic stiffness matrix reads as

$$[D_s] = -\omega [M] = -\omega \begin{bmatrix} m''\Delta S & \cdots & 0 \\ \vdots & m''\Delta S & \vdots \\ & & \ddots \\ 0 & \cdots & m''\Delta S \end{bmatrix} \tag{8.101}$$

Using the above formulation, the total stiffness matrix is

$$[D_{\text{tot}}] = [D_s] + \left[D_{\text{dir}}^{(1)}\right] + \left[D_{\text{dir}}^{(2)}\right] \tag{8.102}$$

and we get the result as shown in Figure 8.21. We see that the wavenumber solution is not in line with the results due to the infinite assumptions of the wave number approach. The differences are highest at low frequencies when the panel is small compared to the wavelength, and the radiation efficiency of the finite panel is much smaller than the infinite half space.

From comparison with experiments, Beranek (1988) found that the best match between experiment and the mass law is found when not the full half space angle is used for integration. The averaging formula (8.14) is modified by a maximum angle ϑ_{max} as an empirical upper integration limit

$$\langle\tau\rangle_E = \frac{\int_0^{\vartheta_{max}} \tau(\vartheta)\cos\vartheta\sin\vartheta d\vartheta}{\int_0^{\vartheta_{max}} \cos\vartheta\sin\vartheta d\vartheta} \quad \langle\tau\rangle_E = \frac{2}{\sin^2\vartheta_{max}} \int_0^{\vartheta_{max}} \tau(\vartheta)\cos\vartheta\sin\vartheta d\vartheta \tag{8.103}$$

The best fit is found for $\vartheta = 78°$. Even though it fits best to the high frequency results, the 78° approximation does not comply well with the discrete results at lower frequencies.

A simple step to integrate the bending plate dynamics without setting up a full FE model of the plate is to use the discrete plate radiation stiffness formula (8.76). So, the total stiffness matrix is built up as follows.

$$[D_{\text{tot}}] = \left[D_{\text{dir}}^{(plate)}\right] + \left[D_{\text{dir}}^{(1)}\right] + \left[D_{\text{dir}}^{(2)}\right] \tag{8.104}$$

The results are shown in Figure 8.22. Those results can be efficiently calculated using the above formula and are much more precise than estimations from wavenumber space. We chose two aluminium plates of thicknesses $h = 4$ mm and $h = 6$ mm. Three different values for the damping loss are considered.

For the two plates this leads to coincidence frequencies $\omega_c \approx 18\,700\ \text{s}^{-1}$ and $\omega_c \approx 12\,500\text{s}^{-1}$. In Figure 8.22 several phenomena can be observed. First, the high mass of the thick plate leads to generally larger transmission loss; second, the higher thickness shifts the coincidence dip to lower frequencies. Thus, the increasing stiffness of thicker plates may jeopardize the positive mass effect. Third, structural damping may weaken the coincidence dips. The difference between $\eta = 0.001$ and $\eta = 0.01$ is small. The reason for this is that the damping loss from radiation (8.91) is dominating the damping, and small structural dissipation does not have great effect. Globally the coincidence dip is not as sharp as would be predicted by the wavenumber equation (8.95).

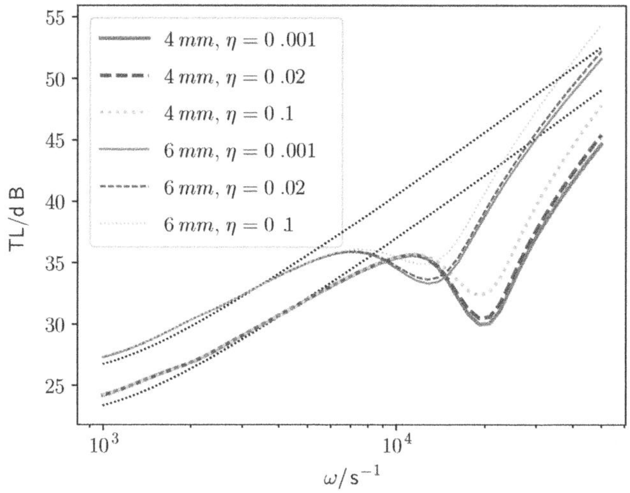

Figure 8.22 Transmission loss of 4 mm and 6 mm aluminium plate at different damping losses. *Source:* Alexander Peiffer.

8.2.5 Line Radiation Stiffness

For line connections the situation is complex even in wavenumber space, because there is a combination of four different wave types in plates. We recall the different wave types in plates from Chapter 3. This multi wave situation leads to specific challenges for the application of the hybrid coupling loss factor in the context of plate line junctions that must be worked out in detail. For homogeneous plates the in-plane wave motion is decoupled from the out-of-plane motion. The formula for in-plane waves (3.184) affects the displacement components u and v

$$\begin{Bmatrix} \boldsymbol{u} \\ \boldsymbol{v} \end{Bmatrix} = \boldsymbol{\Psi}_S \begin{Bmatrix} k_{yS} \\ -k_{xS} \end{Bmatrix} e^{-j\mathbf{k}_S\mathbf{x}} + \boldsymbol{\Psi}_L \begin{Bmatrix} k_{xL} \\ k_{yL} \end{Bmatrix} e^{-j\mathbf{k}_L\mathbf{x}}$$

with two different wavenumbers for shear and longitudinal motion. The longitudinal wavenumber was given by (3.167)

$$k_L = \frac{\omega}{c_L} = \sqrt{\frac{\omega^2(1-\nu^2)\rho_0}{E}}$$

and the shear wavenumber by (3.171)

$$k_T = \frac{\omega}{c_T} = \sqrt{\frac{\omega^2\rho_0}{G}} \tag{8.105}$$

Out-of-plane waves are called bending waves, and their wavenumber is usually higher than both in-plane waves. In principle we have the superposition of two waves

$$\boldsymbol{w} = \boldsymbol{w}_1 e^{-jk_{xB1}x - jk_{yB1}y} + \boldsymbol{w}_2 e^{-jk_{xB2}x - jk_{yB2}y}$$

with

$$k_{B1} = j\sqrt[4]{\frac{\omega^2\rho_0 h}{B}} \quad k_{B2} = \sqrt[4]{\frac{\omega^2\rho_0 h}{B}}$$

In wavenumber based SEA, a plate is described by wave fields as discussed in Chapter 6. In most cases the bending wave is determining the random behavior because the wavelength of the in-plane waves is too high to reach high modal density. However, for the definition of the radiation stiffness, the in-plane waves are very important, especially when the connections are rectangular and the bending motion of one plate acts against the in-plane stiffness of the second plate. The wave formulation shows some limits of the hybrid coupling loss factor that will be revealed later. However, we will here focus on the hybrid coupling loss factor but comparing the results to those from Langley and Heron (1990).

8.2.5.1 Plates - Wavenumber Domain

There are several papers on the subject of plate edge wave transmission, all of them based on the plane wave transmission and mode conversion. The method was first introduced by Langley and Heron (1990), adapted and implemented by Johansson and Comnell (2010) and many other authors applying this important relationship to SEA. For a detailed derivation including beams as deterministic component in the junction, see Langley and Heron (1990). Langley used the reaction forces that lead to a reverse sign in the stiffness matrices, as shown in the appendix B.2. This does not affect the transmission theory but does not comply with the hybrid coupling loss factor theory.

In Figure 8.23 the arrangement of degrees of freedom and wavenumbers is shown. Langley and Heron considered four degrees of freedom: $\boldsymbol{u}_e, \boldsymbol{v}_e, \boldsymbol{w}_e$, and $\boldsymbol{\beta}_{xe}$, and the related forces per unit length $\boldsymbol{F}'_x, \boldsymbol{F}'_y, \boldsymbol{F}'_z$, and moment $\boldsymbol{M}'_x$. For the derivation of the edge radiation stiffness we consider the local edge coordinate system denoted by the index e. The edge is supposed to be of infinite length, and the harmonic motion is given by the local edge coordinate system

$$\begin{Bmatrix} u_e \\ v_e \\ w_e \\ \beta_{xe} \end{Bmatrix}(x,t) = \begin{Bmatrix} \boldsymbol{u}_e \\ \boldsymbol{v}_e \\ \boldsymbol{w}_e \\ \boldsymbol{\beta}_{xe} \end{Bmatrix} e^{-jk_x x + j\omega t} = \{\boldsymbol{q}_e\} e^{-jk_x x + j\omega t} \tag{8.106}$$

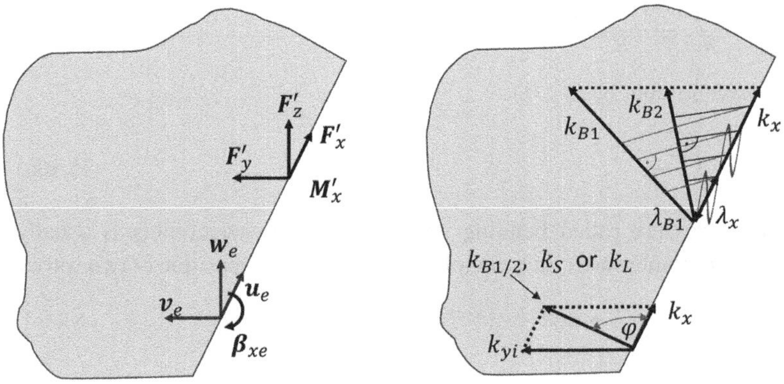

Figure 8.23 Semi infinite plate with edge forces, moments and wavenumber vectors. *Source:* Alexander Peiffer.

$$\begin{Bmatrix} F'_x \\ F'_y \\ F'_z \\ M'_x \end{Bmatrix}(x,t) = \begin{Bmatrix} \boldsymbol{F}'_x \\ \boldsymbol{F}'_y \\ \boldsymbol{F}'_z \\ \boldsymbol{M}'_x \end{Bmatrix} e^{-jk_x x + j\omega t} = \{\boldsymbol{F}'_e\} e^{-jk_x x + j\omega t} \tag{8.107}$$

With $\boldsymbol{q}_e$ as the generic edge displacement degrees of freedom. The free edge boundary conditions at $y = 0$ are according to Langley and Heron (1990) or given by explicit derivation in Ventsel and Krauthammer (2001).

$$F'_x = -N'_x = -S\left(\frac{\partial u}{\partial y} + \frac{\partial v}{\partial x}\right) \tag{8.108a}$$

$$F'_y = -N'_{xy} = -C\left(\frac{\partial v}{\partial y} + \nu\frac{\partial u}{\partial x}\right) \tag{8.108b}$$

$$F'_z = B\left(\frac{\partial^3 w}{\partial y^3} + (2-\nu)\frac{\partial^3 w}{\partial x^2 \partial y}\right) \tag{8.108c}$$

$$M'_x = -B\left(\frac{\partial^2 w}{\partial y^2} + \nu\frac{\partial^2 w}{\partial x^2}\right) \tag{8.108d}$$

with constants B, C, and S as defined in section 3.7. Those are the bending stiffness, the in-plane longitudinal stiffness, and the in-plane shear stiffness, respectively.

$$B = \frac{Eh^3}{12(1-\nu)} \qquad C = \frac{Eh}{(1-\nu^2)} \qquad S = \frac{Eh}{2(1+\nu)}$$

The first two equations (8.108a),(8.108b), and the last (8.108d) follow directly from plate stress resultants in section 3.7, namely equations (3.154), (3.155), and (3.197). The external forces are in equilibrium with the internal stress resultants, thus requiring different signs as in Langley's paper. For example, a positive force in the y-direction leads to a negative slope in displacement v requiring the negative sign in equation (8.108b).

The third equation (8.108c) dealing with forces in the z-direction is derived in more detail from the boundary conditions. A stress resultant at the edge is not only a resultant from the shear force Q'_y but also a second force resulting from the twisting moment M'_{yx} as depicted in Figure 8.24.

Imagine two elements of length dx with a twisting moment resulting from the shear stress σ_{xy} as shown in 8.24a. The moment per per length on the left element is $M'_{yx}dx$ and on the right element $(M'_{yx} + \frac{\partial M'_{yx}}{\partial x}dx)dx$. These moments can be represented by pairs of vertical forces M'_{yx} and $M'_{xy} + \frac{\partial M'_{yx}}{\partial x}dx$ both acting on a lever of dx as shown in 8.24b.

Finally, the right force of the left element and the left force of the right element add up to $\frac{\partial M'_{yx}}{\partial x}dx$ (8.24c)) along the line A–B. Thus, the effective shear force V'_y is the combination of both shear forces and supposed to be zero. Boundary condition (8.108c) is now modified to

$$F'_z = -F'_{\text{eff},z} = -Q'_y - \frac{\partial M'_{yx}}{\partial x} = 0 \qquad M'_y = 0 \tag{8.109}$$

With equations (3.199) and (3.202) we get (8.108c). In addition the relationship

$$\frac{Gh^3}{6} = B(1-\nu) \tag{8.110}$$

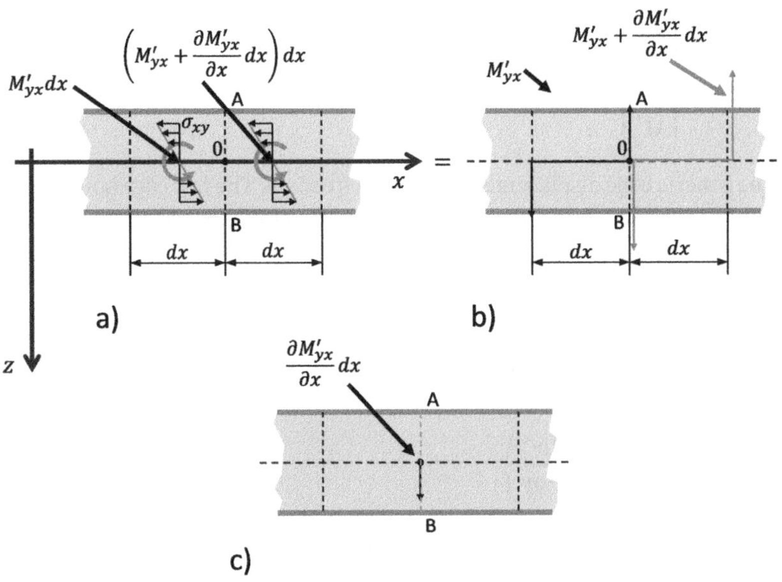

Figure 8.24 Replacement of twisting moment by effective shear force. *Source:* Alexander Peiffer.

was used. In the above derived equations (8.108a)–(8.108d) the solution of the in-plane and bending waves of plates from section 3.7 are inserted under the condition that the edge wavenumber in the x-direction is given. Thus, for all wave types we have $k_x = k_{xS} = k_{xL} = k_{xB1/2}$ with the given time and x-dependency $e^{-jk_x x + j\omega t}$ as shown in Figure 8.25. The individual wavenumber in the y-direction is given by

$$k_x^2 + k_{y_i}^2 = k_i^2 \text{ with} \qquad i = S, L, B1, B2 \tag{8.111}$$

For convenience and in order to follow the convention from Langley's paper, we use the propagation constant $\mu_i = -jk_{yi}$ instead of the wavenumber. Consequently, a complex propagation constant means wave propagation; a real constant denotes a decaying wave mode

$$\mu_y = \pm\sqrt{k_x^2 - k^2} \tag{8.112}$$

For bending waves we have potentially four wavenumbers, so we get as solution

$$\mu_{yB1} = \pm\sqrt{k_x^2 + k_B^2} \qquad \mu_{yB2} = \pm\sqrt{k_x^2 - k_B^2} \tag{8.113}$$

and for shear and longitudinal waves

$$\mu_{yS} = \pm\sqrt{k_x^2 - k_S^2} \qquad \mu_{yL} = \pm\sqrt{k_x^2 - k_L^2} \tag{8.114}$$

The waves must propagate in the positive y-direction and vanish for large y. This is only possible when negative values are considered. When, for example, $k_B^2 > k_x^2$ the roots become complex, and the edge motion cannot radiate into the wave mode of the plate. From section 3.7 it is known that the displacement in u and v-directions is a superposition of in-plane waves as shown in Figure 8.25. According to (3.184) the u

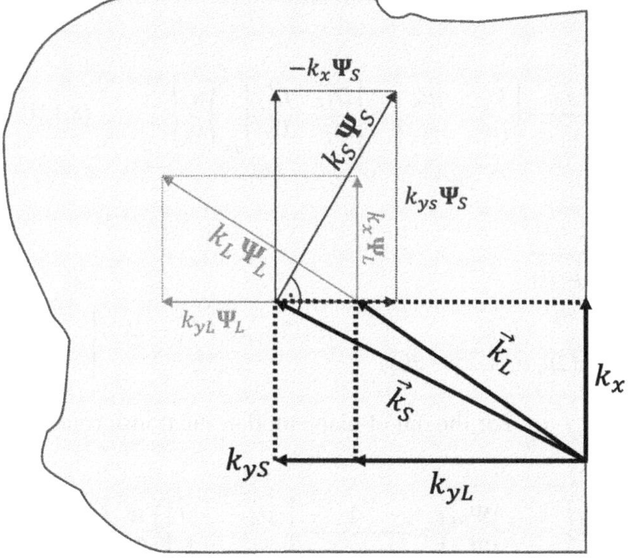

Figure 8.25 Combined vector of propagation and displacement for in-plane waves. *Source:* Alexander Peiffer.

and v-displacement expressed with the propagation constant is

$$\boldsymbol{u} = (\boldsymbol{\Psi}_L k_x e^{\mu_L y} + j\mu_S \boldsymbol{\Psi}_S e^{\mu_S y}) e^{-jk_x x + j\omega t} \tag{8.115}$$

$$\boldsymbol{v} = (j\boldsymbol{\Psi}_L \mu_L e^{\mu_L y} - k_x \boldsymbol{\Psi}_S e^{\mu_S y}) e^{-jk_x x + j\omega t} \tag{8.116}$$

The out-of-plane displacement is given by $\boldsymbol{w}$ and $\boldsymbol{\beta}_x$ with

$$\boldsymbol{w} = (\boldsymbol{\Psi}_{B1} e^{\mu_{B1} y} + \boldsymbol{\Psi}_{B2} e^{\mu_{B2} y}) e^{-jk_x x + j\omega t} \tag{8.117}$$

$$\boldsymbol{\beta}_x = \frac{\partial \boldsymbol{w}}{\partial y} = (\mu_{B1} \boldsymbol{\Psi}_{B1} e^{\mu_{B1} y} + \mu_{B2} \boldsymbol{\Psi}_{B2} e^{\mu_{B2} y}) e^{-jk_x x + j\omega t} \tag{8.118}$$

For $y = 0$ we name the local edge motion by the coordinates $\{\boldsymbol{u}_e, \boldsymbol{v}_e, \boldsymbol{w}_e, \beta_{x,e}\}^T$, and the wave coordinates are $\{\boldsymbol{\Psi}_L, \boldsymbol{\Psi}_S, \boldsymbol{\Psi}_{B1}, \boldsymbol{\Psi}_{B2}\}^T$. We get the following block matrix for the transformation between both coordinate systems of the in plane motion

$$\begin{Bmatrix} \boldsymbol{u}_e \\ \boldsymbol{v}_e \end{Bmatrix} = \begin{bmatrix} k_x & j\mu_S \\ j\mu_L & -k_x \end{bmatrix} \begin{Bmatrix} \boldsymbol{\Psi}_L \\ \boldsymbol{\Psi}_S \end{Bmatrix} \tag{8.119a}$$

$$\begin{Bmatrix} \boldsymbol{\Psi}_L \\ \boldsymbol{\Psi}_S \end{Bmatrix} = \frac{1}{k_x^2 - \mu_L \mu_S} \begin{bmatrix} k_x & j\mu_S \\ j\mu_L & -k_x \end{bmatrix} \begin{Bmatrix} \boldsymbol{u}_e \\ \boldsymbol{v}_e \end{Bmatrix}$$

Equations (8.115) and (8.116) are entered into the boundary conditions (8.108a)–(8.108d) to derive the edge dynamics in wave coordinates

$$\begin{Bmatrix} \boldsymbol{F}'_x \\ \boldsymbol{F}'_y \end{Bmatrix} = [\boldsymbol{S}']_{LS,\text{dir}} \begin{Bmatrix} \boldsymbol{\Psi}_L \\ \boldsymbol{\Psi}_S \end{Bmatrix} = S \begin{bmatrix} -2\mu_L k_x & j(k_S^2 - 2k_x^2) \\ j(k_S^2 - 2k_x^2) & -k_x \mu_S \end{bmatrix} \begin{Bmatrix} \boldsymbol{\Psi}_L \\ \boldsymbol{\Psi}_S \end{Bmatrix} \tag{8.120}$$

Using (8.119) for the replacement of the wave coordinates, we get the stiffness block matrix for the in plane motion

$$\begin{Bmatrix} \boldsymbol{F}'_x \\ \boldsymbol{F}'_y \end{Bmatrix} = \frac{[\boldsymbol{S}']_{LS,dir}}{k_x^2 - \mu_L\mu_S} \begin{bmatrix} k_x & j\mu_S \\ j\mu_L & -k_x \end{bmatrix} = \begin{bmatrix} \boldsymbol{D}'_{11} & \boldsymbol{D}'_{12} \\ \boldsymbol{D}'_{21} & \boldsymbol{D}'_{22} \end{bmatrix}_{\text{dir}} \begin{Bmatrix} \boldsymbol{u}_e \\ \boldsymbol{v}_e \end{Bmatrix} \tag{8.121}$$

$$\begin{bmatrix} \boldsymbol{D}'_{11} & \boldsymbol{D}'_{12} \\ \boldsymbol{D}'_{21} & \boldsymbol{D}'_{22} \end{bmatrix}_{LS,dir} = \frac{S}{k_x^2 - \mu_L\mu_S} \cdots$$

$$\cdots \begin{bmatrix} -\mu_L k_S^2 & -jk_x(2\mu_L\mu_S + k_S^2 - 2k_x^2) \\ +jk_x(2\mu_L\mu_S + k_S^2 - 2k_x^2) & -\mu_S k_S^2 \end{bmatrix}$$

with $k_S^2 = k_x^2 - \mu_S^2$ and $k_L^2 = k_x^2 - \mu_L^2$. For the out-of-plane motion the transformation matrices at the edge are

$$\begin{Bmatrix} \boldsymbol{w}_e \\ \boldsymbol{\beta}_{x,e} \end{Bmatrix} = \begin{bmatrix} 1 & 1 \\ \mu_{B1} & \mu_{B2} \end{bmatrix} \begin{Bmatrix} \boldsymbol{\Psi}_{B1} \\ \boldsymbol{\Psi}_{B2} \end{Bmatrix} \qquad \begin{Bmatrix} \boldsymbol{\Psi}_{B1} \\ \boldsymbol{\Psi}_{B2} \end{Bmatrix} = \frac{1}{\mu_{B2} - \mu_{B1}} \begin{bmatrix} \mu_{B2} & -1 \\ -\mu_{B1} & 1 \end{bmatrix} \begin{Bmatrix} \boldsymbol{w}_e \\ \boldsymbol{\beta}_{x,e} \end{Bmatrix} \tag{8.122a}$$

We repeat the procedure for the in-plane waves here using equations (8.117) and (8.118) in the boundary conditions (8.108c) and (8.108d) at $y = 0$

$$\begin{Bmatrix} \boldsymbol{F}'_z \\ \boldsymbol{M}'_y \end{Bmatrix} = [\boldsymbol{S}']_{B,dir} \begin{Bmatrix} \boldsymbol{\Psi}_{B1} \\ \boldsymbol{\Psi}_{B2} \end{Bmatrix} = B \begin{bmatrix} \mu_{B1}^3 - (2-\nu)\mu_{B1}k_x^2 & \mu_{B2}^3 - (2-\nu)\mu_{B2}k_x^2 \\ \nu k_x^2 - \mu_{B1}^2 & \nu k_x^2 - \mu_{B2}^2 \end{bmatrix} \tag{8.123}$$

With the transformation matrix (8.122), the radiation stiffness matrix for bending reads:

$$\begin{Bmatrix} \boldsymbol{F}'_z \\ \boldsymbol{M}'_y \end{Bmatrix} = \frac{[\boldsymbol{S}']_{B,dir}}{\mu_{B2} - \mu_{B1}} \begin{bmatrix} \mu_{B2} & -1 \\ -\mu_{B1} & 1 \end{bmatrix} = \begin{bmatrix} \boldsymbol{D}'_{33} & \boldsymbol{D}'_{34} \\ \boldsymbol{D}'_{43} & \boldsymbol{D}'_{44} \end{bmatrix}_{\text{dir}} \begin{Bmatrix} \boldsymbol{w}_e \\ \boldsymbol{\beta}_{xe} \end{Bmatrix} \tag{8.124}$$

$$\begin{bmatrix} \boldsymbol{D}'_{33} & \boldsymbol{D}'_{34} \\ \boldsymbol{D}'_{43} & \boldsymbol{D}'_{44} \end{bmatrix}_{\text{dir}} = B \begin{bmatrix} -\mu_{B1}^2\mu_{B2} + \mu_{B1}\mu_{B2}^2 & \mu_{B1}\mu_{B2} - \nu k_x^2 \\ \mu_{B1}\mu_{B2} + \nu k_x^2 & -\mu_{B1} - \mu_{B2} \end{bmatrix}$$

In contrast to Langley and Heron (1990) and Johansson and Comnell (2010), the equation of motion for external forces and not for the internal tractions are used. In wave transmission theory this has no impact on the result, but it would lead to negative radiated power in the hybrid formulation. The antisymmetry results from the complex base function from equation (8.106). Finally, the total radiation stiffness matrix in wavenumber space for the edge coordinates of the mth plate is given by

$$\{\boldsymbol{F}_{em}\} = [\boldsymbol{D}']_{\text{dir}}^{(m)} \{\boldsymbol{q}_{em}\} \tag{8.125}$$

$$[\boldsymbol{D}'(k_x)]_{\text{dir}}^{(m)} = \begin{bmatrix} \boldsymbol{D}'_{11} & \boldsymbol{D}'_{12} & 0 & 0 \\ \boldsymbol{D}'_{21} = -\boldsymbol{D}'_{12} & \boldsymbol{D}'_{22} & 0 & 0 \\ 0 & 0 & \boldsymbol{D}'_{33} & \boldsymbol{D}'_{34} \\ 0 & 0 & \boldsymbol{D}'_{43} = \boldsymbol{D}'_{34} & \boldsymbol{D}'_{44} \end{bmatrix}$$

The extra index m denotes that the radiation stiffness is given in edge coordinates of the mth plate with the specific material constants $\mu_{B1}, \mu_{B2}, k_S, k_L$, e.t.c. of each plate.

8.2.6 Line Junctions

In a junction configuration several plates may be connected to the line with angle Θ_m in relation to the global coordinate system. The radiation stiffness is defined in the local plate coordinate system (8.106). For the calculation of the global radiation stiffness the local coordinates of the mth plate must be rotated by the angle Θ_m. The coordinates u and β_x are not affected by the rotation, so the transformation matrix is given by

$$\left[T^{(m)}\right] = \begin{bmatrix} 1 & 0 & 0 & 0 \\ 0 & \cos\Theta_m & -\sin\Theta_m & 0 \\ 0 & \sin\Theta_m & \cos\Theta_m & 0 \\ 0 & 0 & 0 & 1 \end{bmatrix} \tag{8.126}$$

Using this transformation the local edge coordinates are converted into the degrees of freedom of the global system $\boldsymbol{q}_{e0}$ or vice versa by

$$\{\boldsymbol{q}_{e0}\} = \left[T^{(m)}\right]\{\boldsymbol{q}_{em}\} \qquad \{\boldsymbol{q}_{em}\} = \left[T^{(m)}\right]^T\{\boldsymbol{q}_{e0}\} \tag{8.127}$$

The radiation stiffness matrix of plate m that is connected with angle ϑ_m reads in global coordinates

$$\left[\boldsymbol{D}_{\text{dir}}^{\prime(m)}\right]_{e0} = \left[T^{(m)}\right]\left[\boldsymbol{D}_{\text{dir}}^{\prime(m)}\right]_{e}\left[T^{(m)}\right]^T \tag{8.128}$$

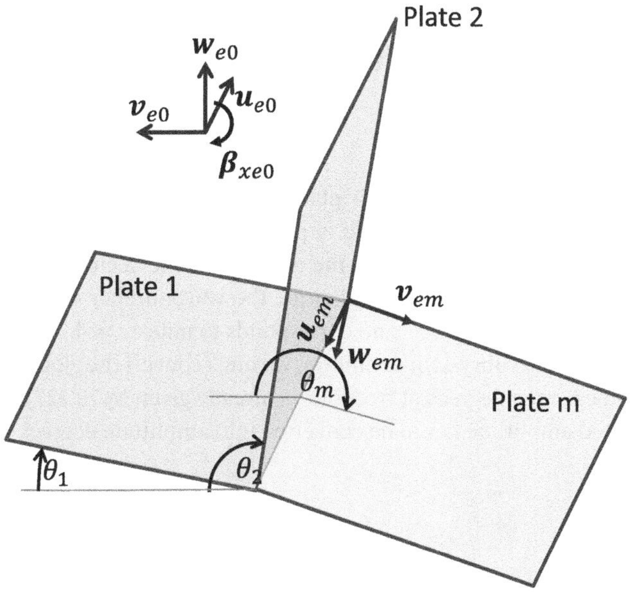

Figure 8.26 Flat plates connected via line junctions. *Source:* Alexander Peiffer.

For a multiple plate junction the total stiffness sums up to:

$$[\boldsymbol{D}'_{\text{tot}}(k_x)]_{e0} = \sum_m \left[\boldsymbol{D}'^{(m)}_{\text{dir}}\right]_{e0} \tag{8.129}$$

Langley and Heron (1990) derived the coupling loss factors based on wave transmission coefficients assuming one incoming wave and calculating all transmitted and reflected waves. A detailed derivation and some further explanation are given in the appendix section B.2. Here we stay as far as possible with the stringent application of the coupling loss factor definition from (7.27), because this keeps the doors open for the derivation of hybrid line junctions.

The total junction radiation stiffness and the radiation stiffness of each plate is given by equation (8.128) and is defined in the global coordinates. Thus, in the global system we have all the details collected to calculate the coupling loss factor using (7.27)

$$\eta_{mn}(k_x) = \frac{2}{\pi n_m \omega} \sum_{i,j} Im\left\{\left[\boldsymbol{D}'^{(n)}_{e0,dir}\right]_{ij}\left(\left[\boldsymbol{D}'_{e0,tot}\right]^{-1} Im\left[\boldsymbol{D}'^{(m)}_{e0,dir}\right]\left[\boldsymbol{D}'_{e0,tot}\right]^{-H}\right)_{ij}\right\} \tag{8.130}$$

The sum runs over the four displacement degrees of freedom. The imaginary is over the full expression of the sum because the radiation stiffness here is not symmetric.

Equation (8.130) doesn't distinguish between the different wave types, but random plate systems are often described by the three diffuse wave fields of the plate: longitudinal, shear, and bending waves with very different wavelength and modal densities, especially between bending and in-plane waves.

So, each plate comprises three systems that can be deterministic and random independently because of different modal density. Thus, we should put more effort into a theory for the coupling of each system and wave field.

The wavenumber coordinate k_x is the lead coordinate for the infinite line junction and can be considered as the projected wave-number of any incoming wave, as shown in Figure 8.27. When the incoming wave is hosted by the mth plate the wavenumber in x depends on the angle φ as follows

$$k_x = k_v^{(m)} \sin\varphi \tag{8.131}$$

The definition of different random fields in each plate requires a new indexing scheme for each SEA system. We name the wave field of plate m by the combination of the indexes v, w thus ${}^{(m)}_{v}$. The wave index follows the order as in the vector $\{\boldsymbol{q}_\Psi\}$, so $v, w = 1, 2, 3, 4$ corresponds to $L, S, B1, B2$. For example the wavenumber of the bending wave of the mth plate is denoted by $k_B^{(m)}$ and corresponds to index $v = 4$.

In order to cope with this wave field subsystem definition, we must convert the global degrees of freedom into the local edge degrees of freedom as already given by (8.127). Once this is known the radiated amplitude has to be converted into amplitude degrees of freedom defined by

$$\{\boldsymbol{q}_{\Psi m}\} = \begin{Bmatrix} \boldsymbol{\Psi}_L^{(m)} \\ \boldsymbol{\Psi}_S^{(m)} \\ \boldsymbol{\Psi}_{B1}^{(m)} \\ \boldsymbol{\Psi}_{B2}^{(m)} \end{Bmatrix} \tag{8.132}$$

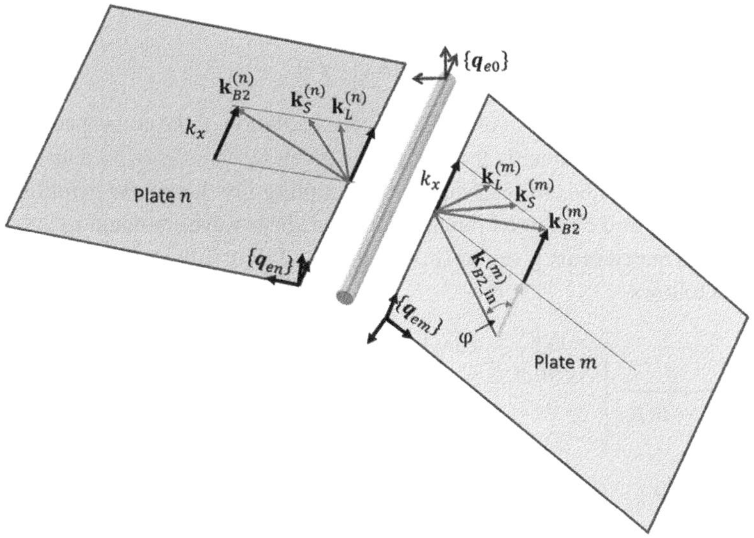

Figure 8.27 Transmission of plane bending wave in *m*th plate to multiple waves in *n*th plate. *Source:* Alexander Peiffer.

The required transformation was already derived in the radiation stiffness section 8.2.5 but block-wise in equations (8.119) and (8.122). For the full set the conversion from wave amplitude to edge coordinates reads

$$\begin{Bmatrix} u_{em} \\ v_{em} \\ w_{em} \\ \beta_{x,em} \end{Bmatrix} = \begin{bmatrix} k_x & j\mu_S^{(m)} & & \\ j\mu_L^{(m)} & -k_x & & \\ & & 1 & 1 \\ & & \mu_{B1}^{(m)} & \mu_{B2}^{(m)} \end{bmatrix} \begin{Bmatrix} \Psi_L^{(m)} \\ \Psi_S^{(m)} \\ \Psi_{B1}^{(m)} \\ \Psi_{B2}^{(m)} \end{Bmatrix} \tag{8.133}$$

The pair of transformations is then given by

$$\{q_{em}\} = \left[T_\Psi^{(m)}\right]\{q_{\Psi m}\} \qquad \{q_{\Psi m}\} = \left[T_\Psi^{(m)}\right]^{-1}\{q_{em}\} \tag{8.134}$$

We have to use the inverse of the above matrix, because it is not orthonormal, in contrast to the transformation matrix in (8.127).

8.2.6.1 Coupling Loss Factor of in-plane and out-of-plane Waves

For the application of the hybrid coupling loss factor and keeping the approach of different wave fields, it would be necessary to find a radiation stiffness definition that provides the cross correlation of the blocked forces of longitudinal, shear, and bending waves. The diffuse field reciprocity (7.19) is valid for bending waves but not for single in-plane waves. The argumentation is beyond the scope of a text book, but the details are given in the appendix B.2. It is shown there that the diffuse field reciprocity is valid for the combination of both in-plane waves

$$\left[S'_{ff}\right]_B = \frac{4E_m}{\pi\omega n(\omega)} Im\left[D_{\text{dir}}^{(m)}\right]_B \qquad \left[S'_{ff}\right]_{LS} = \frac{4E_m}{\pi\omega n(\omega)} Im\left[D_{\text{dir}}^{(m)}\right]_{LS} \tag{8.135}$$

but not for single in-plane waves

$$Im\left[S'_{ff}\right]_L \neq 0 \qquad\qquad Im\left[S'_{ff}\right]_S \neq 0 \tag{8.136}$$

because the cross spectral density is not real. The direct calculation of the cross spectral density leads to imaginary components that cannot be given by cross spectral density of diffuse field reciprocity. Thus, we will derive the coupling loss factor for bending wave fields denoted by B and combined longitudinal and shear waves denoted by LS. The radiation stiffness matrices are given by (8.121) and (8.124). In full matrix notation they would read as follows

$$[\boldsymbol{D}'_{\text{dir}}]_{LS} = \frac{S}{k_x^2 - \mu_L\mu_S}\begin{bmatrix} -\mu_L k_S^2 & -jk_x(2\mu_L\mu_S + k_S^2 - 2k_x^2) & 0 & 0 \\ Asymm. & -\mu_S k_S^2 & 0 & 0 \\ 0 & 0 & 0 & 0 \\ 0 & 0 & 0 & 0 \end{bmatrix} \tag{8.137}$$

$$[\boldsymbol{D}'_{\text{dir}}]_{B} = B\begin{bmatrix} 0 & 0 & 0 & 0 \\ 0 & 0 & 0 & 0 \\ 0 & 0 & -\mu_{B1}^2\mu_{B2} + \mu_{B1}\mu_{B2}^2 & \mu_{B1}\mu_{B2} - \nu k_x^2 \\ 0 & 0 & Symm. & -\mu_{B1} - \mu_{B2} \end{bmatrix} \tag{8.138}$$

For the application of the above radiation stiffnesses, the global edge motion must be transformed into the local edge motion. For the derivation it is helpful to recommend the fact that the coupling loss factor formula is based on the random response (expressed as cross correlation) of general displacement coordinates to random excitation (7.21).

$$\left[\boldsymbol{S}^{(n)*}_{e0,qq}\right] = \left[\boldsymbol{D}'_{e0,tot}\right]^{-1}\left[\boldsymbol{S}^{(m)*}_{e0,ff}\right]\left[\boldsymbol{D}'_{e0,tot}\right]^{-H} \tag{7.21}$$

The task of transformation from global edge coordinates to local edge coordinates is carried out by converting the displacement on the LHS of the above equation into the edge system of the nth plate (receiving system) and the force in the central cross spectral density matrix into the local edge system of the mth plate (exciting system).

$$\{\boldsymbol{q}_{en}\} = [T^{(n)}]^T\{\boldsymbol{q}_{e0}\} \tag{8.139}$$

Using (8.139) the cross spectral displacement matrix is converted into the nth local edge system by

$$\left[\boldsymbol{S}^{(n)*}_{e,qq}\right] = \left\langle \boldsymbol{q}_{en}\boldsymbol{q}_{en}^H\right\rangle_E = [T^{(n)}]^T\underbrace{\left\langle \boldsymbol{q}_{e0}\boldsymbol{q}_{e0}^H\right\rangle_E}_{\left[\boldsymbol{S}^{(n)*}_{e0,qq}\right]}[T^{(n)}]^* \tag{8.140}$$

The transformation from the local edge system into the global edge system is done using

$$\{\boldsymbol{f}_{e0}\} = \left[\boldsymbol{T}^{(m)}\right]\{\boldsymbol{f}_{em}\} \tag{8.141}$$

The cross spectral force matrix can be written as

$$\left[\boldsymbol{S}^{(m)*}_{e0,ff}\right] = \left\langle \boldsymbol{f}_{e0,m}\boldsymbol{f}_{e0,m}^H\right\rangle_E = [T^{(m)}]\underbrace{\left\langle \boldsymbol{f}_{em}\boldsymbol{f}_{em}^H\right\rangle_E}_{\left[\boldsymbol{S}^{(m)*}_{e,ff}\right]}[T^{(m)}]^H \tag{8.142}$$

Entering this in (7.21) and putting the result into (8.140) reads finally as

$$\left[S^{(n)*}_{e,qq}\right] = \underbrace{\left[T^{(n)}\right]^T \left[D_{e0,\mathrm{tot}}\right]^{-1} \left[T^{(m)}\right]}_{\left[D_{\mathrm{tot},mn}\right]^{-1}} \underbrace{\left[S^{(m)*}_{\Psi,ff}\right]}_{\left[S^{(m)*}_{\Psi,ff}\right]} \underbrace{\left[T^{(m)}\right]^H \left[D_{e0,\mathrm{tot}}\right]^{-H} \left[T^{(n)}\right]^*}_{\left[D_{\mathrm{tot},mn}\right]^{-H}} \tag{8.143}$$

with

$$\left[D_{\mathrm{tot},mn}\right] = \left[T^{(m)}\right]^T \left[D_{\mathrm{tot}}\right] \left[T^{(n)}\right] \tag{8.144}$$

The above matrix is the total stiffness matrix with the nth wave displacement coordinates as input and the reaction forces of the mth plate as output. With this matrix the one-dimensional coupling loss factor is given by:

$$\eta_{mv,nw}(k_x) = \frac{2}{\pi n_{m,v}\omega} \sum_{i,j} Im\left\{\left[D'^{(n)}_{e,v,\mathrm{dir}}\right]_{ij} \left(\left[D'_{\mathrm{tot},mn}\right]^{-1} Im\left[D'^{(m)}_{e,\mathrm{dir}}\right]\left[D'_{\mathrm{tot},mn}\right]^{-H}\right)_{ij}\right\} \tag{8.145}$$

With $u, v = LS$ or $u, v = B$, so $\eta_{mB,nSL}$ is the coupling loss factor of the bending wave of the mth plate to the nth plate in-plane waves. Note that a coupling loss factor for bending to in-plane is also possible in one single plate. With (8.8a) the transmission coefficient reads

$$\tau_{mv,nw}(k_x) = 4\sum_{i,j} Im\left\{\left[D'^{(n)}_{e,v,\mathrm{dir}}\right]_{ij} \left(\left[D'_{e0,\mathrm{tot}}\right]^{-1} Im\left[D'^{(m)}_{e,\mathrm{dir}}\right]\left[D'_{e0,\mathrm{tot}}\right]^{-H}\right)_{ij}\right\} \tag{8.146}$$

For further considerations we chose an example of two aluminium plates[2] of thickness $t = 10$ cm as shown in Figure 8.28, connected in an L-shape with $\theta = 90°$.

See the results in Figures 8.29 and 8.30. In both figures the transmission is compared to Langley's reference solution using plane-wave transmission theory. For wavenumbers smaller than k_L and larger than k_S, both solutions agree well, because the combined radiation stiffness provides the correct power. But, there is a difference for $k_L < k_x <= k_B$. In this wavenumber range the radiated power is not correctly determined, because there is only the shear wave existing as shown in Figure 8.31.

A new approach is required to calculate the radiated power correctly in this wavenumber range. We switch to wave amplitude coordinates to determine the radiated power and use equation (8.134) to transform the cross spectral matrix in edge coordinates further to wave amplitudes

$$\left[S^{(n)*}_{\Psi,qq}\right] = \left[T^{(n)}_{\Psi}\right]^{-1} \left[S^{(n)*}_{\Psi,qq}\right] \left[T^{(n)}_{\Psi}\right] \tag{8.147}$$

In the diagonal we find the squared wave amplitude for each specific wave type with value $\Psi_w \Psi_w^*$ with w as index for each wave amplitude. We use the power amplitude relationships of each wave type to calculate it from the squared wave amplitude. For in-plane waves the power per length in the wave direction is the average kinetic energy

2 The author is aware of the fact that 10 cm is not thin. This number is chosen for better wavenumber visibility.

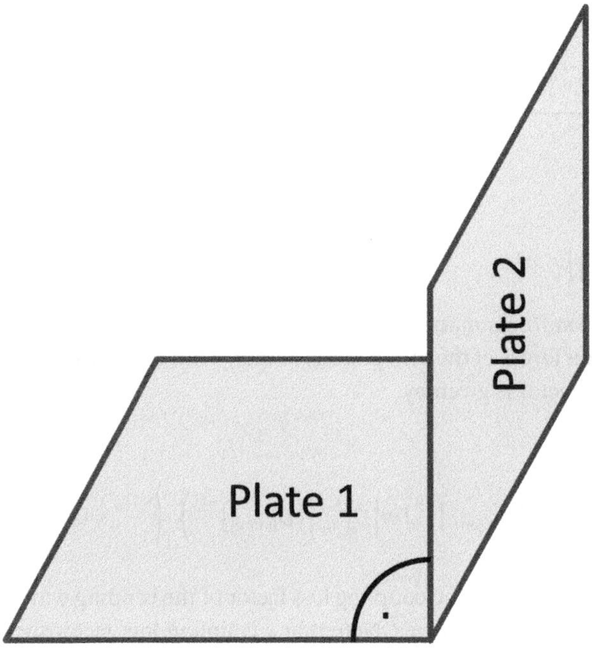

Figure 8.28 L-plate example for line junction. Frequency $\omega = 2000$ Hz. Both plates are made of aluminium of thickness $t = 10$ cm. *Source:* Alexander Peiffer.

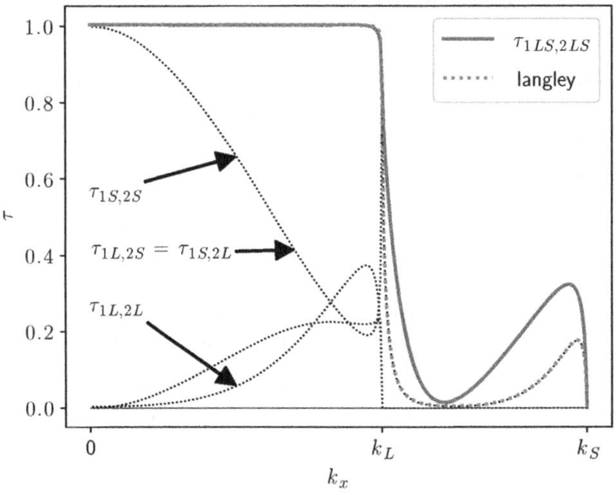

Figure 8.29 In-plane to in-plane transmission coefficient for L-shaped line junction using the radiation stiffness for radiated power calculation. *Source:* Alexander Peiffer.

times the speed of sound

$$\Pi'_L = e_s c_L = \frac{1}{2}\rho_0 h\omega^2 c_L \hat{u}_L^2 = \frac{1}{2}\rho_0 h\omega^3 k_L \Psi_L^2 \tag{8.148}$$

$$\Pi'_S = \frac{1}{2}\rho_0 h\omega^3 k_S \Psi_S^2 \tag{8.149}$$

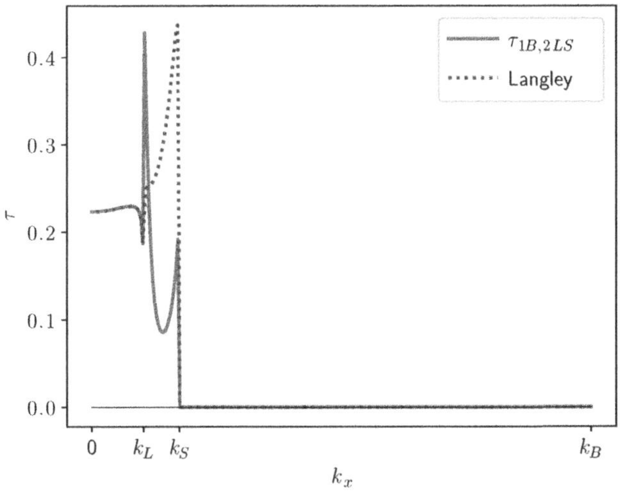

Figure 8.30 Bending to in-plane transmission coefficient for L-shaped line junction using the radiation stiffness for radiated power calculation. *Source:* Alexander Peiffer.

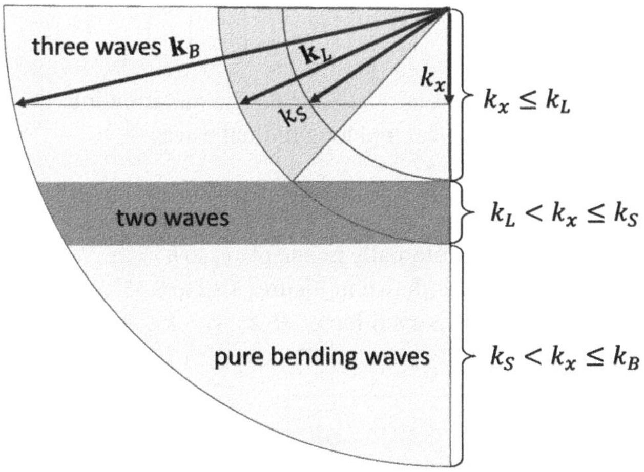

Figure 8.31 Wavenumber regimes and available wave types. *Source:* Alexander Peiffer.

A similar expression is derived from (3.228)

$$\Pi'_B = \rho_0 h\omega^3 / k_B \Psi_B^2 \tag{8.150}$$

For the orthogonal radiated power per length, each term must be multiplied by $\sin(\varphi_w^{(n)})$ with

$$\sin\varphi_w^{(n)} = \frac{k_x}{k_w^{(n)}} \tag{8.151}$$

There are values of k_x without a possible radiating angle. If k_x is too large for some wave type, there is no solution available and thus no power transmitted (Figure 8.31). This formulation enables us to define a generic scalar for $D_{im,w}^{(n)}$ that represents the imaginary stiffness and provides the radiated power per length and perpendicular to the line:

$$\Pi'^{(n)}_{\perp,w} = \frac{\omega}{2} D^{(n)}_{im,w} \hat{\Psi}^2 \tag{8.152}$$

with

$$D^{(n)}_{im,L} = \rho_0 h\omega^2 k_L \sin\varphi_L \tag{8.153}$$

$$D^{(n)}_{im,S} = \rho_0 h\omega^2 k_S \sin\varphi_S \tag{8.154}$$

$$D^{(n)}_{im,B} = 2\rho_0 h\omega^2/k_B \sin\varphi_B \tag{8.155}$$

leading to the radiated power of each wave type and system

$$\Pi'_{mv,nw}(k_x) = \frac{2D^{(n)}_{im,w}}{\pi n_{mv}\omega}\left(\left[\boldsymbol{T}^{(n)}_{\Psi}\right]^{-1}\left[\boldsymbol{D}'_{\text{tot}}\right]^{-1}_{mn} Im\left[\boldsymbol{D}'^{(m)}_{e,dir}\right]\left[\boldsymbol{D}'_{\text{tot}}\right]^{-H}_{mn}\left[\boldsymbol{T}^{(n)}_{\Psi}\right]^{-H}\right)_{ww} \tag{8.156}$$

or for the transmission coefficient

$$\tau_{mv,nw}(k_x) = 4D^{(n)}_{im,w}\left(\left[\boldsymbol{T}^{(n)}_{\Psi}\right]^{-1}\left[\boldsymbol{D}'_{\text{tot}}\right]^{-1}_{mn} Im\left[\boldsymbol{D}'^{(m)}_{e,dir}\right]\left[\boldsymbol{D}'_{\text{tot}}\right]^{-H}_{mn}\left[\boldsymbol{T}^{(n)}_{\Psi}\right]^{-H}\right)_{ww} \tag{8.157}$$

Please note that v is either LS or B and that $w = L, S$, or B. The radiation into the in-plane waves is determined by the sum of shear and longitudinal waves

$$\tau_{mn,vLS}(k_x) = \tau_{mn,vL}(k_x) + \tau_{mn,vS}(k_x) \tag{8.158}$$

Different wave types can exchange energy internally in one plate, so $n = m$.

The transfer solutions (plate 1 to plate 2) are shown in Figures 8.32 to 8.35, and they are in accordance with the reference solutions even for $k_L < k_x <= k_S$. For further

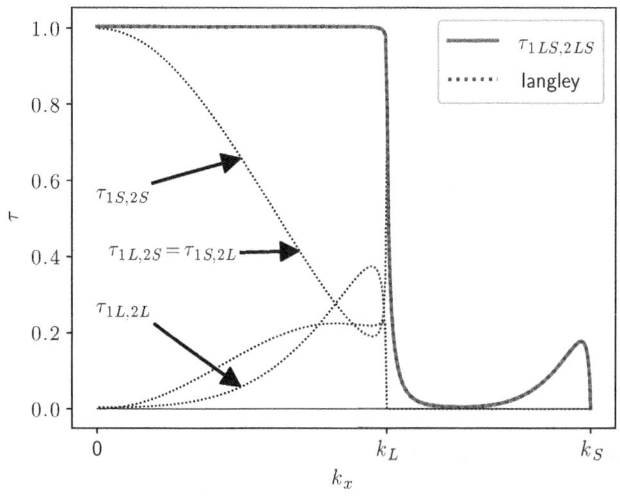

Figure 8.32 In-plane to in-plane transmission coefficient for L-shaped line junction using wave amplitude for radiated power. *Source:* Alexander Peiffer.

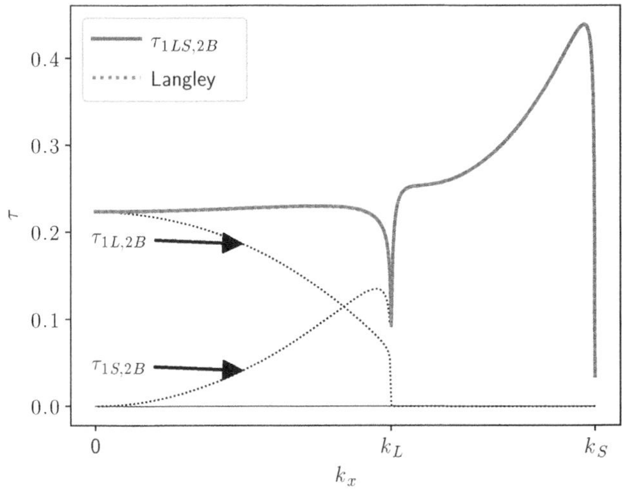

Figure 8.33 In-plane to bending transmission coefficient for L-shaped line junction using wave amplitude for radiated power. *Source:* Alexander Peiffer.

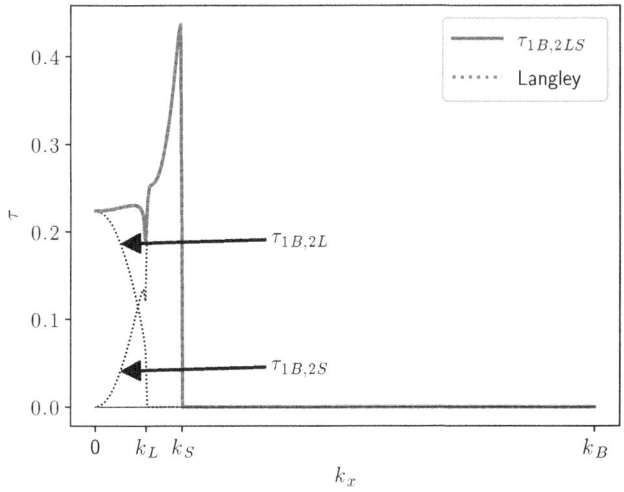

Figure 8.34 Bending to in-plane transmission coefficient for L-shaped line junction using wave amplitude for radiated power. *Source:* Alexander Peiffer.

insight the detailed reference solutions of Langley's single wave results are also presented. The specific L and S-results must be added in order to get the full in-plane solution.

The results for the conversion in the same plate 1 are shown in Figures 8.37 and 8.36. There is a steady plate–internal energy exchange between the in-plane and bending waves.

8.2.6.2 Diffuse Field Coupling Loss Factor

The diffuse sound coupling loss factor is derived by equation (8.10)

$$\langle\tau\rangle_E = \frac{1}{2}\int_0^{\pi} \tau(\varphi)\sin(\varphi)d\varphi$$

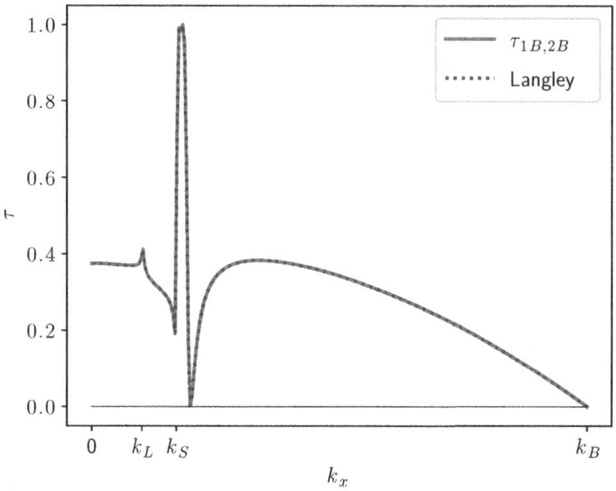

Figure 8.35 Bending to bending transmission coefficient for L-shaped line junction using wave amplitude for radiated power. *Source:* Alexander Peiffer.

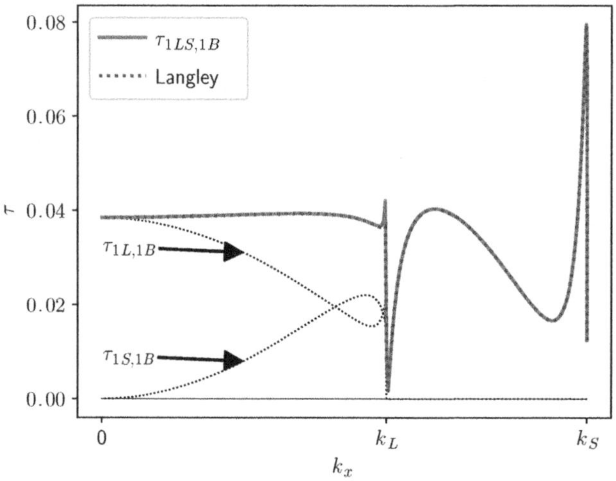

Figure 8.36 Internal in-plane to bending transmission coefficient for plate 1 connected to the L-shaped line junction. *Source:* Alexander Peiffer.

We stay with the integration over φ and use equation $k_x = k_{mv}\cos(\varphi)$ determined by the incoming wave and system. For in-plane excitation the maximum wavenumber at $\varphi = \pi/2$ is $k_{x,max} = k_{mS}$. So, the diffuse transmission coefficient is

$$\langle \tau_{mv,nw} \rangle_E = \frac{1}{2}\int_0^{\pi} \tau_{mv,nw}(k_v^{(m)}\cos(\varphi))\sin(\varphi)d\varphi$$

The coupling loss factor follows from (8.8b)

$$\eta_{mv,nw} = \frac{k_v^{(m)} L_j}{4\pi^2 n_{mv}(\omega)\omega}\int_0^{\pi} \tau_{mv,nw}(k_v^{(m)}\cos(\varphi))\sin(\varphi)d\varphi \tag{8.159}$$

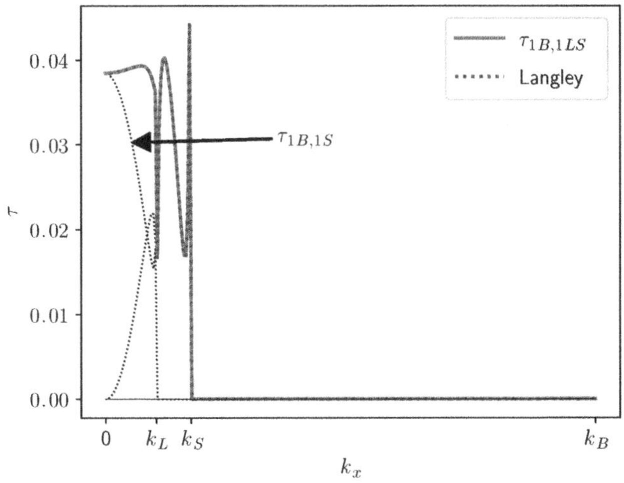

Figure 8.37 Internal bending to in-plane transmission coefficient for plate 1 connected to the L-shaped line junction. *Source:* Alexander Peiffer.

The angle integration in equation (8.10) is motivated by the plane wave model of diffuse fields. When we try to apply this equation for the diffuse in-plane wave field, we cannot specify a clear wavenumber $k_v^{(m)}$, because longitudinal and shear wavenumbers are different. When the integration variable is changed to $k_x = k_v^{(m)} \cos(\varphi)$, the wavenumber is cancelled, and we get the following expression

$$\eta_{mv,nw} = \frac{L_j}{4\pi^2 n_{mv}(\omega)\omega} \int_{-k_x}^{+k_x} \tau_{mv,nw}(k_x) dk_x \tag{8.160}$$

that can be used for in-plane waves, too. The modal densities of longitudinal and shear waves are added to get the in-plane waves modal density:

$$n_{m,LS}(\omega) = n_L(\omega) + n_S(\omega) \tag{8.161}$$

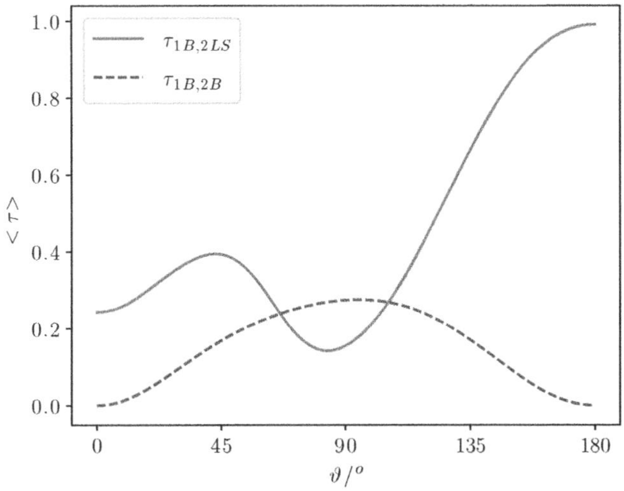

Figure 8.38 Diffuse transmission coefficient for a V-plate junction with subtended angle ϑ and the wavenumber ratio $k_L/k_B = 0.3$. *Source:* Alexander Peiffer.

How strong the conversion between the wave types is depends on the junction angle ϑ, as can be seen, for example, in Figure 8.38 that shows the reference solution taken from figure 2 of Langley and Heron (1990).

8.2.7 Summary

With this collection of junctions, many practical problems can be solved, and this set of junctions provides a general overview of typical coupling phenomena. A weak coupling is usually desired, and the transmission coefficients provide means to calculate the coupling and how to design connections that provide such weak couplings. We have learned that the determination of the radiation stiffness is not always simple. Especially the two- and one-dimensional junctions require complex mathematical analysis. In many cases the derivation must be performed in the wavenumber domain and converted into discrete or diffuse field coefficients by angle- or wavenumber integration.

In addition the discussions of the line junction coupling loss factor show that there are two options for considering specific wave fields as SEA subsystem. We either choose a combination of all wave types in one plate subsystem and use (8.130) for the determination of the coupling loss factor, or we stay with in-plane and bending waves and apply the above described methods. Note, that with wave transmission theory, a separation between shear and longitudinal waves is also possible, but the elegant formulation of the coupling loss factor based on diffuse field reciprocity does not apply to this wave-type selection.

Bibliography

Leo L. Beranek. *Noise and Vibration Control.* McGraw-Hill, Cambridge, MA, second edition, 1988. ISBN 0-9622072-0-9.

Lothar Cremer, Manfred Heckl, and Björn Petersson. *Structure-Borne Sound: Structural Vibrations and Sound Radiation at Audio Frequencies.* Springer Verlag, Berlin, Germany, 3rd edition edition, December 2005. ISBN 978-3-540-26514-6.

Daniel Johansson and Peter Comnell. Statistical Energy Analysis software. Master's thesis, Chalmers University, Göteborg, 2010.

R. S. Langley. Numerical evaluation of the acoustic radiation from planar structures with general baffle conditions using wavelets. *The Journal of the Acoustical Society of America*, 121(2):766–777, 2007.

R. S. Langley and P. J. Shorter. The wave transmission coefficients and coupling loss factors of point connected structures. *The Journal of the Acoustical Society of America*, 113(4):1947–1964, 2003.

R.S. Langley and K.H. Heron. Elastic wave transmission through plate/beam junctions. *Journal of Sound and Vibration*, 143(2):241–253, December 1990. ISSN 0022460X.

F. G. Leppington, E. G. Broadbent, and K. H. Heron. The Acoustic Radiation Efficiency of Rectangular Panels. *Proceedings of the Royal Society A: Mathematical, Physical and Engineering Sciences*, 382(1783):245–271, August 1982. ISSN 1364-5021, 1471-2946.

R.H. Lyon and R.G. DeJong. *Theory and Application of Statistical Energy Analysis, Second Edition*. Butterworth Heinemann, second edition, 1995. ISBN 0-7506-9111-5.

Eduard Ventsel and Theodor Krauthammer. *Thin Plates and Shells: Theory: Analysis, and Applications*. CRC Press, August 2001. ISBN 978-0-8247-0575-6.

9

Deterministic Applications

In technical acoustics deterministic systems are usually treated by numeric methods like the finite element method or the boundary element method. Those methods require complex and powerful solvers as far as pre and post-processors to handle large and detailed models. Even though such models are extremely useful for the simulation of vibroacoustic systems, it is hard to develop a deep understanding of the dynamic phenomena with these numeric models and to draw the right conclusions.

In this book we will treat deterministic systems as far as possible by analytical approaches or by models that consist of sub-elements that can be described by analytical formulas. This allows the reader to follow and understand the details of the theory and may help to provide a deeper understanding of typical vibroacoustic systems. However, even with such constraints, the examples in this chapter are about several deterministic subsystems that are used in real technical systems and create the basement for later SEA or hybrid FEM/SEA examples.

9.1 Acoustic One-Dimensional Elements

One-dimensional acoustic elements are used in the simulation of mufflers, ventilation systems, or hydraulics. The wavelength is assumed to be much larger than the dimension of the cross section. Such systems can become very complex, and they are also used as a designed network in audio systems, for example, the housing and resonators of loudspeakers. In the context of this book, these elements are presented as typical deterministic applications in order to explain the different effects of filters, resonators, and absorbers.

9.1.1 Transfer Matrix and Finite Element Convention

When dealing with one-dimensional systems, the literature often refers to the transfer matrix theory (Pierce, 1991; Mechel, 2002). This approach is useful when the full system is also one-dimensional, meaning that it is a linear chain of subsystems without additional branches.

In Figure 9.1 the convention for both is shown. When the pressure (or the velocity) is used as state variable the mobility matrix reads as (4.11)

$$\begin{bmatrix} \boldsymbol{Y}_{11} & \boldsymbol{Y}_{12} \\ \boldsymbol{Y}_{21} & \boldsymbol{Y}_{22} \end{bmatrix} \begin{Bmatrix} \boldsymbol{p}_1 \\ \boldsymbol{p}_2 \end{Bmatrix} = \begin{Bmatrix} \boldsymbol{v}_{x1} \\ \boldsymbol{v}_{x2} \end{Bmatrix} \tag{9.1}$$

Vibroacoustic Simulation: An Introduction to Statistical Energy Analysis and Hybrid Methods, First Edition. Alexander Peiffer.

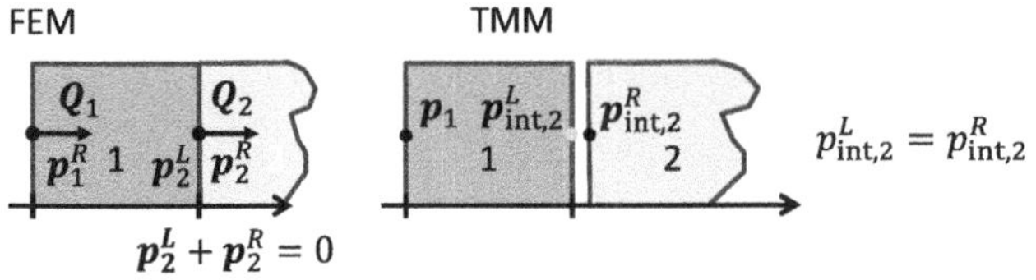

Figure 9.1 Convention for stiffness matrix and TMM. *Source:* Alexander Peiffer.

or the impedance matrix

$$\begin{bmatrix} z_{11} & z_{12} \\ z_{21} & z_{22} \end{bmatrix} \begin{Bmatrix} v_{x1} \\ v_{x2} \end{Bmatrix} = \begin{Bmatrix} p_1 \\ p_2 \end{Bmatrix} \tag{9.2}$$

The velocity in equation (9.2) is a shared internal degree of freedom, and the pressure corresponds to an external pressure as discussed in Xue (2003). Thus, due to the continuity of pressure (or force), we assume for the external pressure $p_2^L = -p_2^R$ on the right hand side.

In the transfer matrix method theory the situation is different – here the pressure is the *internal* pressure or the state variable, and $p_{int,2}^L = p_{int,2}^R$.

To conclude, when we switch from the transfer matrix method presentation to FE, the right hand side *internal* pressure is the negative *internal* pressure: $p_2^R = -p_{int,2}^R$. In the following we will use p_2 in the mobility presentation and p_2' in the transfer matrix approach. The transfer matrix representation of the same system is:

$$\begin{Bmatrix} p_1 \\ v_{x1} \end{Bmatrix} = \begin{bmatrix} T_{11} & T_{12} \\ T_{21} & T_{22} \end{bmatrix} \begin{Bmatrix} p_2' \\ v_{x2} \end{Bmatrix} = \begin{Bmatrix} -p_2 \\ v_{x2} \end{Bmatrix} \tag{9.3}$$

Both representations can be easily exchanged. Solving the above equation for each different state variable gives

$$\frac{1}{T_{12}} \begin{bmatrix} T_{22} & \det([T]) \\ 1 & T_{11} \end{bmatrix} \begin{Bmatrix} p_1 \\ p_2 \end{Bmatrix} = \begin{Bmatrix} v_{x1} \\ v_{x2} \end{Bmatrix} \tag{9.4}$$

$$\begin{Bmatrix} p_1 \\ v_{x1} \end{Bmatrix} = \frac{1}{Y_{21}} \begin{bmatrix} Y_{22} & 1 \\ \det[Y] & Y_{11} \end{bmatrix} \begin{Bmatrix} p_2' \\ v_{x2} \end{Bmatrix} \tag{9.5}$$

When the system consists of a cascade of one-dimensional systems, the transfer matrix method is very convenient

$$\begin{Bmatrix} p_1 \\ v_1 \end{Bmatrix} = [T]_{12} \begin{Bmatrix} p_2 \\ v_1 \end{Bmatrix} \quad \begin{Bmatrix} p_2 \\ v_2 \end{Bmatrix} = [T]_{23} \begin{Bmatrix} p_3 \\ v_3 \end{Bmatrix} \ldots \begin{Bmatrix} p_{N-1} \\ v_{N-1} \end{Bmatrix} = [T]_{N-1,N} \begin{Bmatrix} p_N \\ v_N \end{Bmatrix} \tag{9.6}$$

because the total transfer matrix is the product of all transfer matrices

$$[T]_{1N} = [T]_{12} [T]_{23} \cdots [T]_{N-1,N} \tag{9.7}$$

This makes the calculation fast and simple, because no matrix inversion is involved. However, the FE approach is more straightforward and allows for branches in the total system.

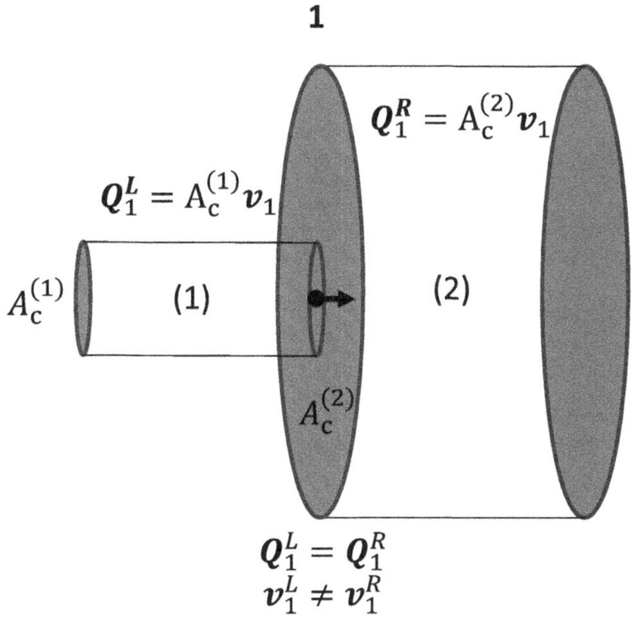

Figure 9.2 Connection of one-dimensional acoustic systems of different cross sections. *Source:* Alexander Peiffer.

9.1.2 Acoustic One-Dimensional Networks

Acoustic networks consist of systems with specific cross sections $A_c^{(i)}$ at both ends. Thus, it is useful in accordance with the finite element formulation from section 4.3.2 to switch from velocity to volume flow. This is the continuous quantity at connections as shown in Figure 9.2

The appropriate matrix equation to describe this would be the radiation mobility matrix with the pressure as internal state variable and the volume flow as external excitation quantity.

$$\begin{bmatrix} \boldsymbol{Y}_{a,11} & \boldsymbol{Y}_{a,12} \\ \boldsymbol{Y}_{a,21} & \boldsymbol{Y}_{a,22} \end{bmatrix} \begin{Bmatrix} \boldsymbol{p}_1 \\ \boldsymbol{p}_2 \end{Bmatrix} = \begin{Bmatrix} Q_1 \\ Q_2 \end{Bmatrix} \tag{9.8}$$

An example for an acoustic network is shown in Figure 9.2. Here the net flow into the nodes is zero when no external volume source is applied. In the final matrix equation, the state variables $\boldsymbol{Q}_i$ are calculated and must be derived from the impedance of the connected and cut-free subsystems. So, the flow $\boldsymbol{Q}_i$ into each node i determines the nodal pressure as state solution. We use the following convention: The flow Q_{in} denotes the volume flow into the node i from the system n.

The equation of motion for the acoustic system as shown in Figure 9.3 is

$$\begin{bmatrix} \boldsymbol{Y}_{a,11} & \boldsymbol{Y}_{a,12} & \cdots & & \boldsymbol{Y}_{a,15} \\ \boldsymbol{Y}_{a,21} & \boldsymbol{Y}_{a,22} & \cdots & & \boldsymbol{Y}_{a,25} \\ \vdots & & \ddots & & \\ \vdots & & & \ddots & \\ \boldsymbol{Y}_{a,51} & \boldsymbol{Y}_{a,42} & & \cdots & \boldsymbol{Y}_{a,55} \end{bmatrix} \begin{Bmatrix} \boldsymbol{p}_1 \\ \boldsymbol{p}_2 \\ \boldsymbol{p}_3 \\ \boldsymbol{p}_4 \\ \boldsymbol{p}_5 \end{Bmatrix} = \begin{Bmatrix} Q_{1,\text{ext}} \\ 0 \\ 0 \\ 0 \\ 0 \end{Bmatrix}. \tag{9.9}$$

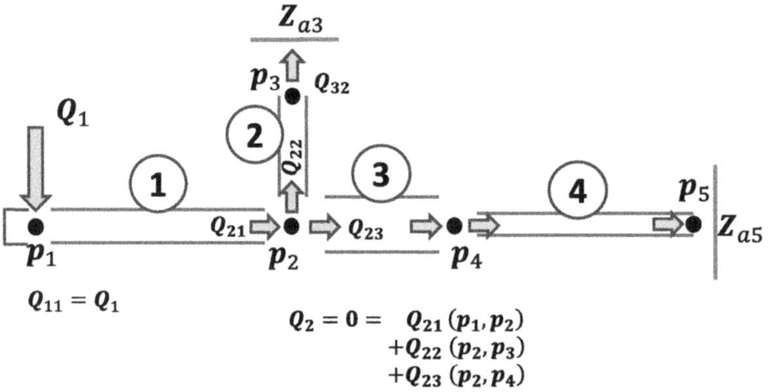

Figure 9.3 Acoustic network with nodal volume flow, volume sources, and radiation impedance at open ends. The numbers in circles denote the system numbers. *Source:* Alexander Peiffer.

Equation (9.9) is derived by using the element mobility from (9.8) and adding each element mobility to the total system matrix similar to the procedure described in section 4.3.1 but with a different source term.

This finite element formulation is efficient when used as a numerical solution but not when analytical expressions are required. The network equation must be solved or inverted to get the system response. Inverting the analytical formulas is not easily done or possible in many cases, and the transfer matrices are more useful.

According to the discussions regarding the transfer matrices in section 9.1.1, we can change easily between the different formulations. The transfer matrix $[T_a]$ reads as:

$$\begin{Bmatrix} p_1 \\ Q_1 \end{Bmatrix} = \begin{bmatrix} T_{a,11} & T_{a,12} \\ T_{a,21} & T_{a,22} \end{bmatrix} \begin{Bmatrix} p_2' \\ Q_2 \end{Bmatrix} = \begin{Bmatrix} -p_2 \\ Q_2 \end{Bmatrix} \tag{9.10}$$

and the conversion between each representation is

$$\frac{1}{T_{a,12}} \begin{bmatrix} T_{a,22} & \det([T_a]) \\ 1 & T_{a,11} \end{bmatrix} \begin{Bmatrix} p_1 \\ p_2 \end{Bmatrix} = \begin{Bmatrix} Q_1 \\ Q_2 \end{Bmatrix} \tag{9.11}$$

$$\begin{Bmatrix} p_1 \\ Q_1 \end{Bmatrix} = \frac{1}{Y_{a,21}} \begin{bmatrix} Y_{a,22} & 1 \\ \det[Y_a] & Y_{a,11} \end{bmatrix} \begin{Bmatrix} p_2' \\ Q_2 \end{Bmatrix} \tag{9.12}$$

Thus, for one-dimensional acoustic networks with changing cross section, equation (9.10) may be the best choice.

9.1.2.1 Properties of the System Matrices

From the reciprocity principle some useful properties can be derived. Reciprocity states

$$\frac{Q_1}{p_2} = \frac{Q_2}{p_1} \tag{9.13}$$

Entering this into the two port equation gives for this mobility matrix

$$Y_{a,11} = Y_{a,22} \qquad\qquad Y_{a,12} = Y_{a,21} \tag{9.14}$$

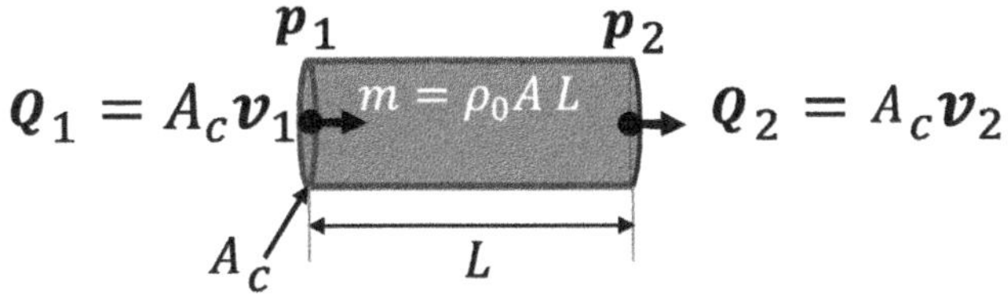

Figure 9.4 Properties of an acoustic pipe. *Source:* Alexander Peiffer.

Hence, they are symmetric. The transfer matrix is obviously not symmetric which can be seen from equations (9.5) and (9.12). From the same equations it can be derived that the determinant of the transfer matrix equals 1

$$\det\left[\boldsymbol{T}\right] = 1 \tag{9.15}$$

9.1.3 The Acoustic Pipe

We take the acoustic pipe from section 4.1 and switch to the volume flow using $\boldsymbol{Q} = A_c\boldsymbol{v}$; we get the radiation mobility matrix formulation with pressure as state variable and the volume flow as external source from equation (4.11):

$$\frac{A_c}{z}\begin{bmatrix} \dfrac{1}{j\tan(\boldsymbol{k}L)} & \dfrac{1}{j\sin(\boldsymbol{k}L)} \\ \dfrac{1}{j\sin(\boldsymbol{k}L)} & \dfrac{1}{j\tan(\boldsymbol{k}L)} \end{bmatrix}\begin{Bmatrix} \boldsymbol{p}_1 \\ \boldsymbol{p}_2 \end{Bmatrix} = \begin{Bmatrix} \boldsymbol{Q}_1 \\ \boldsymbol{Q}_2 \end{Bmatrix} \tag{9.16}$$

The transfer matrix representation reads as:

$$\begin{Bmatrix} \boldsymbol{p}_1 \\ \boldsymbol{Q}_1 \end{Bmatrix} = \begin{bmatrix} \cos(\boldsymbol{k}L) & j\frac{\rho_0 c_0}{A_c}\sin(\boldsymbol{k}L) \\ j\frac{A_c}{\rho_0 c_0}\sin(\boldsymbol{k}L) & \cos(\boldsymbol{k}L) \end{bmatrix}\begin{Bmatrix} \boldsymbol{p}_2' \\ \boldsymbol{Q}_2 \end{Bmatrix} \tag{9.17}$$

9.1.4 Volumes and Closed Pipes

The closed volume was derived in section 4.3.1 leading to equation (4.98). We divide this equation by $j\omega$ in order to get the mobility version with $\boldsymbol{Q}$ as source term

$$j\omega\frac{V}{K}\boldsymbol{p} = \boldsymbol{Q} \tag{9.18}$$

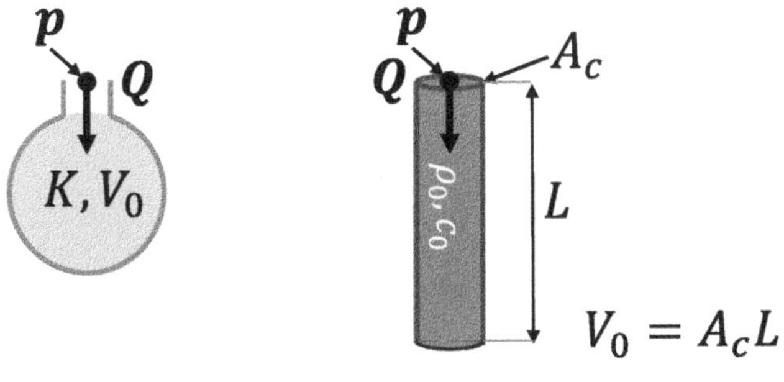

Figure 9.5 Closed volume and pipe. *Source:* Alexander Peiffer.

The bulk modulus can be replaced using (2.18), giving

$$j\omega \frac{V}{\rho_0 c_0^2} \boldsymbol{p} = \boldsymbol{Q} \tag{9.19}$$

The volume formulation assumes a volume extension to be much smaller than the wavelength. For thin cylinder shaped volumes, the one-dimensional pipe formulation with rigid end (4.16) leads to

$$j\frac{A_c \tan(\boldsymbol{k}L)}{z_0} \boldsymbol{p} = \boldsymbol{Q} \tag{9.20}$$

This equation is more appropriate for volumes that have only cross section dimensions that are small compared to the wavelength but can have large dimensions in the direction of sound propagation. However, when $kL = \frac{2\pi\omega L}{c_0} \ll 1$, the tangent can be approximated by $\tan x \approx x$, and equation (9.20) leads to

$$j\frac{A_c kL}{z_0} \boldsymbol{p} = j\omega \frac{A_c L}{\rho_0 c_0^2} \boldsymbol{p} = \boldsymbol{Q} \tag{9.21}$$

This is exactly corresponding to expression (9.19). We conclude with the mobility of the volume and tube, namely

$$\text{volume: } \boldsymbol{Y}_{a,1D} = \boldsymbol{p} = j\omega \frac{A_c L}{\rho_0 c_0^2} \qquad \text{pipe: } \boldsymbol{Y}_{a,1D} = j\frac{A_c \tan(\boldsymbol{k}L)}{z_0} \tag{9.22}$$

and the according impedances from the reciprocal.

9.1.5 Limp Layer

This generic model describes a lumped element in the pipe flow, representing mass, stiffness, or damping effects. The condition for the validity of lumped elements is that the wavenumber must be much larger than the dimension of the element. Thus, we assume $\boldsymbol{v}_1 = \boldsymbol{v}_2$.

As shown in Figure 9.6, the lumped element presentation has in common that the volume flow or velocity is equal on both sides. Thus, the dynamic behavior can be

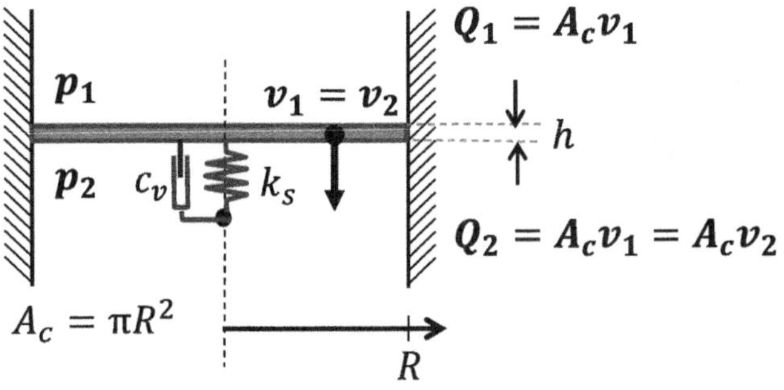

Figure 9.6 Mass and stiffness element in a pipe of cross setion A_c. *Source:* Alexander Peiffer.

described by transfer impedances

$$z_{1D} = \frac{\boldsymbol{p}_2' - \boldsymbol{p}_1'}{\boldsymbol{v}} = \frac{\Delta \boldsymbol{p}}{\boldsymbol{v}} \qquad \boldsymbol{Z}_{a,1D} = \frac{\boldsymbol{p}_2' - \boldsymbol{p}_1'}{A_c \boldsymbol{v}} = \frac{\Delta \boldsymbol{p}}{\boldsymbol{Q}} \tag{9.23}$$

The transfer impedance can be written in the following way:

$$z_{1D} = R_{1D} + jX_{1D} \qquad \boldsymbol{Z}_{a,1D} = R_{a,1D} + jX_{a,1D} \qquad \text{with } \boldsymbol{Z}_{a,1D} = \frac{z_{1D}}{A_c} \tag{9.24}$$

The real and imaginary parts represent the reactive and dissipative parts, respectively. The transfer matrix of such an element is then determined by reshuffling the above equations to an appropriate set

$$\boldsymbol{p}_1' = \boldsymbol{p}_2' + \boldsymbol{Z}_{1D}\boldsymbol{v}_2 \tag{9.25}$$

$$\boldsymbol{v}_1 = \qquad \boldsymbol{v}_2 \tag{9.26}$$

and the transfer matrix of the generic layer reads as:

$$\boldsymbol{T}_{1D} = \begin{bmatrix} 1 & z_{1D} \\ 0 & 1 \end{bmatrix} \qquad \boldsymbol{T}_{a,1D} = \begin{bmatrix} 1 & \boldsymbol{Z}_{a,1D} \\ 0 & 1 \end{bmatrix} \tag{9.27}$$

According to the transformation in (9.11) the radiation mobility matrix representation is

$$\frac{1}{\boldsymbol{Z}_{a,1D}} \begin{bmatrix} 1 & 1 \\ 1 & 1 \end{bmatrix} \begin{Bmatrix} \boldsymbol{p}_1 \\ \boldsymbol{p}_2 \end{Bmatrix} = \begin{Bmatrix} \boldsymbol{Q}_1 \\ \boldsymbol{Q}_2 \end{Bmatrix} \tag{9.28}$$

In the literature the transfer matrix is often used to describe the acoustics of specific layers.

9.1.5.1 Mass

Mass layers can be limp membranes that are closing the pipe or plates that are in the pipe where friction can be neglected. For example, a thin fluid layer of small thickness can also be approximated by a mass layer. Following Newton's law $F = m\ddot{x}$, the equation of motion is

$$A_c(\boldsymbol{p}_1' - \boldsymbol{p}_2') = j\omega m \boldsymbol{v} \Longleftrightarrow z_{1D} = \frac{j\omega m}{A_c} = j\omega m'' \tag{9.29}$$

Similar to the discussion concerning volume and pipe end the mobility expressions can also be derived as an approximation of equation (9.16) for small values of $|\boldsymbol{k}L|$.

9.1.5.2 Stiffness

A stiffness in the pipe can be thought of as an infinitely stiff plate supported by a spring of stiffness k_s or a specific stiffness $k_s'' = k_s/A_c$. The equation of motion for a spring $F = -kx$ leads to

$$A_c(\boldsymbol{p}_1' - \boldsymbol{p}_2') = -k_s \frac{\boldsymbol{v}}{j\omega} \Rightarrow z_{1D} = -\frac{k_s}{j\omega A_c} = -\frac{k_s''}{j\omega} \tag{9.30}$$

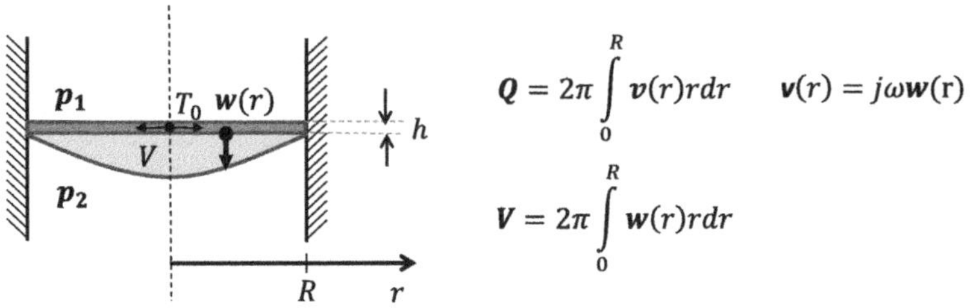

Figure 9.7 Cylindrical membrane exposed to pressure. *Source:* Alexander Peiffer.

9.1.5.3 Viscous Damping

The viscous damping is determined by $F = -c_v v$, and hence

$$A_c(\boldsymbol{p}_1 - \boldsymbol{p}_2') = -c_v \boldsymbol{v} \Rightarrow \boldsymbol{z}_{1D} = -\frac{c_v}{A_c} = -c_v'' = R_{1D} \tag{9.31}$$

We see that the damping results in the real resistance R, whereas the imaginary part considers mass and stiffness effects X. The total transfer impedance is hence

$$\boldsymbol{z}_{1D} = c_v'' + j(\omega m'' + \frac{k_s''}{\omega}) \tag{9.32}$$

9.1.6 Membranes

A practical implementation of a system with mainly mass and stiffness is a membrane with tension T_0 mounted in a circular cross section A_c. From the membrane equation of motion (3.139)

$$T_0 \Delta w = P_1 - P_2 \tag{9.33}$$

and using the cylindrical Laplace operator $\Delta = \frac{1}{r}\frac{\partial}{\partial r} + \frac{\partial^2}{\partial r^2}$ we get

$$\frac{1}{r}\frac{\partial w}{\partial r} + \frac{\partial^2 w}{\partial r^2} = P_1 - P_2 = \Delta P \tag{9.34}$$

When we assume a circular membrane with boundary condition $w(R) = 0$, the solution is

$$w(r) = w_0(1 - \frac{r^2}{R^2}) \text{ with } w_0 = \frac{\Delta P R^2}{4T_0} \tag{9.35}$$

When we derive the limp parameter, we have to use the surface averaged quantities. The stiffness k_s can be defined based on the volume change due to the static pressure:

$$k_s = -\frac{F}{w_{\text{eff}}} = -\frac{A_c \Delta P}{w_{\text{eff}}} = -\frac{A_c^2 \Delta P}{V} \text{ with } w_{\text{eff}} = V/A_c \tag{9.36}$$

The area specific stiffness is defined by $k_s'' = A_c \Delta P / V$ and the volume is given by

$$V = 2\pi \int_0^R w(r) r dr = 2\pi \int_0^R w_0 \left(1 - \frac{r^2}{R^2}\right) r dr = w_0 \frac{\pi R^2}{2} = \frac{\pi R^2}{8T_0} \Delta P \tag{9.37}$$

Thus, the stiffness of the membrane is

$$k_s'' = 8T_0 \tag{9.38}$$

The membrane is displaced non-uniformly. For the efficient mass estimation we use the displacement $w_{\text{eff}} = V/A_c = w_0/2$ and calculate the kinetic energy based on this

$$E_{\text{kin}} = \frac{\omega^2}{2} m_{\text{eff}} w_{\text{eff}}^2 = \frac{\omega}{2} m_{\text{eff}} \frac{w_0}{4} \tag{9.39}$$

The velocity maximum occurs at zero displacement position

$$v_z(r) = Re(j\omega \mathbf{w}(r) e^{j\omega t}) = \omega w(r)$$

This energy must be equal to the kinetic energy of the membrane movement integrated over the given velocity shape

$$\begin{aligned} E_{\text{kin}} &= \frac{1}{2}\rho_0 h 2\pi \int_0^R \hat{v}_z(r)^2 r dr = \frac{1}{2}\rho_0 h 2\pi\omega^2 \int_0^R w_0^2 (1 - \frac{r^2}{R^2}) r dr \\ &= \rho_0 h \omega^2 w_0^2 \frac{\pi R^2}{6} \end{aligned} \tag{9.40}$$

Setting both kinetic energies equal leads to the efficient mass

$$m_{\text{eff}} = \frac{4\rho_0 h \pi R^2}{3} \tag{9.41}$$

or the specific mass

$$m''_{\text{eff}} = \frac{4\rho_0 h}{3} \tag{9.42}$$

which corresponds to $\frac{4}{3}$ of a mass layer. Thus, the final reactance of the membrane is:

$$X_{1D,membrane} = \frac{4}{3}\rho_0 h\omega + \frac{8T_0}{\omega} \tag{9.43}$$

9.1.7 Perforated Sheets

Perforated sheets are a major device for acoustic treatment in engineering acoustics. Consider for example the ceiling of typical offices showing perforated surfaces or surfaces with a regular grid of holes. Such systems provide control over the parameters mass, stiffness, and damping. As a first attempt one might consider the volume in the hole as a mass with $m = \rho_0 L S_{hole}$ plus a specific end correction resulting from a fluid volume partition that is moving on both ends of the channel. But, the channels are supposed to be so small that the wave motion is affected by friction at the walls, and we have a certain flow profile in the hole. A detailed treatment of this theory would go too far, but an established model is derived by Maa (1998) and extended with further details by Fuchs and Zha (1995).

In order to get more insight, only the basic concepts of Maa's model are given here. The pore is assumed to be cylindrical and so small that friction is affecting the flow profile in the fluid. Maa derived that the velocity profile is given by

$$v(r) = -\frac{1}{j\omega\rho_0}\frac{\partial p}{\partial z}\left(1 - \frac{J_0(k_{sh}r\sqrt{-j})}{J_0(k_{sh}R\sqrt{-j})}\right) \tag{9.44}$$

with

$$k_{sh} = \frac{\omega\rho_0}{\eta_{sh}} \tag{9.45}$$

and η_{sh} as shear or dynamic viscosity. In order to get the average or efficient velocity we have to integrate over the cross section

$$v_{\text{eff,hole}} = \frac{2\pi\int_0^R v(r)rdr}{\pi R^2} = -\frac{1}{j\omega\rho_0}\frac{\partial p}{\partial z}\left(1 - \frac{2}{k_{sh}R\sqrt{-j}}\frac{J_1(k_{sh}R\sqrt{-j})}{J_0(k_{sh}R\sqrt{-j})}\right) \tag{9.46}$$

When we assume small hole depth h compared to the wave length we can assume $\frac{\partial p}{\partial z} \approx \delta p/h$ and thus:

$$z_{1D,\text{hole}} = j\omega\rho_0 h\left(1 - \frac{2}{k_{sh}R\sqrt{-j}}\frac{J_1(k_{sh}R\sqrt{-j})}{J_0(k_{sh}R\sqrt{-j})}\right)^{-1} \tag{9.47}$$

So, we found the right expression for a single hole. The holes are covering only part of the surface. Thus, when averaging the velocity over the surface we have to consider this. The ratio of hole surface to total surface is the surface porosity $\sigma' = S_{\text{holes}}/S$, and the efficient velocity related to the total surface S is $v_{\text{eff}} = \sigma' v_{\text{eff, hole}}$
and finally we get

$$z_{1D} = \frac{j\omega\rho_0 h}{\sigma'}\left(1 - \frac{2}{k_{sh}\sqrt{-j}}\frac{J_1(k_{sh}\sqrt{-j})}{J_0(k_{sh}\sqrt{-j})}\right)^{-1}. \tag{9.48}$$

The surface porosity can be derived from the distance of the hole in a square grid as shown in Figure 9.8.

$$\sigma'_{\text{square}} = \frac{\pi R^2}{d^2} \tag{9.49}$$

In principle we have found the transfer impedance that is required to describe the dynamics of a perforated layer. But, there are some modifications required:

1. The flow outside in the nearfield of the pore must be considered.
2. The Bessel functions are quite unwieldy and should be simplified.

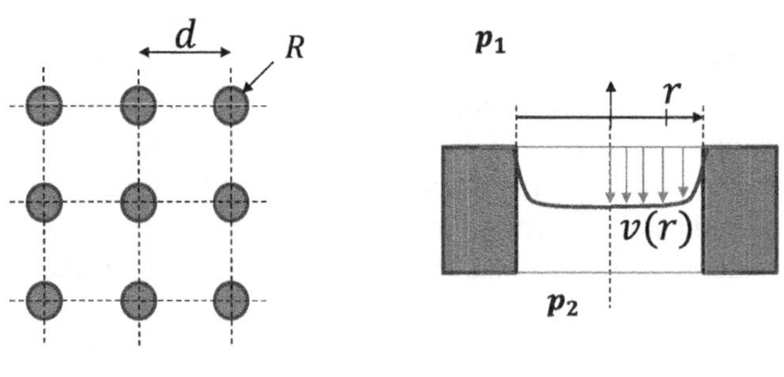

Figure 9.8 Geometry of a perforated plate with square grid. *Source:* Alexander Peiffer.

The first item is addressed by the so called end corrections. They consider the mass of the fluid above and below the pores and the additional friction at the pore edges. The second item is address by an approximation of the Bessel functions that leads to a maximal error of 5% (Fuchs and Zha, 1995).

Finally, the approximation derived by Maa and extended by the nearfield corrections in FreeFieldTechnologies (2015) is:[1]

$$\boldsymbol{Z}_{1D} = R_{1D} + jX_{1D} \tag{9.50}$$

$$R_{1D} = \frac{8\eta h}{\sigma' R^2}\left(\sqrt{1+\frac{(k_{sh}R)^2}{8}} + \frac{\alpha\sqrt{2}k_{sh}R^2}{8h}\right)$$

$$X_{1D} = \frac{\omega\rho_0 h}{\sigma'}\left(1 + \frac{1}{\sqrt{9+\frac{(k_{sh}R)^2}{2}}} + \frac{16R}{3\pi h}(1-f_{int})\right)$$

with α being a constant that is considered to be $\alpha = 4$ for sharp edged holes and $\alpha = 2$ for round edges; f_{int} depends on the porosity σ' and the grid pattern and is given by

$$f_{\text{int}} = (1+\epsilon)\sqrt{\sigma'} - \epsilon\sqrt{\sigma'^3} \tag{9.51}$$

For a squared grid $\epsilon = 0.47$.

9.1.7.1 Example for a Micro Perforated Grid

The task of perforate is to provide specific acoustic mass and flow resistivity for absorber applications. In Figure 9.9 the normalized transfer impedance of a perforate with thickness $h = 2$ mm, hole radius $R = 1$ mm and squared grid distance $d = 2$ mm is shown. The porosity is $\sigma' = 0.007\,2$.

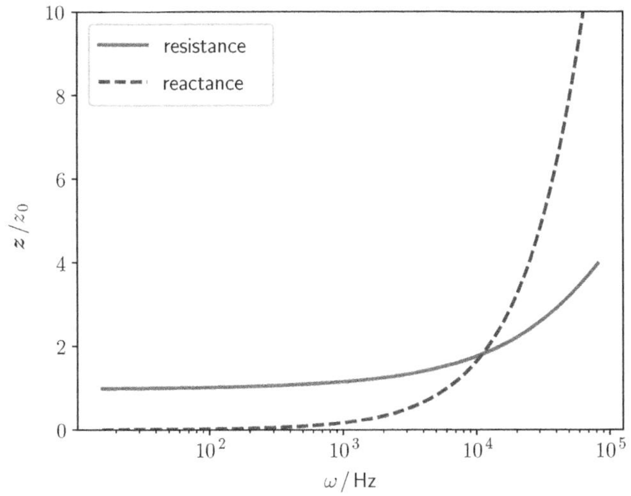

Figure 9.9 Normalized transfer impedance of perforate $h = 2$ mm, $R = 1$ mm, $d = 2$ mm and $\sigma' = 0.007\,2$. *Source:* Alexander Peiffer.

1 The third term corresponds to the moving mass in front of a piston as discussed in section 2.7.3.

The normalized resistance is near 1 thus equal to the impedance of air. This means the perforate is well adapted to plane waves in air. The reactance shows an increase in mass by the nonlinear slope or the reactance curve resulting from viscosity effects in the pores.

9.1.8 Branch Lumped Elements

Additional to the transfer impedance, we define a branch element with radiation impedance $\boldsymbol{Z}_{a,branch}$. The situation is contrary to the transfer impedance. As shown in Figure 9.10, the pressure is constant at both ports, but the volume flow is not.

$$\boldsymbol{p}_1' = \boldsymbol{p}_2' \tag{9.52}$$

$$\boldsymbol{Q}_1 - \boldsymbol{Q}_2 = \frac{\boldsymbol{p}_1}{\boldsymbol{Z}_{a,branch}} \tag{9.53}$$

This leads to the transfer matrix presentation:

$$\boldsymbol{T}_{a,1D} = \begin{bmatrix} 1 & 0 \\ 1/\boldsymbol{Z}_{a,branch} & 1 \end{bmatrix} \tag{9.54}$$

Any expression for this branch equation can be used – for example, the piston radiation impedance or the results from section 9.1.5.

9.1.9 Boundary Conditions

The boundary conditions in the network descriptions are required to define end conditions by the ratio of pressure and velocity. For the volume flow as state variable, an open end corresponds to zero pressure. When pipe systems are given, they may be excited by sources with an inner impedance, and one end might radiate into the free field. Thus, for realistic systems an impedance end condition must be defined. This could be the characteristic impedance of one-dimensional wave propagation or the radiation impedance of a piston (9.95).

The single node expression for the impedance in the finite element expression is

$$[\boldsymbol{Y}_{a,i}]\{\boldsymbol{Q}_i\} = \{\boldsymbol{p}_i\} \qquad [\boldsymbol{Y}_i]\{\boldsymbol{v}_i\} = \{\boldsymbol{p}_i\} \tag{9.55}$$

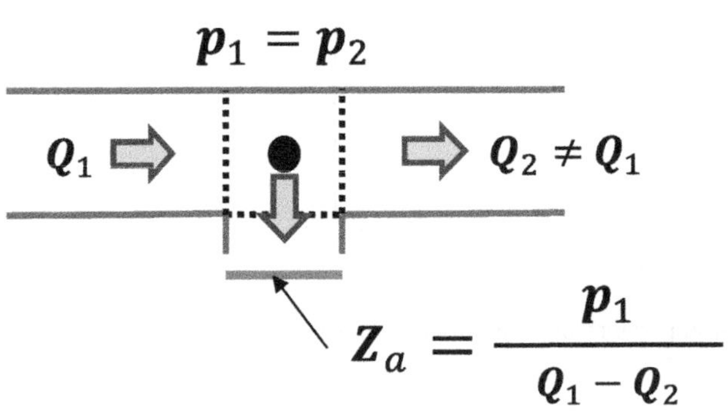

Figure 9.10 Branch impedance configuration of the pipe network. *Source:* Alexander Peiffer.

When we assume the free field boundary condition, the mobility values are

$$\boldsymbol{Y}_{a,i} = \frac{A_c^{(i)}}{z_0} \qquad \boldsymbol{Y}_i = \frac{1}{z_0} \tag{9.56}$$

or in case of the pipe ending in a semi infinite space, the radiating piston would be more appropriate; in this case z_0 must be replaced by the impedance from equation (9.95). For unbaffled configurations the radiating sphere would be more realistic, using equation (2.84) for Z_a.

In case of the transfer impedance description, we set one value of the state to 1, say the velocity. In the cascade transfer matrices, the inputs of the first state vector are

$$\begin{Bmatrix} \boldsymbol{p}_i \\ \boldsymbol{v}_i \end{Bmatrix} = \begin{Bmatrix} z \\ 1 \end{Bmatrix} \qquad \begin{Bmatrix} \boldsymbol{p}_i \\ \boldsymbol{Q}_i \end{Bmatrix} = \begin{Bmatrix} \boldsymbol{Z}_{a,1} \\ 1 \end{Bmatrix} \tag{9.57}$$

$$\begin{Bmatrix} \boldsymbol{p}_i \\ \boldsymbol{v}_i \end{Bmatrix} = \begin{Bmatrix} 1 \\ \boldsymbol{Y}_i \end{Bmatrix} \qquad \begin{Bmatrix} \boldsymbol{p}_i \\ \boldsymbol{Q}_i \end{Bmatrix} = \begin{Bmatrix} 1 \\ \boldsymbol{Y}_{a,1} \end{Bmatrix} \tag{9.58}$$

9.1.10 Performance Indicators

Evaluating such networks aims at a certain effect on the noise propagation. So, we need parameters that describe the effect or performance of such networks. In general the above described systems can be set up and solved for any source and receiver configuration. There are two kinds of indicators: one is a comparative criteria given by the pressure or transmitted power with and without the device, and the other parameter is a local absorption criteria.

9.1.10.1 Transfer and Insertion Coefficients

The performance indicator results from the comparison of a reference system to the newly designed system. The ratio of squared result variables is called the insertion loss as defined in equation (2.169)

$$IL = 10 \log_{10} \frac{p_{\text{out}}^2}{p_{\text{in}}^2} = 20 \log_{10} \frac{p_{\text{out}}}{p_{\text{in}}} \tag{2.169}$$

A further quantity for system performance is the transmission coefficient.

$$\tau = \frac{\Pi_{\text{in}}}{\Pi_{\text{out}}}$$

In Chapter 8 the transmission coefficient definition results from diffuse field reciprocity and the coupling of semi infinite systems. Equations (8.8a) and (7.27) can be used to calculate the transmission factor for a one-dimensional system by adjusting the mobility of impedance degrees of freedom to the stiffness coordinates. In the literature there are several derivations using the transfer matrix approach, for example Allard and Atalla (2009), Tageman (2013), or Jacobsen (2011). Here, the mobility matrix approach will be used.

As can be seen in Figure 9.11, an arbitrary system is extended by the radiation mobilities $\boldsymbol{Y}_s$ and $\boldsymbol{Y}_r$ for sender and receiver, respectively. The input power is given

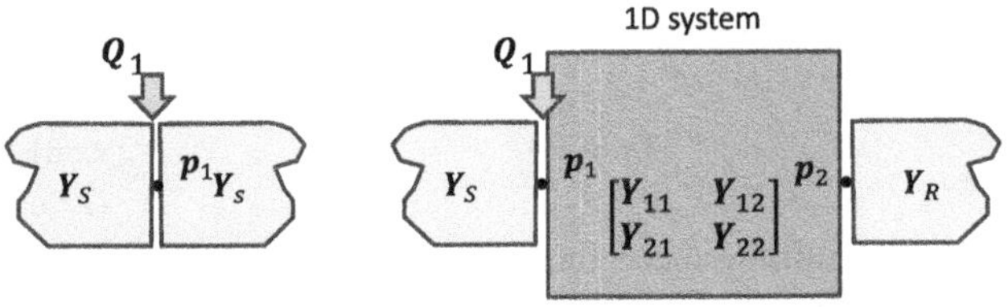

Figure 9.11 Reference and test system configuration for the definition of the transmission. *Source:* Alexander Peiffer.

by

$$\Pi_{\text{in}} = \frac{1}{2}\hat{p}_1^2 Re \boldsymbol{Y}_s \tag{9.59}$$

because the pressure and the (internal) volume flow are given by the entry boundary condition. The reference system is described by two connected equal mobilities $\boldsymbol{Y}_s$, so an external source $\boldsymbol{Q}_1$ leads to the pressure

$$\boldsymbol{p}_1 = \frac{\boldsymbol{Q}_1}{2\boldsymbol{Y}_s} \tag{9.60}$$

and finally:

$$\Pi_{\text{in}} = \frac{1}{2} Re\{\boldsymbol{Y}_s\} \frac{Q_1^2}{4\mathbf{Y}_s^2} \tag{9.61}$$

The power radiated to the receiver mobility $\boldsymbol{Y}_r$ is:

$$\Pi_{\text{out}} = \frac{1}{2} p_2^2 Re \boldsymbol{Y}_r \tag{9.62}$$

In a complex network with many nodes, the network matrix equation (9.9) must be solved to calculate the output pressure of the network and the radiation mobilities. In order to cope with an input and output description, the network matrix must be condensed to the external degrees of freedom. Thus, the assumption of a 2x2 matrix is not a constraint of generality, and the result can be compared to the transfer matrix theory. A general one-dimensional system as shown in Figure 9.11 with radiation impedance endings is described by:

$$\begin{bmatrix} \boldsymbol{Y}_{a,11} + \boldsymbol{Y}_{a,s} & \boldsymbol{Y}_{a,12} \\ \boldsymbol{Y}_{a,21} & \boldsymbol{Y}_{a,22} + +\boldsymbol{Y}_{a,r} \end{bmatrix} \begin{Bmatrix} \boldsymbol{p}_1 \\ \boldsymbol{p}_2 \end{Bmatrix} = [\boldsymbol{Y}_{a,tot}] \begin{Bmatrix} \boldsymbol{p}_1 \\ \boldsymbol{p}_2 \end{Bmatrix} = \begin{Bmatrix} \boldsymbol{Q}_1 \\ \boldsymbol{Q}_2 \end{Bmatrix} \tag{9.63}$$

Matrix inversion gets $\boldsymbol{p}_2$ from $\boldsymbol{Q}_1$, and we consider that there is no source at port 2 hence $\boldsymbol{Q}_2 = 0$:

$$\begin{Bmatrix} \boldsymbol{p}_1 \\ \boldsymbol{p}_2 \end{Bmatrix} = \frac{1}{\det[\boldsymbol{Y}_{a,tot}]} \begin{bmatrix} \boldsymbol{Y}_{a,22} + \boldsymbol{Y}_{a,r} & -\boldsymbol{Y}_{a,21} \\ -\boldsymbol{Y}_{a,12} & \boldsymbol{Y}_{a,11} + +\boldsymbol{Y}_{a,s} \end{bmatrix} \begin{Bmatrix} \boldsymbol{Q}_1 \\ 0 \end{Bmatrix} \tag{9.64}$$

So, the pressure is given by

$$\boldsymbol{p}_2 = \frac{-\boldsymbol{Y}_{a,12}}{det([\boldsymbol{Y}])} \boldsymbol{Q}_1 \tag{9.65}$$

Entering this into (9.62) gives

$$\Pi_{out} = \frac{1}{2}p_2^2 Re\boldsymbol{Y}_r \frac{\boldsymbol{Y}_{a,12}^2}{det([\boldsymbol{Y}])^2}\boldsymbol{Q}_1^2 \tag{9.66}$$

This reads for the transmission factor

$$\tau = \frac{\Pi_{out}}{\Pi_{in}} = 4\frac{Re\boldsymbol{Y}_r}{Re\boldsymbol{Y}_s}\frac{\boldsymbol{Y}_{a,12}^2\boldsymbol{Y}_{a,s}^2}{\left|\det\left[\boldsymbol{Y}_{a,tot}\right]\right|^2} \tag{9.67}$$

When we have condensed the network, this expression defines the transmission through the network. We rearrange the details of the total matrix determinant

$$\begin{aligned}
\tau &= \frac{Re\boldsymbol{Y}_r}{Re\boldsymbol{Y}_s}\frac{4\boldsymbol{Y}_{a,12}^2\boldsymbol{Y}_{a,s}^2}{\left|[(\boldsymbol{Y}_{a,11}+\boldsymbol{Y}_{a,s})(\boldsymbol{Y}_{a,22}+\boldsymbol{Y}_{a,r}) - \boldsymbol{Y}_{a,12}+\boldsymbol{Y}_{a,21}]\boldsymbol{Y}_{a,s}\right|^2} \\
&= \frac{Re\boldsymbol{Y}_r}{Re\boldsymbol{Y}_s}\frac{4\boldsymbol{Y}_{a,12}^2\boldsymbol{Y}_{a,s}^2}{\left|\det(\boldsymbol{Y}_a)+\boldsymbol{Y}_{a,11}\boldsymbol{Y}_{a,r}+\boldsymbol{Y}_{a,22}\boldsymbol{Y}_{a,s}+\boldsymbol{Y}_{a,s}\boldsymbol{Y}_{a,r}\right|^2} \\
&= \frac{Re\boldsymbol{Y}_r}{Re\boldsymbol{Y}_s}\frac{4}{\left|\frac{\det(\boldsymbol{Y}_a)}{\boldsymbol{Y}_{a,12}\boldsymbol{Y}_{a,s}}+\frac{\boldsymbol{Y}_{a,11}\boldsymbol{Y}_{a,r}}{\boldsymbol{Y}_{a,12}\boldsymbol{Y}_{a,s}}+\frac{\boldsymbol{Y}_{a,22}}{\boldsymbol{Y}_{a,12}}+\frac{\boldsymbol{Y}_{a,r}}{\boldsymbol{Y}_{a,12}}\right|^2}
\end{aligned} \tag{9.68}$$

and convert this into the transmission values using equation (9.12)

$$\tau = \frac{Re\boldsymbol{Y}_r}{Re\boldsymbol{Y}_s}\frac{4}{\left|\boldsymbol{T}_{a,11}+\boldsymbol{T}_{a,12}\boldsymbol{Y}_{a,r}+\frac{\boldsymbol{T}_{a,21}}{\boldsymbol{Y}_{a,s}}+\boldsymbol{T}_{a,22}\frac{\boldsymbol{Y}_{a,r}}{\boldsymbol{Y}_{a,s}}\right|^2} \tag{9.69}$$

corresponding to expressions in the literature. When both media are the same, $\boldsymbol{Y}_{a,s} = \boldsymbol{Y}_{a,r}$, then

$$\tau = \frac{4}{\left|\boldsymbol{T}_{a,11}+\boldsymbol{T}_{a,12}\boldsymbol{Y}_{a,r}+\frac{\boldsymbol{T}_{a,21}}{\boldsymbol{Y}_{a,s}}+\boldsymbol{T}_{a,22}\right|^2} \tag{9.70}$$

9.1.10.2 Absorption

The absorption is also derived in Chapter 2, namely by equation (2.105) for $\vartheta = 0$. The relevant quantity in this context is the input impedance. In the network context we have to solve the FE equation for loads at the input, say $\boldsymbol{Q}_1$. The input impedance follows from the resulting pressure $\boldsymbol{p}_1$

$$\boldsymbol{Z}_a = \frac{\boldsymbol{p}_1}{\boldsymbol{Q}_1} \tag{9.71}$$

leading to the selected absorption depending on the impedance of the connected system.

$$\alpha_s = (1 - |\boldsymbol{R}|^2) \text{ and } \boldsymbol{R} = \frac{\boldsymbol{Z}_a - Z_{a,s}}{\boldsymbol{Z}_a + Z_{a,s}} \tag{9.72}$$

9.2 Coupled One-Dimensional Systems

In Section 9.1 the finite element representation of typical systems was developed to simulate acoustic networks. In the following we will elaborate some examples created by such subsystems.

9.2.1 Change in Cross Section

The changes in cross section motivated the selection of the volume flow as state variable. As a consequence the cross section element is supposed to be very simple, and the transfer matrix is the unit matrix

$$[\boldsymbol{T}]_{\text{change}} = \begin{bmatrix} 1 & 0 \\ 0 & 1 \end{bmatrix} \tag{9.73}$$

Assuming the same fluid on both sides, the transfer coefficient reads, with $\boldsymbol{Y}_s = \frac{A_c^{(1)}}{\rho_0 c_0}$ and $\boldsymbol{Y}_r = \frac{A_c^{(2)}}{\rho_0 c_0}$:

$$\tau = \frac{A_c^{(2)}}{A_c^{(1)}} \frac{4}{\left|1 + \frac{A_c^{(2)}}{A_c^{(1)}}\right|^2} = \frac{4A_c^{(1)}A_c^{(2)}}{(A_c^{(1)} + A_c^{(2)})^2} \tag{9.74}$$

9.2.2 Impedance Tube

The impedance tube is a device to measure the surface impedance of a specimen, for example a layer of foam, an absorber, or natural wall surfaces. In Figure 9.12 the impedance tube set-up is shown.

We search for the dependency of the two measured pressures at positions x_1 and x_2 from the contact impedance at the end. In principle equation (4.15) can be applied directly, but in order to present the use of the transfer matrix method, we use it here.

The transfer matrices between x_1 and $x = 0$ as far as x_1 and x_2 are given by

$$[T_i] = \begin{bmatrix} \cos(kL_i) & j\frac{z_0}{A_c}\sin(kL_i) \\ j\frac{A_c}{z_0}\sin(kL_i) & \cos(kL_i) \end{bmatrix} \quad \text{with } L_1 = x_1 \quad L_2 = x_2 - x_1 \tag{9.75}$$

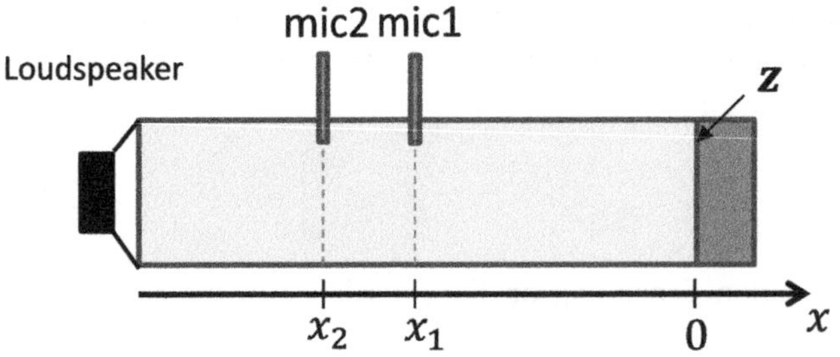

Figure 9.12 Impedance tube for measuring the surface impedance of a flat probe. *Source:* Alexander Peiffer.

The end condition is given by (9.57) with $\boldsymbol{Z}_a = \boldsymbol{z}/A_c$. The pressure $\boldsymbol{p}_1 = \boldsymbol{p}(x_1)$ follows from

$$\begin{Bmatrix} \boldsymbol{p}_1 \\ \boldsymbol{Q}_1 \end{Bmatrix} = [T_1] \begin{Bmatrix} \boldsymbol{Z}/A_c \\ 1 \end{Bmatrix} \tag{9.76}$$

and $\boldsymbol{p}_2 = \boldsymbol{p}(x_2)$ from

$$\begin{Bmatrix} \boldsymbol{p}_2 \\ \boldsymbol{Q}_2 \end{Bmatrix} = [T_2] \begin{Bmatrix} \boldsymbol{p}_1 \\ \boldsymbol{Q}_1 \end{Bmatrix} \tag{9.77}$$

Using the pressure ratio $\boldsymbol{H} = \boldsymbol{p}_1/\boldsymbol{p}_2$ and solving for $\boldsymbol{Z}$ gives:

$$\boldsymbol{z} = jz_0 \frac{\boldsymbol{H} \sin(kx_2) - \sin(kx_1)}{\boldsymbol{H} \cos(kx_2) - \cos(kx_1)} \tag{9.78}$$

9.2.3 Helmholtz Resonator

The Helmholtz resonator is the acoustic network representation of a tuned vibration absorber. Spring, mass, and damper are realised as fluid devices. The construction is as shown in Figure 9.13 on the left hand side.

This cascade of involved subsystems can be readily described by transfer matrices. The neck of length L_1 works as a mass of $m = \rho_0 A_c L_1$ and thus with transfer impedance $\boldsymbol{Z}_{a,1D} = j\omega\rho_0 l_1/A_c$ and resulting in the transfer matrix

$$[\boldsymbol{T}]_{\text{neck}} = \begin{bmatrix} 1 & j\omega\rho_0 \frac{\text{eff}}{A_c} \\ 0 & 1 \end{bmatrix} \tag{9.79}$$

For the neck we follow the same argumentation as for the pores in the perforate absorber. We may have different situations as shown in figure 9.14. When there is no perforate at the opening, the resonator may use an end correction representing a fluid cylinder of length L_c on both sides if the jump in cross section is very high, thus $L_{\text{eff}} = L_1 + 2L_{\text{eff}}$. As shown for the low frequency approximation for the piston in the wall (2.156), this length is $L_c = 0.85R$ with R being the radius.

If the opening is covered by a perforate, this end correction can be replaced by the transfer impedance of the perforate. The correction length at the connection of the volume is taken into account by the length correction adjusting L_{eff}. The end correction at the opening is included in the general transfer impedance term.

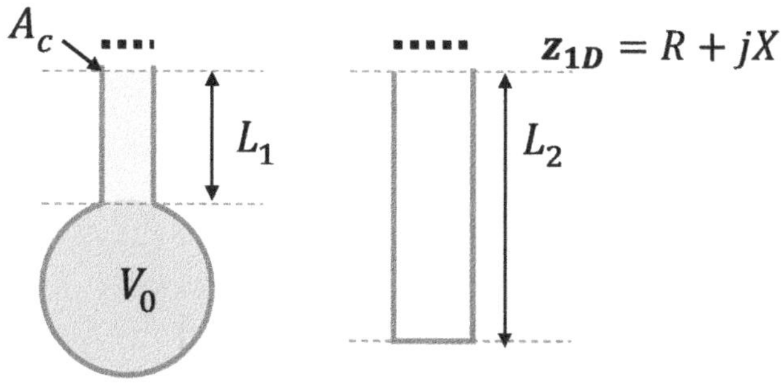

Figure 9.13 Helmholtz (LHS) or quarter wave resonator (RHS). *Source:* Alexander Peiffer.

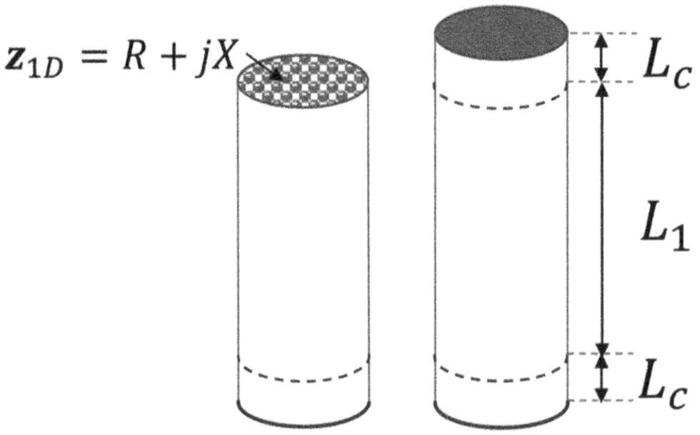

Figure 9.14 Neck tube with different end corrections. *Source:* Alexander Peiffer.

The end condition at the degree of freedom 2 is given by the volume form equation (9.22) and (9.57):

$$\begin{Bmatrix} \boldsymbol{P}_2 \\ \boldsymbol{Q}_2 \end{Bmatrix} = \begin{Bmatrix} \frac{\rho_0 c_0^2}{j\omega V_0} \\ 1 \end{Bmatrix} \tag{9.80}$$

The classical Helmholtz resonator is open, but in some applications the neck is covered by a perforate. In order to keep the option free, we represent the perforate by the generic transfer impedance $R_a + jX_a$

$$[\boldsymbol{T}]_{\text{perf}} = \begin{bmatrix} 1 & R_a + jX_a \\ 0 & 1 \end{bmatrix} \tag{9.81}$$

In case of an open neck, the reactance is determined by the end correction $2L_c$ as discussed before:

$$X_a = \omega\rho_0 \frac{L_c}{A_c} \qquad R_a = 0 \tag{9.82}$$

When covered by a perforate, the transfer impedance is given by equation (9.50) converted into a radiation impedance

$$\boldsymbol{Z}_a = \frac{\boldsymbol{z}_{\text{perf}}}{A_c} \tag{9.83}$$

The total transfer matrix follows from

$$[\boldsymbol{T}]_{\text{helm}} = [\boldsymbol{T}]_{\text{perf}} [\boldsymbol{T}]_{\text{neck}} \tag{9.84}$$

and the surface impedance is given by

$$\begin{Bmatrix} \boldsymbol{P}_1 \\ \boldsymbol{Q}_1 \end{Bmatrix} = [\boldsymbol{T}]_{\text{perf}} [\boldsymbol{T}]_{\text{neck}} \begin{Bmatrix} \boldsymbol{Z}_{a,\text{volume}} \\ 1 \end{Bmatrix} = \begin{bmatrix} 1 & j\omega\rho_0 \frac{l_1}{A_c} + (R + jX) \\ 0 & 1 \end{bmatrix} \begin{Bmatrix} \frac{\rho_0 c_0^2}{j\omega V_0} \\ 1 \end{Bmatrix} \tag{9.85}$$

Thus,

$$\boldsymbol{Z}_{a,helm} = R_a + j\left(\omega\rho_0\frac{L_1}{A_c} + X - \frac{\rho_0 c_0^2}{\omega V_0}\right) \tag{9.86}$$

The system is in resonance when the imaginary part is zero, and for the pure resonator without porous sheet we get

$$\omega_0 = \sqrt{\frac{c_0^2}{V_0}\frac{A_c}{L_1 + 2L_c}} \tag{9.87}$$

At resonance the input impedance is $\boldsymbol{Z}_{a,helm} = R_a$. Thus, for specific frequencies the Helmholtz resonator creates a matching end that would not be possible at low frequencies with such small dimensions. In Figure 9.15 the radiation impedance of one example resonator is shown. At resonance ($\omega_0 = 1876.\text{s}^{-1}$) the reactance curve crosses zero creating a purely resistive impedance at resonance frequency.

When we need a perfectly matched end at a specific frequency, the Helmholtz resonator is the best choice for restricted space. Using the perforate of section 9.1.7.1 at the ending, the resonance moves to higher frequencies. This is because the upper mass is missing, but the resistance matches perfectly $Z_a = \rho_0 c_0/A_c$. The results are shown in Section 9.2.4 in combination with the quarter wave absorber.

9.2.4 Quarter Wave Resonator

The quarter wave resonator is similar to the Helmholtz resonator except the fact that the resonator does not clearly separate between the mass (neck) and spring (volume) part. It consists of a tube with length l_1 with rigid end and the same end correction as for the Helmholtz resonator neck. From section 4.1.1 and equation (4.17), we know the

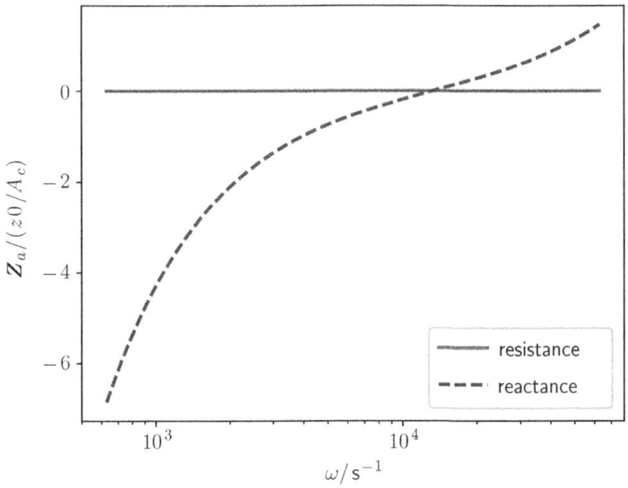

Figure 9.15 Radiation impedance of Helmholtz resonator of parameters $R = 2$ mm, $V_{0=1}$ cm^3 and $L_{1=5}$mm. *Source:* Alexander Peiffer.

input impedance of the resonator.

$$\boldsymbol{Z}_{a,pipe} = \frac{z_0}{jA_c \tan(\boldsymbol{k}L_1)} \text{ with } \boldsymbol{k} = k(1 - j\frac{\eta}{2}) \tag{9.88}$$

The impedance is zero for $L_1 = \frac{2n+1}{4}\lambda$ with $n = 0, 1, 2, 3..$ so the resonance frequencies are

$$\omega_n = \frac{(2n+1)\pi c_0}{l_1} \tag{9.89}$$

With the transfer impedance for end correction of perforate consideration

$$\begin{Bmatrix} \boldsymbol{p}_2 \\ \boldsymbol{Q}_2 \end{Bmatrix} = \begin{bmatrix} 1 & R_a + jX_a \\ 0 & 1 \end{bmatrix} \begin{Bmatrix} \boldsymbol{Z}_{a,pipe} \\ 1 \end{Bmatrix} \tag{9.90}$$

Deriving the input impedance from this we get

$$\boldsymbol{Z}_{a,quarter} = \frac{\boldsymbol{p}_2}{\boldsymbol{Q}_2} = R_a + jX_a + \frac{z_0}{jA_c \tan(\boldsymbol{k}L_1)} \tag{9.91}$$

We may consider the transfer matrix of both end correction (9.82) and (9.83). Quarter wave resonators are rarely used without cover. They are covered in most cases by perforate, for example in the liner absorber of turbofan engines. The perforate from the last section can be used together with a resonator length of $L_2 = 10$ mm.

In Figure 9.16 the impedances of both resonators are shown. The resonance frequency of the Helmholtz resonator is lower as for the quarter wave version. This results in better low frequency absorption as shown in Figure 9.17.

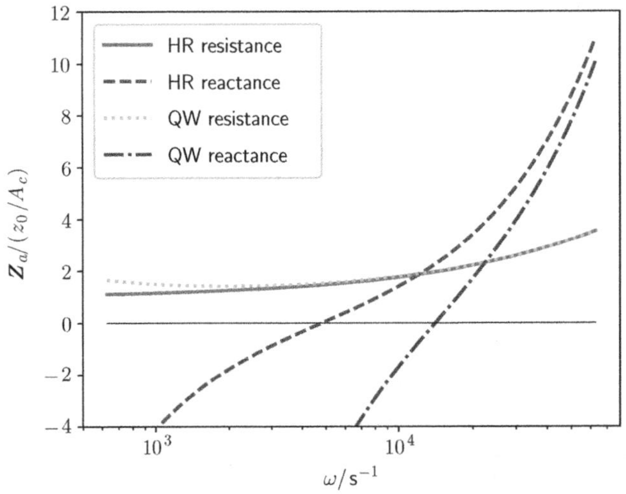

Figure 9.16 Radiation Impedance of Helmholtz resonator of parameters $R = 2$ mm, $V_{0=1}$ cm^3, and $l_{1=5}$ mm and quarter wave resonator $l_{2=10}$ mm and perforated sheet. *Source:* Alexander Peiffer.

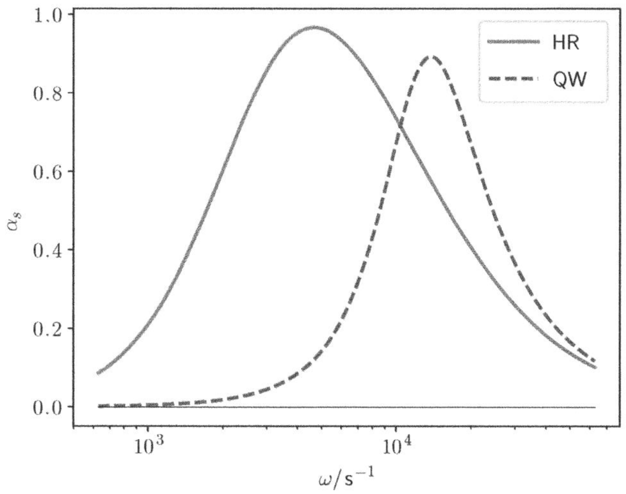

Figure 9.17 Absorption coefficient of Helmholtz resonatorand quarter wave resonator. *Source:* Alexander Peiffer.

9.2.5 Muffler System

Mufflers are applied to reduce, for example, the noise of combustion engines or other machinery that creates pulsating volume flow in the audible frequency range. Here, we neglect any flow component by assuming that the flow speed is much lower than the speed of sound. More details for realistic mufflers can be found in Munjal (1987).

9.2.5.1 Expansion Chamber

A simple muffler consists of a combination of three tubes as shown in Figure 9.18. A pipe of cross section $A_c^{(1)}$ is expanded in the middle to a specific cross section $A_c^{(2)}$; this

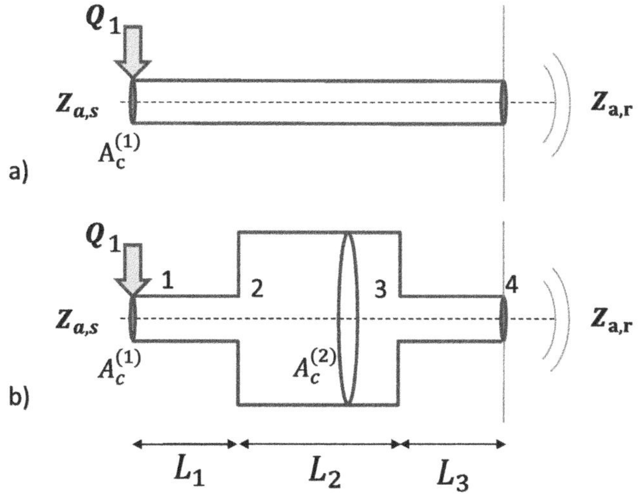

Figure 9.18 Expansion chamber arrangement a) reference system b) expansion system. *Source:* Alexander Peiffer.

leads to 3 tube sections. As this is a cascaded set-up the transfer matrix method can be applied.

The system is described by the matrix product of all sub pipes using equation (9.17) with according length and cross section

$$[T]_{\mathrm{EC}} = [T]_{\mathrm{pipe1}} [T]_{\mathrm{pipe2}} [T]_{\mathrm{pipe3}} \tag{9.92}$$

or we use the mobility form setting up the system matrix.

9.2.5.2 Open end Conditions

We start with the case that source impedance and end impedance are considered as one-dimensional free field $\boldsymbol{Z}_{a,s} = \boldsymbol{Z}_{r,s} = \rho_0 c_0 / A_c^{(1)}$. In this situation the first and third pipe in Figure 9.18b can be neglected, and the total system performance is defined by the expanded chamber in the middle. From equation (9.70) we get

$$\begin{aligned} \tau &= \frac{4}{\left| \cos(\boldsymbol{k}L_2) + j\frac{\rho_0 c_0}{A_c^{(2)}} \frac{A_c^{(1)}}{\rho_0 c_0} \sin(\boldsymbol{k}L_2) + j\frac{\rho_0 c_0}{A_c^{(1)}} \frac{A_c^{(2)}}{\rho_0 c_0} \sin(\boldsymbol{k}L_2) + \cos(\boldsymbol{k}L_2) \right|^2} \\ &= \frac{1}{1 + \frac{1}{4}\left(\frac{A_c^{(2)}}{A_c^{(1)}} + \frac{A_c^{(1)}}{A_c^{(2)}} \right)^2 \sin^2(\boldsymbol{k}L_2)} \end{aligned} \tag{9.93}$$

or expressed as transmission loss

$$TL = 10 \log_{10} \left(\frac{1}{\tau} \right) = 10 \log_{10} \left[1 + \frac{1}{4} \left(\frac{A_c^{(2)}}{A_c^{(1)}} + \frac{A_c^{(1)}}{A_c^{(2)}} \right)^2 \right] \tag{9.94}$$

In Figure 9.19 different transmission losses are shown. The higher the change in cross section, the better the performance. We see that for sources in a well defined frequency range, a transmission loss of more than 30 dB can be achieved. However, when the center chamber is in resonance the transmission loss is zero, and the muffler does not work.

9.2.5.3 Realistic End Conditions

Every muffler pipe radiates finally into the three-dimensional space. Thus, the pipe end will have the piston radiation impedance. In this case the first and third pipe of the system must be included, because reflections will occur at both ports.

The radiation impedance of the piston, according to equation (2.152), is:

$$\boldsymbol{Z}_{a,r} = \frac{\rho_0 c_0}{A_c^{(1)}} \left(1 - \frac{J_1(2kR)}{kR} + j\frac{H_1(2kR)}{kR} \right) \text{ with } A_c^{(1)} = \pi R^2 \tag{9.95}$$

When the matrix is built up as explained in section 9.1.2, we can consider each detail of the setup. We still assume an open condition at the source and compare the muffler to the reference pipe of same total length. The parameters of the expansion chamber are given in Table 9.1; the given center radius corresponds to an area ratio of 10.

The pressure magnitude at the end of the reference tube and expansion chamber is shown on figures 9.20 and 9.21. In the reference chamber the piston end condition

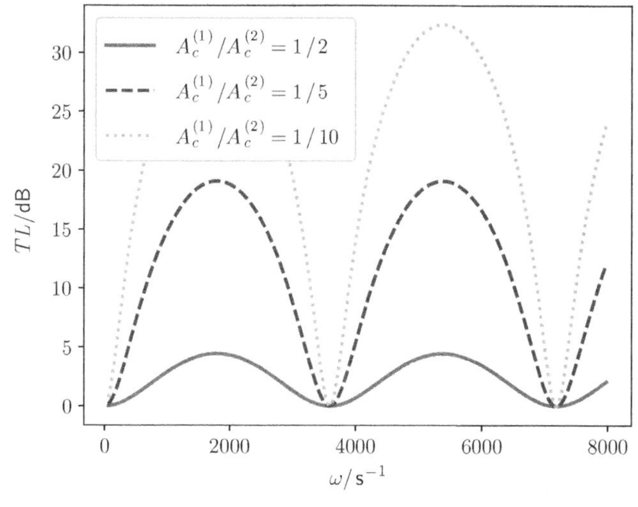

Figure 9.19 Transmission loss of expansion chamber with various cross section ratios; $l_{2=30}$ cm. *Source:* Alexander Peiffer.

Table 9.1 Expansion chamber geometry parameter.

Tube	Length	Radius
1	$L_1 = 20$ cm	5 cm
2	$L_2 = 30$ cm	15.8 cm
3	$L_3 = 20$ cm	5 cm

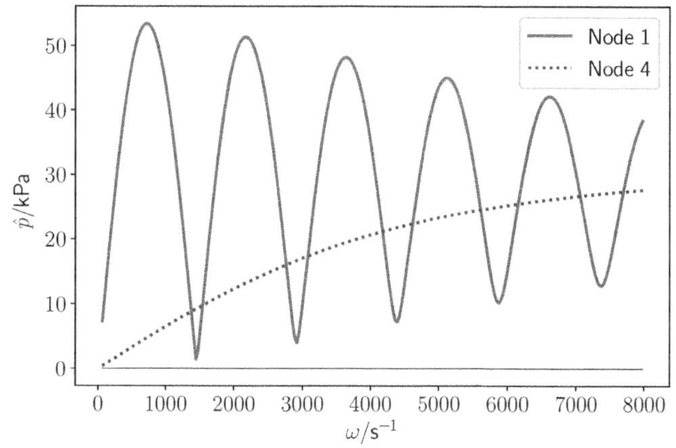

Figure 9.20 Pressure at entry (node1) and end (node4) of the reference tube with flanged end. *Source:* Alexander Peiffer.

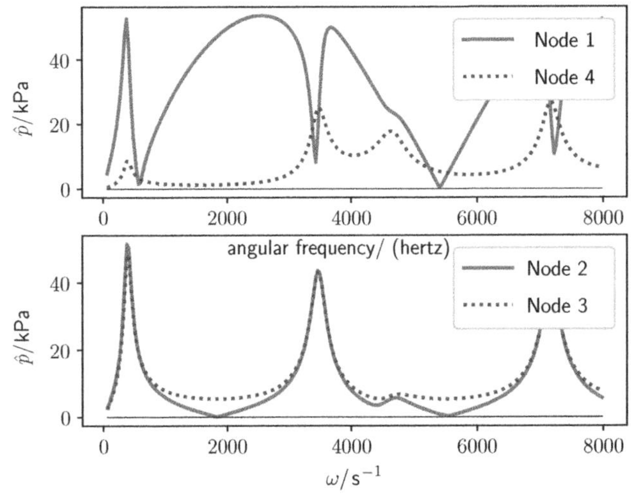

Figure 9.21 Pressure magnitude of expansion chamber at entry (node 1) and end (node 4), and at entry (node 2) and end (node 3) of the expansion chamber. *Source:* Alexander Peiffer.

leads to reflections that are seen at the entry. The expansion chamber shows several resonances coming from the the different pipe sections. Globally the pressure is reduced, but there are still some resonances where the muffler has a weak performance.

The performance is qualified by the insertion loss comparing the pressure of reference and muffler system. The result is shown in figure 9.22. We see that at some frequencies, there are negative values, meaning that the reference system is more efficient than the muffler. Thus, real mufflers require additional damping, realized for example by steel wool, to take care of the resonances.

9.2.6 T-Joint

This system makes use of the Helmholtz resonator as a resonant damper in the one-dimensional propagation path. Such devices are called T-joints and are applied in

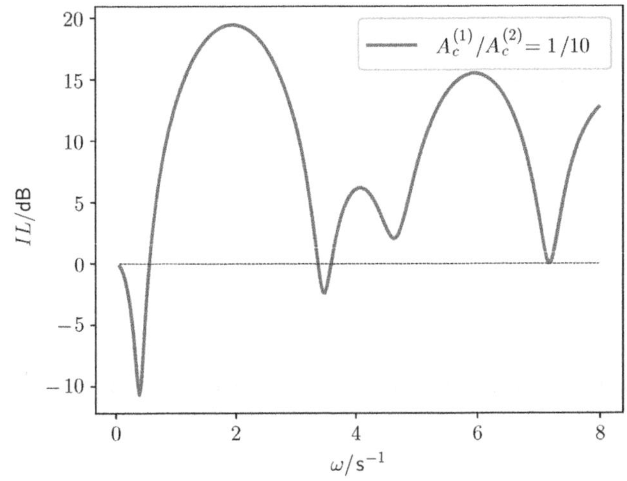

Figure 9.22 Insertion loss of expansion chamber. *Source:* Alexander Peiffer.

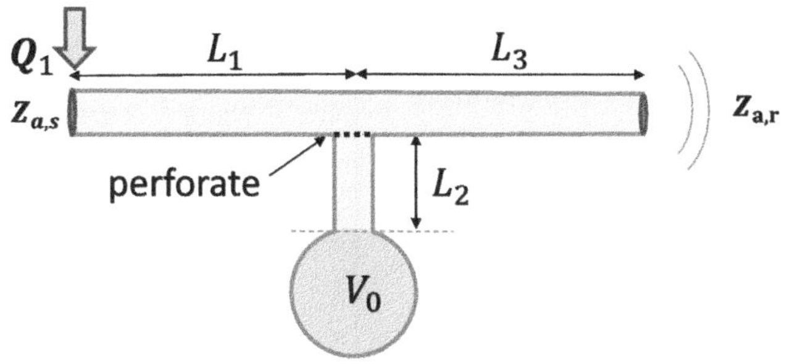

Figure 9.23 Pipe with t-joint and connected Helmholtz resonator. *Source:* Alexander Peiffer.

hydraulic pipes to fight pulsation from hydraulic pumps or to reduce noise in the engine air intake. In Figure 9.23 a typical set-up is shown. The Helmholtz resonator is located in the middle, hence $L_1 = L_3 = 20$ cm, the source impedance is open, and the end dynamics is given by the piston radiation impedance. The system can be described both ways: either the FE method using the Helmholtz resonator impedance as boundary condition at the center node, or by applying the transfer matrix method with a branch impedance.

In this example we use Helmholtz resonators of the following parameters: $R_{HR} = 1$ cm, $V_0 = 100$ cm^3, and $l_2 = 2$ cm. We use the pure Helmholtz resonator configuration and with a perforate cover in order to show the effect of damping. The perforate parameters are thickness $h = 0.2$ mm, hole radius $R = 0.2$ mm, and porosity $\sigma = 0.05$. Such radiation impedance of both resonators is shown in Figure 9.24; we see a relative moderate resistivity of approximately 0.1, and the resonance of both is at $\omega_0 = 2235$ Hz.

With the pure Helmholtz resonator, the insertion loss can be very high, as can be seen in Figure 9.25, but showing some negative loss at some resonances. The perforate version is not as effective but avoids the negative insertion loss.

9.2.7 Conclusions of 1D-Systems

In the last section a set of system and tools was developed to deal with one-dimensional fluid systems, namely pipes or tubes. Besides the practical use of such systems, the idea

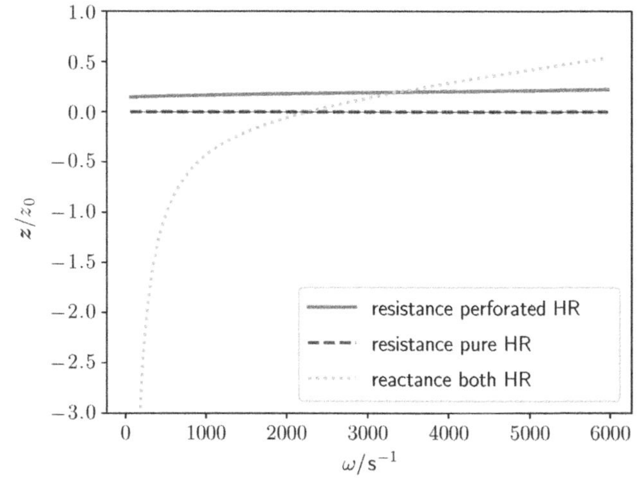

Figure 9.24 Transfer impedance of T-joint perforate $h = 0.2$ mm, $R_{=0.2}$ mm, and $\sigma = 0.05$. *Source:* Alexander Peiffer.

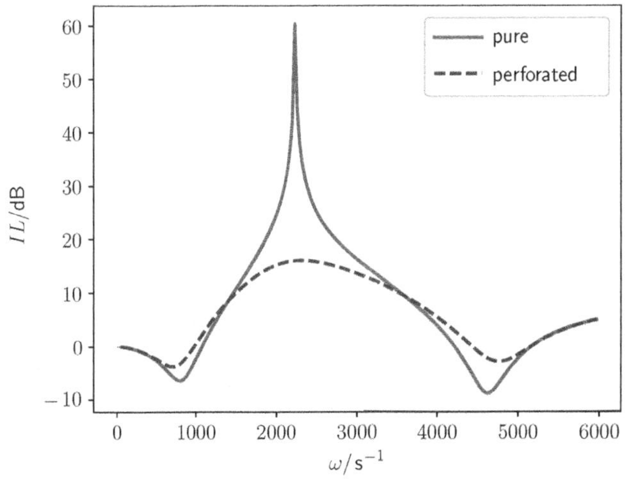

Figure 9.25 Insertion loss of T-joint system with and without perforate. *Source:* Alexander Peiffer.

was to present means of noise control that are based on deterministic and coherent devices. From the point of view of the vibroacoustics engineer, the design and layout means adjusting the different resonances correctly.

9.3 Infinite Layers

Infinite layers are one-dimensional systems which are infinite in the other two dimensions. The infinity guaranties that plane waves can propagate in all directions, and the projection of the wave propagation normal to the plane is a one-dimensional wave motion. From the space vector definition in section 8.2.3, it is clear that the components k_x, k_y of the wavenumber $\boldsymbol{k}$ determines the in-plane state, whereas k_z defines the

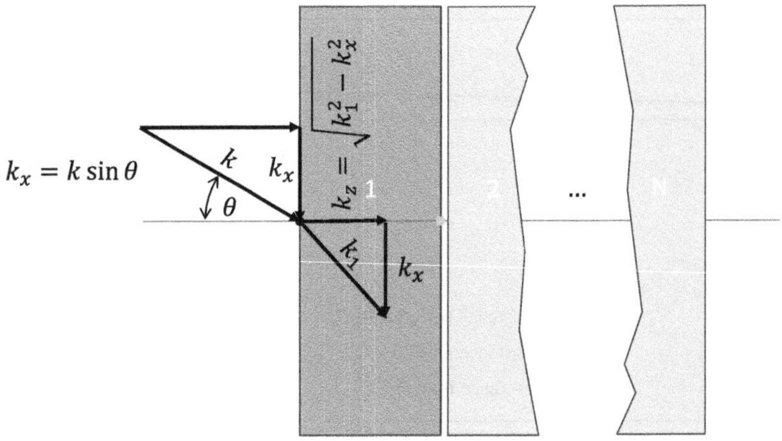

Figure 9.26 Wavenumbers in *x* and *z*-directions for infinitely extended layers. *Source:* Alexander Peiffer.

propagation through the layer. For homogeneous layers we set $k_y = 0$ without loss of generality. We take an infinite layer of fluids and a plane wave impinging at angle θ; the wavenumber parallel to the surface is $k_x = k \sin\theta$. The coordinate in the x-direction is given in wavenumber space by k_x. In case of infinite layers – and only then – k_x remains constant in each layer, and only k_z varies. The in-depth wavenumber k_z depends on the speed of sound in the layer. Thus, it is $\boldsymbol{k}_{z,i} = \sqrt{\boldsymbol{k}_i^2 - k_x^2}$. The transfer matrix method is converted to an infinite layer by adding a wavenumber argument to the transfer matrix.

$$\begin{Bmatrix} \boldsymbol{p}_1(k_x) \\ \boldsymbol{v}_{z1}(k_x) \end{Bmatrix} = \begin{bmatrix} \boldsymbol{T}_{11}(k_x) & \boldsymbol{T}_{12}(k_x) \\ \boldsymbol{T}_{21}(k_x) & \boldsymbol{T}_{22}(k_x) \end{bmatrix} \begin{Bmatrix} \boldsymbol{p}_2(k_x) \\ \boldsymbol{v}_{z1}(k_x) \end{Bmatrix} \tag{9.96}$$

9.3.1 Plate Layer

The mass layer can be enriched by bending stiffness effects using the results from section 8.2.4.3. Using the Ansatz of plane waves again this reads

$$\Delta \boldsymbol{p}(x) = \Delta \boldsymbol{p}(k_x) e^{-j(k_x)} \tag{9.97}$$

For infinite plates the displacement must have the same wavenumber

$$\boldsymbol{v}_z(x) = j\omega \boldsymbol{w}(k_x x, k_y y) e^{-j(k_x x + k_y y)} \tag{9.98}$$

entering this into the wave equation of the plate (3.206) leads to

$$B\left(k_x^4 - k_B^4\right) \boldsymbol{w}(k_x, k_y) = \Delta \boldsymbol{p}(k_x, k_y) \tag{9.99}$$

and the transfer impedance is

$$\boldsymbol{z}(k_x) = \frac{\Delta \boldsymbol{p}(k_x)}{\boldsymbol{v}_z(k_x)} = \frac{\Delta \boldsymbol{p}(k_x)}{j\omega \boldsymbol{w}(k_x)} = j\omega B\left[\left(k_x x^2 + k_y y^2\right)^2 - k_B^4\right] \tag{9.100}$$

After some modification the transfer impedance reads as:

$$\boldsymbol{z}(k_x) = j\omega m''\left(1 - \frac{k_x^4}{\boldsymbol{k}_B^4}\right) \text{ with } \boldsymbol{k}_B(\omega) = \frac{m''\omega^2}{B} \tag{9.101}$$

For zero bending stiffness $B \to 0$, the transfer impedance is $\boldsymbol{Z}(k_x) = j\omega m''$ showing that the limp behavior does not depend on the wavenumber k_x. As there is no in-depth wave propagation, the transfer matrix has the same form as (9.27) with the above given transfer impedance.

$$[\boldsymbol{T}]_{\text{plate}} = \begin{bmatrix} 1 & \boldsymbol{z}(k_x) \\ 0 & 1 \end{bmatrix} \tag{9.102}$$

9.3.2 Lumped Elements Layers

In general, the parameters of lumped elements don't change due to the wavenumber argument. So, mass layers and perforates can be used as is. The given transfer impedances according to (9.27) or specifically for the mass layer (9.29) or the perforate

(9.50) are still valid.

$$[\boldsymbol{T}]_{\text{lumped}} = \begin{bmatrix} 1 & z_{1D} \\ 0 & 1 \end{bmatrix}$$

9.3.3 Fluid Layer

The transfer matrix method of one layer excited by a plane wave of k_x, k_z follows from the solutions of the wave equation in the z-direction. We slightly modify equation (4.1)

$$\boldsymbol{p}_i(z, k_x) = \boldsymbol{A}e^{-j\boldsymbol{k}_{z,i}z} + \boldsymbol{B}e^{j\boldsymbol{k}_{z,i}z} \tag{9.103}$$

and (4.2) using (2.35)

$$\boldsymbol{v}_{z_i}(z, k_x) = -\frac{1}{j\omega\boldsymbol{\rho}_i}\frac{\partial \boldsymbol{p}_i}{\partial z} = \frac{\boldsymbol{k}_{z,i}}{\omega\boldsymbol{\rho}_i}(\boldsymbol{A}e^{-j\boldsymbol{k}_{z,i}z} - \boldsymbol{B}e^{j\boldsymbol{k}_{z,i}z}) \tag{9.104}$$

keeping in mind that

$$\boldsymbol{k}_{z,i}(k_x) = \sqrt{\boldsymbol{k}_i^2 - k_x^2} \tag{9.105}$$

is a function of k_x. We get in accordance with the one-dimensional fluid transfer matrix method

$$[\boldsymbol{T}]_{i,fluid}(k_x) = \begin{bmatrix} \cos(\boldsymbol{k}_{z,i}h) & j\frac{\omega\boldsymbol{\rho}_i}{\boldsymbol{k}_{z,i}}\sin(\boldsymbol{k}_{z,i}h) \\ j\frac{\boldsymbol{k}_{z,i}}{\omega\boldsymbol{\rho}_i}\sin(\boldsymbol{k}_{z,i}h) & \cos(\boldsymbol{k}_{z,i}h) \end{bmatrix} \tag{9.106}$$

9.3.4 Equivalent Fluid – Fiber Material

Fiber materials are used for acoustic isolation and absorption. The acoustic fluid motion in the fiber network leads to losses in the flow and wave propagation. The deceleration of the fluid motion acts on the fiber matrix and accelerates the fibers. A model for such porous materials is mandatory for noise control application, but the required theory is out of scope for this book. Allard et al. (2005) provides a comprising overview about the acoustic theory and application of porous material.

We use the model of the *limb equivalent fluid* that was developed by Champoux for the rigid or fixed matrix Champoux and Stinson (1992) and extended by the limb frame model as described, for example, by Paneton Panneton (2007). The useful thing with this model is that the acoustics in the fiber material are still described by acoustic fluid parameters. They become complex and frequency dependent, but the existing models

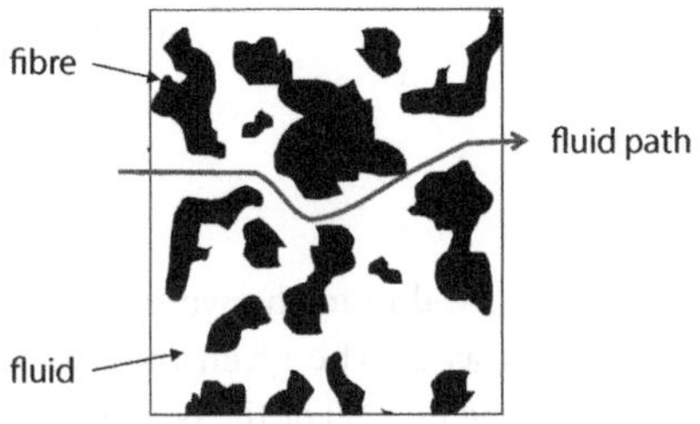

Figure 9.27 Sketch of fiber absorber. *Source:* Alexander Peiffer.

of fluid layers or tubes can still be used. A thorough description by complex models – as for example Biot models Biot (1962) – considering the two coupled waves in the solid and fluid phase, require a completely different approach handling several degrees of freedom and connectivity conditions. In order to keep the efforts reasonable, we leave it with the simple presentation of the final formulas:

$$\boldsymbol{G}_j(\omega) = \left(1 + \frac{4j\alpha_\infty^2 \eta_V \rho_0 \omega}{\sigma^2 \Lambda^2 \Phi^2}\right)^{1/2} \tag{9.107a}$$

$$\boldsymbol{\rho}_{\text{eq}} = \frac{\alpha_\infty \rho_0}{\Phi}\left(1 + \frac{\sigma\Phi}{j\omega\rho_0\alpha_\infty}\boldsymbol{G}_j(\omega)\right) \tag{9.107b}$$

$$\boldsymbol{\rho}_{\text{limp}} = \frac{\boldsymbol{\rho}_{\text{eq}}\rho_{\text{bulk}} - \rho_0^2}{\rho_{\text{bulk}} + \boldsymbol{\rho}_{eq} - 2\rho_0} \tag{9.107c}$$

$$\boldsymbol{K} = \frac{\rho_0 c_0^2}{\kappa - (\kappa - 1)\left[1 + \frac{8\eta}{j\Lambda'^2 Pr\omega\rho_0}\sqrt{1 + j\rho_0 \frac{\omega Pr \Lambda'^2}{16\eta}}\right]^{-\frac{1}{2}}} \tag{9.107d}$$

$$\boldsymbol{z} = \sqrt{\boldsymbol{K}\boldsymbol{\rho}_{\text{limp}}} \tag{9.107e}$$

$$\Gamma = j\omega\sqrt{\boldsymbol{\rho}_{\text{limp}}/\boldsymbol{K}} \tag{9.107f}$$

$$\boldsymbol{c} = j\omega/\Gamma \tag{9.107g}$$

There is a confusing variety of parameters that are not all independent as shown by Horoshenkov et al. (2019). A heuristic explanation of the parameters is given in tables 9.2 and 9.3.

The soft fibrous material from Panneton (2007) is taken as an example. In figures 9.29 and 9.28 the results for the complex speed of sound and the density show a strong dependency on frequency. The are two major versions of equivalent fluid models: the rigid frame model, assuming a fixed fiber matrix with no participation of the fibers to the motion, and the limp frame model that considers forced motion of the frame inertia.

Table 9.2 Fiber parameters of the equivalent fluid model.

Symbol	Description
$\Phi = \frac{V_{\text{fiber}}}{V_{\text{all}}}$	Volume porosity, fraction of fiber and total volume
α_∞	Tortuosity, ratio of average path through the absorber to straight path, a measure for the diversion of the fluid
Λ	Viscous characteristic length
Λ'	Thermal characteristic length
σ	Static air flow resistivity
ρ_{bulk}	Apparent total density of fluid and fiber
$\boldsymbol{\rho}_{\text{eq}}$	Equivalent density of the fluid in a rigid/fixed fiber matrix
$\boldsymbol{\rho}_{\text{limp}}$	Equivalent density of the limp equivalent fluid

Table 9.3 Fluid parameters.

Symbol	Description
ρ_0	Fluid density
η_V	Dynamic viscosity
Pr	Prandtl number

Table 9.4 Material parameters of soft fiber material. *Source:* Panneton (2007).

Symbol	Value	Units
Φ	0.98	
σ_{flow}	25 000	N s/m^4
α_∞	1.02	
Λ^1	90	μm
ρ_0	1.208	kg/m^3
ρ_{bulk}	31.1	kg/m^3

[1] For fibrous material $\Lambda' = 2\Lambda$ can be assumed.

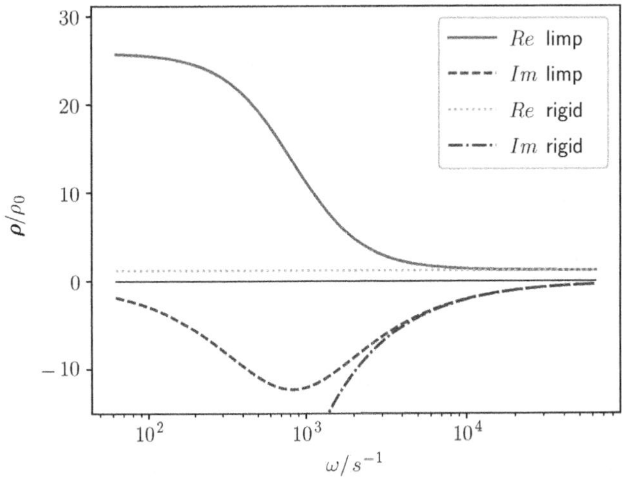

Figure 9.28 Dynamic complex density of limp and rigid fiber material. *Source:* Alexander Peiffer.

One can easily switch from the rigid to the limp model by using equation (9.107c). For low frequencies the real density of the limp model reaches values near the bulk density of the material, because the material moves with the inertia of the bulk material. The rigid frame model does not catch this effect. In the high frequency limit, both models provide the same result. According to Panneton the high frequency limit is

$$\lim_{\omega \to \infty} \rho_{eq} = \frac{\alpha_\infty \rho_0}{\Phi} \tag{9.108}$$

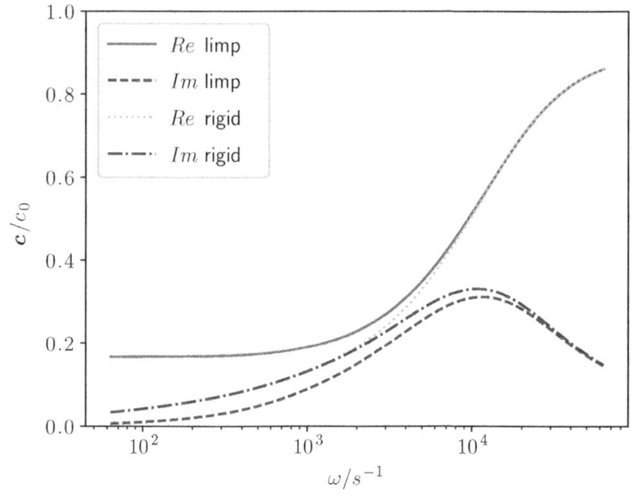

Figure 9.29 Dynamic complex sound speed of limp and rigid fiber material. *Source:* Alexander Peiffer.

A similar effect can be seen for the speed of sound. For high frequencies the real value of the limp model reaches the real value of the rigid model. At low frequencies the speed of sound of the fiber material is below the fluid sound speed because of the additional density from the bulk material. In Panneton (2007) the results are compared to tests, showing that the rigid frame model fails for low frequencies.

To conclude, for the calculation of fiber material sound propagation, we presented a useful model that is, for example, used frequently in aerospace applications (Moeser et al., 2008).

9.3.5 Performance Indicators

The performance indicators from the last section are reused here but with wavenumber or angle as argument and now formulated for the impedance instead of radiation mobility. The source and receiver impedances of the fluids are also dependant on the angle of attack due to equation (2.116). With $\boldsymbol{z}_{s/r} \rightarrow \boldsymbol{z}_{s/r}/\cos(\vartheta)$ we get

$$\tau(\vartheta) = \frac{Re\boldsymbol{z}_s}{Re\boldsymbol{z}_r} \frac{4}{\left|\boldsymbol{T}_{11}(k_x) + \frac{\boldsymbol{T}_{12}(k_x)\cos(\vartheta)}{\boldsymbol{z}_r} + \frac{\boldsymbol{T}_{21}(k_x)\boldsymbol{z}_s}{\cos(\vartheta)} + \boldsymbol{T}_{22}(k_x)\frac{\boldsymbol{z}_r}{\boldsymbol{z}_s}\right|^2} \tag{9.109}$$

The diffuse field performance is derived by equations (8.13) or (8.14). The above formula is often used with an empirical maximum angle ϑ_{max} leading to the formula shown in (8.103). When we are interested in the absorption, we have similar modifications for (9.72)

$$\alpha_s(\vartheta) = (1 - |\boldsymbol{R}(\vartheta)|^2) \text{ and } \boldsymbol{R}(\vartheta) = \frac{\boldsymbol{z}_s(\vartheta) - \frac{z_0}{\cos(\vartheta)}}{\boldsymbol{z}_s(\vartheta) + \frac{z_0}{\cos(\vartheta)}} \tag{9.110}$$

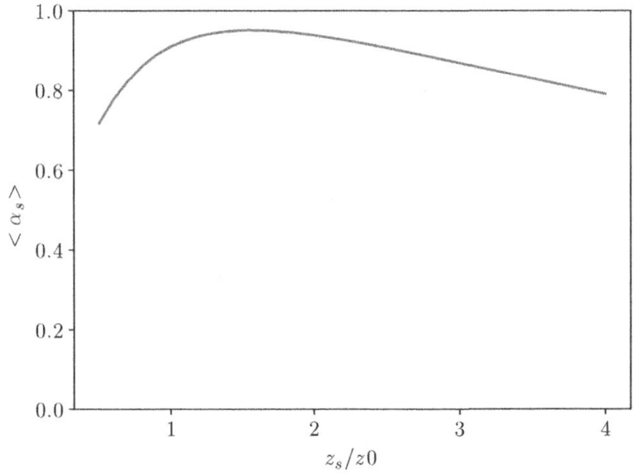

Figure 9.30 Diffuse field absorption for varying values of z_s. *Source:* Alexander Peiffer.

getting an according diffuse field absorption by angular averaging

$$\langle \alpha_s \rangle = 2 \int_0^{\pi/2} \alpha_s(\vartheta) \cos(\vartheta) \sin(\vartheta) d\vartheta \tag{6.72}$$

For normal irradiation the perfect absorption is achieved for $\mathbf{z}_s = \rho_0 c_0$, thus a perfectly matching impedance with zero reactance and resistance equal to the characteristic impedance of air. However, this condition cannot be met for all angles of incidence. In the diffuse field integration the most important angle is $\vartheta = 45°$, because the $\sin\vartheta \cos\vartheta$-term in equation (6.70) has its maximum at this angle.

In Figure 9.30 the diffuse field absorption is calculated for different values of the real surface impedance. We see that highest diffuse field absorption can be achieved by $\mathbf{z}_s = a z_0$ with $a \in [1, 1.5]$. So, the design goal for a best diffuse field absorbing device is given by this rule.

9.3.6 Conclusions on Layer Formulation

The above derived transfer matrices are useful for the determination of transmission (and therefore also coupling) of large area junctions. Energy is removed from the reverberant field by absorption taking place at every reflection. The approximation of infinite layer is the more valid the larger the areas are. In any case this formulation is a powerful tool to derive properties of so-called acoustic treatments, layers of different materials that are used as noise control treatment in many applications.

9.4 Acoustic Absorber

Acoustic absorbers are used to reduce the noise levels in cavities and rooms. The task is to create a lay-up that maximizes the absorption in a specified frequency range with the least space and weight requirements. The target value is $\mathbf{z}_s = \sqrt{2}\rho_0 c_0$, because this leads to perfect matching at an angle of 45° that contributes most to the diffuse absorption coefficient. We start with a simple absorber created by a single layer of fiber material. The next step will be a combined absorber consisting of a fiber absorber plus perforated sheet.

9.4.1 Single Fiber Layer

When placing an absorbing fiber layer in front of a rigid wall, the transfer matrix is connected to a rigid wall, meaning that $v_2 = 0$ in equation (9.96)

$$\begin{Bmatrix} \boldsymbol{p}_1(k_x) \\ \boldsymbol{v}_{z1}(k_x) \end{Bmatrix} = \begin{bmatrix} \boldsymbol{T}_{11}(k_x) & \boldsymbol{T}_{12}(k_x) \\ \boldsymbol{T}_{21}(k_x) & \boldsymbol{T}_{22}(k_x) \end{bmatrix} \begin{Bmatrix} \boldsymbol{p}_2(k_x) \\ 0 \end{Bmatrix} = \begin{Bmatrix} \boldsymbol{T}_{11}(k_x) \\ \boldsymbol{T}_{21}(k_x) \end{Bmatrix} \boldsymbol{p}_2(k_x) \tag{9.111}$$

For the surface impedance this leads to

$$\boldsymbol{z}(k_x) = \frac{\boldsymbol{p}_1(k_x)}{\boldsymbol{v}_z(k_x)} = \frac{\boldsymbol{T}_{11}(k_x)}{\boldsymbol{T}_{21}(k_x)} = \frac{\boldsymbol{z}_0}{j\tan(\boldsymbol{k}h)} \tag{9.112}$$

For first evaluation of our fiber material described in section 9.3.4, we use the fluid transfer matrix (9.106) with the material parameters from (9.107a)–(9.107g). In figure 9.31 the impedance spectrum for $\vartheta = 0$ is shown for materials of different thicknesses.

The reactance of the 20 cm layer shows a first $\lambda/2$ resonance at $\omega = 650\,Hz$ but with no clear peak. The 10 cm version has a local maximum around the double frequency. As damping increases with frequency, the 10 cm resonance is weaker and does not reach zero reactance. When we consider equation (2.104), at higher frequency both layers coincide, because the high damping in the material prevents any feedback from reflected waves.

We see that zero reflection requires a real $\boldsymbol{z}_s = z_s = \rho_0 c_0$ for normal incidence. So, the material shown before seems to be a performing fiber absorber, as resistance is near the impedance of air. In Figure 9.32 the absorption due to normal wave incidence is shown for both layers. The early resonance of the 20 cm layer leads to better performance at

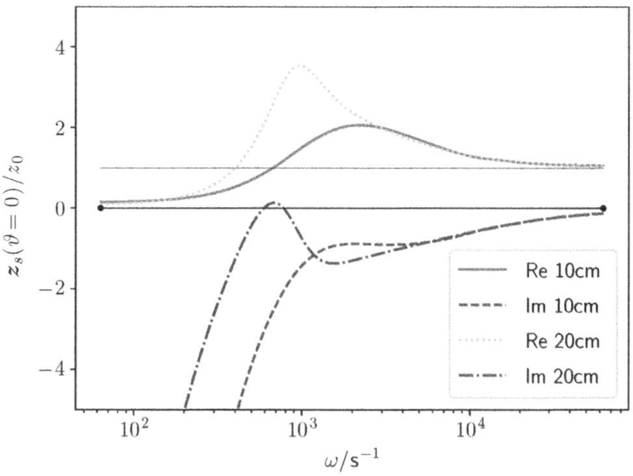

Figure 9.31 Perpendicular surface impedance of single fiber layer in front of a rigid wall. *Source:* Alexander Peiffer.

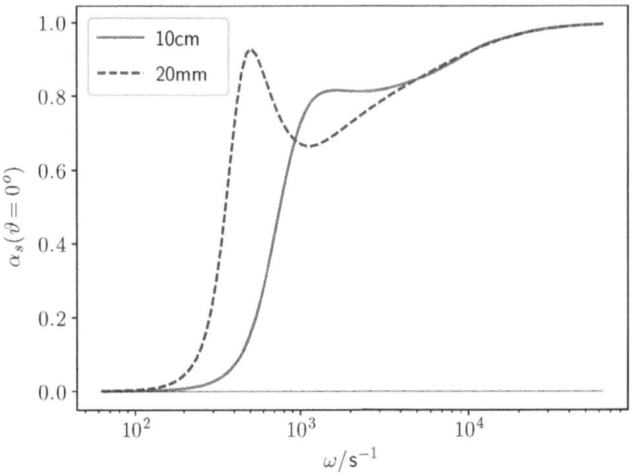

Figure 9.32 Perpendicular surface absorption of a single fiber layer in front of a rigid wall. *Source:* Alexander Peiffer.

lower frequencies with a dip afterwards. At $\omega = 400\mathrm{s}^{-1}$, there is weak absorption for the 10 cm layer, whereas the 20 cm layer provides already $\alpha \approx 0.75$. However, a 20 cm layer requires a lot of space that is not always available. Thus, we look for an option to reduce the thickness but keep the low frequency performance.

The low speed of sound in the fiber material leads to a large wavenumber in the fiber. The consequence is that even oblique waves are diffracted to the surface normal, and the angle dependence to the impedance is quite low. Thus, the normal and diffuse absorption are not that different (figure 9.33).

9.4.2 Multiple Layer Absorbers

In many cases there is not enough space for large thickness absorbers available; or, the environment does not allow for porous materials, because the surface will be exposed

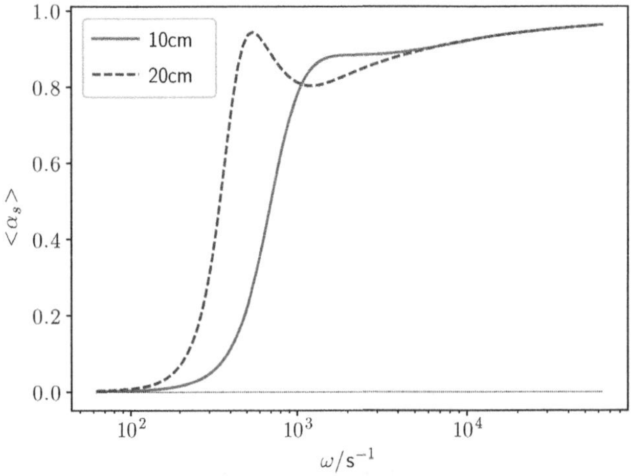

Figure 9.33 Diffuse absorption of a single layer in front of a rigid wall. *Source:* Alexander Peiffer.

to dirt and humidity. we must protect the absorber by a thin layer (thin and soft plates) or use perforate that can be cleaned.

9.4.3 Absorber with Perforate

A perforate with micro absorption may have a transfer impedance with resistance on the order of magnitude of air and with additional mass generated by the neck effect as described in section 9.1.7. This mass can be used to lower the resonance of the total system and therefore to reduce the thickness of the absorber.

The advantage of this concept is that we can separately tune the resistance of the front sheet to air and the thickness of fluid or fiber layer to create zero reactance. Let us assume that the layer generates a surface impedance of z_{s2}. The state vector is then given by

$$\begin{Bmatrix} p_2 \\ v_2 \end{Bmatrix} = \begin{Bmatrix} z_{s2} \\ 1 \end{Bmatrix} v_2 \tag{9.113}$$

The state vector of the front layer follows from matrix multiplication with the limp layer transfer matrix

$$\begin{Bmatrix} p_1 \\ v_1 \end{Bmatrix} = \begin{bmatrix} 1 & z_{\text{perf}} \\ 0 & 1 \end{bmatrix} \begin{Bmatrix} z_{s2} \\ 1 \end{Bmatrix} v_2 = \begin{Bmatrix} z_{\text{perf}} + z_{s2} \\ 1 \end{Bmatrix} v_2 \tag{9.114}$$

so the final surface impedance reads

$$z_{s1} = \frac{p_1}{v_1} = z_{\text{perf}} + z_{s2} \tag{9.115}$$

This is why limb elements that are characterised by $v_1 = v_2$ are well represented by a transfer impedance that simply adds to the backing impedance. Coming back to our design problem, we can select an air layer that provides zero reactance at thickness $h = n\lambda/2$ and choose a perforate with the matching resistivity. In Fuchs and Zha (1995) some micro-perforate absorbers with the desired quantities are given. We use the absorber of figure 6 of Fuchs and Zha and the parameters shown in Table 9.5.

In figures 9.34 and 9.35 the resistance and reactance of the perforates are shown. One can recognize that the resistance of the second mesh fits best to our above condition for best absorption. The resistance of mesh 1 is too low and mesh 3 too high. Mesh 2 nearly meets the requirements over a large frequency range.

From the three reactance curves, we conclude that the high inertia from mesh 1 and 2 might reduce the frequency range of the absorber, because the reactance of the mesh is added to the reactance of the air spring as shown in Figure 9.35. A pure resistive absorber would have best absorption for $X = 0$; with the mass effect, the resonance is

Table 9.5 Absorbers with different perforate plates and constant surface porosity. *Source:* Fuchs and Zha (1995).

Mesh	t/mm	R/mm	d/mm	h/mm	σ'
1	3.0	1.5	22.5	50	0.0140
2	3.0	0.225	3.37	50	0.0140
3	3.0	0.075	1.13	50	0.0138

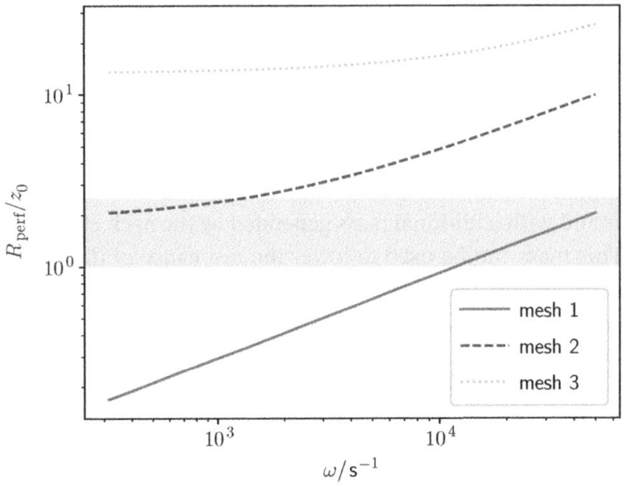

Figure 9.34 Transfer resistance of three different micro-perforates. *Source:* Alexander Peiffer.

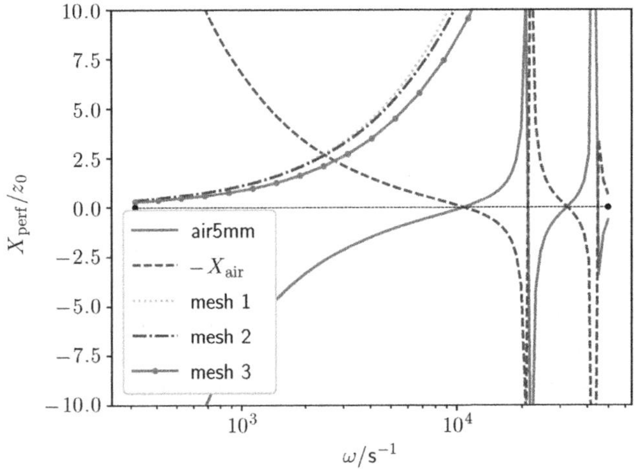

Figure 9.35 Transfer reactance of three different micro-perforates. *Source:* Alexander Peiffer.

lower. This is because the stiffness of the air spring sees the mass of the perforate, and the resonance frequency is reduced. In Figure 9.35 this can be seen by the intersection point of X_{perf} and $-X_{\text{air}}$.

The normal impedance of the full absorber is shown in Figure 9.36. The normal and diffuse field absorption is shown in figures 9.37 and 9.38. The result does not fit perfectly to Fuchs' test results, but the global tendency and resonance frequency is well met.

9.4.4 Single Degree of Freedom Liner

The micro-perforate absorber is a perfect candidate for building absorbers. In aerospace applications the absorber must withstand strong aerodynamic load, pressure, and heat. Thus, the perforate is fixed to a carrier structure. This is usually a honeycomb material

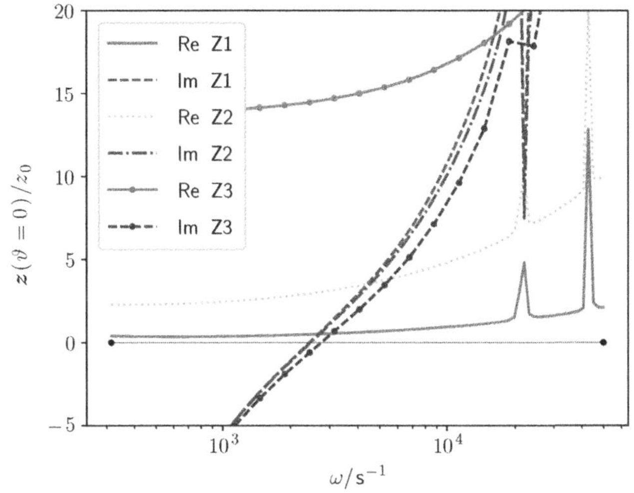

Figure 9.36 Surface impedance of micro-perforate absorber. *Source:* Alexander Peiffer.

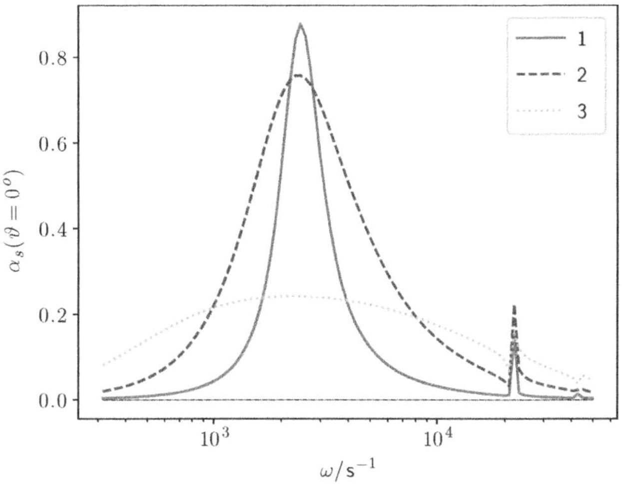

Figure 9.37 Normal incidence absorption of micro-perforate absorber. *Source:* Alexander Peiffer.

made out of aluminium or aramid fiber paper (Figure 9.39). Such liners are used in the inlet section of turbo-fan nacelles but also inside engines in cold and hot streams. In these absorbers the waves cannot propagate in in-plane directions, and we must use the normal incidence impedance of an air layer as surface impedance of the honeycomb layer. This means we set the wave number in the x-direction $k_x = 0$ in equation (9.105) and neglect the volume fraction of the honeycomb wall material.

With the configuration from Table 9.6, we get the transfer impedances as shown in figure 9.40. We see that the resistance condition $R = z_0$ and the $\lambda/2$ resonance occur at the same frequency. In Figure 9.41 we see that the absorption is nearly 1 at a certain frequency band around this regime. This is acceptable, because in turbofan engines the absorption is optimized for specific tones in the engine. It must be noted

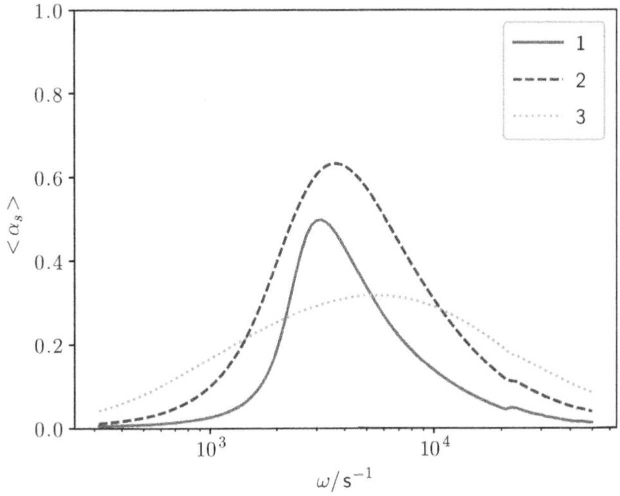

Figure 9.38 Diffuse field absorption of micro-perforate absorber. *Source:* Alexander Peiffer.

Table 9.6 Aerospace liner set-up.

t/mm	R/mm	d/mm	h/mm	σ'
1.5	0.2	3	30	0.022

that in real aircraft liner designs, the airflow must be considered, because the high Mach number current changes the neck correction (Hubbard and Acoustical Society of America, 1995).

9.5 Acoustic Wall Constructions

9.5.1 Double Walls

Inspecting wall configurations that aim at high acoustic isolation at reasonable weight leads to the conclusion that many engineering systems are enhancing the acoustic isolation of a single wall by simply adding a second wall with some absorbing material in the middle. Examples are the interior lining in aircraft covering layers of glass fiber blankets (Peiffer et al., 2007, 2013), gypsum plasterboards in buildings, or mass-spring layers in the automotive industry used, for example, to increase the isolation of the firewall in cars.

The generic setup is shown in Figure 9.42. The *walls* consist of panels that may be considered as limp mass or plates. In both cases they are described by thin layer transfer equation (9.27) with different transfer impedances $\boldsymbol{z}_1$ and $\boldsymbol{z}_2$

$$[\boldsymbol{T}]_1 = \begin{bmatrix} 1 & \boldsymbol{z}_1 \\ 0 & 1 \end{bmatrix} \qquad [\boldsymbol{T}]_2 = \begin{bmatrix} 1 & \boldsymbol{z}_2 \\ 0 & 1 \end{bmatrix} \tag{9.116}$$

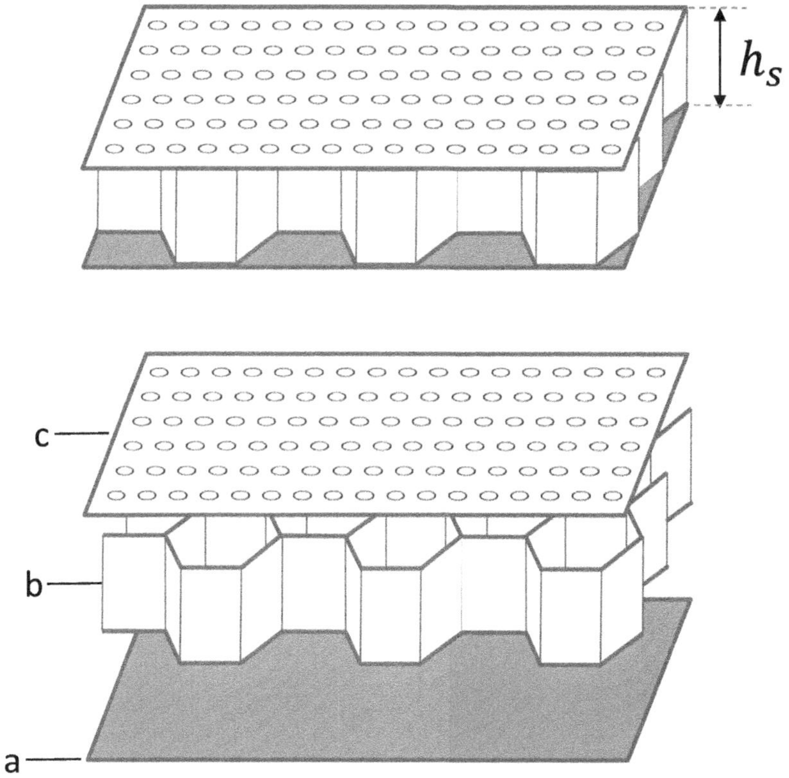

Figure 9.39 Single degree of freedom liner with (a) hard backing, (b) honeycomb core and (c) perforated sheet. *Source:* Alexander Peiffer.

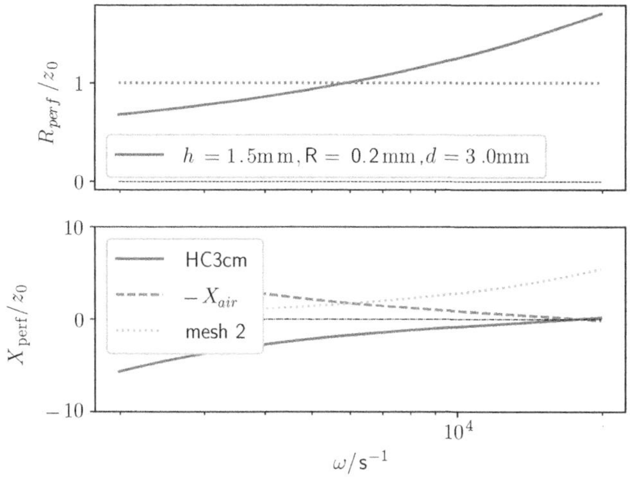

Figure 9.40 Transfer resistance and reactance of liner perforate. *Source:* Alexander Peiffer.

The cavity or fluid layer is given by (9.106).

The transfer matrix of the full system is given by

$$[\boldsymbol{T}]_{DW} = [\boldsymbol{T}]_1 [\boldsymbol{T}]_{\text{fluid}} [\boldsymbol{T}]_2 \tag{9.117}$$

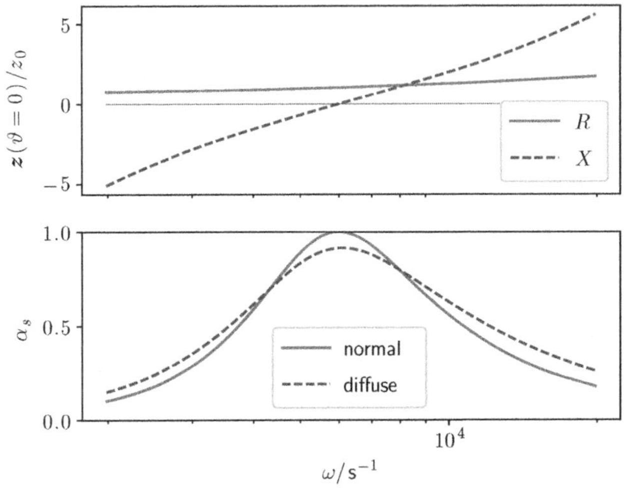

Figure 9.41 Surface impedance and absorption of liner. *Source:* Alexander Peiffer.

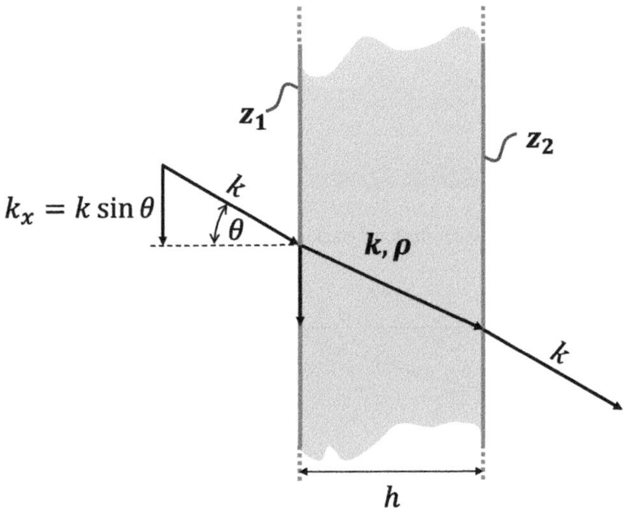

Figure 9.42 Double wall lay-up of panel–cavity–panel. *Source:* Alexander Peiffer.

In this case the result can be derived analytically and we get after some math

$$[\boldsymbol{T}]_{DW} = \begin{bmatrix} \boldsymbol{T}_{11} & \boldsymbol{T}_{12} \\ \boldsymbol{T}_{21} & \boldsymbol{T}_{22} \end{bmatrix} \tag{9.118}$$

$$\boldsymbol{T}_{11} = \frac{jz_1 \boldsymbol{k}_z \sin(h\boldsymbol{k}_z)}{\omega\rho} + \cos(h\boldsymbol{k}_z)$$

$$\boldsymbol{T}_{12} = (z_1 + z_2)\cos(h\boldsymbol{k}_z) + j\left(\frac{z_1 z_2 \boldsymbol{k}_z}{\omega\rho} + \frac{\omega\rho}{\boldsymbol{k}_z}\right)\sin(h\boldsymbol{k}_z)$$

$$\boldsymbol{T}_{21} = \frac{j\boldsymbol{k}_z \sin(h\boldsymbol{k}_z)}{\omega\rho}$$

$$\boldsymbol{T}_{22} = \frac{jz_2 \boldsymbol{k}_z \sin(h\boldsymbol{k}_z)}{\omega\rho} + \cos(h\boldsymbol{k}_z)$$

Applying equation (9.109) gives the transmission coefficient for the double leaf configuration

$$\tau = \left| \frac{-2j\cos(\vartheta)/\cos(h\boldsymbol{k}_z)}{\frac{\boldsymbol{k}_z}{\omega\rho}(z_0 + \boldsymbol{z}_1 + \boldsymbol{z}_2)\cos(\vartheta)\tan(h\boldsymbol{k}_z) - 2j\cos(\vartheta)\cdots \\ \cdots + \left[\left(\frac{\boldsymbol{z}_1\boldsymbol{z}_2\boldsymbol{k}_z}{\omega\rho z_0} + \frac{\omega\rho}{\boldsymbol{k}_z z_0}\right)\tan(h\boldsymbol{k}_z) - j\frac{\boldsymbol{z}_1+\boldsymbol{z}_2}{z_0}\right]\cos^2(\vartheta)} \right|^2 \tag{9.119}$$

For low frequencies $|\boldsymbol{k}_z h| \ll 1$ the wavelength in the double wall cavity is much larger than the thickness, and equation (9.118) can be simplified to

$$\lim_{k_z h\to 0} [\boldsymbol{T}]_{DW} = \begin{bmatrix} 1 & \boldsymbol{z}_1 + \boldsymbol{z}_2 \\ 0 & 1 \end{bmatrix} \tag{9.120}$$

Thus, at low frequencies the layers are supposed to be stiffly connected.

9.5.1.1 Double Wall of Limp Mass

For further insight we use the ideal case of two limp mass layers with transfer impedance $\boldsymbol{z}_1 = jm_1''\omega$ and $\boldsymbol{z}_2 = jm_2''\omega$. The fluid gap can be a fluid like air, other gases, or an equivalent fluid, hence $\boldsymbol{k}_z = \boldsymbol{k}\cos(\vartheta)$. The double wall transmission loss reads with this assumption

$$\tau = \left| \frac{-2j\cos(\vartheta)/\cos(h\boldsymbol{k}_z)}{\frac{\boldsymbol{k}_z}{\omega\rho}(z_0 + j(m_1'' + m_2'')\omega)\cos(\vartheta)\tan(h\boldsymbol{k}_z) - 2j\cos(\vartheta)\cdots \\ \cdots + \left[\left(-\frac{m_1''m_2''\omega\boldsymbol{k}_z}{\rho z_0} + \frac{\omega\rho}{\boldsymbol{k}_z z_0}\right)\tan(h\boldsymbol{k}_z) + \frac{(m_1''+m_2'')\omega}{z_0}\right]\cos^2(\vartheta)} \right|^2 \tag{9.121}$$

We can show that at low frequencies $|\boldsymbol{k}_z h \ll 1|$, eq. (9.121) gives the mass law of infinite single walls (8.96) with total mass $m_t'' = m_1'' + m_2''$. The same conclusion follows from (9.120) when we assume the mass transfer impedance $\boldsymbol{z} = jm_t''\omega$. In figure 9.43 the transmission coefficients are shown for $\vartheta = 0°$ and $\vartheta = 50°$. There is a first minimum of the transmission loss that is called the double wall resonance and will be dealt with later. Below this resonance the result coincides with the mass law curve of a single wall of the same total mass.

In the regime above the double wall resonance, there are several dips that correspond to resonances of the cavity for $k_z h = n\pi$, and the tangent in equation (9.121) becomes infinite. In other words the thickness of the cavity equals integer multiples of the half wavelength. The cavity resonances are denoted by ω_i in figure 9.43. When damping in the cavity is not too high (which is definitely the case for air), the anti-resonances for

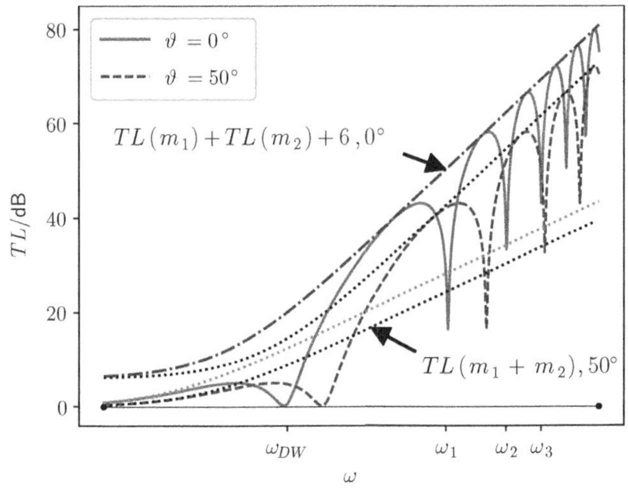

Figure 9.43 Transmission coefficient for plane waves transmission. Double wall of two limp layers of $m''_1 = m''_2 = 1$ kg/m^2 and air cavity of $h = 10$ cm thickness. At the axis the resonance frequencies of perpendicular waves are denoted. *Source:* Alexander Peiffer.

$k_z h = \pi(2n-1)/2$ lead to a decoupling and equation (9.121) is approximated for high frequencies by:

$$\tau = \left| \frac{j2\omega\rho z_0}{\omega^2 m''_1 m''_2 \cos^2\vartheta \boldsymbol{k}} \right|^2 = \left(\frac{2z_0}{\omega m''_1 \cos\vartheta} \right)^2 \left(\frac{2z_0}{\omega m''_2 \cos\vartheta} \right)^2 \left(\frac{z}{2z_0} \right)^2 \tag{9.122}$$

Note that high damping means a complex wavenumber and the condition $k_z h = \pi \sin((2n-1)/2) \neq 1$ cannot be met, and the above estimation is not valid. Writing the above expression as transmission loss, we see that above the double wall resonance, the single transmission loss of each wall is summed up, plus an extra term

$$TL = TL(m''_1) + TL(m''_2) + 6\,\text{dB} + 10\log_{10}\left|\frac{z_0}{z}\right| \tag{9.123}$$

The best use of available mass for high transmission loss is given for a distribution of two walls. This is the reason why double leaf constructions are the power weapons for acoustic isolation in many fields of technical acoustics.

However, at the resonances the isolation performance of the double wall can be low, lower than the mass law. Thus, the resonances must be reduced by damping in the cavity, and we must understand what determines the frequency of the double wall resonance. This can be found by determining the minimum of the denominator of equation (9.121) as shown, for example, by Fahy (1985). We apply the transfer matrix expression (9.118). Perfect transmission is given when the transfer matrix becomes a unit-matrix, thus $\boldsymbol{T}_{11} = \boldsymbol{T}_{22} = 1$ and $\boldsymbol{T}_{12} = \boldsymbol{T}_{21} = 0$. In general the thickness of a double wall is much smaller than the wavelength at low frequencies, so $hk_z \ll 1$ and we use $\sin(\boldsymbol{k}_z h) \approx \boldsymbol{k}_z h$ and $\cos(\boldsymbol{k}_z h) \approx 1$.

The diagonal components of (9.118) are, with $\boldsymbol{z}_{ii} = jm''_i\omega$

$$\boldsymbol{T}_{ii} \approx 1 - \frac{m''_i \omega^2 h}{\rho c^2 \cos^2\vartheta} \tag{9.124}$$

and are close to unity for typical wall dimensions. The approximation of $\boldsymbol{T}_{12}$ reads

$$\boldsymbol{T}_{12} \approx j\left[(m_1'' + m_2'')\omega - \frac{m_1''m_2''\omega^3}{\rho c^2}\cos^2\vartheta + \omega\rho h\right] \tag{9.125}$$

This matrix coefficient is zero at

$$\omega_{DW} = \sqrt{\frac{\rho c^2}{h}\frac{m_1'' + m_2'' + \rho h}{m_1''m''}\frac{1}{\cos\vartheta}} \tag{9.126}$$

$$\underset{\rho h \ll m_1''+m_2''}{\approx} \sqrt{\frac{\rho c^2}{h}\frac{m_1'' + m_2''}{m_1''m_2''}\frac{1}{\cos\vartheta}}$$

The result is the natural frequency of a spring with masses at both ends. The approximation $\rho h \ll m_1'' + m_2''$ corresponds to a massless spring. However, in double wall construction, the filling of a double wall is lightweight. Figure 9.44 shows the diffuse field transmission loss determined from angle integration. The cavity resonances are smeared out due to the integration but the double wall resonance is still visible.

9.5.2 Limp Double Walls with Fiber

As the excellent performance of a double wall is partly compensated by the cavity or double wall resonances, those must be damped by absorbing material, and due to equation (9.122), the characteristic impedance should not be too high. The fiber material of section 9.3.4 is an appropriate candidate for such an application. The air of the double wall cavity is replaced by the fiber material by using the material data of the equivalent fluid. First, we see in Figure 9.45 that the double wall resonance is lower than in air. This results from the lower speed of sound in the fiber material. Second, the transmission loss is much higher than for the air filled cavity, and the notches due

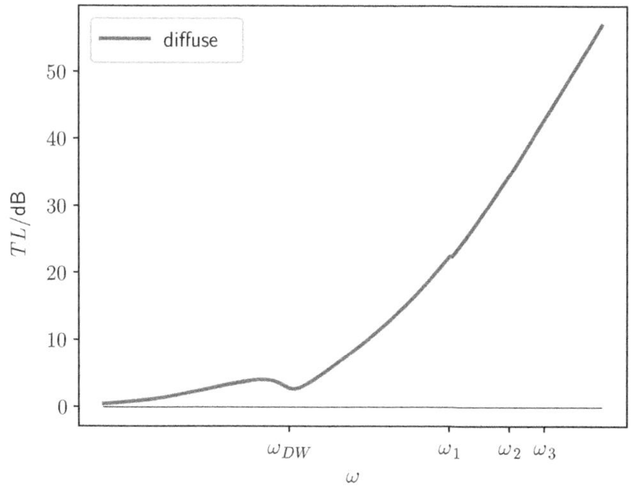

Figure 9.44 Diffuse field transmission coefficient. Double wall of two limp layers of $m_1'' = m_2'' = 1$ kg/m^2 and air cavity of $h = 10$ cm thickness. *Source:* Alexander Peiffer.

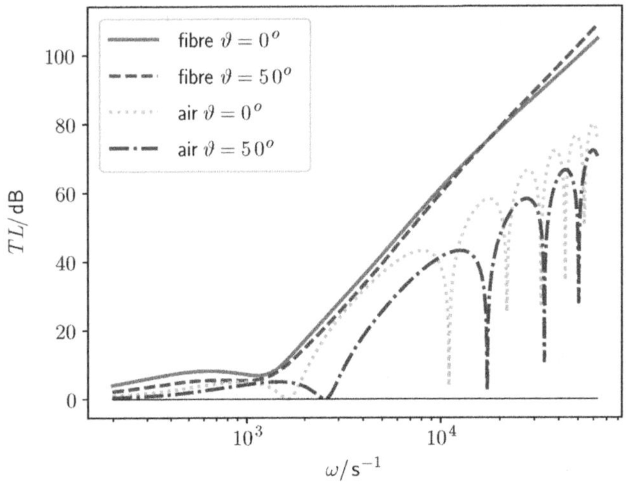

Figure 9.45 Transmission coefficient for plane waves transmission. Double wall of two limp layers of $m''_1 = m''_2 = 1$ kg/m^2 and fiber cavity of $h = 10$ cm thickness. *Source:* Alexander Peiffer.

to the thickness resonances are not existing. This is a result of the high damping in the fiber material.

It must be kept in mind that this is infinite layer theory. Thus, the calculated transmission loss is slightly overestimated compared to real systems. Every wall must be connected to the other wall by mounts that jeopardize the high isolation; secondly, the finite size of panel and absorber is not taken into account as it is already shown for single walls in section 8.2.4.5. However, for a principle understanding of double wall phenomena, the infinite layer theory is excellent. In Chapters 10 and 11, we will see how the finite size can be considered.

9.5.3 Two Plates with Fiber

Large areas cannot be realized by limp heavy layers, because a certain stiffness is necessary to mechanically hold the plate. The consideration of bending stiffness is done by using the transfer impedance of plates (9.101) in equation (9.119). Figure 9.46 shows a similar shape as for the limp double wall with fiber material. In addition there are the coincidence peaks from the bending waves of the aluminium plates. It is clear that plate materials with similar coincidence frequencies should be avoided, so that only one plate coincides with the exciting or radiating fluid wave.

9.5.4 Conclusion on Double Walls

The results from Section 9.5 are some of the most relevant results for passive noise control and acoustic isolation of enclosures or walls. Therefore, we recapitulate the most important conclusions.

Below the double wall resonance, the system behaves as a single wall of the same mass with no benefit regarding noise isolation. Above the double wall resonance, the

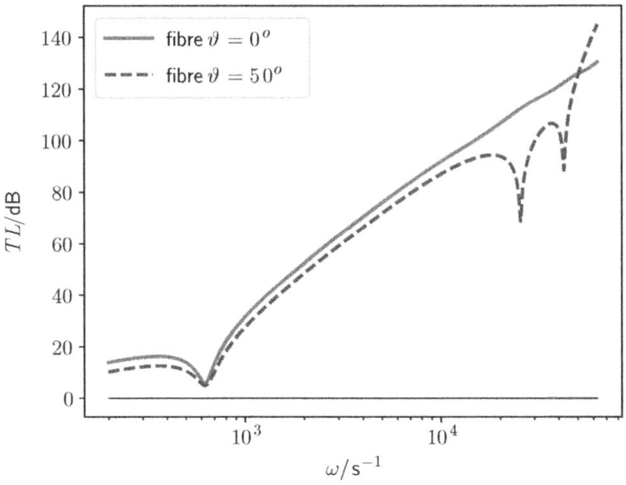

Figure 9.46 Transmission coefficient for plane wave transmission. Double wall of two aluminium plates of thicknesses 3 and 5 mm and fiber cavity of $h = 5$ cm thickness. *Source:* Alexander Peiffer.

system performs much better, likely 20 dB and more, because the single transmission coefficients are multiplied and not added.

Thus, design ***rule number one*** is: Try to keep the double wall resonance as low as possible. For further details we introduce the reduced mass by

$$m''_{\text{red}} = \frac{m''_1 + m''_2}{m''_1 m''_2} \tag{9.127}$$

and the frequency is then given by

$$\omega_{DW} = \sqrt{\frac{\rho c^2}{h\, m''_{\text{red}}} \frac{1}{\cos\vartheta}} \tag{9.128}$$

Weight is a critical parameter in most technical systems. So, we need the best mass distribution for given total mass per area $m''_{\text{tot}} = m''_1 + m''_2$. Let us assume that μ is the fraction of the total mass giving the first mass.

$$m''_1 = \mu \quad m''_{\text{tot}} \qquad\qquad m''_1 = (1 - \mu) m''_{\text{tot}}$$

and the reduced mass reads for μ

$$m''_{\text{red}} = m''_{\text{tot}} \left(\frac{1}{4} - (\mu - 0.5)^2 \right) \tag{9.129}$$

The maximum is at $\mu = 0.5$ so the lowest double wall frequency is given for a symmetric mass distribution. This ideal condition cannot always be achieved, because the main structure has to provide certain stability and is the heavy part of the wall, for example fuselage panels (Peiffer et al., 2013), walls of buildings, or the body-in-white of a car. In those cases the weight of the second wall is lower than for the main structure

$m''_2 \ll m''_1$. In this case the reduced mass is approximated by

$$m''_{\text{red}} \approx m''_2 \tag{9.130}$$

Thus, the light part determines the frequency. As a further consequence it does not make much sense to use a much heavier structure as the first structure, because the resonance is no longer efficiently decreased. So, rule number one under these constraints means to distribute the mass symmetrically if possible. If one wall structure is given, don't use a mass that is much higher than the first leaf, because this will not be efficient.

For high frequencies the performance according to equation (9.122) is

$$\tau^{-1} \sim m''^4_{tot}\mu^2(1-\mu)^2 = m''^4_{tot}\left(\left(\mu - \frac{1}{2}\right)^2 - \frac{1}{4}\right)^2 \tag{9.131}$$

that becomes also maximal at $\mu = 0.5$.

The next important parameters are the cavity thickness and stiffness, given by $\boldsymbol{\rho c^2}$. Thus, a possible trade off can be space to weight. However, this can be difficult for lower frequencies, because the thickness can, for example, reach more than 15 cm in case of the first blade, passing the frequency of turbo-props around 100 Hz for fuselage noise control.

The stiffness of the centre cavity or layer can be reduced by using fiber material with low speed of sound, but this is partly compensated by the effective higher density. Double walls are also often realized by soft foams with Young's modulus around $E = 90-200$ kN/m^2. Those systems are called mass-spring treatments. We should keep in mind that the lowest reasonable Young's modulus is near 90 kN/m^2, because this is the stiffness per area of the air spring in the material, as $\rho_0 c_0^2 \approx 100$ kN/m^2.

Rule number two: The space between the layers must be damped. At least there must be *some* damping; more damping will further increase the performance, but not much. The necessity for damping depends on how much transmission is allowed near the double wall resonance.

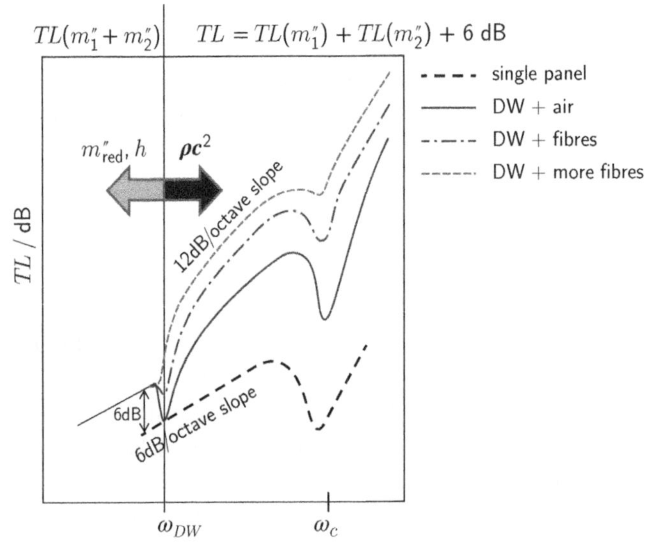

Figure 9.47 General shape of the double wall transmission and the relationship to main parameters. *Source:* Alexander Peiffer.

Rule number three: Don't forget the coincidence. If possible use wall properties that lead to different frequencies, or use a lining that is limp.[2]

To conclude, an efficient double wall system is designed by two panels of similar weight and thick space filled with soft and damped material. One remark concerns the impermeability of the walls. This must be guaranteed to exploit the advantages of double walls. Due the high performance of the total system, small leakages may have a tremendous effect. So, if leakage cannot be avoided due to technical reasons, eg. cable, ventilation, or pipe cut-outs, put special care on the design of sealing systems at those critical areas.

Bibliography

Jean-F. Allard and Noureddine Atalla. *Propagation of Sound in Porous Media*. Wiley, second edition, 2009. ISBN 978-0-470-74661-5.

Jean F. Allard, Michel Henry, Laurens Boeckx, Philippe Leclaire, and Walter Lauriks. Acoustical measurement of the shear modulus for thin porous layers. *The Journal of the Acoustical Society of America*, 117(4):1737–1743, April 2005. ISSN 0001-4966.

M. A. Biot. Generalized Theory of Acoustic Propagation in Porous Dissipative Media. *The Journal of the Acoustical Society of America*, 34(9A):1254–1264, September 1962. ISSN 0001-4966.

Yvan Champoux and Michael R. Stinson. On acoustical models for sound propagation in rigid frame porous materials and the influence of shape factors. *The Journal of the Acoustical Society of America*, 92(2):1120–1131, August 1992. ISSN 0001-4966.

Frank Fahy. *Sound and Structural Vibration: Radiation, Transmission and Response*. Academic Press, London, The United Kingdom, 1985. ISBN 0-12-247670-0.

FreeFieldTechnologies. *Actran 16.0 User's Guide*. Mont-Saint-Guibert, Belgium, two thousand, sixteenth edition, October 2015.

H. V. Fuchs and X. Zha. Einsatz mikro-perforierter Platten als Schallabsorber mit inhärenter Dämpfung. *Acta Acustica united with Acustica*, 81(2):107–116, March 1995.

Kirill V. Horoshenkov, Alistair Hurrell, and Jean-Philippe Groby. A three-parameter analytical model for the acoustical properties of porous media. *The Journal of the Acoustical Society of America*, 145(4):2512–2517, April 2019. ISSN 0001-4966.

Harvey H. Hubbard and Acoustical Society of America, editors. *Aeroacoustics of Flight Vehicles: Theory and Practice*. Published for the Acoustical Society of America through the American Institute of Physics, Woodbury, NY, 1995. ISBN 978-1-56396-407-7 978-1-56396-404-6 978-1-56396-406-0.

Finn Jacobsen. PROPAGATION OF SOUND WAVES IN DUCTS. Technical Note 31260, Technical University of Denmark, Lynby, Denmark, September 2011.

Dah-You Maa. Potential of microperforated panel absorber. *The Journal of the Acoustical Society of America*, 104(5):2861–2866, November 1998. ISSN 0001-4966.

Fridolin P. Mechel, editor. *Formulas of Acoustics*. Springer, Berlin ; New York, 2002. ISBN 978-3-540-42548-9.

---

2 This is why the lining is often called trim, because in historic aircraft the second wall was realized by a trimmed membrane

Clemens Moeser, Alexander Peiffer, Stephan Brühl, and Stephan Tewes. FEM Schalldurchgangsrechnungen einer Doppelwandstruktur. In *Fortschritte der Akustik*, pages 593–594, Dresden, Germany, March 2008.

M. L. Munjal. *Acoustics of Ducts and Mufflers with Application to Exhaust and Ventilation System Design*. Wiley, New York, 1987. ISBN 978-0-471-84738-0.

Raymond Panneton. Comments on the limp frame equivalent fluid model for porous media. *The Journal of the Acoustical Society of America*, 122(6):EL217–EL222, 2007.

Alexander Peiffer, Stephan Tewes, and Stephan Brühl. SEA Modellierung von Doppelwandstrukturen. In *Fortschritte der Akustik*, Stuttgart, Germany, March 2007.

Alexander Peiffer, Clemens Moeser, and Arno Röder. Transmission loss modelling of double wall structures using hybrid simulation. In *Fortschritte Der Akustik*, pages 1161–1162, Merano, March 2013.

Allan D. Pierce. *Acoustics - An Introduction to Its Physical Principles and Applications*. Acoustical Society of America (ASA), Woodbury, New York 11797,U.S.A., one thousand, nine hundred eighty-ninth edition, 1991. ISBN 0-88318-612-8.

Karin Tageman. Modelling of sound transmission through multilayered elements using the transfer matrix method. Master's thesis, Chalmers University of Technology, Gothenburg, Sweden, 2013.

Huiyu Xue. A combined finite element–stiffness equation transfer method for steady state vibration response analysis of structures. *Journal of Sound and Vibration*, 265(4): 783–793, August 2003. ISSN 0022-460X.

10

Application of Random systems

The random description of vibroacoustic systems is well suited for large and complex systems with high Helmhotz number. Thus, large technical systems are excellent applications for SEA. These are buildings providing large cavities and walls, ships, trains, and aircraft. Even the acoustics of cars can be calculated by SEA when we keep in mind that the statistical conditions are not perfectly met, and one can accept a certain uncertainty.

However, this chapter explains the principle procedures to perform a SEA simulation, the way of subsystem description, and the determination of the coupling loss factor on realistic examples. We try to find a compromise between presenting realistic examples and being not too sophisticated.

10.1 Frequency Bands for SEA Simulation

In Chapter 9 we used narrow band spectra, because we dealt with deterministic excitations and responses. In general the nature of the excitation signal determines the result, and this can be an amplitude, an rms-value, or a spectral density. This is also true for SEA systems. Here, the input is power, and the output is energy.

Due to the statistic nature of the reverberant fields, SEA simulation uses third-octave bands, because frequency averaging can further improve the statistics. The perception of human hearing is linked to octaves. A doubling of the frequency is called an increase of one octave. According to this and in order to facilitate the comparison of test and simulation results, there are standardized frequency intervals: the so-called octave or third-octave bands.

Those bands have center frequencies $f_{n,c}$ that are given by repeated multiplication of a main frequency by roots of 2 according to the ANSI Standard S1.11:

$$f_{n,c} = 1000 \cdot 2^{n/N} \text{for } 1/N - \text{th octave bands} \tag{10.1}$$

The system is based on the main frequency at $f_{0,c} = 1000$ Hz with $n = 0$, and for third-octaves $N = 3$ is used. The frequency limits of each band are determined by a square root factor $d = \sqrt{2^{1/N}}$. Hence, the left and right limits are given by:

$$f_{n,L} = f_{n,c}/d \qquad f_{n,R} = f_{n,c}d \tag{10.2}$$

Vibroacoustic Simulation: An Introduction to Statistical Energy Analysis and Hybrid Methods,
First Edition. Alexander Peiffer.

Table 10.1 ANSI S1.11 center frequencies.

n	ANSI	f_c
−10	100 Hz	99.213 Hz
−9	125 Hz	125.000 Hz
−8	160 Hz	157.490 Hz
−7	200 Hz	198.425 Hz
−6	250 Hz	250.000 Hz
−5	315 Hz	314.980 Hz
−4	400 Hz	396.850 Hz
−3	500 Hz	500.000 Hz
−2	630 Hz	629.961 Hz
−1	800 Hz	793.701 Hz
0	1 kHz	1000.000 Hz
1	1.25 kHz	1259.921 Hz
2	1.6 kHz	1587.401 Hz
3	2 kHz	2000.000 Hz
4	2.5 kHz	2519.842 Hz
5	3.15 kHz	3174.802 Hz
6	4 kHz	4000.000 Hz
7	5 kHz	5039.684 Hz
8	6.3 kHz	6349.604 Hz
9	8 kHz	8000.000 Hz

The resulting frequency values are written as rounded values for better readability, but the bands always refer to precise frequencies.[1]

Unlike the sections before we will not use the angular frequency ω because of this industry standard, and use third-octave bands. In addition, as SEA is energy based, all results of the engineering units are rms-values. When we speak about power or energy, it is the averaged power or energy in one band.

10.2 Fluid Systems

The random property of cavities is given by the modal density

$$n_{3D}(\omega) = \frac{V\omega^2}{2\pi^2 c_0^3} + \frac{S\omega}{8\pi c_0^2} + \frac{P}{16\pi c_0} \tag{4.71}$$

1 Besides the presented base 2 formulas, there is also a base 10 definition that leads to slightly different frequencies

where the second and third summand is often neglected. The relationship between energy and physical units is given by

$$p_{m,rms}^2 = \rho_0 c_{gr}^2 \frac{E_m}{V} \tag{6.116}$$

The load or excitation is in most cases described as power input.

10.2.1 Twin Chamber

The twin chamber set-up is a test bench for the determination of the transmission loss described e.g. in detail in the norm ISO 10140-2.

One typical arrangement is shown in Figure 10.1. Two large rooms of volumes V_1 and V_2 are connected exclusively via the test panel. This panel is of size S_j. The aim is to measure the transmission loss of this panel under diffuse field excitation. Thus, this is by nature an excellent application for SEA, because the cavities are supposed to be random due to specification in the norm. We don't need to determine the coupling loss factor or transmission coefficient here, because this is the intrinsic result of this test bench.

As can be seen in Figure 10.1, the rooms are confined by slightly oblique walls in order to avoid regularly shaped modes. The first room is a sending room with a loudspeaker of source power Π_1. For this simple two system configuration, the detailed analytical equations can be readily worked out. The SEA equation (6.113) for two systems reads

$$\omega \begin{bmatrix} n_1(\eta_{11} + \eta_{21}) & -n_2\eta_{21} \\ -n_1\eta_{12} & n_2(\eta_{22} + \eta_{12}) \end{bmatrix} \begin{Bmatrix} E_1/n_1(\omega) \\ E_2/n_2(\omega) \end{Bmatrix} = \begin{Bmatrix} \Pi_1 \\ \Pi_2 \end{Bmatrix} \tag{10.3}$$

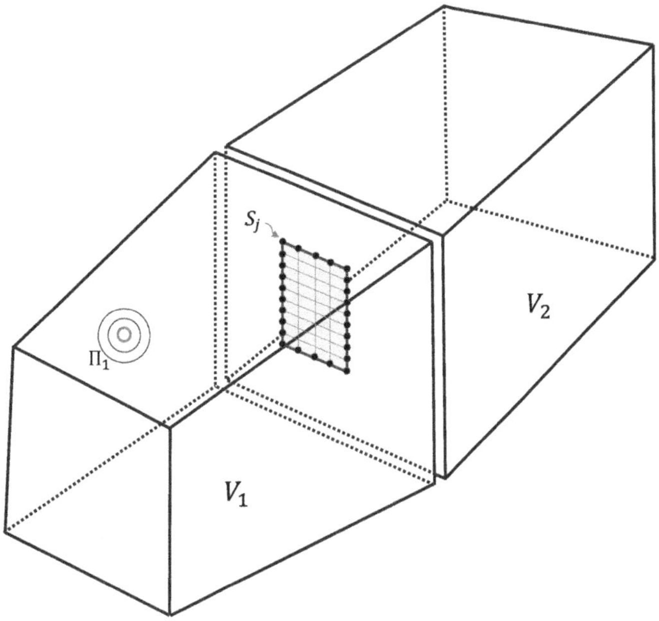

Figure 10.1 Twin chamber arrangement for TL tests of panels. *Source:* Alexander Peiffer.

and the solution using the reciprocity relationship is

$$\begin{Bmatrix} \frac{E_1}{n_1} \\ \frac{E_2}{n_2} \end{Bmatrix} = -\frac{1}{\omega[\eta_{12}^2 n_1 - n_2(\eta_{11}+\eta_{12})(\eta_{12}\eta_{22})]} \begin{bmatrix} \frac{n_2}{n_1}(\eta_{22}+\eta_{12}) & \eta_{12} \\ \eta_{12} & \eta_{11}+\eta_{12} \end{bmatrix} \begin{Bmatrix} \Pi_1 \\ \Pi_2 \end{Bmatrix} \tag{10.4}$$

With source in the sending room (1) the solution is

$$E_1 = -\frac{n_2(\omega)}{\omega n_1(\omega)} \frac{\eta_{12}+\eta_{22}}{\eta_{12}^2 n_1 - n_2(\eta_{11}+\eta_{12})(\eta_{12}\eta_{22})} \Pi_1 \tag{10.5}$$

$$E_2 = -\frac{1}{\omega} \frac{\eta_{12}}{\eta_{12}^2 n_1 - n_2(\eta_{11}+\eta_{12})(\eta_{12}\eta_{22})} \Pi_1 \tag{10.6}$$

As the rooms must be large enough to be considered as diffuse, we assume that the modal density is well approximated by the first summand in (4.71). Expressing the energy by the pressure (6.116) gives

$$p_{1,rms} = -\frac{n_2(\omega)\omega\rho_0}{2\pi^2 c_0 n_1(\omega)} \frac{\eta_{12}+\eta_{22}}{\eta_{12}^2 n_1 - n_2(\eta_{11}+\eta_{12})(\eta_{12}\eta_{22})} \Pi_1 \tag{10.7}$$

$$p_{2,rms} = -\frac{\omega\rho_0}{2\pi^2 c_0} \frac{\eta_{12}}{\eta_{12}^2 n_1 - n_2(\eta_{11}+\eta_{12})(\eta_{12}\eta_{22})} \Pi_1 \tag{10.8}$$

According to the test norm, the level difference of the two rooms $\Delta L_p = L_{p2} - L_{p1}$ is measured. So, we need the rms pressure ratio

$$\frac{p_{2,rms}^2}{p_{1,rms}^2} = \frac{n_1(\omega)\eta_{12}}{n_2(\omega)(\eta_{12}+\eta_{22})} \tag{10.9}$$

The coupling loss factor depends only on the damping loss of the receiving room and the squared pressure ratio. Solving for η_{12} gives

$$\eta_{12} = \frac{n_2(\omega)\eta_{22}\frac{p_{2,rms}^2}{p_{1,rms}^2}}{n_1(\omega) - n_2(\omega)\frac{p_{2,rms}^2}{p_{1,rms}^2}} \approx \eta_{22}\frac{p_{2,rms}^2}{p_{1,rms}^2} \tag{10.10}$$

The approximation is true if the pressure level in the receiving room is much lower than in the sending room. In other words the norm requires a certain minimum isolation so that the power flow from receiving room back into the sending room can be neglected. This back flow of energy is a natural part of the SEA analysis.

This is good news for a test set-up, because it should depend on as few parameters as possible. The damping is determined from equation (6.72), where we neglect the fluid damping

$$\eta_{22} = \frac{A_s c_0}{4V\omega} \tag{10.11}$$

and we express the coupling loss factor by the two-dimensional transmission coefficient (8.7) using the modal density

$$\langle \tau_{12} \rangle_E = \frac{4V_2\omega}{c_0 S_j}\eta_{12} = \frac{p^2_{2,rms}}{p^2_{1,rms}}\frac{A_s}{S_j} \tag{10.12}$$

Writing the above equation as transmission loss (2.168) leads to

$$TL = L_{p1} - L_{p2} + 10\log_{10}\left(\frac{S_j}{A_s}\right) \tag{10.13}$$

A_s and S_j are the absorption area of the receiving room and panel or junction area, respectively. The absorption area is usually measured by reverberation time measurements. This expression is in full accordance with the norm ISO/DIS 10140-2 considering the assumptions from equation (10.10).

10.3 Algorithms of SEA

In contrast to the twin chamber example, it does not make sense to provide analytical expression for multiple subsystems. All following examples are calculated by numerical methods.

The symmetric SEA equation (6.113)

$$\omega \begin{bmatrix} n_1(\omega)\sum_{n=1}^{N}\eta_{1n} & -n_2(\omega)\eta_{21} & \cdots & -n_N(\omega)\eta_{N1} \\ \vdots & n_2(\omega)\sum_{n=1}^{N}\eta_{2n} & & \vdots \\ & & \ddots & \\ sym & \cdots & & n_N(\omega)\sum_{n=1}^{N}\eta_{Nn} \end{bmatrix} \begin{Bmatrix} \frac{E_1}{n_1(\omega)} \\ \frac{E_2}{n_2(\omega)} \\ \vdots \\ \frac{E_N}{n_N(\omega)} \end{Bmatrix} = \begin{Bmatrix} \Pi^{(1)}_{in} \\ \Pi^{(2)}_{in} \\ \vdots \\ \Pi^{(N)}_{in} \end{Bmatrix}$$

requires the modal densities $n_m(\omega)$, the damping loss factors η_{mm}, the power inputs Π_m, and the coupling loss factors η_{mn}. All quantities are given by the physical properties of the systems. The complicated part is the coupling loss factor calculation.

Each junction couples reverberant wave fields. In the case of two plates, these are the bending and in-plane waves of each plate. The wave fields are denoted by a combined system and wave field index. Let us take an SEA-matrix with six wave fields and a junction that connects wave field 1,2, and 4. The junction part that must be added to the SEA matrix is

$$[J]' = \begin{bmatrix} n_1(\eta_{12}+\eta_{13}) & -n_2\eta_{21} & -n_4\eta_{41} \\ & n_2(\eta_{21}+\eta_{24}) & -n_4\eta_{42} \\ sym & & n_4(\eta_{41}+\eta_{42}) \end{bmatrix} \tag{10.14}$$

Because of the the matrix symmetry, it is sufficient to calculate only the upper triangle of the matrix. Each junction contributes to connected subsystems and leads to entries in the rows and columns of the global matrix. As we are calculating only the upper matrix, we add the coupling loss factor twice. For example the term $n_1(\omega)\eta_{21}$ contributes to the

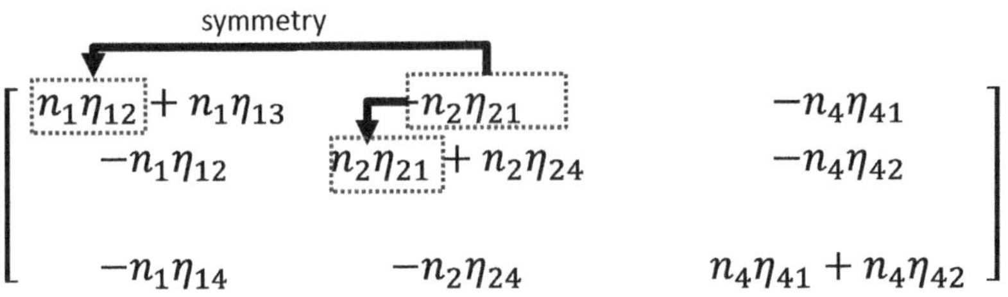

Figure 10.2 Schematics of junction *element* algorithm. *Source:* Alexander Peiffer.

sum at the matrix diagonal at index 11 and 22.

$$[J] = \begin{bmatrix} n_1(\eta_{12} + \eta_{13}) & -n_2\eta_{21} & -n_3\eta_{41} & \cdots & \cdots \\ sym & n_2(\eta_{21} + \eta_{24}) & -n_4\eta_{42} & & \\ & & \ddots & & \\ sym & sym & n_4(\eta_{41} + \eta_{42}) & & \vdots \\ \cdots & & & \ddots & \\ & \cdots & & & \ddots \end{bmatrix} \tag{10.15}$$

In Figure 10.2 the procedure is depicted. The algorithm loops over all upper triangular coefficients $-n_m(\omega)\eta_{mn}$ at index n, m. Next, the coefficient is added as a positive value to the diagonal at two positions: index n, n and index m, m. Each must be used twice to recover the part from the symmetric lower triangular matrix.

10.4 Coupled Plate Systems

Typical structure systems that become random at audible frequencies are plates. As shown in section 6.4.1, in many cases only the bending has wavelengths at audible frequencies that allow the description of the wave field as a reverberant field. We apply the system descriptions that were derived in section 8.2.4.3 and the coupling loss factor calculation from section 8.2.6. We use the discussed separation of in-plane and bending wave to describe the SEA subsystem. Thus, one plate consists of two SEA subsystems of particular wave fields. Hence, there are two energies for two reverberant fields in each plate system.

10.4.1 Two Coupled Plates

We will investigate the L-plate set-up as shown in Figure 10.3. For clarity we present the indexes of this problem. The Energy solution vector reads

$$\left\{ \frac{E_{1B}}{n_{1B}(\omega)} \quad \frac{E_{1LS}}{n_{1LS}(\omega)} \quad \frac{E_{2B}}{n_{2B}(\omega)} \quad \frac{E_{2B}}{n_{2LS}(\omega)} \right\}^T \tag{10.16}$$

For the inspection of the system and whether it is suitable for SEA simulation, we have a first look at the modal densities. In figure 10.4 it becomes obvious that the in-plane waves are not random, because the modal density is several orders of magnitude lower than for the bending waves. A more explicit quantity is the number of modes in band that counts the modes in one frequency band. For a reverberant field

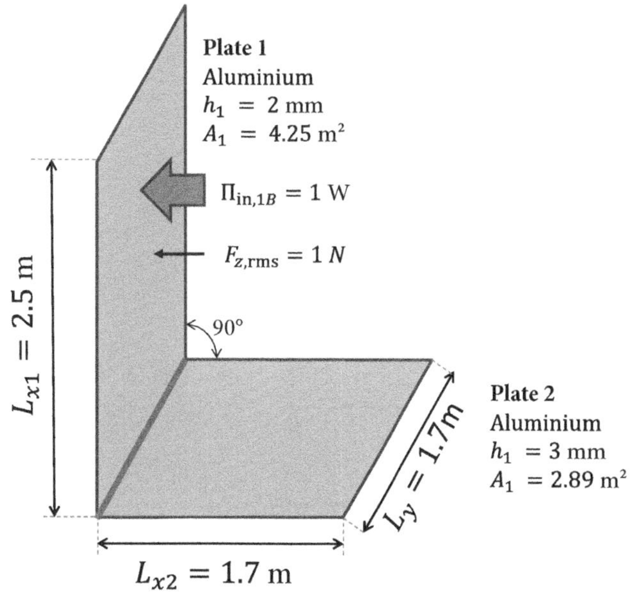

Figure 10.3 Twin plates connected by a rectangular angle. *Source:* Alexander Peiffer.

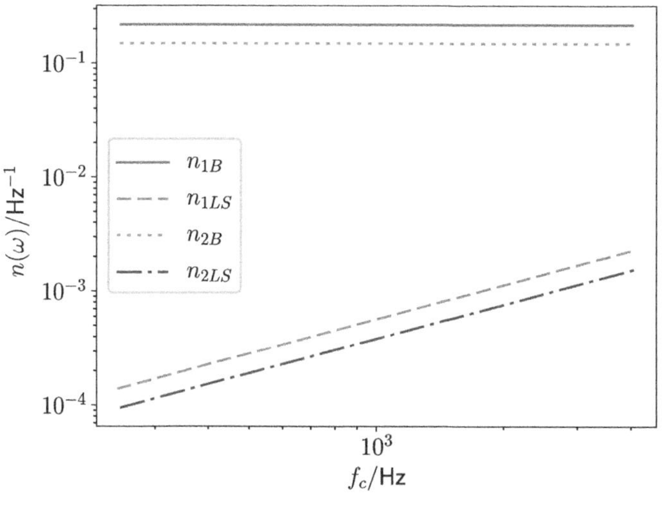

Figure 10.4 Modal densities of both plates. *Source:* Alexander Peiffer.

there should be a minimum of 10 modes according to Lyon and DeJong (1995). In Figure 10.5 the in-plane waves hardly reach 10, but the mode count for bending waves is nearly above 100 in all bands.

In our example we assume a global damping loss of $\eta = 0.01$. In comparison to the coupling loss factor, the damping loss must be higher to fulfil the requirement of low boundary impact. In Figure 10.6 we see that the coupling is below 0.004. Hence the coupling condition is fulfilled.

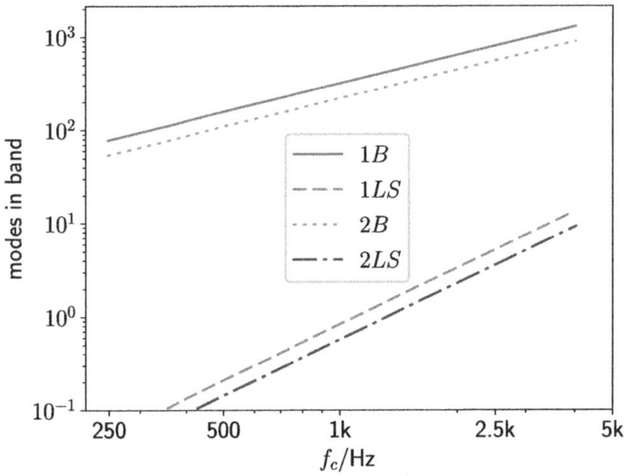

Figure 10.5 Modes in band of both plates. *Source:* Alexander Peiffer.

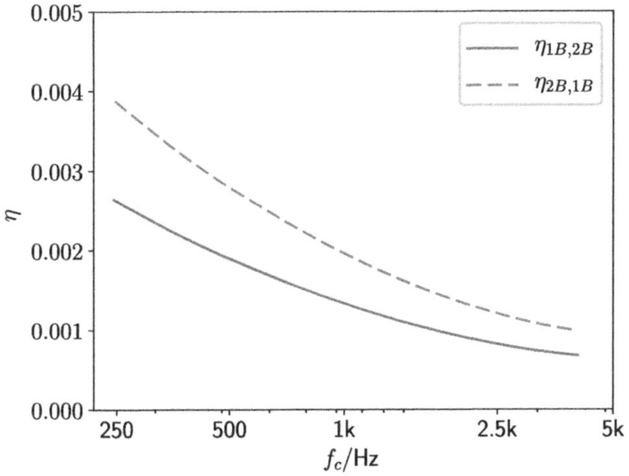

Figure 10.6 Coupling loss factors of bending waves. *Source:* Alexander Peiffer.

We consider a power source of $\Pi_{\text{in},1B} = 1$ W (per band), set up the SEA matrix, and solve the system of equations numerically. The energy results are shown in Figure 10.7. Both bending waves are four orders of magnitude higher than the in-plane waves. The energy of plate 1 is higher, because this is the excited subsystem.

The energy has to be converted into the field quantity of plate systems; the root mean square velocity is shown in Figure 10.8.

Usually, we don't introduce a dedicated amount of power into a subsystem. A more technical source is a force excitation normal to the plate. The power of a point force follows from the point force expression of an infinite plate from section 6.4.2.1. We modify equation (6.61) by introducing the rms-force $F_{z,\text{rms}} = \hat{F}_z/\sqrt{2}$

$$\Pi_{\text{in},F} = \frac{1}{8\sqrt{Bm''}} F_{z,rms}^2 \tag{10.17}$$

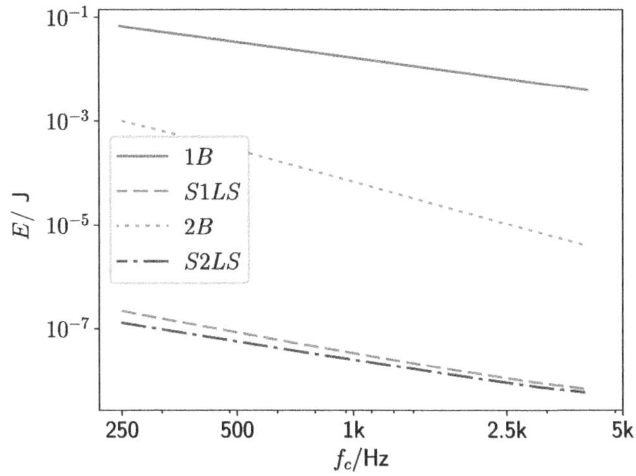

Figure 10.7 Energy results of two plate system. *Source:* Alexander Peiffer.

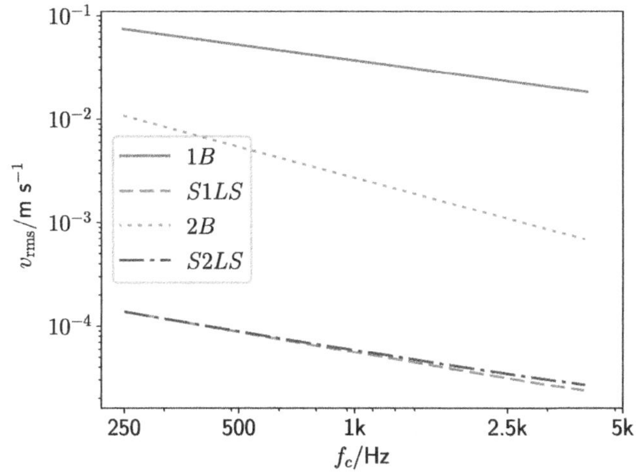

Figure 10.8 Velocity results of the two plates. *Source:* Alexander Peiffer.

With the given data of plate 1, the velocity result of unit power must be multiplied by $\sqrt{\Pi_{\text{in},F}}$ to get the results for 1 N rms force.

10.5 Fluid-Structure Coupled Systems

Cavities that are indirectly connected via plate area junctions constitute a critical case for SEA. With the methods of hybrid theory, we can treat the plate as deterministic subsystem. But in this case the plate is not modelled as SEA subsystem even though it is a random subsystem in this frequency range. One way out of this dilemma is the concept of resonant and non-resonant transmission. The separation between resonant and non-resonant transmission can be well explained in the modal model. Resonant transmission occurs when the modes of the plate are in the same frequency range as

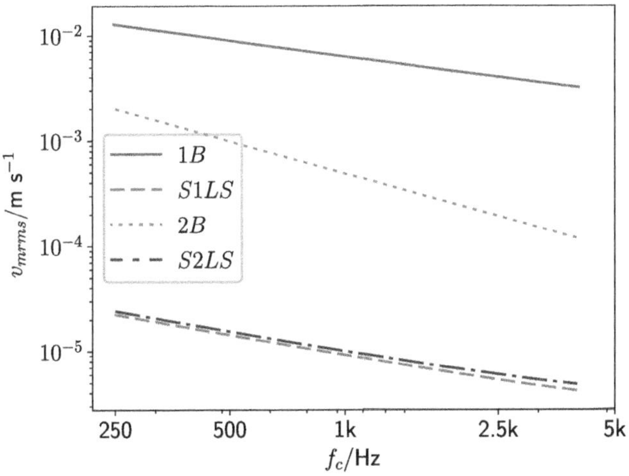

Figure 10.9 Velocity results of the two plates excited by 1 N force. *Source:* Alexander Peiffer.

the excitation. Non-resonant transmission comes from the excitation of modes whose modal frequency is not in the frequency band of excitation. In the wave field model the excitation of a reverberant field requires the existence of enough waves in this frequency range to create a reverberant field.

For the explanation of this effect the wavenumber result for an area junction in equation (8.95) is useful

$$\tau = \left[\left(\frac{m''\omega}{2\rho_0 c_0} \right)^2 \left(\frac{k_a^4 \sin^4 \vartheta}{k_B^4} - 1 \right)^2 \cos^2 \vartheta + 1 \right]^{-1} \tag{8.95}$$

The term in the inner bracket shows two components; the first is related to the wave propagation in the plate, the second leads to the mass law of plane wave transmission. Hence, the first term describes the excitation of *resonating* waves in the plate, and therefore energy can be stored in this wavefield.

When the projected wavenumber of the fluid does not coincide well with the plate wavenumber, there is only litle excitation of resonant waves in the plate. Thus, the coupling between the reverberant fields of cavity and plate is low. In this case only the forced motion of the limp plate remains, which is described by the mass law. The expression limp is determined by the ratio of structure and air wavenumbers. If the air wavenumber is much smaller than the structural wave number (or $\lambda_p < \lambda_{air}$), the plate is considered as limp in view of the fluid wave.

How can we consider this in SEA theory, namely by the junction formulation? The trick is to introduce multiple paths. The resonant part is described by direct junctions between the cavities and the bending wave field of the plate.

The coupling loss factor from plate (bending) to cavity is given by equation (8.91)

$$\eta_{12} = \frac{\rho_0 c_0 \sigma}{\omega m''}$$

with σ calculated by (8.92). The in-plane waves are obviously not coupled to the reverberant pressure waves, so the coupling loss factor of these waves is zero.

In addition a non-resonant path is introduced to allow energy transport via mass law. This path describes the energy transport between the cavities due to forced excitation. The plate wavenumber k_B is larger than for the fluid wavenumber k_a, so the plate wave is not *fast* enough to follow the fluid waves. The plate motion is forced by the external fluid pressure as would be the case for a limp mass layer (with infinite wavenumber) and not resonantly exciting a wave field in the plate. Therefore, the non-resonant path is also called mass-law and considered by evaluating equation (8.103) numerically or using the transfer matrix theory. The cavity–cavity coupling loss factor is then given by (8.8c):

$$\eta_{13} = \frac{k^2 S_j}{8\pi^2 n_1(\omega)\omega} \langle \tau_{13} \rangle_E \tag{10.18}$$

This multiple path construct as depicted in Figure 10.10 may seem inconsistent, but when the mass law is neglected and only resonant transfer is taken into account, the most important path of power transfer would not be considered correctly.

Note that in a hybrid formulation with two cavities as SEA subsystems and the plate modelled as a deterministic system, all paths are considered automatically.

This formulation leads to one issue. When the engineering result of the subsystems is calculated, only the resonant energy of the plate is considered. Though the forced motion of the non-resonant path is considered in the power flow, it doesn't occur in the energy of the reverberant field. So, when inspecting the velocity results, we must keep in mind that this is only the reverberant component. In other words, when we would compare the reverberant plate result to tests, the agreement would be low, because the test sensor would measure both the resonant and the non-resonant part. As described by Wang (2015), the SEA result can be corrected by the additional consideration of the forced excitation of both reverberant pressure fields by

$$v^+_{2,rms} = \frac{2}{m''\omega}(p_{1,rms} + p_{3,rms}). \tag{10.19}$$

10.5.1 Twin Chamber

For the application of the area junction we use the twin chamber set-up from Section 10.2.1. In contrast to section 10.2.1, we use the area junction instead of the transmission coefficient for the cavity coupling. Figure 10.11 shows a realization of such a

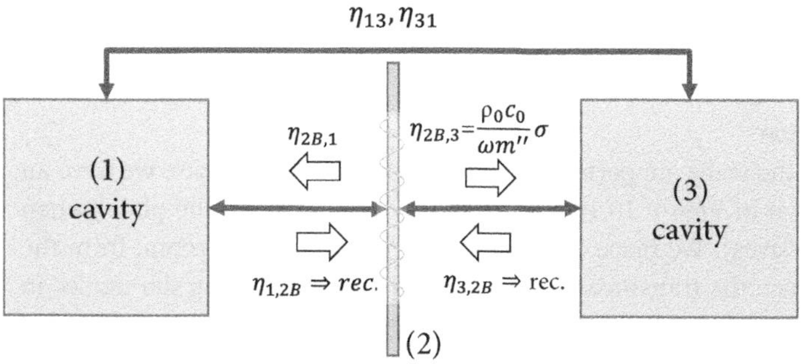

Figure 10.10 Resonant and non-resonant paths of cavity–plate–cavity area junctions. *Source:* Alexander Peiffer.

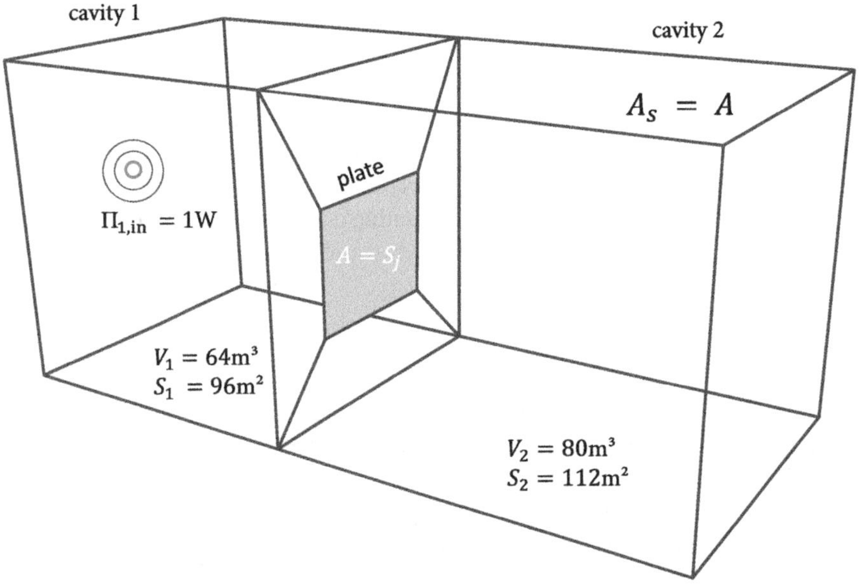

Figure 10.11 Two cavity configuration of transmission loss facility. *Source:* Alexander Peiffer.

configuration. Let us assume as a first step that both cavities are exclusively coupled by the plate in the center of the separating wall. Thus, we have four reverberant fields: two in the plate and one in each cavity; in this case the in-plane waves are fully uncoupled to the pressure waves. So, effectively there are three subsystems. A power source for $\Pi_{1,in} = 1$ W excites cavity 1. The plate is rectangular with lengths $L_x = 2$ m and $L_y = 3$ m and made of 4 mm thick aluminium. Details of the cavities are given in the figure. The damping loss of cavity 1 and the plate is $\eta = 0.01$. The damping loss of cavity 2 is determined by (6.72), and the absorption area $A_s = S_j$ is equal to the plate area.

In this case $10\log_{10}(S_j/A_s) = 0$ in equation (10.13), and the transmission loss follows directly from the pressure level difference.

In Figure 10.11 such a configuration is shown. Both rooms are large enough for a reverberant field at lower frequencies. The walls are made of 20 cm thick concrete (plus reinforcement), and the size of the test window is 3 m^3. The separating wall is connected to both cavities similar to a test plate. The idea is to use SEA to determine if such a construction is good enough for quality tests of the transmission loss.

10.5.1.1 Ideal Situation

First we assume that the walls are perfectly rigid and isolated. In this case we have an area junction as shown in Figure 10.10 connecting two cavities and the plate (when we neglect in-plane waves). We place a power source in the sending room; from the pressure level difference the transmission loss is calculated. Inspecting the modes in band in Figure 10.12, we see that there are more than 10 modes starting from the 250 Hz third-octave band for all relevant wave fields. For demonstration, the in-plane mode count is given. Even for such a large plate, the modes in band reach 10 above 8 kHz, and the in-plane waves cannot be considered as reverberant.

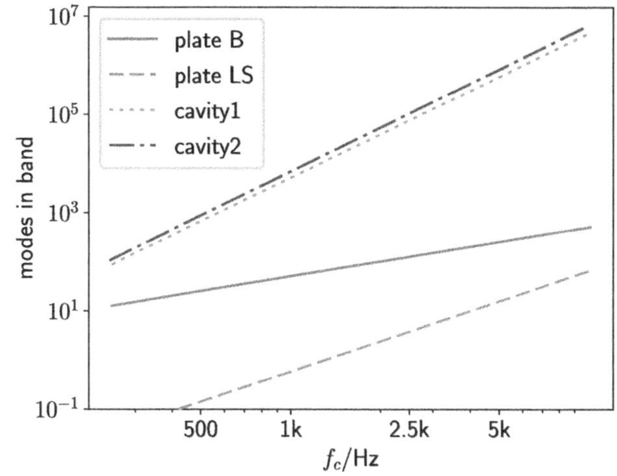

Figure 10.12 Modes in band of twin chamber SEA systems. *Source:* Alexander Peiffer.

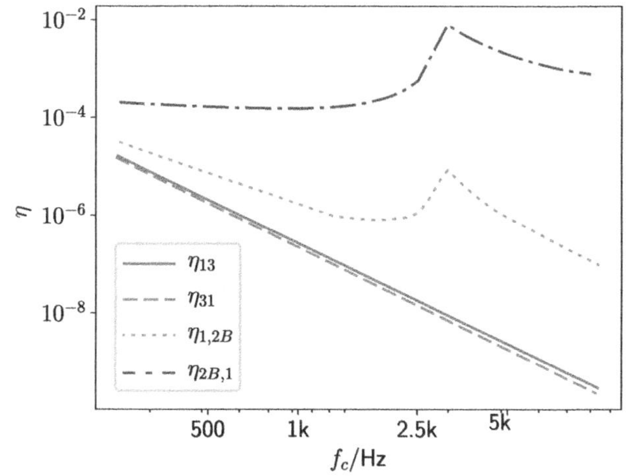

Figure 10.13 Coupling loss factors of the twin chamber arrangement. *Source:* Alexander Peiffer.

The coupling loss factors (Figure 10.13) have values below $\eta = 0.01$, and thus the systems are weakly coupled. In the plate-cavity coupling the coincidence peak can be recognized. The coincidence frequency of 4 mm aluminium is $f_c \approx 3000$ Hz. Running the simulation we get the rms-pressure and velocity results as shown in Figure 10.14.

10.5.1.2 Concrete Walls

For further details we model all walls, the roof, and the floor by concrete walls. The walls are coupled by line junctions with perpendicular connections except the four walls that surround the window. Note that the separating wall could be modelled by a large wall with a hole as window. We split the wall into four subsystems to guarantee a concave plate geometry. The material parameters for concrete are $E = 23$ GPa, $\rho = 2300$ kg/m^3, and $\nu = 0.25$; the wall thickness is $h = 10$ cm.

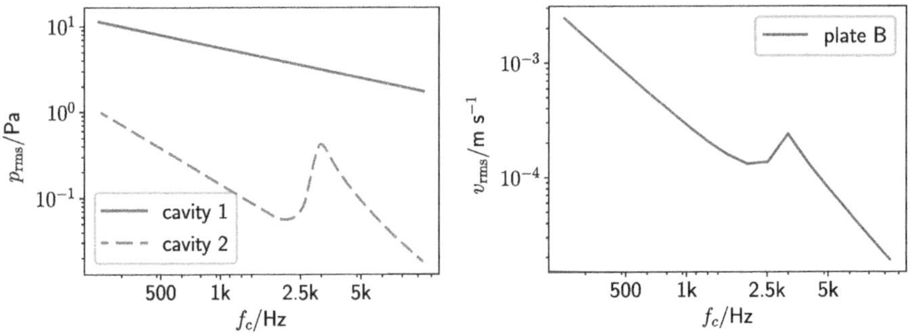

Figure 10.14 Pressure and velocity results. *Source:* Alexander Peiffer.

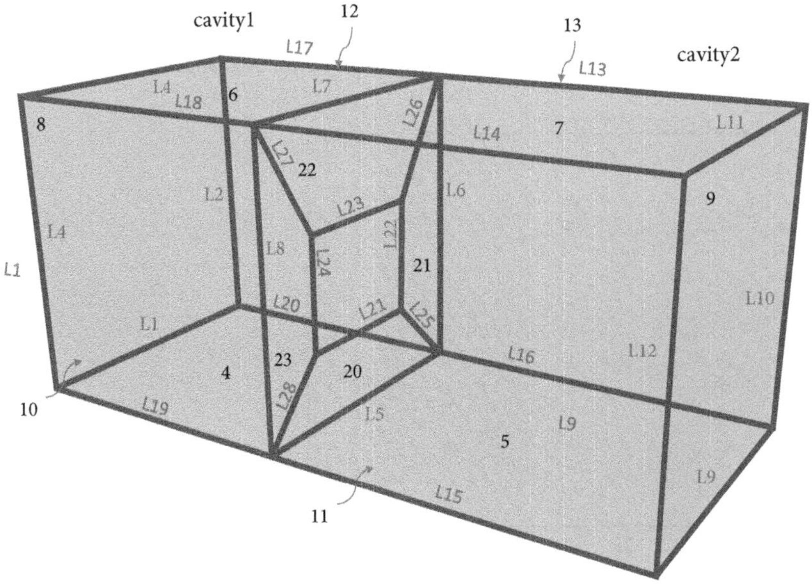

Figure 10.15 Concrete walls forming the chambers and related subsystem IDs. *Source:* Alexander Peiffer.

In Figure 10.15 it is clearly visible that the number of junctions becomes high even for such a simple two room configuration. Therefore, an implementation in a software environment that deals automatically with the junctions between the subsystems is helpful. We will see that there is a parasitic power flow over the walls that contributes to the power flow into cavity 2 and will therefore lead to an error in the transmission loss result. In the standard this so-called side-path power must be at least 10 times lower than the power radiated from the panel.

When we consider equation (7.26), we see that the second term provides the power inflow from each connected subsystem and thus full insight in unwanted side paths.

$$\Pi_{\mathrm{in}}^{(n)} = \Pi_{\mathrm{in,ext}}^{(n)} + \sum_m \omega \eta_{mn} E_m$$

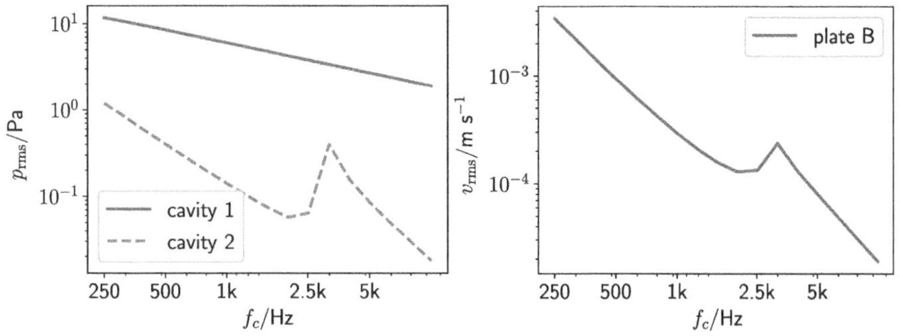

Figure 10.16 Pressure and velocity results. *Source:* Alexander Peiffer.

Thus, from the energy results and the coupling loss factors, we can derive the input power and evaluate if the side paths must be considered. If there is no external power, each contribution $\omega\eta_{mn}E_m$ provides the input power from that subsystem. In Figure 10.17 one can see that the total input power from the separating walls (resonant and non-resonant) is on the same order of magnitude as the resonant radiation from the surrounding walls. Thus, even the flanking paths via the structure borne noise transfer via the concrete walls cannot be neglected. In this example the flanking paths are at least 100 times lower than the test path. However, for test specimen with better performance such as a single aluminium plate, this concrete wall set-up is not sufficient. Therefore, modern test chambers are created by metal sheets filled with heavy glass wool or rock wool in order to avoid the coincidence effects and the structure borne flanking paths of concrete walls.

In Figure 10.18 the difference between the perfect and concrete wall chamber transmission loss is shown. Both rooms give the same results, except at low frequencies where the low transmission loss of the concrete walls leads to erroneous results due to flanking paths.

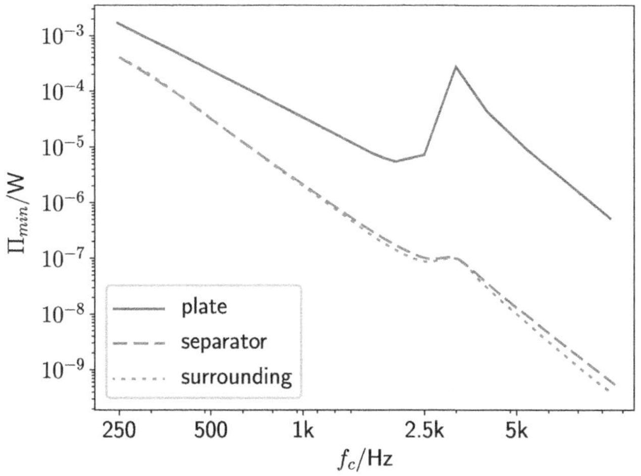

Figure 10.17 Power input into cavity2 from plate, separating and surrounding walls of cavity2. *Source:* Alexander Peiffer.

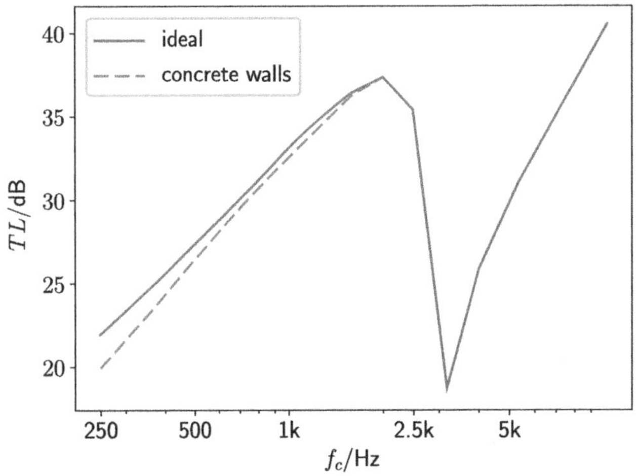

Figure 10.18 Transmission loss of idealistic and real twin chamber configuration. *Source:* Alexander Peiffer.

10.5.2 Noise Control Treatments

A simulation method without practical implementation of absorbing and isolating materials is useless, because the simulation and estimation of the efficiency of noise control treatments is mandatory for acoustic engineering. Therefore, the results of the multilayer simulations from Chapter 9 are adapted to SEA simulation. The parameters that are affected by the noise control treatment are the coupling and damping loss factors, and the treatment has an impact on cavity and plate subsystems.

10.5.2.1 Damping Loss due to Noise Control Treatment

We will not deal with the increase of plate damping due to an extra layer of visco-elastic material or a so-called constrained layer patch. This would increases the damping very efficiently (Crighton, 1996).

We focus on the loss factors related to cavity subsystems, the loss factor of the cavity itself due to treatment of the cavity surface. A specific treatment of area A_T with diffuse field absorption coefficient $\alpha_s(\omega)$ leads to an additional absorption area that must be taken into account in the cavity damping loss factor using equation (6.27):

$$\eta_m = \left(\eta_{m,\text{field}} + \frac{A_T\alpha_s(\omega)c_0}{4V\omega}\right)$$

and (6.70) from the theory of section 9.3.

10.5.2.2 Coupling Loss due to Noise Control Treatment

The coupling loss factor impact of noise control treatment is given by the transmission properties. In Figure 10.19 the resonant and non-resonant paths are shown with additional treatment.

The non-resonant path is considered by applying the transfer matrix method to a limp mass layer plus trim. For the resonant path the impact of noise control treatment is considered by the insertion loss coefficient IC. The idea is to keep the consideration of the finite panel dimensions using Leppington's theory and consider the

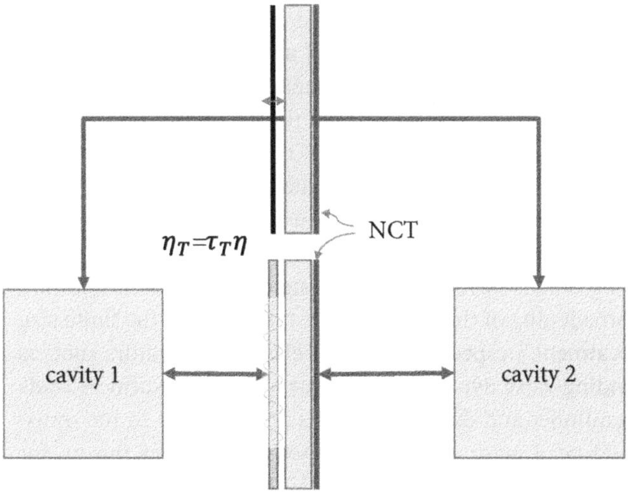

Figure 10.19 Area junction with noise control treatment. *Source:* Alexander Peiffer.

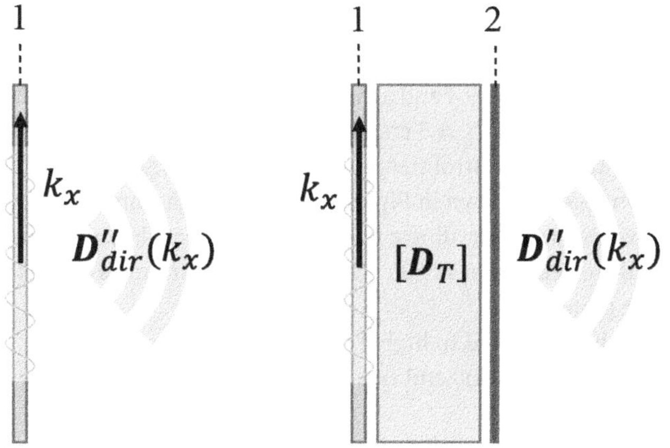

Figure 10.20 Radiated power configuration for untrimmed and trimmed plate. *Source:* Alexander Peiffer.

difference in the radiation efficiency between untrimmed and trimmed configuration. In Figure 10.20 the radiated power due to panel bending waves of the bare and trimmed plate is depicted.

The insertion loss is the correction factor for the untrimmed panel coupling loss factor to the version with trim

$$\eta_{T,12} = IC\,\eta_{12} \tag{10.20}$$

Due to (8.8c) the linear relationship between τ and η means that this insertion loss can be derived from the ratio between the transmission coefficient of the untrimmed panel and the trimmed panel (Allard and Atalla, 2009; Aut, 2003)

$$IC = \frac{\langle \tau_{\text{plate+trim}} \rangle_E}{\langle \tau_{\text{plate}} \rangle_E} \text{ with } \langle \tau_m \rangle_E = \frac{2}{\sin^2 \vartheta_{max}} \int_0^{\vartheta_{max}} \tau_m(\vartheta) \cos\vartheta \sin\vartheta d\vartheta \tag{10.21}$$

We know that the transfer matrix theory is not precise, especially around coincidence, but as we only use the ratio between trimmed and un-trimmed configuration, the effect is supposed to be low. Thus, the coupling loss factor from the radiation efficiency is kept, and the infinite layer is used to calculate the difference between the treated and untreated configurations. However, the choice of the maximal angle to fit the experimental values of transmission loss tests stays quite arbitrary.

Two further things must be noted: First, the resonant wave field of the plate consists of only one wavenumber. So, we might conclude that only the wave transmission at this wavenumber has to be taken. Nevertheless, the angle average is carried out, because this considers the broadening of the wavenumber range due to the finite size of the panel. Second, the treatment – especially when of elastic solid nature such as foams – will change the bending wave dynamics. This extra mass and stiffness leads to changes in bending wavenumber and damping. This is not included in the transfer matrix method. We would need more advanced methods to consider this as, for example, the wave–finite-element theory from Cotoni et al. (2005).

10.5.3 Transmission Loss of Trimmed Plate

In this example the aluminium plate is covered by noise control treatment of two layers: A fiber layer of the fiber material from Table 9.4 and a mass-layer of $m'' = 1\ \text{kg/m}^2$. The thickness of the fiber layer is $t_1 = 2$ cm and $t_2 = 5$ cm. We use the SEA model from the ideal twin chamber and apply the noise control treatment to the receiving room side. The results for the transmission loss are shown in Figure 10.21. The high transmission loss of the trimmed plate above the double wall resonances can be seen. Note that for the 2 cm trim, the double wall resonance is at 300 Hz, and the performance at this frequency is lower than for the plate.

The coincidence dip remains but is shifted to higher values. The SEA model agrees well with the transfer matrix method results, and one may ask why SEA is required

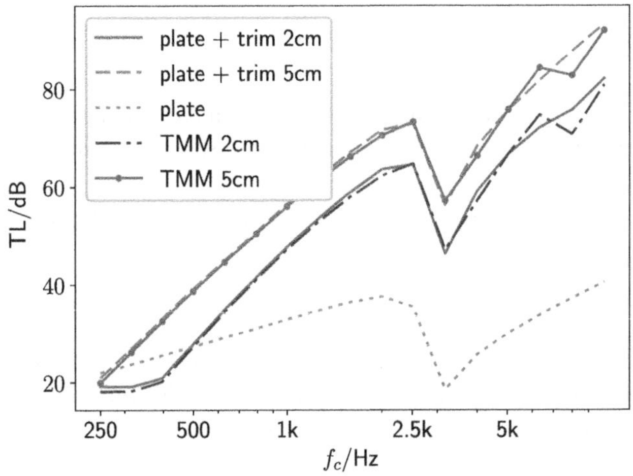

Figure 10.21 Transmission loss of untrimmed aluminium plate ($t = 4$ mm) and mass-fiber layers of different thicknesses. The trimmed results are compared to transfer matrix results (TMM). *Source:* Alexander Peiffer.

here. With the SEA approach even point force excitations on the plate can be considered, and SEA allows more complex setup of a complete system that is not possible with the simple transfer matrix method.

10.5.4 Free Field Radiation into Half Space

Many systems are radiating sound to the exterior space. For example, houses, cars, or ships are always connected to the surrounding space. Thus, we need a system description that simulates the free space in the SEA context. The semi infinite fluid (SIF) half space is an energy sink, because the acoustic energy is radiated to the free field but not reflected. Therefore, the semi infinite fluid can be included into the power flow of the SEA matrix by adding an extra damping loss to every subsystem that is connected to the semi infinite fluid.

The semi infinite fluid plays the role of a cavity in the junction logic, so one cavity is replaced by the semi infinite fluid. The power flows from the cavity and the plate to the semi infinite fluid. The arrows to the semi infinite fluid are going only in one direction, as there is no energy transport from the semi infinite fluid back to the cavity.

The bending wave field damping loss due to the radiation is

$$\eta_{B,SIF} = \frac{\rho_0 c_0 \sigma}{\omega m''} \tag{10.22}$$

and the cavity damping loss is

$$\eta_{C,SIF} = \frac{k^2 S_j}{8\pi^2 n_1(\omega)\omega} \langle \tau_{C,SIF} \rangle_E \tag{10.23}$$

The total radiated power can be calculated from the energy and the damping loss factor related to the semi infinite fluid and using (7.26) in accordance to the power input.

10.5.5 Isolating Box

Imagine a sound source, for example a compressor, a transformer, or an electric engine. These acoustic sources can be annoying when located in your neighborhood. Thus, we would like to design a box to isolate the source from the environment. In Figure 10.23 the setup is shown.

The box is placed on a rigid floor. So, the box is built up by five rectangular plates. We use steel plates of 2 mm thickness to make use of reasonable area mass $m'' = 15.6\ \mathrm{kg/m^2}$. Three configurations will be investigated.

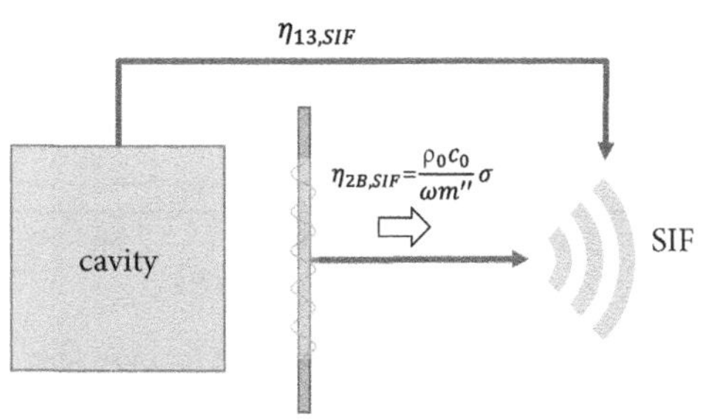

Figure 10.22 Area junction configuration with a semi infinite fluid. *Source:* Alexander Peiffer.

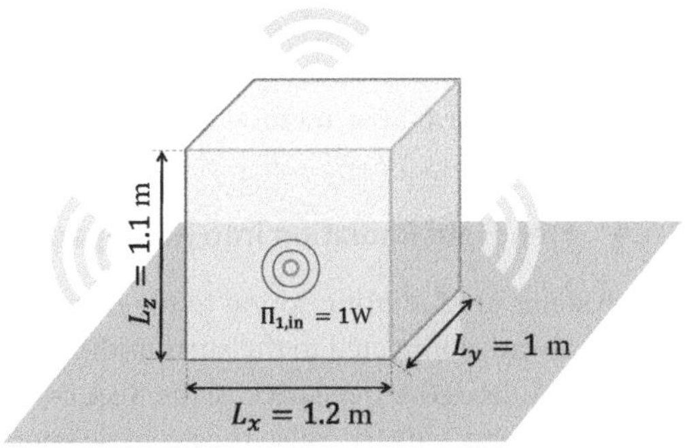

Figure 10.23 Box cover for source isolation. *Source:* Alexander Peiffer.

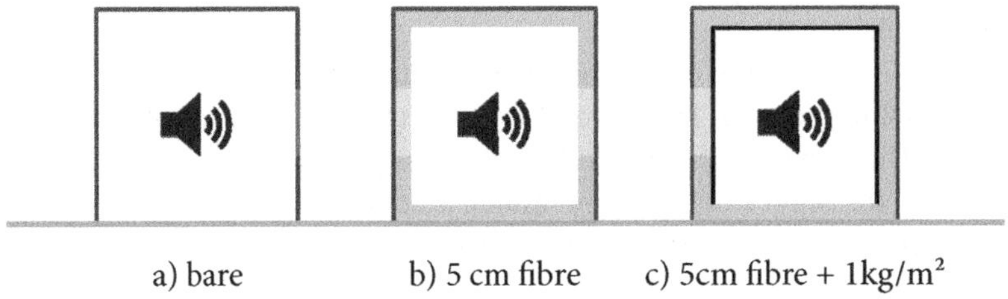

Figure 10.24 Different concepts of noise control with box housing. *Source:* Alexander Peiffer.

With SEA the different concepts can be evaluated regarding their efficiency. With 10.24a, the source is just isolated from the environment, and 10.24b provides a similar isolation. This is because pure fibers don't enhance the transmission loss too much, but there is extra absorption in the inner cavity that leads to lower pressure levels. Concept 10.24c brings high isolation due to the mass-spring system, but the cavity absorption is less than in 10.24b.

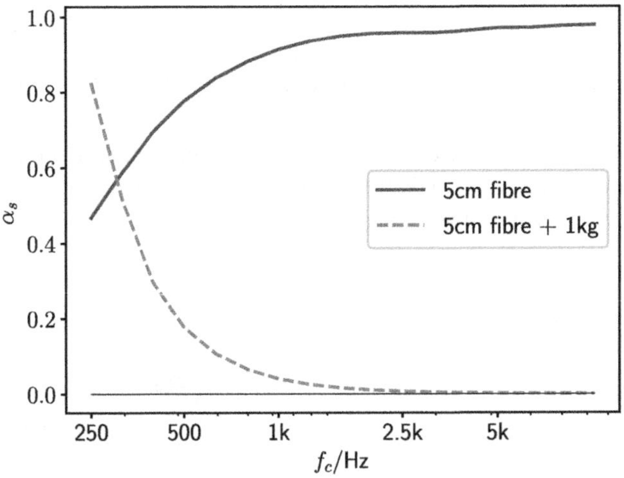

Figure 10.25 Absorption coefficient for pure fiber and mass spring system. *Source:* Alexander Peiffer.

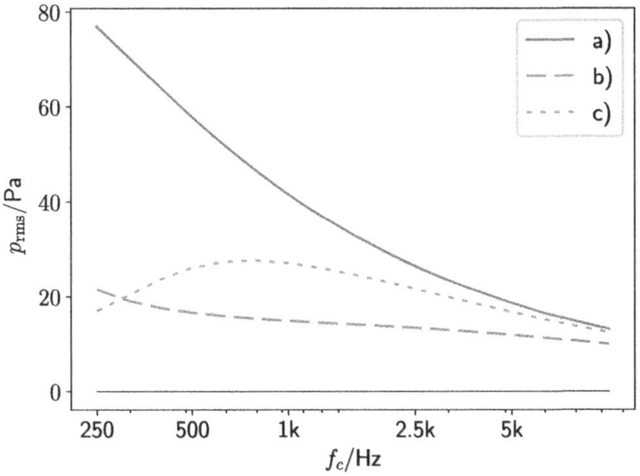

Figure 10.26 Reverberant pressure in the different box configurations. *Source:* Alexander Peiffer.

In Figure 10.25 one can see that the high isolating mass-spring system does not provide much absorption over a wide frequency range. This is the reason why at least the floor of the box is covered with pure fiber to keep some absorption on the cavity system for case 10.24c.

The reverberant pressure level is shown in Figure 3.2. We see that for 10.24a, the pressure in the cavity is quite high, 10.24b tremendously reduces the sound level, whereas the pressure level in 10.24c is higher due to less absorption in the cavity. In Figure 10.27 the noise reductions are shown. Even the simple steel cover provides good isolation, slightly suffering from the coincidence around $f = 6000$ Hz. This isolation improves due to the absorption in the cavity. Note that the fiber does not improve the isolation very much, so the effect results from the absorption in the cavity. With increasing isolation but less absorption, the performance is even better due to the much higher transmission loss of the mass spring system.

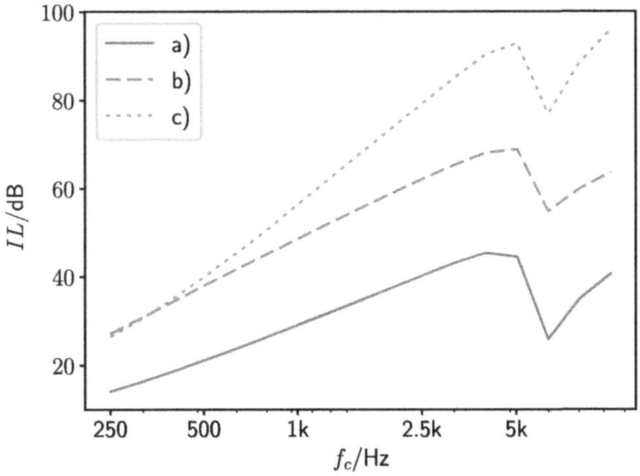

Figure 10.27 Noise reduction for the different box configurations. *Source:* Alexander Peiffer.

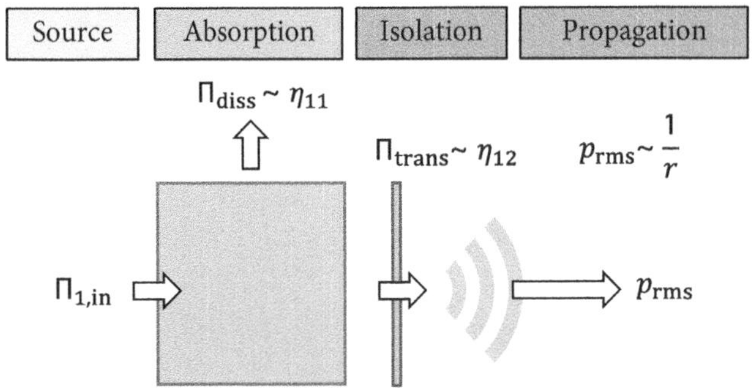

Figure 10.28 Cascade of noise control options from source to receiver. *Source:* Alexander Peiffer.

One may conclude that it is always better to focus on isolation; but, we must keep in mind that practical systems will have weak spots, because the box may not be completely hermetic due to cabling, power supply, or cooling. Thus, for better robustness of the concept, a mix of isolation and absorption is recommended. Note that we have neglected the direct field of the source. So, for a precise layout, it is required to check the ratio of reverberant and direct field pressure at the enclosure walls.

10.5.6 Rules of Noise Control

The housing example in combination with SEA can be applied to formulate a strategy for noise control. As can be seen in Figure 10.28, noise control has to be performed in several steps.

First, the best and most important noise control is at the source. If the source level of noise can be reduced, this is the first means to improve the acoustics of a technical system. This step works obviously without housing.

Second, apply damping and introduce absorption in the surrounding systems. In the example of Section 10.5.5, this means absorption treatment. In the SEA context, this increases the power dissipated in the first system connected to the source.

Third, maximize isolation, meaning that we increase the decoupling of the source or reverberant field from the exterior world.

The last means is distance. If all means of noise control are not sufficient, it remains only to increase distance. For some technical sources such as, for example, rocket launchers, this might be the only way (also for other reasons).

Bibliography

AutoSEA2 Foam2003 User's Guide, Theory @ QA. Technical report, ESI (VASci), 2003.

Jean-F. Allard and Noureddine Atalla. *Propagation of Sound in Porous Media*. Wiley, second edition, 2009. ISBN 978-0-470-74661-5.

V. Cotoni, P. J. Shorter, A. Charpentier, and Bryce Gardner. Efficient models of the acoustic radiation and transmission properties of complex trimmed structures. In *Proceedings Internoise 2005*, Rio, Brasil, August 2005.

D. G. Crighton, editor. *Modern Methods in Analytical Acoustics: Lecture Notes*. Springer, London, 3. printing edition, 1996. ISBN 978-3-540-19737-9.

R.H. Lyon and R.G. DeJong. *Theory and Application of Statistical Energy Analysis, Second Edition*. Butterworth Heinemann, second edition, 1995. ISBN 0-7506-9111-5.

Zhiyi Wang. Correlation between SEA Simulation and Test Results for a Double Wall Structure. Master's thesis, TU-München, München, Germany, June 2015.

11

Hybrid Systems

Basically, we have already applied the hybrid theory in preceding sections, for example Section 7.4. In many cases there were deterministic systems in the junction, for example a plate that we have considered as deterministic connection between the two cavity SEA systems and using the deterministic transfer matrix of the plate.

The twin chamber is an excellent test case for the hybrid method, as it constitutes a clear separation between random and deterministic subsystems and provides all the details of how to deal with each single step of hybrid FEM/SEA simulation.

11.1 Hybrid SEA Matrix

For pure SEA models the matrix from equation (6.113) is applied. The hybrid SEA matrix is similar except one additional damping contribution caused by the damping from the connected FEM-subsystems.

$$[L'] = \begin{bmatrix} n_1\left(\eta_{11}^{\alpha} + \sum_{n=1}^{N} \eta_{1n}\right) & -n_2\eta_{21} & \cdots & -n_N\eta_{N1} \\ -n_1\eta_{12} & n_2\left(\eta_{22}^{\alpha} + \sum_{n=1}^{N} \eta_{2n}\right) & & \vdots \\ \vdots & & \ddots & \\ -n_1\eta_{1N} & \cdots & & n_N\left(\eta_{NN}^{\alpha} + \sum_{n=1}^{N} \eta_{Nn}\right) \end{bmatrix} \tag{7.35}$$

All other impacts of the FE parts of the hybrid model are hidden in the hybrid coupling loss factors or an additional power input that may come from radiation of FE-subsystems into the SEA-subsystems.

11.2 Twin Chamber

The twin chamber set-up is such a perfect example for the hybrid method, because both chambers are designed to be random by the large size and irregular shape. So, the two cavities can be considered as random subsystems for the complete frequency range. The

Vibroacoustic Simulation: An Introduction to Statistical Energy Analysis and Hybrid Methods,
First Edition. Alexander Peiffer.

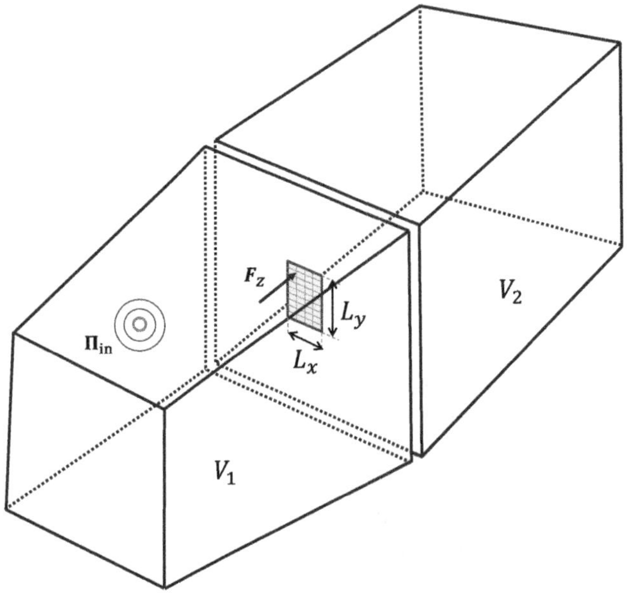

Figure 11.1 Twin chamber arrangement for TL tests of panel. *Source:* Alexander Peiffer.

panel that should be tested is supposed to be smaller than the panel from section 10.2.1. In this case it is not random and must be treated as a deterministic subsystem. We use the analytical solution of rectangular plate normal modes in the discrete version (5.41). As described in section 1.4.4.1 the normal modes are given as vectors and assembled in a matrix

$$[\Phi] = \left[\{\Phi\}_1 \quad \{\Phi\}_1 \quad \cdots \quad \{\Phi\}_N\right]$$

and the deterministic equations of motion are transformed into diagonal form. For the flat plate we use the mode shapes from equation (5.46) so that we can write the deterministic dynamic stiffness matrix in the following form

$$[\boldsymbol{D}_s'] = [\Phi]^H [\boldsymbol{D}_s] [\Phi] = \begin{bmatrix} \omega_1^2(1+j\eta)-\omega^2 & & & \\ & \ddots & & \\ & & \omega_n^2(1+j\eta)-\omega^2 & \\ & & & \omega_N^2(1+j\eta)-\omega^2 \end{bmatrix} \tag{11.1}$$

The radiation stiffness is calculated using (8.42) and also transformed into the modal space by

$$\left[\boldsymbol{D}_{\text{dir}}^{\prime(m)}\right] = [\Phi]^H \left[\boldsymbol{D}_{\text{dir}}^{(m)}\right] [\Phi] \tag{11.2}$$

The global form of the coupling loss factor and the transmission coefficient remains unchanged

$$\eta_{mn} = \frac{2}{\pi n_m \omega} \sum_{i,j} Im\{\boldsymbol{D}_{\text{dir},ij}^{\prime(n)}\} \left([\boldsymbol{D}_{tot}]^{\prime -1} Im\left[\boldsymbol{D}_{\text{dir}}^{(m)}\right]^{\prime} [\boldsymbol{D}_{tot}]^{\prime -H} \right)_{ij} \tag{7.50}$$

$$\tau_{mn} = \frac{8\pi^2 \omega n_m}{k_m^2 A} \eta_{mn} \qquad TL = 10 \log_{10} \frac{1}{\tau}$$

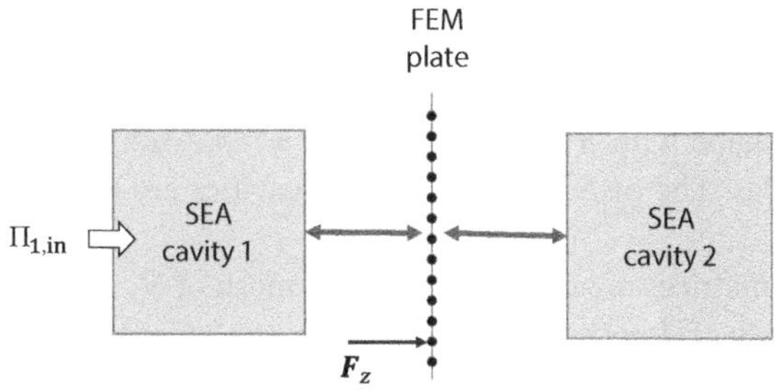

Figure 11.2 System and loads setup for the hybrid twin chamber model. *Source:* Alexander Peiffer.

The modal total stiffness matrix

$$[\boldsymbol{D}'_{tot}] = [\boldsymbol{D}'_s] + \sum_m \left[\boldsymbol{D}'^{(m)}_{\mathrm{dir}}\right] \tag{11.3}$$

requires some comment. The modal form of the direct radiation stiffness is a fully populated but symmetric matrix. Thus, the hybrid method destroys the diagonal form of the dynamic stiffness matrix. Though the introduction of random systems reduces the degrees of freedom tremendously, the computational amount can be higher than for a full deterministic method if the connecting regions are too large. See Peiffer (2012) for details. In Figure 11.2 the subsystem configuration is shown. We chose a rather small flat aluminium plate of size $L_x = 0.5$ m and $L_y = 0.8$ m of 4 mm thickness. At this size we will have a deterministic behaviour until 2 kHz.

The damping in the SEA systems due to the connected FE subsystems is given by (7.52)

$$\eta^{\alpha}_{mm} = \frac{2}{\pi\omega n_m(\omega)} \sum_i \eta\omega_i^2 \left(\left[\boldsymbol{D}_{tot}\right]'^{-1} Im\left[\boldsymbol{D}^{(m)}_{\mathrm{dir}}\right]' \left[\boldsymbol{D}_{tot}\right]'^{-H} \right)_{ii} \tag{7.52}$$

This damping loss corresponds to the dissipation in the cavities because of energy dissipated in the connected FEM plate. In comparison with the situation when the plate is an SEA subsystem, this part replaces the dissipation in the reverberant field of the plate.

For the execution of the full hybrid simulation process, we apply a random power load to the first SEA cavity and a deterministic out-of-plane point force at the plate near the corner. This creates the typical setup where we need both the random modelling of the reverberant fields in the cavity and the deterministic model of the plate. The random point force (3.230) will not provide a good estimation of the real input power, because the plate is a deterministic subsystem. Thus, the modal frequency response to this force excitation must be used. To summarize, two load cases are simulated:

Power 1W	Power source in cavity 1
Force 10 N rms	Point force on plate.

The global set-up of cavities is the same as in Chapter 10. The plate shall be deterministic for a large frequency range. In order to cover the coincidence region, we aim at calculating the frequency response up to f=4000 Hz. The hybrid modelling example is also useful for further clarification when and under which conditions the SEA assumptions become valid. Thus, all results will be compared to the results of an SEA model with the same data.

11.2.1 Step 1 – Setting up System Configurations

The normal modes of the plate are taken from equation 5.46. The modes are sorted in frequency order. The first mode shape with $n = (1, 1)$ is at $f_1 = 54.2$ Hz; until $f_{max} = 4780.1$Hz there are 140 modes. The SEA properties of the cavities are taken from Chapter 10.

The direct field radiation stiffness matrices are calculated using a mesh of $N_x = 40$ and $N_y = 25$ nodes. This leads to a mesh of element length $\Delta x \approx \Delta y \approx 0.02$ m, which is sufficient for this frequency. When the structural plate modes shall be calculated using a FE-mesh, this must be fine enough for the bending wavelength and can thus be much finer than the mesh that is required for a precise radiation stiffness calculation in the desired frequency range. However, in this example it is not necessary, because we map the solution onto our mesh directly from analytical results.

When using fine FE meshes for the structure, it is better to use two meshes, each appropriate for structure and fluid, for the better use of (8.42).

11.2.2 Step 2 – Setting up System Matrices and Coupling Loss Factors

The total system matrix is created according to (11.3). All degrees of freedom are shared by the connected systems and no specific treatment of connecting region is required.

The SEA matrix is set up with the same principles as in section 7.3.2.1 and using equation (7.50). In the SEA matrix the plate does not occur as subsystem. Hence, for the energy degrees of freedom, we have a two system SEA matrix.

The result for the hybrid transmission loss of the plate is shown in Figure 11.3. In addition the SEA transmission loss is also included. We recognize the typical shape of the transmission loss as described, for example, in Bies et al. (2018). At low frequencies below the first resonances, the transmission is stiffness controlled and very high. The transmission loss drops down at the first panel resonances, increases, and reaches mass law slope above 500 Hz. At this frequency the wavelength in the fluid becomes similar to

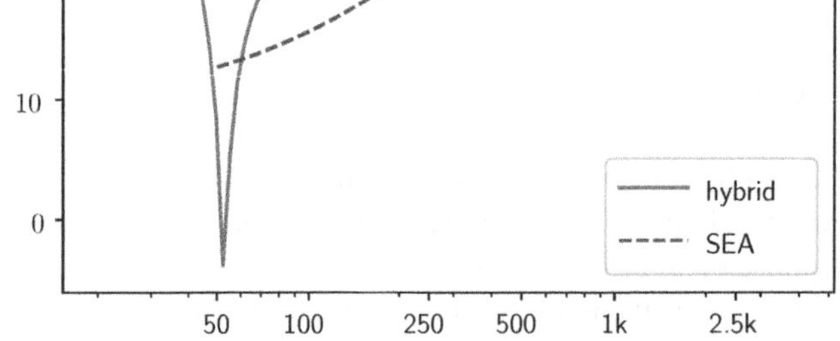

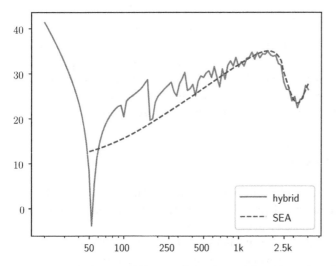

Figure 11.3 Hybrid and SEA transmission loss of aluminium plate. *Source:* Alexander Peiffer.

the panel dimensions. There is a good agreement between the SEA and the hybrid result in the mass law regime. In the coincidence area both curves also coincide well, despite small variations around the ensemble average of the SEA curve. Thus, the reverberant field model of the plate and area coupling loss factors are in good accordance to the hybrid method.

In addition to the inert damping of the SEA systems, there is the damping due to the connected FEM subsystems given in modal form by (7.51). The damping loss for both cavities is shown in Figure 11.4, revealing that this dissipation is relatively low. The reason for the low dissipation is the small area of the panel in relation to the total surface of the room. When all walls of the room would be FEM subsystems, the situation would be different. The highest dissipation occurs at the resonances of the plate.

11.2.3 Step 3 - External Loads

First, the SEA cavity 1 is excited by a power load of $\Pi_{1,ext} = 1$ W. Second, the plate is excited by a point force in the z-direction of the plate with $\hat{F}_{z,rms} = 10$ N that is located at $x = 0.31$ m and $y = 0.11$ m in plate coordinates assuming one corner as origin. Both cases are calculated separately.

11.2.4 Step 4 - Solving System Matrices

11.2.4.1 Step 4.1 - Power Input due to FEM System Excitation

The cross spectral density of the point force in modal coordinates must be calculated with

$$\left[\boldsymbol{S}_{qq}^{\prime(n)}\right] = \left[\boldsymbol{D}_{\text{tot}}^{\prime}\right]^{-1}\left[\boldsymbol{S}_{ff}^{\prime(m)}\right]\left[\boldsymbol{D}_{\text{tot}}^{\prime}\right]^{-H} \tag{11.4}$$

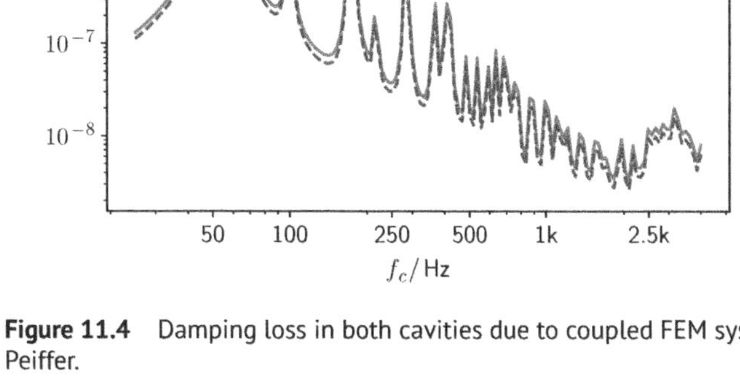

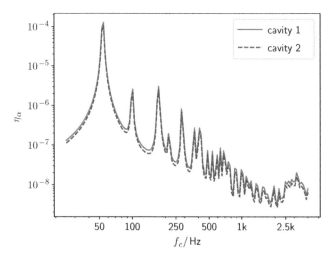

Figure 11.4 Damping loss in both cavities due to coupled FEM system. *Source:* Alexander Peiffer.

and the radiated power with

$$\left\langle \Pi_{\text{in}}^{(n)} \right\rangle_E = \frac{\omega}{2} \sum_{i,j} Im\boldsymbol{D}'_{\text{dir},ij} \left[\boldsymbol{S}'_{qq}\right]_{ij} \tag{11.5}$$

Even in modal form the triple matrix multiplication of (11.4) is computationally expensive. As the point force is fully coherent (to itself), it is faster to calculate the cross spectral density of the plate from the deterministic response, hence solving

$$\left[\boldsymbol{D}'_{\text{tot}}\right]\{\boldsymbol{q}'\} = \{\boldsymbol{F}'\} \text{with} \{\boldsymbol{F}'\} = [\Phi]^H \{\boldsymbol{F}\} \tag{11.6}$$

calculating the cross spectral density directly from the deterministic response, and determining the power input into the SEA subsystems from external excitation that is not related to SEA. The rms-value $F_{z,rms} = 10$ N of the point force corresponds to an amplitude of $\hat{F}_z = \sqrt{2}\ 10$ N. From the above equation we calculate a power radiation from the plate into the SEA systems as given in Figure 11.5. Compared to the source in cavity 1, the power can be neglected. However, at the odd resonances and near coincidence the panel becomes an efficient radiator, leading to high power radiation into the cavities. As both cavities are directly connected to the vibrating plate, the power radiation is the same for both rooms.

The velocity rms-value is reconstructed from the modal coordinates by

$$\{\boldsymbol{v}\} = j\omega\{\boldsymbol{q}\} = j\omega\,[\Phi]\{\boldsymbol{q}'\} \tag{11.7}$$

$$v_{\text{rms}}^2 = \left\langle |\boldsymbol{v}_i|^2 \right\rangle \tag{11.8}$$

The resulting velocity due to the force excitation is shown in Figure 11.7 in combination with the velocity caused by excitation from the reverberant fields.

11.2.4.2 Setp 4.2 – Solve the SEA Equations with SEA and FEM Power Input from Step 4.1

The result for each single case can be seen in Figure 11.6. In comparison the power load in cavity 1 leads to higher pressure levels at low frequencies, but at high frequencies the

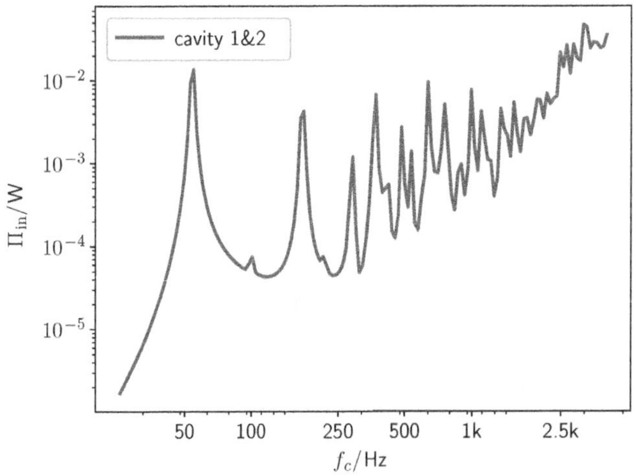

Figure 11.5 Power input to both rooms from deterministic force excitation at the plate. *Source:* Alexander Peiffer.

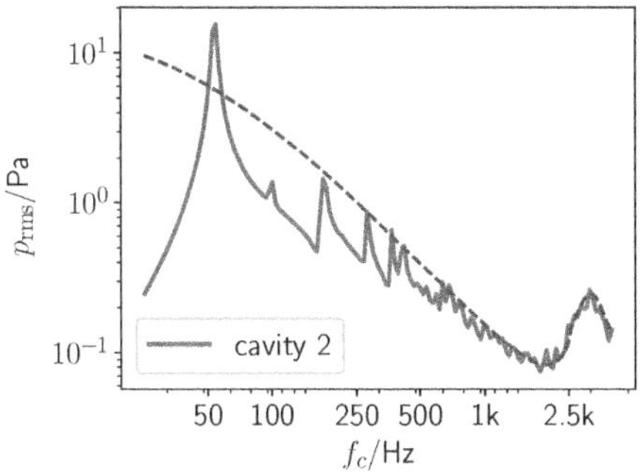

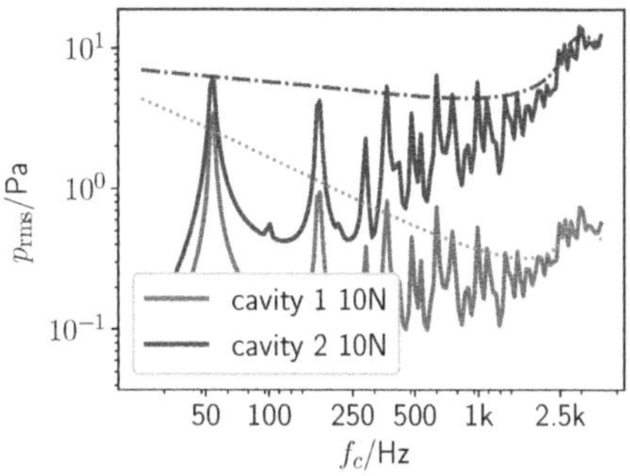

Figure 11.6 Pressure in both rooms due to power excitation in cavity 1 (LHS) and force excitation on plate (RHS). *Source:* Alexander Peiffer.

radiation from the plate becomes more efficient and radiates more energy into cavity 2 than is transmitted from cavity 1 into cavity 2 through the plate. The SEA results of the corresponding model show that the random modelling gives reliable results above 600 Hz for the power excitation in cavity 1 and above 2 kHz for the force excitation. In case of force excitation, the point force does not lead to equal excitation of all modes, and the conditions for SEA are met later in frequency. In addition we must consider that the mass law provides a correct value for the coupling of both cavities, and therefore the results agree earlier in frequency with the hybrid result.

11.2.4.3 Step 4.3 – Calculate FE Response due to Energies in SEA Subsystems from Step 4.2

As the external forces were treated by step 4.1, we finally calculate the response of the FEM systems resulting from the reverberant field excitation in each subsystem using (7.43), omitting the deterministic external loads

$$\left[\boldsymbol{S}'_{qq,rev}\right] = \left[\boldsymbol{D}'\right]_{\text{tot}}^{-1}\left(\sum_m \frac{4E_m}{\pi\omega n_m(\omega)} Im\left[\boldsymbol{D}'\right]_{\text{dir}}^{(m)}\right)\left[\boldsymbol{D}'\right]_{\text{tot}}^{-H} \tag{11.9}$$

As the response is calculated in the modal base, the cross spectral density matrix must be converted into the mesh space by

$$\left[S_{qq,rev}\right] = [\Phi]\left[S'_{qq,rev}\right][\Phi]^H \tag{11.10}$$

and the rms-velocity results from the diagonal of the above matrix

$$v_{\text{rms}}^2 = \left\langle \text{diag}\left[S_{qq,rev}\right]\right\rangle \tag{11.11}$$

11.2.5 Step 5 – Adding the Results

In the final step all results are added together for the deterministic systems. The cross spectral density of the plate results from the external excitation and the reverberant load of the connected SEA systems. In Figure 11.7 the rms-velocity of the plate is shown in combination with the SEA result. For the power load case there is no external excitation of the plate, so the reverberant load is the exclusive excitation. The SEA simulation covers only the resonant energy in the plate. This is the reason why both curves don't agree

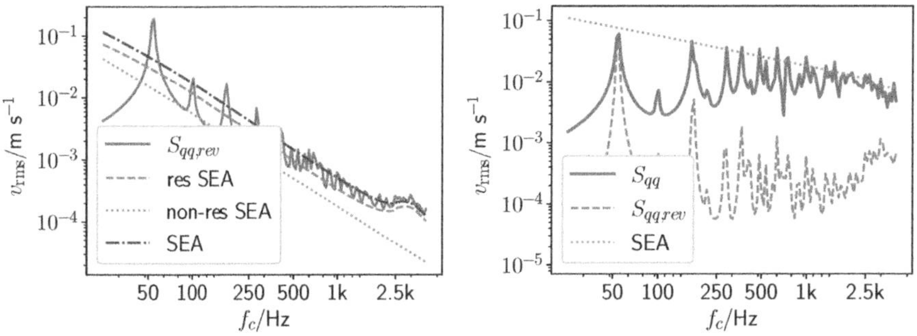

Figure 11.7 Velocity of the plate due to power excitation in room 1 (LHS) and force excitation on plate (RHS). *Source:* Alexander Peiffer.

even at high frequencies, and the resonant SEA solution underestimates the velocity level compared to the FEM system response. When we add the nonresonant motion to the engineering result of the plate using (10.19), both curves agree well above 500 Hz.

In case of the force excitation, the external load response is dominant compared to the reverberant response $S_{qq,rev}$. This is obvious, because the reverberant load from the cavities is the indirect feedback of the force exciting the plate. Globally, the results from section 6.4.2.1 are still valid. A point force requires a large modal overlap to fulfil the conditions of a smooth ensemble average. However, above 1kHz the SEA solution is valid.

11.2.5.1 Conlusions

We can perform a Monte Carlo analysis with our test case by generating an ensemble average of different realizations of the plate, and we would get a smoothly averaged result as shown in Shorter and Langley (2005). The example indicates how useful the hybrid theory is, especially for transmission loss calculations, as these cases provide nearly perfect separation between the random systems (cavities) and the deterministic plate. In addition the hybrid theory illustrates the limits of SEA. The modal and resonant dynamics of the plate cannot be modelled by SEA, whereas the high frequency dynamics can be calculated very efficiently without large losses in precision.

This is a rather academic case with simple cavities and a simple flat plate. When real systems or plates with irregularities, beadings, etc. are dealt with there may be more uncertainty in the system, and thus, the random approach becomes a more valid option in comparison to the natural uncertainty of the technical system.

11.3 Trim in Hybrid Theory

For the consideration of trim layers in hybrid theory, the degrees of freedom have to be reorganized in such a way that there are structural and fluid half space degrees of freedom as far as those of the two trim surfaces. In order to keep the derivation simple we consider the plate to be fully covered with noise control treatment.

As the FEM subsystem is modelled by a FEM matrix, a discrete form of the trim layer stiffness is needed. One option is to apply the porous finite element method (PEM)

(Allard and Atalla, 2009). We will approximate the stiffness matrix of the noise control treatment by converting the wavenumber transfer matrix into the discrete regular meshed space by an inverse Fourier transform. This assumption is valid for large areas with constant lay-up over the area.

11.3.1 The Trim Stiffness Matrix

In order to cope with the hybrid theory, the acoustic treatment must also be modelled by a a stiffness matrix. This is rather unusual, as, for example, FE implementations are using displacement degrees of freedom for the structure side and pressure degrees of freedom for the fluid side (Doutres et al., 2007) due to the fact that the fluid is represented by pressure degrees of freedom in most solvers.

The trim dynamic stiffness matrix has different sets of degrees of freedom for the left (1) and right (2) side. This suggests writing the matrix in blocked form

$$\begin{bmatrix} \boldsymbol{D}_{11}^{SP} & \boldsymbol{D}_{12}^{SP} \\ \boldsymbol{D}_{21}^{SP} & \boldsymbol{D}_{22}^{SP} \end{bmatrix} \begin{Bmatrix} \boldsymbol{w}_1 \\ \boldsymbol{w}_2 \end{Bmatrix} = \begin{Bmatrix} \boldsymbol{F}_1 \\ \boldsymbol{F}_2 \end{Bmatrix} \tag{11.12}$$

The calculation of the above stiffness matrix is done in two steps: First, the transfer matrix of the infinite layer from section 9.3 is transformed into the stiffness matrix in wavenumber domain.

$$\begin{bmatrix} \boldsymbol{D}_{11}^{\prime\prime SP}(k_x) & \boldsymbol{D}_{12}^{\prime\prime SP}(k_x) \\ \boldsymbol{D}_{21}^{\prime\prime SP}(k_x) & \boldsymbol{D}_{22}^{\prime\prime SP}(k_x) \end{bmatrix} \begin{Bmatrix} \boldsymbol{w}_1(k_x) \\ \boldsymbol{w}_2(k_x) \end{Bmatrix} = \tag{11.13}$$

$$\left[\boldsymbol{D}^{\prime\prime SP}(k_x)\right] \begin{Bmatrix} \boldsymbol{w}_1(k_x) \\ \boldsymbol{w}_2(k_x) \end{Bmatrix} = \begin{Bmatrix} \boldsymbol{F}_1^{\prime\prime}(k_x) \\ \boldsymbol{F}_2^{\prime\prime}(k_x) \end{Bmatrix} = \begin{Bmatrix} \boldsymbol{p}_1(k_x) \\ \boldsymbol{p}_2(k_x) \end{Bmatrix}$$

Second, this stiffness matrix must be converted into the discrete space. For the approximation of this discrete stiffness matrix, we assume a regular mesh with constant element lengths $\Delta x,\Delta y$. The transfer matrix of infinite layers (9.96)

$$\begin{Bmatrix} \boldsymbol{p}_1(k_x) \\ \boldsymbol{v}_{z1}(k_x) \end{Bmatrix} = \begin{bmatrix} T_{11}(k_x) & T_{12}(k_x) \\ T_{21}(k_x) & T_{22}(k_x) \end{bmatrix} \begin{Bmatrix} \boldsymbol{p}_2'(k_x) \\ \boldsymbol{v}_{z1}(k_x) \end{Bmatrix} \tag{9.96}$$

is to be converted into the stiffness matrix (11.13). The internal pressure $\boldsymbol{p}_2$ must be replaced by the external pressure $\boldsymbol{p}_2' = -\boldsymbol{p}_2$

$$\frac{1}{\boldsymbol{T}_{21}} \begin{bmatrix} \boldsymbol{T}_{11} & -1 \\ -1 & \boldsymbol{T}_{22} \end{bmatrix} \begin{Bmatrix} \boldsymbol{v}_{z1} \\ \boldsymbol{v}_{z2} \end{Bmatrix} =$$

$$\frac{j\omega}{\boldsymbol{T}_{21}} \begin{bmatrix} \boldsymbol{T}_{11} & -1 \\ -1 & \boldsymbol{T}_{22} \end{bmatrix} \begin{Bmatrix} \boldsymbol{w}_1 \\ \boldsymbol{w}_2 \end{Bmatrix} = \begin{Bmatrix} \boldsymbol{p}_1 \\ \boldsymbol{p}_2 \end{Bmatrix} \tag{11.14}$$

and the coefficients of the stiffness matrix in wavenumber domain are hence

$$D''^{SP}_{11}(k_x) = \frac{j\omega T_{11}}{T_{21}} \tag{11.15}$$

$$D''^{SP}_{12}(k_x) = D''^{SP}_{12}(k_x) = -\frac{j\omega}{T_{21}} \tag{11.16}$$

$$D''^{SP}_{22}(k_x) = \frac{j\omega T_{22}}{T_{21}} \tag{11.17}$$

In Figure 11.8 such a treatment connected to a structure is shown. Tournour et al. (2007) proposed a local approximation that assumes that only the opposite nodes are connected. This assumption is valid for very thin layers ($h \ll \lambda$), and the block matrices in equation (11.12) are diagonal in that case. It is shown by Peiffer (2018) that this assumption is too strict and a nonlocal approach is required for most treatments.

This matrix must be converted into the space domain by inverse Fourier transform. Using the jinc-function approach from Langley (2007) and assuming isotropy in each layer, the transformation becomes simple and implies a finite wavenumber integration.

$$D^{SP}_{ab}(x_{ij}, \omega) = \frac{\Delta A}{2\pi} \int_0^{k_s} D''^{SP}_{ab}(k) J_0(k x_{ij}) k dk \tag{11.18}$$

The transformation is applied to each coefficient of the stiffness matrix with wavenumber arguments

$$a, b = 1, 2 \qquad x_{ij} = \sqrt{x_i^2 - x_j^2} \text{ and } k_s = \frac{2\pi}{\sqrt{\Delta x^2 + \Delta y^2}} \tag{11.19}$$

being the projected distance x_{ij} between nodes, and the maximum wavenumber k_s that can be represented by thespacial sampling of the mesh, respectively. The integral in

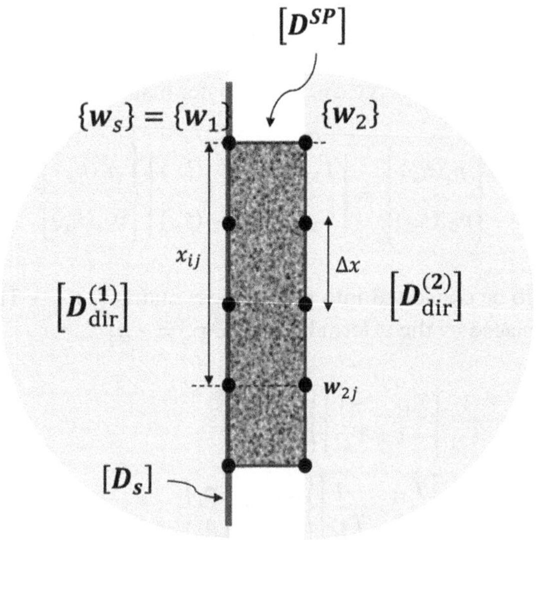

Figure 11.8 2D sketch of plate with trim. *Source:* Alexander Peiffer.

(11.18) is solved numerically, and the stiffness matrix of the trim is available for further considerations.

11.3.2 Hybrid Modal Formulation of Trim and Plate

The trim constitutes an additional deterministic layer that must be considered in the total stiffness matrix. When a layer of noise control treatment is applied to the right surface, the plate degrees of freedom are shared with those from left side of the trim. The right side of the noise control treatment is connected to the fluid as shown in figure 11.8.

This configuration gives rise to the following block matrix configuration

$$\left(\begin{bmatrix} \boldsymbol{D}_s & \\ & \end{bmatrix} + \begin{bmatrix} \boldsymbol{D}_{\text{dir}}^{(1)} & \\ & \boldsymbol{D}_{\text{dir}}^{(2)} \end{bmatrix} + \begin{bmatrix} \boldsymbol{D}_{11}^{SP} & \boldsymbol{D}_{12}^{SP} \\ \boldsymbol{D}_{21}^{SP} & \boldsymbol{D}_{22}^{SP} \end{bmatrix}\right) \begin{Bmatrix} \boldsymbol{w}_1 \\ \boldsymbol{w}_2 \end{Bmatrix} = \begin{Bmatrix} \boldsymbol{F}_1 \\ \boldsymbol{F}_2 \end{Bmatrix} \tag{11.20}$$

adding up to the total stiffness matrix with trim

$$\left[\boldsymbol{D}_{\text{tot}}\right] = \begin{bmatrix} \boldsymbol{D}_s + \boldsymbol{D}_{11}^{SP} + \boldsymbol{D}_{\text{dir}}^{(1)} & \boldsymbol{D}_{12}^{SP} \\ \boldsymbol{D}_{21}^{SP} & \boldsymbol{D}_{22}^{SP} + \boldsymbol{D}_{\text{dir}}^{(2)} \end{bmatrix} \tag{11.21}$$

The coupling loss factor can be calculated when we write (7.27) in block matrix form

$$\eta_{12} = \frac{2}{\pi n_1 \omega} \sum_{i,j} Im \begin{bmatrix} \boldsymbol{D}_{\text{dir}}^{(1)} & \\ & \end{bmatrix}_{ij} \left(\begin{bmatrix} \boldsymbol{D}_{11}^{SP} & \boldsymbol{D}_{12}^{SP} \\ \boldsymbol{D}_{21}^{SP} & \boldsymbol{D}_{22}^{SP} \end{bmatrix}_{\text{tot}}^{-1} Im \begin{bmatrix} & \\ & \boldsymbol{D}_{\text{dir}}^{(2)} \end{bmatrix} \begin{bmatrix} \boldsymbol{D}_{11}^{SP} & \boldsymbol{D}_{12}^{SP} \\ \boldsymbol{D}_{21}^{SP} & \boldsymbol{D}_{22}^{SP} \end{bmatrix}_{\text{tot}}^{-H} \right)_{ij} \tag{11.22}$$

The above equation can be used as is but leads to double size matrices in this case. The idea is to stay with the plate displacement coordinates and condense the above equation to pure structural coordinates, here $\boldsymbol{w}_1$. Writing equation (11.20) using the block matrix gives

$$\left([\boldsymbol{D}]_s + \left[\boldsymbol{D}_{11}^{SP}\right] + \left[\boldsymbol{D}_{\text{dir}}^{(1)}\right]\right)\{\boldsymbol{w}_1\} \qquad + \left[\boldsymbol{D}_{12}^{SP}\right]\{\boldsymbol{w}_2\} = \{\boldsymbol{F}_1\} \tag{11.23}$$

$$\left[\boldsymbol{D}_{21}^{SP}\right]\{\boldsymbol{w}_1\} \qquad + \left(\left[\boldsymbol{D}_{22}^{SP}\right] + \left[\boldsymbol{D}_{\text{dir}}^{(2)}\right]\right)\{\boldsymbol{w}_2\} = \{\boldsymbol{F}_2\} \tag{11.24}$$

We eliminate $\boldsymbol{w}_2$ in order to express the total stiffness in structural coordinates. Doing this with equations (11.23) and (11.24) leads to

$$\left(\left\{[\boldsymbol{D}_s] + \left[\boldsymbol{D}_{11}^{SP}\right] + \left[\boldsymbol{D}_{\text{dir}}^{(1)}\right]\right\} - \left[\boldsymbol{D}_{12}^{SP}\right]\left\{\left[\boldsymbol{D}_{22}^{SP}\right] + \left[\boldsymbol{D}_{\text{dir}}^{(2)}\right]\right\}^{-1}\left[\boldsymbol{D}_{21}^{SP}\right]\right)\{\boldsymbol{w}_1\}$$

$$= \{\boldsymbol{F}_1\} - \left[\boldsymbol{D}_{12}^{SP}\right]\left(\left[\boldsymbol{D}_{22}^{SP}\right] + \left[\boldsymbol{D}_{\text{dir}}^{(2)}\right]\right)^{-1}\{\boldsymbol{F}_2\} \tag{11.25}$$

There is a modified version of the total stiffness matrix

$$\left[\boldsymbol{D}_{\text{tot}}^{SP}\right] = \left\{[\boldsymbol{D}_s] + \left[\boldsymbol{D}_{11}^{SP}\right] + \left[\boldsymbol{D}_{\text{dir}}^{(1)}\right]\right\} - \left[\boldsymbol{D}_{12}^{SP}\right]\left\{\left[\boldsymbol{D}_{22}^{SP}\right] + \left[\boldsymbol{D}_{\text{dir}}^{(2)}\right]\right\}^{-1}\left[\boldsymbol{D}_{21}^{SP}\right] \tag{11.26}$$

which is valid for coordinates $\{\boldsymbol{w}_1\}$. Due to the condensation $\{\boldsymbol{F}_2\}$ is transformed to $\{\boldsymbol{w}_1\}$ degrees of freedom by

$$\begin{aligned}\left[\boldsymbol{D}_{\text{tot}}^{SP}\right]\{\boldsymbol{w}_1\} &= \{\boldsymbol{F}_1\} - \left[\boldsymbol{D}_{12}^{SP}\right]\left(\left[\boldsymbol{D}_{22}^{SP}\right] + \left[\boldsymbol{D}_{\text{dir}}^{(2)}\right]\right)^{-1}\{\boldsymbol{F}_2\} \\ &= \{\boldsymbol{F}_1\} + \left[\boldsymbol{D}_{trans}^{SP}\right]\{\boldsymbol{F}_2\}\end{aligned} \tag{11.27}$$

using the transformation matrix

$$\left[\boldsymbol{D}_{trans}^{SP}\right] = -\left[\boldsymbol{D}_{21}^{SP}\right]\left(\left[\boldsymbol{D}_{22}^{SP}\right] + \left[\boldsymbol{D}_{\text{dir}}^{(2)}\right]\right)^{-1} \tag{11.28}$$

Equation (11.27) provides the deterministic equation of motion for displacement degrees of freedom $\{\boldsymbol{w}_1\}$ of the plate. For use in hybrid coupling loss factor equation (7.27), we replace the direct radiation stiffness by an expression that includes the trim. Let us first consider that the trimmed side is the radiating side. The power radiated by the degrees of freedom from side two $\{\boldsymbol{w}_2\}$ is given by

$$\Pi = \frac{\omega}{2} Im\left\{\{\boldsymbol{F}_2\}\{\boldsymbol{w}_2\}^*\right\} = \frac{\omega}{2} Im\left\{\{\boldsymbol{w}_2\}^T\left[\boldsymbol{D}_{\text{dir}}^{(2)}\right]\{\boldsymbol{w}_2\}^*\right\} \tag{11.29}$$

Reordering of (11.24) and assuming no external forces $\boldsymbol{F}_2 = 0$ allows transforming $\{\boldsymbol{w}_1\}$ to $\{\boldsymbol{w}_2\}$

$$\begin{aligned}&-\left(\left[\boldsymbol{D}_{22}^{SP}\right] + \left[\boldsymbol{D}_{\text{dir}}^{(2)}\right]\right)^{-1}\left[\boldsymbol{D}_{21}^{SP}\right]\{\boldsymbol{w}_1\} = \\ &\left[\boldsymbol{D}_{\text{trans}}^{SP}\right]^T\{\boldsymbol{w}_1\} \overset{\text{symm}}{=} \left[\boldsymbol{D}_{\text{trans}}^{SP}\right]\{\boldsymbol{w}_1\} = \{\boldsymbol{w}_2\}\end{aligned} \tag{11.30}$$

and we get for the radiated power

$$\Pi = \frac{\omega}{2} Im\left\{\{\boldsymbol{w}_1\}^T\left[\boldsymbol{D}_{\text{trans}}^{SP}\right]^T\left[\boldsymbol{D}_{\text{dir}}^{(2)}\right]\left[\boldsymbol{D}_{\text{trans}}^{SP}\right]^*\{\boldsymbol{w}_1^*\}\right\} \tag{11.31}$$

The term between both $\{\boldsymbol{w}_1\}$ is called the reduced radiation stiffness

$$\left[\boldsymbol{D}_{\text{red}}^{SP}\right] = \left[\boldsymbol{D}_{\text{trans}}^{SP}\right]^T\left[\boldsymbol{D}_{\text{dir}}^{(2)}\right]\left[\boldsymbol{D}_{\text{trans}}^{SP}\right]^* \tag{11.32}$$

Thus, the radiated power can be calculated with

$$\begin{aligned}\Pi &= \frac{\omega}{2} Im\left\{\{\boldsymbol{w}_1\}^T\left[\boldsymbol{D}_{\text{red}}^{SP}\right]\{\boldsymbol{w}_1\}^*\right\} = \frac{\omega}{2}\sum_{i,j} Im\left\{\boldsymbol{D}_{\text{red},ij}^{SP}\boldsymbol{w}_{1,j}\boldsymbol{w}_{1,i}^H\right\} \\ &= \frac{\omega}{2}\sum_{i,j} Im\left\{\boldsymbol{D}_{\text{red},ij}^{SP}\right\}\left[\boldsymbol{S}_{ww,1}^*\right]_{ij}\end{aligned} \tag{11.33}$$

and (11.32) replaces $\left[\boldsymbol{D}_{\text{dir}}^{(2)}\right]$ in equation (7.27).

$$\eta_{12}^{SP} = \frac{2}{\pi n_1 \omega}\sum_{i,j} Im\boldsymbol{D}_{\text{red},ij}^{SP}\left(\left[\boldsymbol{D}_{\text{tot}}^{SP}\right]^{-1} Im\left[\boldsymbol{D}_{\text{dir}}^{(1)}\right]\left[\boldsymbol{D}_{\text{tot}}^{SP}\right]^{-H}\right)_{ij} \tag{11.34}$$

From reciprocity (6.112) it follows that the reduced radiation stiffness can also be used for the reverse direction.

$$\eta_{21}^{SP} = \frac{2}{\pi n_2 \omega}\sum_{i,j} Im\boldsymbol{D}_{\text{dir},ij}^{(1)}\left(\left[\boldsymbol{D}_{\text{tot}}^{SP}\right]^{-1} Im\left[\boldsymbol{D}_{\text{red}}^{SP}\right]\left[\boldsymbol{D}_{\text{tot}}^{SP}\right]^{-H}\right)_{ij} \tag{11.35}$$

In addition to the reciprocity argument, the above expression can also be derived using the fact that in the condensed equation of motion (11.27), forces $\{\boldsymbol{F}_2\}$ are converted to $\{\boldsymbol{F}_1\}$ by the transformation matrix $\left[\boldsymbol{D}_{\text{trans}}^{SP}\right]$.

The diffuse field excitation from side two is then given by

$$\begin{aligned}\left[\boldsymbol{S}_{ff,11}^{(2)}\right] &= \left\langle \boldsymbol{F}_1\boldsymbol{F}_1^H\right\rangle_E = \left\langle \left[\boldsymbol{D}_{\text{trans}}^{SP}\right]\boldsymbol{F}_2\boldsymbol{F}_2^H\left[\boldsymbol{D}_{\text{trans}}^{SP}\right]^H\right\rangle_E \\ &= \frac{4E_2}{\pi n_2(\omega)\omega}\left[\boldsymbol{D}_{\text{trans}}^{SP}\right] Im\left[\boldsymbol{D}_{\text{dir}}^{(2)}\right]\left[\boldsymbol{D}_{\text{trans}}^{SP}\right]^H \\ &= \frac{4E_2}{\pi n_2(\omega)\omega} Im\left[\boldsymbol{D}_{\text{red}}^{SP}\right]\end{aligned} \tag{11.36}$$

Entering this expression in (7.21) leads also to (11.35). It can be shown that a trimmed surface can be considered in general by using the total stiffness matrix given by

$$\left[\boldsymbol{D}_{tot}^{SP}\right] = \left[\boldsymbol{D}_s\right] + \sum_m \left[\boldsymbol{D}_{\text{dir}}^{(m)}\right] \tag{11.37}$$

using the different expression for trimmed sufaces

$$\left[\boldsymbol{D}_{\text{dir,red}}^{(m)}\right] = \left[\boldsymbol{D}_{12}^{SP}\right] - \left[\boldsymbol{D}_{12}^{SP}\right]\left(\left[\boldsymbol{D}_{22}^{SP}\right] + \left[\boldsymbol{D}_{\text{dir}}^{(m)}\right]\right)^{-1}\left[\boldsymbol{D}_{21}^{SP}\right]$$

According to equations (11.34) and (11.35), the radiation stiffness must be replaced by the reduced radiation stiffness if there is trim on the surface. The same approach is used for all other expressions used in the hybrid formulation, for example (7.52).

11.3.3 Modal Space

The stiffness matrix must be transformed into the modal space for both sets of degrees of freedom. The transformation reads:

$$\{\boldsymbol{w}_1\} = [\Phi]\{\boldsymbol{w}_1'\} \qquad \{\boldsymbol{w}_2\} = [\Phi]\{\boldsymbol{w}_2'\} \tag{11.38}$$

Or, in matrix form when we write all connecting region degrees of freedom in one column vector

$$\begin{Bmatrix}\boldsymbol{w}_1\\ \boldsymbol{w}_2\end{Bmatrix} = \begin{bmatrix}\Phi & \\ & \Phi\end{bmatrix}\begin{Bmatrix}\boldsymbol{w}_1'\\ \boldsymbol{w}_2'\end{Bmatrix} \tag{11.39}$$

The transformation matrix is a block matrix with twice the modal transformation matrix. All system matrices are transformed into modal space by this matrix meaning to apply the modal transformation matrix to every block matrix. All above given matrices are converted to modal space with the usual transformation.

The same is true for the block matrices of the trim leading to:

$$\left[\boldsymbol{D}_{ab}^{\prime SP}\right] = [\Phi]^H\left[\boldsymbol{D}_{ab}^{SP}\right][\Phi] \tag{11.40}$$

From the conversion to modal space follows a useful property for evaluating the radiation: the modal radiation efficiency given by

$$\sigma_n = \frac{1}{\omega\Delta A\rho_2 c_2}\frac{Im\boldsymbol{D}_{\text{dir}}^{(2)}}{\Phi_n^H\Phi_n} \tag{11.41}$$

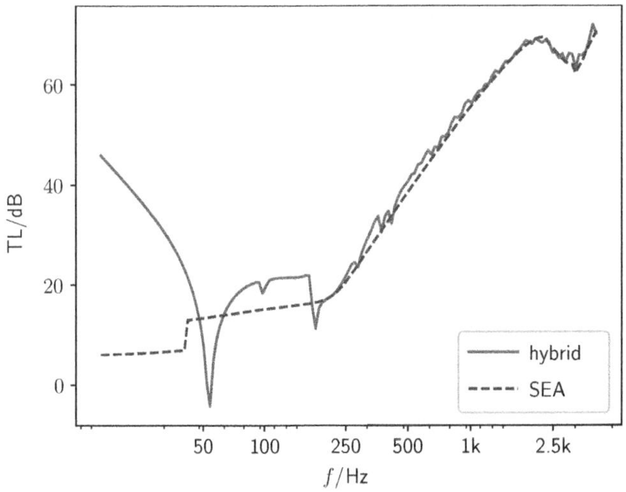

Figure 11.9 Transmission loss of small panel with trim, hybrid and SEA solution.
Source: Alexander Peiffer.

For configurations with trim the reduced radiation stiffness from equation (11.32) shall be considered.

11.3.4 Plate Example with Trim

We apply this theory to the plate from section 11.2. The panel surface connected to cavity 2 is treated with the 5 cm fiber-mass system from section 10.5.3. With the given theory the transmission loss is calculated and shown in Figure 11.8 in comparison with the SEA result. The agreement between hybrid and SEA is even better than for the untrimmed case. Both approaches start to coincide above 250 Hz in contrast to figure 11.3 where the agreement started around 1 kHz. This is the reason why SEA is considered often as valid even at low frequencies for airborne problems. The dissipation by the noise treatment introduces damping into the system that results in an earlier fulfillment in frequency of the requirements of SEA.

Bibliography

Jean-F. Allard and Noureddine Atalla. *Propagation of Sound in Porous Media.* Wiley, second edition, 2009. ISBN 978-0-470-74661-5.

David Alan Bies, Colin H. Hansen, and Carl Q. Howard. *Engineering Noise Control.* CRC Press, Boca Raton, fifth edition edition, 2018. ISBN 978-1-138-30690-5 978-1-4987-2405-0.

Olivier Doutres, Nicolas Dauchez, and Jean-Michel Génevaux. Porous layer impedance applied to a moving wall: Application to the radiation of a covered piston. *The Journal of the Acoustical Society of America*, 121(1):206, 2007. ISSN 00014966.

R. S. Langley. Numerical evaluation of the acoustic radiation from planar structures with general baffle conditions using wavelets. *The Journal of the Acoustical Society of America*, 121(2):766–777, 2007.

Alexander Peiffer. Comparison of the computational expense of hybrid FEM/SEA calculation. In *Proceedings NOVEM 2012*, Sorrento, Italy, April 2012.

Alexander Peiffer. Hybrid modelling of transmission loss with acoustic treatment. In *Fortschritte Der Akustik*, pages 50–53, München, Germany, March 2018.

P. J. Shorter and R. S. Langley. Vibro-acoustic analysis of complex systems. *Journal of Sound and Vibration*, 288(3):669–699, 2005. ISSN 0022-460X.

Michel A. Tournour, Fumihiko Kosaka, and Hirotaka Shiozaki. Fast Acoustic Trim Modeling using Transfer Admittance and Finite Element Method. In *Proceedings SAE 2007 Noise and Vibration Conference*, pages 2007–01–2166, May 2007.

12

Industrial Cases

In the preceding chapters all examples were rather simple and academic cases so that they can be retraced by the reader. This chapter is about industrial examples dealing with practical and complex cases. Thus, the models presented here rely on commercial software tools, namely NASTRAN™ for pure FE models or VAOne™ from ESI-Group and Wave6™ from Dassault Systems as hybrid FEM/SEA solvers. All software tools have various materials, properties, and subsystem formulations implemented that allow the consideration of realistic technical systems.

The examples are presented briefly but will outline the basic ideas and point out typical challenges. A detailed description and derivation of the vibroacoustic models is out of the scope of this textbook. In fact, some of the presented examples are subject to complete masters and PhD theses, which are referenced for further reading.

However, the aim is to give a summary about simulation strategy for important applications. This means that we need to decide which subsystems are to be considered as deterministic and which are random for the interesting frequency range and how should the global system be divided into subsystems. Furthermore, it is important to develop a strategy for what is not included in the model, because not every detail can be included.

Special thanks goes to Ulf Orrenius[1] who wrote the train and motivation section of this chapter. With his experience and knowledge, this chapter could be enriched by one further industrial sector, the railway industry.

12.1 Simulation Strategy

12.1.1 Motivation

Before getting started with the choice of simulation methodology and the substructuring of the system to be analyzed, it is wise to reflect somewhat on the modelling targets. Why do we spend time and money on modelling? Apart from a general motivation of the work for the modeller, as well as for the people allocating the resources, it is important to know the targets before setting up the model for a successful result. Also, the validation strategy must support the modelling target. For example, if the model is

1 ulf@akustikdoktorn.se

Vibroacoustic Simulation: An Introduction to Statistical Energy Analysis and Hybrid Methods,
First Edition. Alexander Peiffer.

mainly going to be used to compare different designs, it can be argued that the absolute levels of the source(s) are less important. Rather, the elements that are to be compared, e.g. different noise control lay-ups of an aircraft fuselage, should be modelled accurately enough so that the physics of the vibroacoustic transmission is sufficiently well described for the purpose of the modelling. However, when the transmission paths are different for two significant sources, for example when both air- and structure borne excitation mechanisms contribute to the total, also the relative strength of the source mechanisms need to be well described, to understand the effect of different noise control treatments. One fundamental rule in vibroacoustic modelling is that one should spend effort in describing the dominant paths and sources. In addition, one should pay attention to those elements for which design changes are planned. For example, one may be able to change the bags and trim elements of an aircraft fuselage but not the fuselage structure itself. In this case, effort should be made on the modelling of these elements, in particular if insertion losses due to design changes are being sought. Another example is for structural point excitation, for which the local impedance of the receiving structure determines the power injected. This means that, e.g. for a vibrating compressor that creates noise in a railcar (often tonal), the modelling of the structure where it is mounted must resemble its local impedance well. For this reason, the use of higher order modal data of the structure is typically required. When it comes to optimizing the design with respect to parameters like weight, structural strength, stiffness, noise, etc., it is generally useful to also include the structural design in the optimization, in addition to noise control treatments and trim. When vibroacoustic parameters can be integrated in the functional targets of the structure, there is a greater chance to fulfill the targets at lowered weight and cost.

12.1.2 Choice of Simulation Method

From the academic examples it becomes clear that for a full audio frequency simulation there is usually a transition from FEM, over SEA to geometrical acoustics over the frequency range. Depending on the size and properties of the subsystems, this transition occurs at different frequencies. Based on this fact and in view of the author's experience, a global simulation strategy can be developed.

This strategy is illustrated in a two-dimensional axes system, with the Helmholtz number on the x-axis and the dynamic complexity or type of system on the y-axis. In the specific sections of aerospace, automotive, and trains, the global strategy is described with regards to the *frequency* axis because the knowledge of size and system details allows estimating the frequency that corresponds to the Helmholtz number.

The y-axis is roughly categorized into systems of increasing dynamic complexity. All acronyms are explained in this book except MBS which stands for multibody simulation (Blundell and Harty, 2010). In multibody simulations the systems are modelled by various rigid or elastic bodies analogue to the lumped elements approach from Chapter 1.

The above strategy must not be understood as fixed limits. There is an overlap zone where both methods will provide reliable results. In addition the precision of the SEA method depends on the nature of excitation. A distributed source with low correlation results in higher quality of the SEA results.

12.2 Aircraft

The vibroacoustic design of planes, helicopters, and even satellites is a demanding challenge. The target conflict between low noise cabins and light weight design and thus environmentally friendly systems is much stronger than for surface based transportation systems. In addition, powerful turbofans, jet exhausts, rotors, and propeller create a noisy environment and external excitation. The high Mach number flow creates pressure fluctuations that excite the full fuselage surface from cockpit to aft section (Bhat, 1971).

The need for efficient noise control requires simulation tools to design and evaluate acoustic concepts in the early design phase. The focus on simulation is further enhanced by the fact that prototypes are rare and expensive. A first flight occurs late in the project when major noise control decisions are already made, and in-flight conditions are also very difficult to simulate in a transmission loss suit.

The fuselage of a typical single aisle aircraft, as the A320 series from Airbus or the B737 series from Boeing, is about 35 m long with a diameter of 4 m. In terms of Helmholtz number, all ranges of Figure 12.1 occur. Thus, the full range of deterministic finite elements to random methods and even geometrical acoustics is necessary.

Even though helicopter and launcher are of interest for vibroacoustic simulation, we focus on turbofan airplanes because they are well known and used by a large community. Aircraft exterior noise is a major subject of environmental acoustics and thus an important possible job-stopper for civil aviation. However, here we stay with the passenger perspective and focus on the interior noise comfort.

12.2.1 Excitation

Modern aircraft are powered by so-called turbofan engines. The word *turbofan* is a combination of turbine and fan. The turbine is a gas turbine engine that generates mainly mechanical energy driving a ducted fan, that is the main source of propulsion. Even though we are not dealing with exterior aircraft noise, the passenger benefits from the global target of low noise engines.

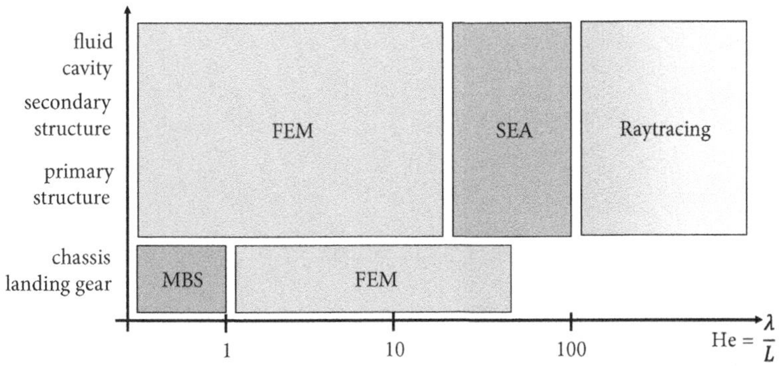

Figure 12.1 Global simulation strategy. Methods related to Helmholtz number and system type. *Source:* Alexander Peiffer.

Consequently this leads to lower noise excitation at the surface of the fuselage but brings forward the turbulent boundary layer (TBL) excitation in the source ranking.

The main sources are depicted in Figure 12.2. The turbofan engine generates a mainly tonal sound called turbofan noise (12.2a) resulting from the rotor stator interactions at the duct inlet and modal effects in the duct. This component occurs often at high thrust levels during take-off phase. The jet-noise (12.2b) is flow noise generated by the turbulent flow sources in the downstream of the engine outlets. This mainly affects the aft cabin. In addition the engines excite structure borne noise that propagates via the pylons and wings into the cabin. The aircraft fuselage is surrounded by a boundary layer that becomes turbulent near the cockpit zone.

The surface pressure excitation generated by the turbulent flow surrounding the fuselage is the turbulent boundary layer excitation. The subject of turbulent boundary layer is a matter of intense research. See for example Klabes et al. (2016); Palumbo (2012); Goody (2004) and one of the first papers on TBL-excitation from Corcos (1963). One advantage of the TBL-excitation is that it is a lowly correlated source and therefore not as efficient as acoustic waves of similar surface pressure. In terms of coherence the TBL is rather a rain-on-the-roof than an acoustic excitation.

Technically the TBL-models used are empirical definitions of $[S_{ff}]$ from equation (1.203) and they require the use of (7.43). The consequence is that even when the excited subsystem is deterministic, a matrix inversion and triple multiplication is required, making the TBL excitation quite unhandy. There are several methods to overcome these issues, as shown for example by Peiffer and Mueller (2019) or Maxit et al. (2015).

The turbofan and jet-noise excitation is also the subject of intense research, not only for exterior noise (Leylekian et al., 2014). However, the determination of the surface pressure from the engine is often performed empirically on flight tests (Mengle et al., 2006).

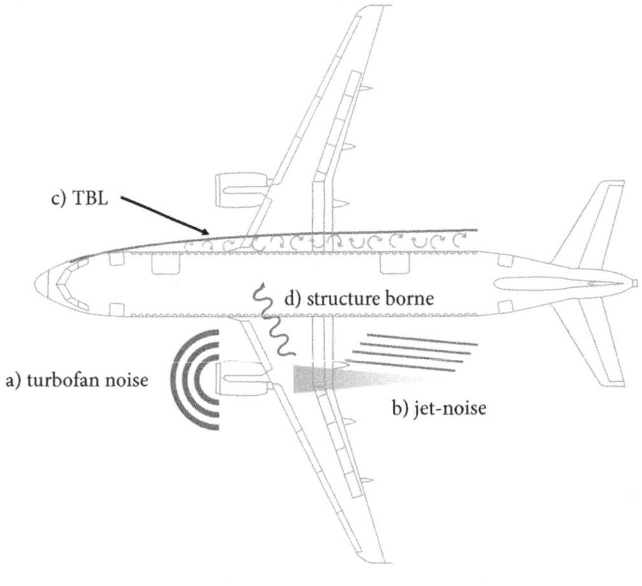

Figure 12.2 Major exterior sources of aircraft interior noise. *Source:* Alexander Peiffer.

Please note that the topics of engine and TBL-excitation are far beyond a textbook on vibroacoustics, so only the basic principle can be addressed here. Please refer to the original papers for more details.

12.2.2 Simulation Strategy

The application of the simulation strategy to twin aisle aircraft leads to the estimated frequency limits and subsystem configuration as shown in Figure 12.3. The deterministic modelling approach can be easily used for deterministic excitation, for example propeller noise and even jet noise that is random in time but deterministic in space. The situation becomes complicated for TBL-excitation where special tricks must be applied to overcome the computational costs of equation (7.43).

Due to the size of aircraft, a full random modelling is suitable above approximately 1000 Hz when precise structural results are required. However, the large interior cavities are reaching even the upper frequency limit of SEA and ray tracing is required. In Figure 12.3 the icons show the implementation of such models. An example for a hybrid model based on a generic test case is given by Peiffer et al. (2011). The author did not succeed in creating a hybrid model for a realistic fuselage section. With the given commercial tools in 2014, the reasons are quite practical: The deterministic models are imported as modal base; because of the thin aluminum construction and the size, the modal base becomes much too large for the current software, and there are several local modes that pollute the modal base and further increase the computational costs. Thus, for a full industrial application of hybrid FEM/SEA models, a computationally powerful tool is mandatory. One further practical limitation is that the large area coupling leads to fully populated radiation stiffness matrices that are computationally expensive and sometime even more expensive as a full FEM calculation. See Peiffer (2012) for more details on that subject.

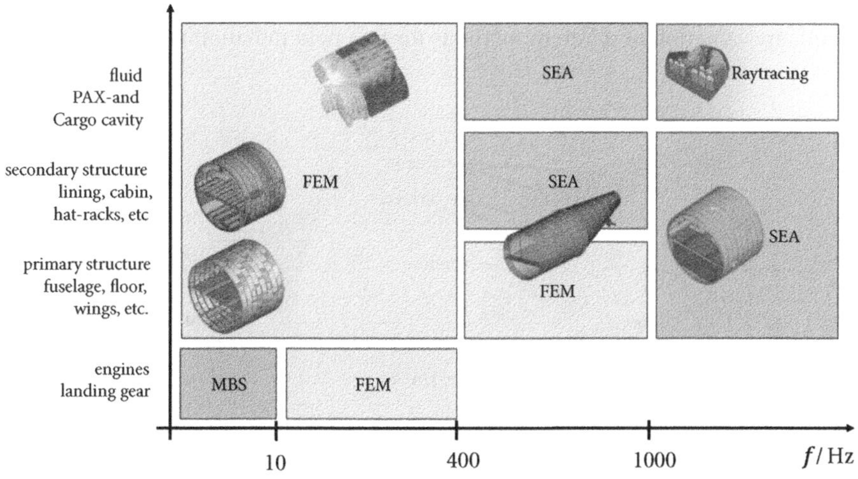

Figure 12.3 Global simulation strategy for twin aisle aircraft. Fuselage diameter $D \approx 6$ m. *Source:* Alexander Peiffer.

12.2.3 Fuselage Sidewall

Aircraft sidewalls are a typical example for double walls. A good acoustic isolation without violating the weight restrictions can only be achieved by such a construction. The outer structure is the fuselage panel covered by glass fiber blankets for thermal and acoustic insulation. The thickness of the aluminum fuselage skin varies from $h \approx 2{-}6$ mm leading to an area weight $m'' > 5.4\,\text{kg/m}^2$ plus the weight of the stiffeners. The volume of the double wall cavity behind the window panels is nearly completely filled with fiber material, and the thickness is determined by the height of the frames, which is about 12 cm. The lining presents the inner plate of the double wall made of sandwich material with an area weight of $m'' \approx 1.5$ kg. Because of the much lower weight of the lining compared to the fuselage, the double wall resonance is determined by the cavity, the sound speed of the fiber material, and the lining.

For the study as presented in Peiffer et al. (2013); Peiffer and Wang (2015), a fuselage panel of the front section of an Airbus A300 was simulated and tested. The naming of structure and fluid components is shown in Figure 12.4 in conjunction with the FE model.

12.2.3.1 Double Wall Simulation Strategy

Various subsystem modelling configurations of FEM, hybrid FEM/SEA and SEA models were investigated by Peiffer et al. (2013). The aim of the study was to compare the results to high frequency FE simulation and tests in order to verify the frequency range of validity of each set-up. In order to get reliable results the reference FE model was designed in such a way that the element size and material models are valid until 1000 Hz. For the calculation of the transmission loss of the FE model, the diffuse field excitation is synthesized by creating an ensemble of plane waves as in section 6.1.4, and the radiated power is determined by free field radiation into the semi infinite half space. More details of the mathematical background of the FE modelling of the transmission loss can be found in (Davidsson, 2004). The process of diffuse field synthesis by plane waves and an evaluation of different methods for free field radiation are described in

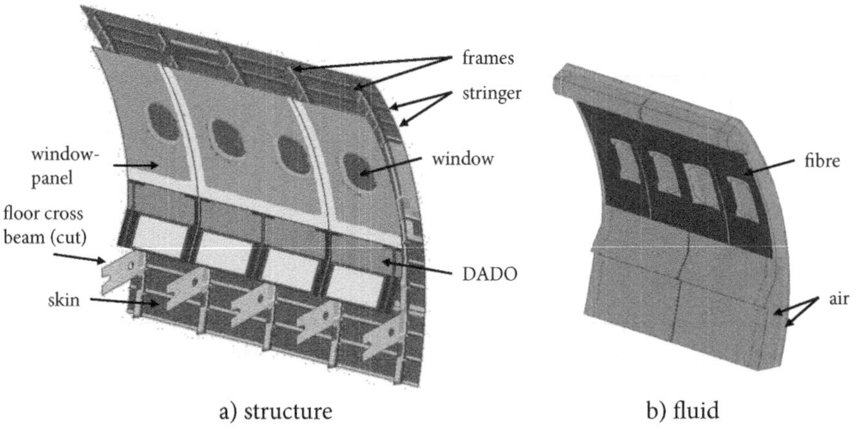

Figure 12.4 FE Model of an A300 sidewall panel with interior lining. *Source:* Alexander Peiffer.

(Peiffer et al., 2015). A description of the FE model and details of the double wall panel are also given by de Matos (2008).

According to the discussions above in sections 7.4 and 12.1.2, decisions must be taken how to subdivide the system into subsystems and what simulation method is applied in which frequency range. To support the decision of which systems can be treated as stochastic, one option is *visual inspection* of the finite element simulation results and the number of modes in the third octave band. The outcome of this inspection and calculation in terms of model set-ups is shown in Figure 12.5.

Though we ended up with four different models to cover the full frequency range there is a certain overlap area between the models. The following set of model was created:

FEM	All structure, fluid, fibers, and lining modelled by FEM.
Hybrid1	The fuselage structure is modelled by FEM, the fluid, fiber and lining are modelled with the transfer matrix method from section 11.3 specifically for the different areas.
Hybrid2	The fuselage structure is modelled by FEM, the fluid, fiber, and lining are modelled as SEA.
SEA	All subsystems are modelled as SEA systems.

12.2.3.2 Fuselage Panels

The aircraft fuselage design is typically a monocoque shell design as depicted in Figure 12.6. That means a singly curved skin of aluminum or carbon fiber reinforced plastic is strengthened by longerons called stringers and circumferential frames. The idea is that the loads are mainly taken by the skin, with the frames and stringers to support against buckling. The stringers are in most cases directly connected to the skin. The

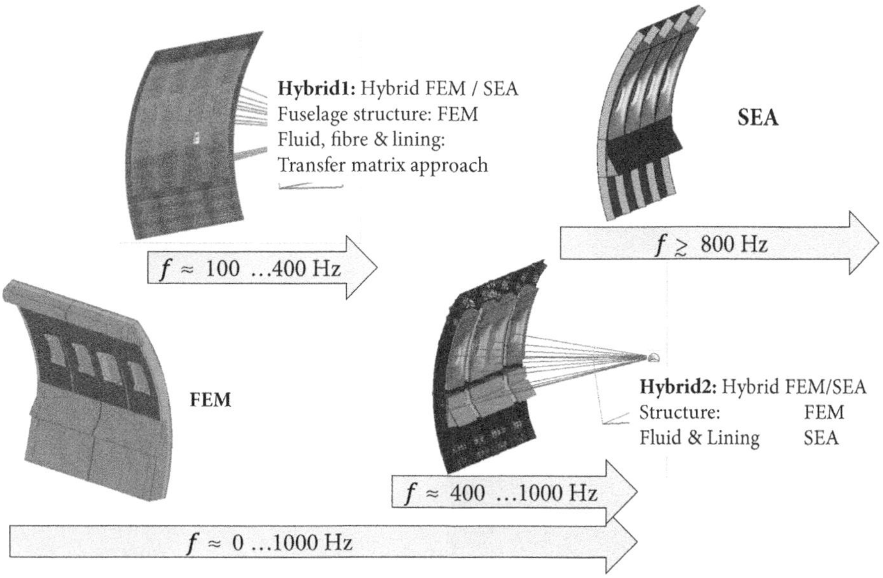

Figure 12.5 Subsystem and simulation strategy for the aircraft double wall over frequency range. *Source:* Alexander Peiffer.

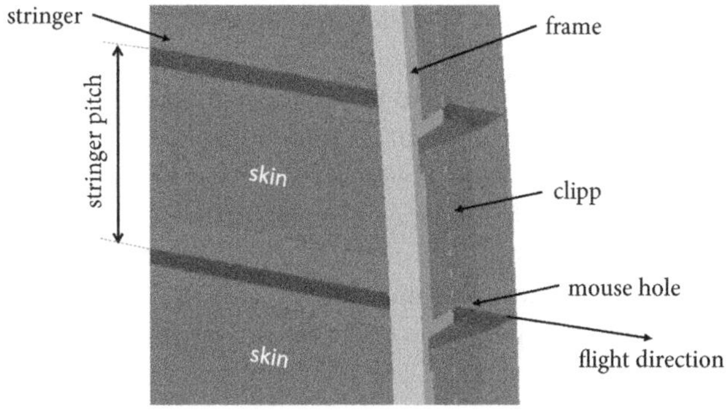

Figure 12.6 Monocoque design of fuselage panel. *Source:* Alexander Peiffer.

frames are sometimes directly connected to the skin or via small clips. These kinds of structures are called rib-stiffened structures.

There are several papers dealing exclusively with vibroacoustics of rib-stiffened structures, for example Maidanik (1962); Efimtsov and Lazarev (2009). Many authors tried to derive a valid analytical formulation for such a structure, but the mathematical effort of those models is sometimes so high that a classical numerical approach is a better option. Moreover, such models are structurally approximative in that the real structure has multiple irregularities such as windows, window frames, varying skin thickness, and structural enhancements. Therefore, the representation of the fuselage panel is the field of finite element simulation possibly in combination with periodic structure theory as proposed by Cotoni et al. (2009).

In Figure 12.7 a few mode shapes of a representative panel are shown. At low frequencies the stiffness and mass of skin and stiffeners are smeared to a smeared bending stiffness and mass density. At high frequencies the skin vibration is decoupled from that of the stiffeners; the stiffened panel can be treated as a cylindrical shell essentially

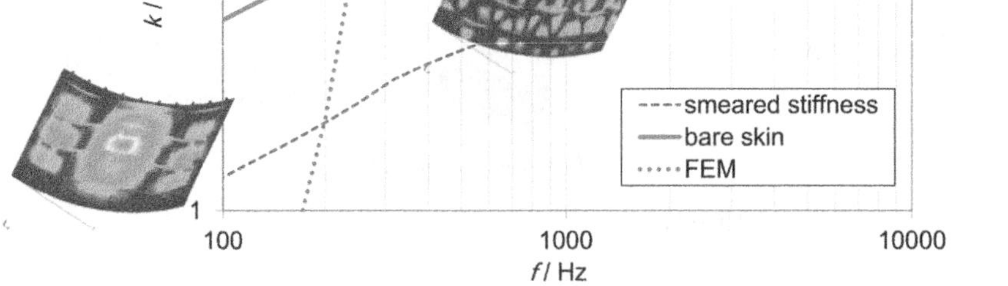

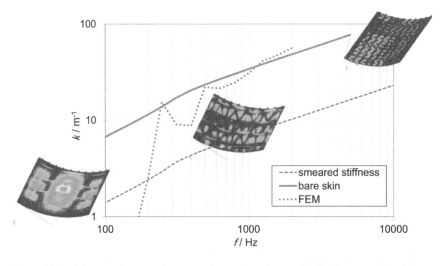

Figure 12.7 Wavenumber over frequency for modes of a regular fuselage panel and three representative shapes. *Source:* Alexander Peiffer.

neglecting the stiffening members, because the wavelength of skin bending waves is much smaller than the skin field dimensions.

However, in the mid frequency range, the coupling between skin and stiffener is still effective and the wave motion is determined by a complex motion of partly coupled stiffeners and skin with wavenumbers oscillating between both asymptotes. To conclude, the fuselage panel should be modelled as an FE subsystem or, alternatively, asymptotic approximations valid for low or high frequencies must be applied.

The graph in Figure 12.7 shows the wavenumber envelope over the modal frequencies. The wavenumber was derived from the two-dimensional Fourier transform of the mode shapes using equation (A.48b). The position $\mathbf{k} = \{k_{x,n}, k_{y,n}\}$ of the maximum peak in the wavenumber spectrum is selected for each modal frequency $\omega_n = 2\pi f_n$; see (Peiffer et al., 2009) for details of this method.

The fuselage panels are a good example for a typical dilemma in SEA. Though the panel has a high mode count, and we may assume a diffuse sound field in the plate structure, there is a lack of valid and simple analytical models that precisely describe the wavenumber and radiation efficiency over frequency.

The ribbed plate limitations are one reason why the SEA model becomes valid from 800 Hz. In the SEA model the frames are modelled as plate subsystems, and the skin is stiffened exclusively by stringers in axial direction. In this frequency regime near the pure skin dynamics, the ribbed plate model as implemented in VAOne™ works satisfactory.

12.2.3.3 Windows

Aircraft windows are acrylic glass panes fixed with rubber sealing in a stable aluminum frame. The windows panes have thickness of $h_1 = 7$ mm and $h_2 = 4$ mm with an air gap of 3 mm. This double pane window is a result of the failsafe design of aircraft. If one sheet breaks the other will do the job until a safe landing. This double window is covered in addition by 2 mm acrylic glass of the double frame window panel.

12.2.3.4 Lining and Insulation

The lining in aircraft consists of sandwich panels with glass fiber reinforced plastic (GFRP) fabric with a honeycomb core. The inner surface is covered with decoration foil. The double wall cavity is filled with fiber blankets that are in bags of thin foil. Near the fuselage skin the stringers prevent the fiber blankets from touching the aluminum directly, so there is always an air gap of about 20 mm. The lay-ups from Figure 12.8 are especially used for the Hybrid1 model that applies the transfer matrix with the specific layup of each zone. The DADO panels are flat sandwich plates of 8–5 mm thickness.[2] The DADO panels are designed to release automatically in case of sudden pressure loss in the cabin.

In the Hybrid2 and SEA model the lining and insulation are modelled as plates and cavities. The equivalent fluid of the double wall cavity is determined by equations (9.107a)–(9.107g).

12.2.3.5 Transmission Loss

The transmission loss results are shown in Figure 12.9; the valid frequency range for each model is denoted by bars. For illustration the calculation is performed for

2 In 18 years of work at Airbus, I didn't find someone who could explain to me the acronym DADO. In English literature some authors call them catwalk panels.

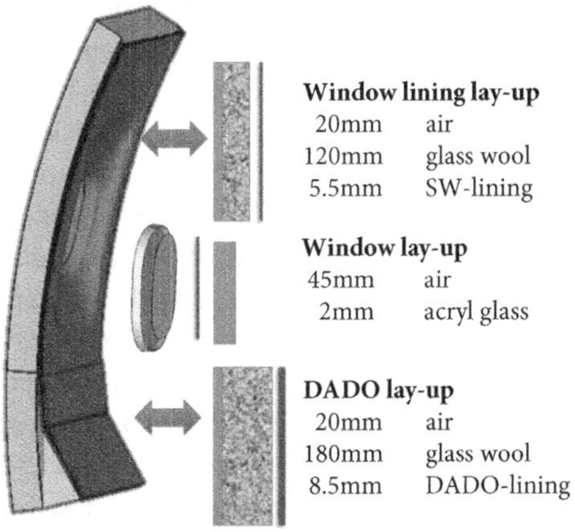

Figure 12.8 Double wall, isolation, window, and lining set-up between the frames. *Source:* Alexander Peiffer.

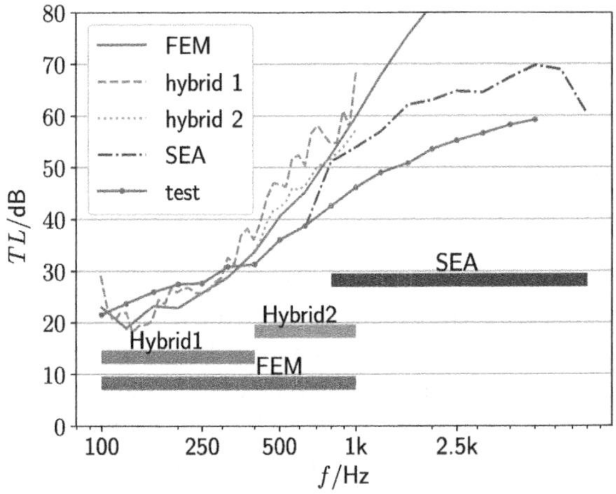

Figure 12.9 Transmission loss results from various simulation models and experiment. The bars denote the range of validity. *Source:* Alexander Peiffer.

frequencies extending the range of validity of each model. It can be seen that generally all methods provide similar results in the overlapping frequency range. From low to mid, all results fit satisfying the test but not above 600 Hz. Further investigation was made as discussed by Peiffer and Wang (2015). One outcome of this study was that the twin-chamber arrangement suffered from some side paths leading to erroneous results of the tests.

In any case the results show that for full frequency simulation sometimes several models are required in accordance with the discussions from section 7.4. As a consequence vibroacoustic simulation software should provide all required methods in one tool in order to avoid inefficient switching between different tools.

12.2.4 SEA Model of a Fuselage Section

The sound pressure level changes only slightly in an aircraft along the aisle. Due to this and for practical reasons mid- and long-range aircraft are rarely modelled using

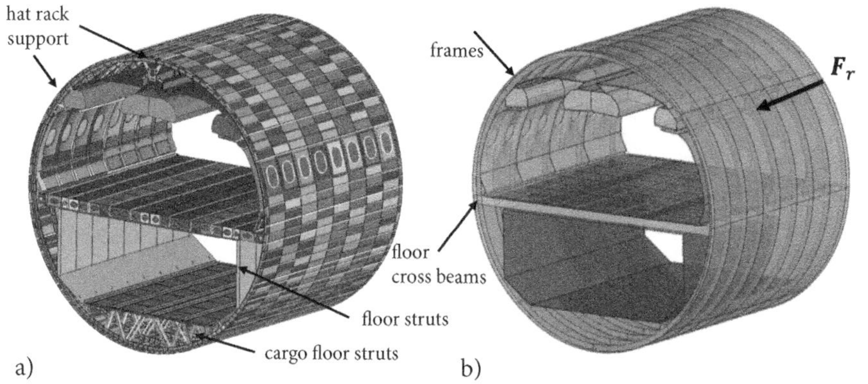

Figure 12.10 A340 barrel stucture. a) FE-model and b) SEA model. Reproduced with friendly permission from Weineisen (2014).

the full fuselage. Usually, a representative section is used as depicted in Figure 12.10. Details and assumptions of model creation are given in detail by Weineisen (2014). Other examples for interior noise predictions can be found in Davis (2004).

12.2.4.1 Fuselage Structure and Lining

Weineisen (2014) used the FE model from Tewes and Peiffer (2013) as input for material data and geometry. In Figure 12.10 the details of the complicated fuselage construction can be seen. The comparison with the SEA model on the right hand side reveals that the creation of an SEA model requires abstraction and simplification. The complicated and deterministic parts that cannot be captured by the simple and canonical formulations of SEA must either be modelled as FEM and included with the hybrid method or neglected. A further option is to apply experimental SEA to determine the coupling loss factors of floor subsystems and fuselage as described by Bouhaj et al. (2017). The modelling strategy of the fuselage and lining are similar to section 12.2.3.1.

This A340 section was investigated in the public funded project AKUCON-RAST funded by the German Federal Ministry of Education and Research. In the final reports from Tewes and Peiffer (2013), the tests and various other modelling approaches are described.

The fuselage aluminum skin was modelled as a rib-stiffened singly curved panel with the stringers as stiffeners in the axial direction as in the double wall case. The frames are modelled as plate systems. Weineisen (2014) applied the energy influence method from MACE and SHORTER (2000) to check if the coupling loss factors between frames and skin fields and the modal density of the frames are correct.

Several further components must be in included in the model, for example the hat racks, ceiling panels, floor panels, cargo panels, the floor structure of crossbeams, and seat rails.

12.2.4.2 Cavities

The cylindrical volume of the fuselage is subdivided into several cavities. The main cavities are the cargo and passenger (PAX) cavities. In addition there are several small

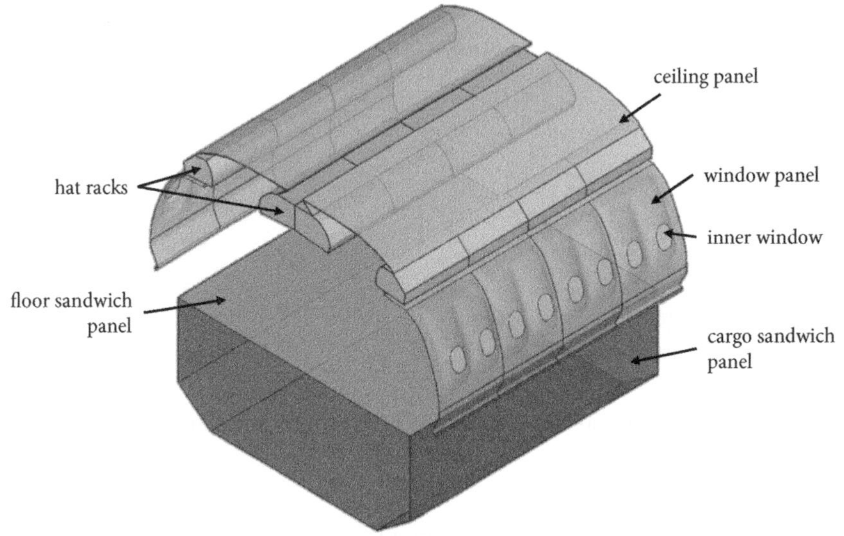

Figure 12.11 Hat rack, floor as far as passenger and cargo lining plate subsystems of the A340 barrel. *Source:* Alexander Peiffer.

cavities created by the hat racks plus the volumes that represent the double wall cavities near window, DADO, ceiling and cargo floor, and side wall. The main cavity is not equipped with seats to represent the test conditions in Tewes and Peiffer (2013). Many authors use an additional subdivision of the PAX cavities into small cavities (Davis, 2004) for getting local pressure values. Please note that this provides a variation of the pressure field inside the cabin but may violate typical SEA assumptions, as for example the required weak coupling. However, this issue is solved by using ray tracing as denoted in the strategy in Figure 12.3.

12.2.4.3 Force excitation

In the process of model validation, precisely given testing parameters like boundary conditions and excitations is a critical factor for a reliable test database. It was therefore decided to use point forces perpendicular to the fuselage surface skin. The advantage

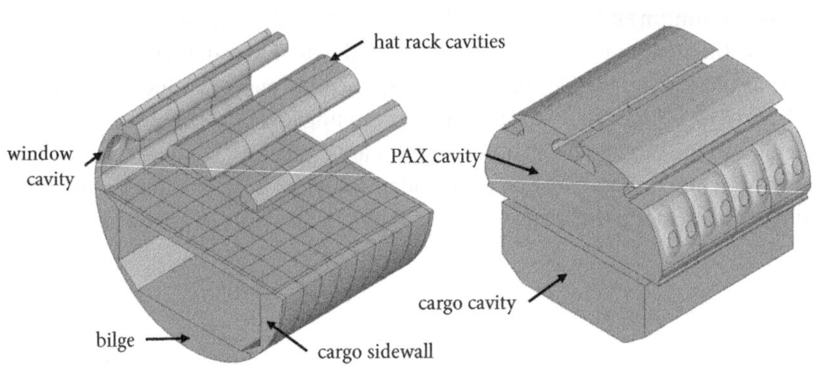

Figure 12.12 SEA cavities of A340 section. *Source:* Alexander Peiffer.

Figure 12.13 Velocity level in the 1600 Hz band for shaker test and SEA model. Reproduced with permission from Weineisen (2014).

of point forces is that they can be precisely described and also used for experimental modal analysis. The disadvantage of such point forces is that they are not a good excitation for generating a random field and therefore a difficult test case for SEA. In section 6.4.2.1 it was shown that even for flat plates, the input point impedance for random waves is only valid at high frequencies. Thus, for the determination of input power, it is recommended to calculate it from measured force and acceleration. The shaker was positioned at the frame between two skin SEA subsystems, and the excitation is therefore located on a junction (Figure 12.10). For a correct power distribution the input power must be calculated from the point stiffness of each excited subsystem. As this information is hard to gather, Weineisen (2014) used a 50 % distribution of input power into each connected skin bending wave subsystem.

In Figure 12.13 the velocity results from tests and SEA simulation are compared for the 1600 Hz band. Near the excitation point the velocity is slightly overestimated by the SEA result. Weineisen performed several model updates to find the best fit. However, the distribution of acoustic energy along the fuselage and especially the velocity levels in the floor are not well predicted by the SEA model. This is in accordance with the expectation that the energy distribution over several subsystems is a critical task for the SEA approach. The poor agreement of the floor levels results from the inaccurate coupling loss factors for floor to fuselage subsystems. The differences at the cargo and PAX lining are not relevant, because in the SEA results in Figure 12.13 the nonresonant components are not taken into account. Their consideration using equation (10.19) would lead to better agreement as shown by Wang (2015). To conclude, a point excitation of complex structures is a critical application case for SEA. If precise prediction is required, at least the excited subsystem should be modelled as an FE subsystem.

12.2.4.4 Turbulent Boundary Layer Excitation

In contrast to the force excitation, TBL excitation constitutes an excellent application case for SEA. The excitation is broadband and distributed over the entire fuselage. Moreover, the computational efforts of a corresponding FEM simulation are very high because of the random character excitation (Peiffer and Mueller, 2019). Weineisen used

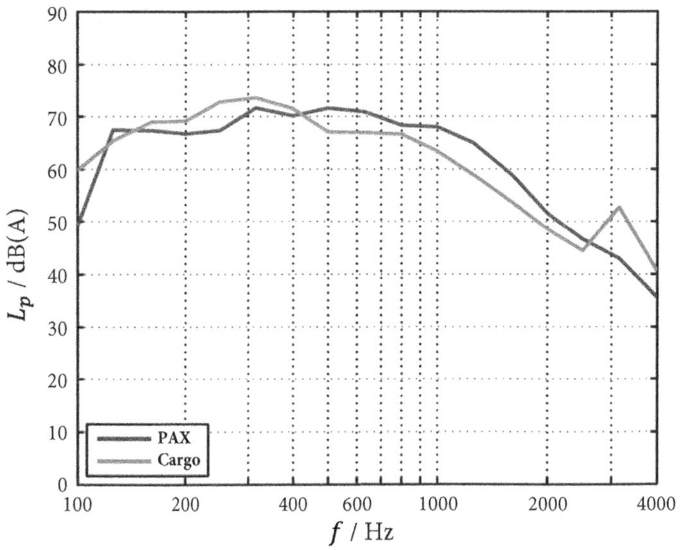

Figure 12.14 Interior pressure level due to TBL excitation. Ma=0.82, h = 35 000 ft, ΔP = 50 kPa. Reproduced with permission from (Weineisen, 2014).

the TBL model from Cockburn and Robertson (1974). The flight conditions are: Height $h = 35000$ ft and cruise Mach number Ma= 0.82. The external atmosphere and cabin internal pressure lead to pressurization of $\Delta P = 50\,000$ Pa. The pressure results from Figure 12.14 match well with flight tests. Thus, for a global design of the cabin noise control and identification of hot spots under cruise conditions, the SEA model is an excellent tool in the early design phase.

12.2.4.5 Conclusions

The large size of aircraft, the nearly regular and periodic structure, and the random nature of an important excitation make the interior noise prediction an excellent field of application for SEA. When the detailed simulation of structure borne paths is required, FEM or hybrid FEM/SEA may become necessary.

12.3 Automotive

Cars provide an excessively complex case for vibroacoustic simulation. There is a large variety of sources, source mechanisms, and structure variations not only driven by model, carbody, engine, suspension, tire, and interior options even for one single model. Neither the simulation nor the test of all combinations is possible. The different source and wave propagation phenomena are reflected in the specific name for vibroacoustics that is used in the automotive industry: Noise, vibration, and harshness (NVH). Nevertheless, SEA models are used early in the vehicle development program to identify potential problems before a pototype has even been built and when changes are still possible

The noise and vibration control of cars would provide more than enough content for many textbooks, for example Wang (2010) in English or Zeller (2009); Genuit (2010) in German. Those textbooks cover many details of acoustic and vibration sources and the related wave propagation and radiation. Within the context of this book, we focus on SEA and hybrid FEM/SEA simulation and present some highlights and issues.

12.3.1 Simulation Strategy

The strategy from section 12.1.2 leads to Figure 12.15 in case of automotive structures. Due to the relatively small cabin, ray tracing is applied lately in frequency for the passenger cavity. In combustion engines, electric motors, and chassis there are many rotating and oscillating parts. Therefore, those parts are modelled by multi-body simulation (MBS). This method allows the simulation of rolling and engine dynamics by simultaneously considering nonlinear effects and combustion sources. For more details see, for example, Blundell and Harty (2010) and García de Jalón and Bayo (1994). Due to the smaller size of cars, the frequencies are shifted to higher values compared to the aerospace plane model.

12.3.2 Excitation

The main excitation of acoustic and structural waves comes from the power train. The combustion engine or electric motor is mounted on bushes to decouple it from the structure. In case of combustion engines, high efforts are made to decrease the noise from the exhaust and air intake. Mufflers are designed to reduce the pulsating flow that enters and leaves the engine. Due to reduced exterior noise levels required by law, the design of mufflers is a peculiar field of acoustics. The tire–road contact is the second source of so-called rolling noise. Vibrations and acoustic waves propagate through the car structure to the drivers ear. In addition we must take care of many other potential sources from hydraulics, HVAC compressors, fans, electric motors, and other equipment.

12.3.3 Rear Carbody

This example deals with the acoustic evaluation of different aft car construction details. The object of interest is the trunk floor structure that separates the electric motor

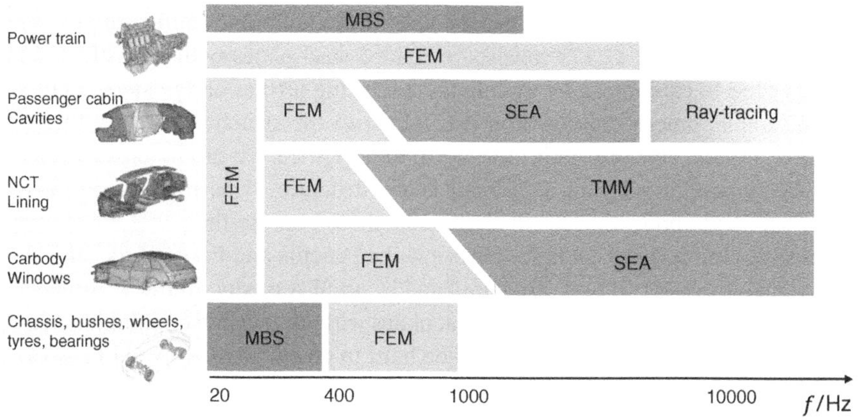

Figure 12.15 Simulation strategy for automotive applications. *Source:* Alexander Peiffer.

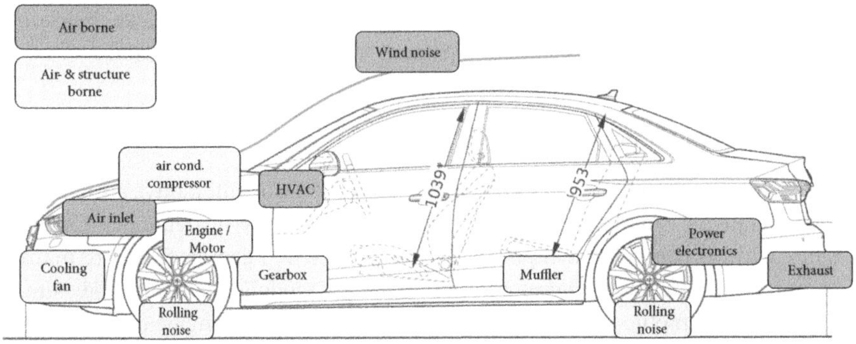

Figure 12.16 A selection of sources and source mechanisms in a car. *Source:* Alexander Peiffer.

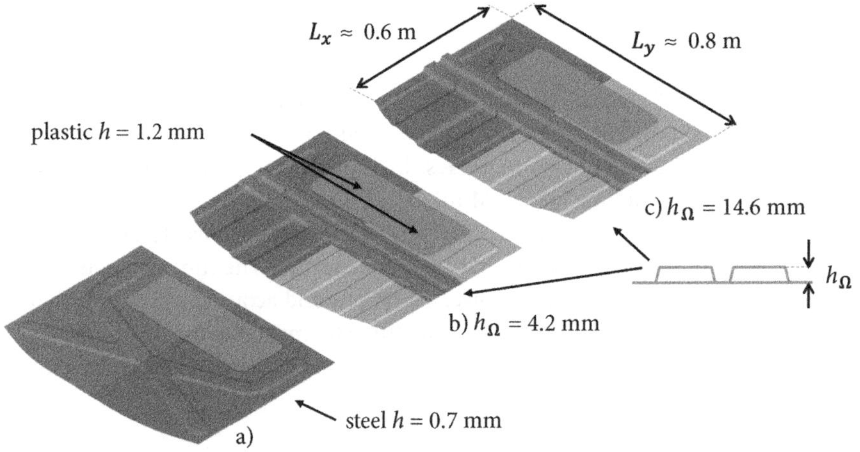

Figure 12.17 Sructural FE models of different design versions of a car trunk floor. *Source:* Alexander Peiffer.

of a rear driven vehicle from the trunk. In station wagon designs this is the only barrier for electric motor noise. In Figure 12.17 three different configurations are shown. The initial design 12.17a consists of beaded steel plates with a plastic cover to provide access to equipment for maintenance. During further design steps, a cross beam with double omega cross-section was added to the structure to carry higher loads. There are two versions with low 12.17b and high 12.17c profile cross beams. What is the impact of such an additional beam structure? This transmission loss case is an excellent application for the hybrid method, because the panel is flat and equations (8.42) can be used. In order to review the benefits and limits of the statistical energy analysis case 12.17a is also simulated by an SEA model. In the SEA model the steel and plastic plate are modelled as flat plates without any line or point junction definition between them. Obviously, it will be hard to create an SEA model that correctly takes the cross beam into account. Therefore, only case 12.17a is given as an SEA model.

In middle and premium class cars, the steel plates are usually covered by noise control treatment with soft foam core and a heavy layer. See Figure 12.18 for details. Foam materials are modelled using Biot's model Biot (1962) and applying a more sophisticated version of the transfer matrix method as presented in Allard and Atalla (2009). Two treatments are applied: a 20 mm and a 30 mm foam core treatment with similar heavy layer.

Simply supported boundary conditions are applied to the edge degrees of freedom of the FE model, and a modal analysis with upper frequency limit $f_{\max} = 2$ kHz is performed. In Figure 12.19 one shape at $f_{683} = 997.4$ Hz is shown. From the shape and related Helmholtz number, we conclude that the dynamics of the structure is still deterministic, but transition to random dynamics has started at this frequency. The plastic cover is connected to the steel plate at six vertex and edge point connections.

12.3.3.1 Transmission Loss

The transmission loss is calculated using equation (7.50) in combination with (8.8c).

In Figure 12.20 the results for all cases are shown. Until 400 Hz there are different resonances from the plate structures visible. Because of the stiff cross beam the resonances go higher in frequency for cases 12.17b and 12.17c. Above 500 Hz all versions are equivalent in terms of transmission isolation performance. The SEA model agrees well with the hybrid result above the resonances. Figure 12.21 shows the results of

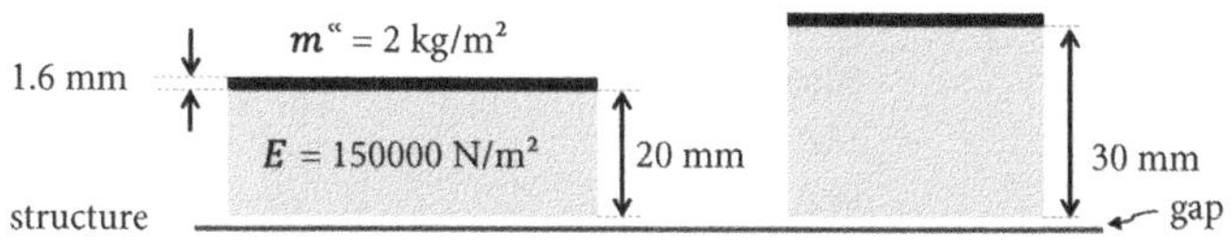

Figure 12.18 Mass-spring treatment with heavy layer and soft foam core.
Source: Alexander Peiffer.

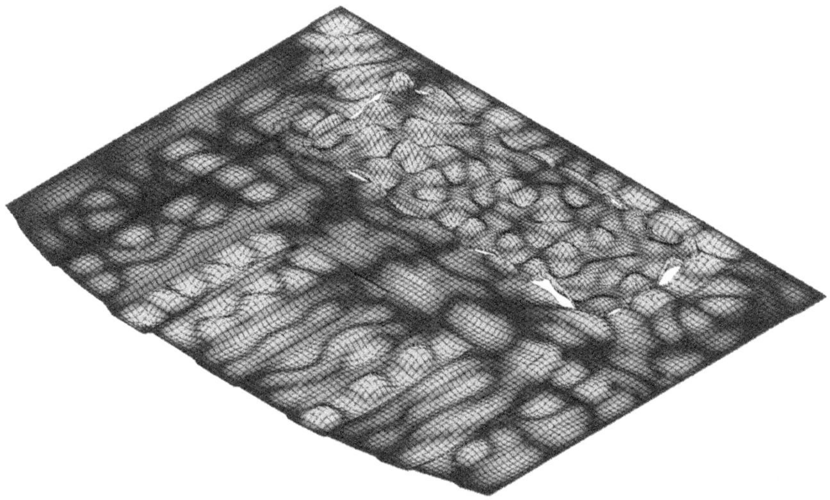

Figure 12.19 Mode shape of number $n = 683$ at $f = 1997$ Hz for version 12.17c.
Source: Alexander Peiffer.

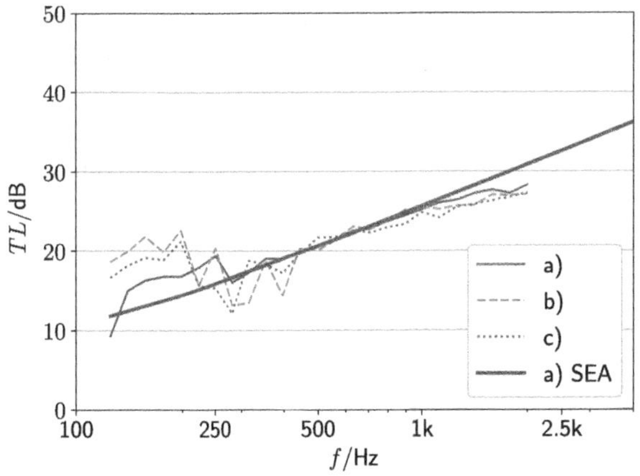

Figure 12.20 Hybrid transmission loss of pure structure plate. *Source:* Alexander Peiffer.

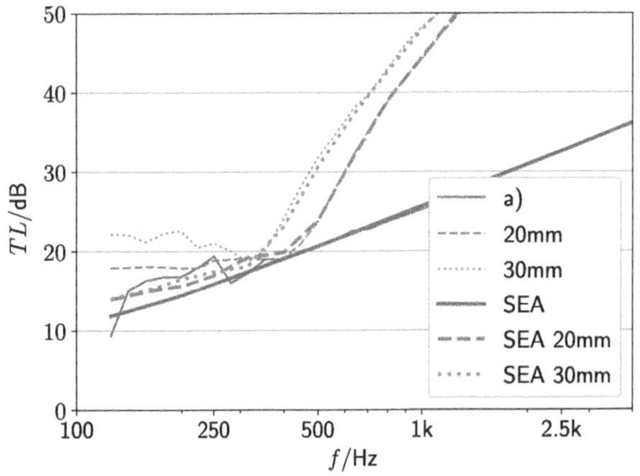

Figure 12.21 Hybrid and SEA transmission loss of case 12.17a with mass-spring trim. *Source:* Alexander Peiffer.

the trimmed structure 12.17a both with trims and without. We see that the differences between hybrid FEM/SEA and SEA are even less when trim is applied. In all cases SEA underestimates the transmission loss. The thicker treatment shifts the double wall resonance frequency to lower values leading to benefits from the mid frequency above 300 Hz. Note that there is a difference of 5 dB between the different thicknesses of the foam core.

The comparison of all cases in Figure 12.22 with treatment reveals that the differences between the cases are smoothed out by the trim. To conclude, from the airborne point of view, all cases are comparable, and there is no major drawback from both cross beam designs.

12.3.3.2 Force Excitation

When the structure is excited by a normal point force, the conclusions from Section 12.3.3.1 must be qualified. In Figure 12.23 the power radiated to the car interior when

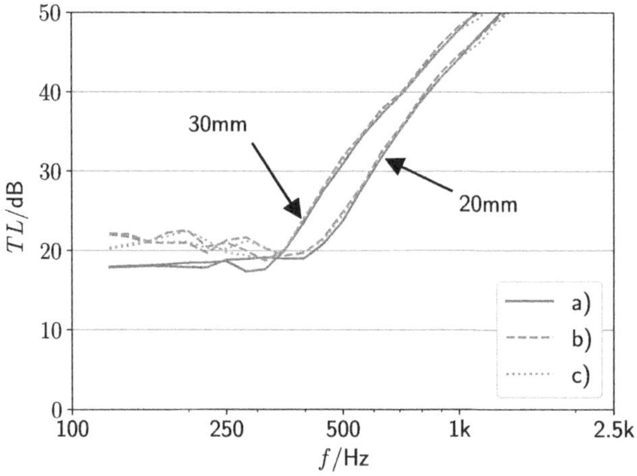

Figure 12.22 Hybrid transmission loss of all cases with mass-spring trim.
Source: Alexander Peiffer.

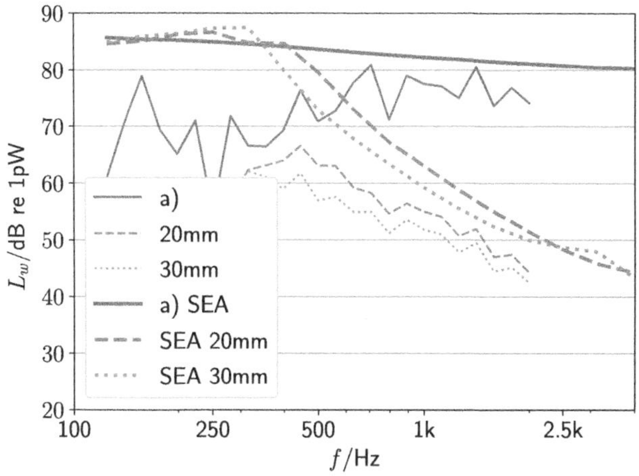

Figure 12.23 Radiated acoustic power of case 12.17a under force excitation.
Source: Alexander Peiffer.

the steel plate is excited by a point force is presented. The hybrid result shows lower values for all trim configurations. Even for high frequencies, the hybrid and SEA results don't agree, and the SEA method overestimates the radiated power. Thus, for structure borne excitation SEA, can be used for comparisons of different treatment efficiency but not for absolute values.

12.3.4 Full Scale SEA Models

The number of sources and excitation mechanisms in combination with the different drive conditions constitutes a vast amount of data that would be required to predict the interior noise level even when only airborne excitation is considered. Thus, in industrial applications, SEA models of cars are less applied for the prediction of noise levels and the achievement of specific target levels. Nevertheless, SEA is used early in the vehicle development program to evaluate the noise and vibration status of cars before a prototype has been built. For this purpose a full SEA model is required that includes all paths and sources of a car. Classical SEA models are introduced by Venor and Burghardt, Marc (2005) and Venor et al. (2003) and include many details as shown in Figure 12.24.

According to the simulation strategy from Figure 12.15, most structural parts have to be modelled by FEM because they are still deterministic at 1000 Hz. The detailed substructuring shown in Figure 12.24 results from the restrictions of the 3D geometry modelling of the used software and leads to even smaller structural subsystems. Every cavity must be constituted by surrounding patches, and in most cases only edges can be connected. Due to this fact many subsystems have less than one mode in the third octave band. Thus, hybrid models would be excellent to fill this mid frequency gap as shown by Charpentier et al. (2008). However, the application of hybrid theory may suffer from the fact that FE and SEA methods are not yet practically implemented (Peiffer, 2016).

When Venor et al. (2003) developed the global simulation strategy, the hybrid theory was not available. Therefore, SEA was used for all components but enriched with many modifications to the coupling and damping loss factors so that the simulation results fits to the measurement results. It is especially due to this fact that SEA has the reputation of requiring a high level of experience.

12.3.4.1 Cavities

The first important step in the modelling strategy is a realistic representation of the exterior noise field and wave propagation around the car.

In Figure 12.25 the shape of the cavities and the sound field at 1000 Hz resulting from the rear tire rolling noise is shown, which follows from the adaption proposed by Venor. The outer cavities are connected to a small semi infinite fluid. The interior cavity

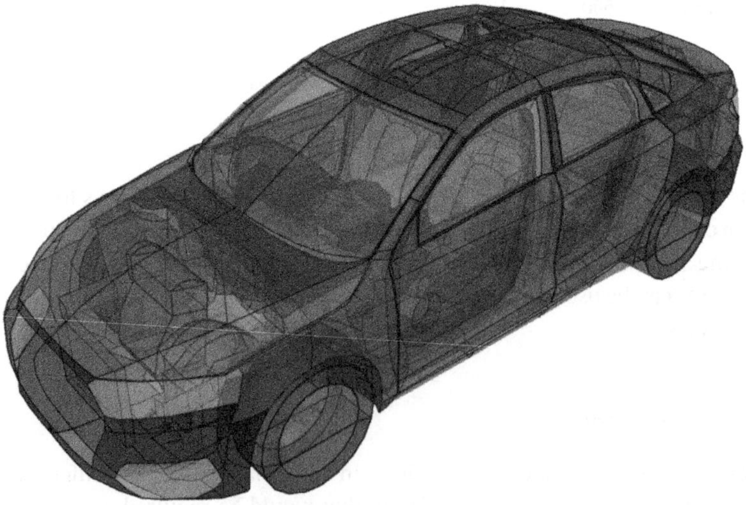

Figure 12.24 Full SEA model of a sedan car with combustion engine. *Source:* Alexander Peiffer.

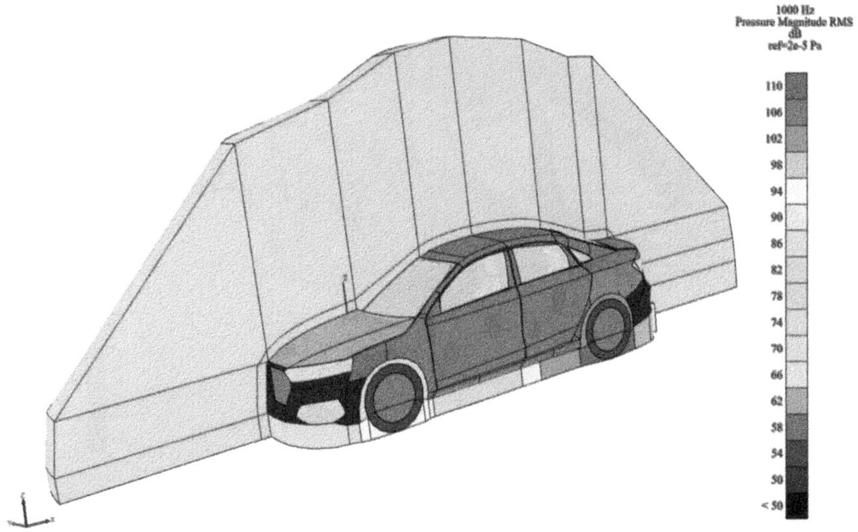

Figure 12.25 Exterior near, under floor, and far field cavities proposed by Venor et al. (2003) to represent sound field and propagation around the car. *Source:* Alexander Peiffer.

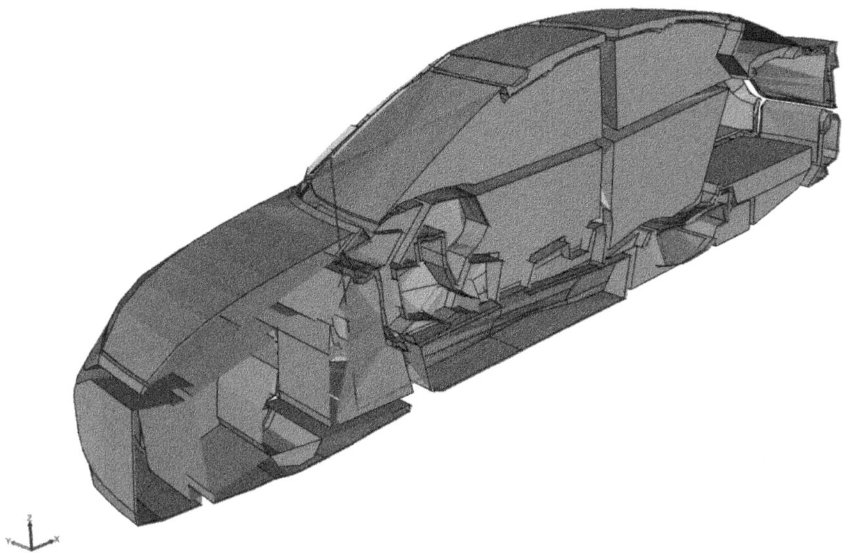

Figure 12.26 Subdivision of passenger cavity, engine bay, trunk, and many other hollow volumes. *Source:* Alexander Peiffer.

subdivision follows from the given shape of most inner volumes, but the passenger cavity is further separated. The idea of this additional subdivision is to get different noise levels at different positions in the car, but it violates the week coupling requirement of SEA. This dilemma can also be solved by ray tracing or other gradient methods as shown by Schell and Cotoni (2015).

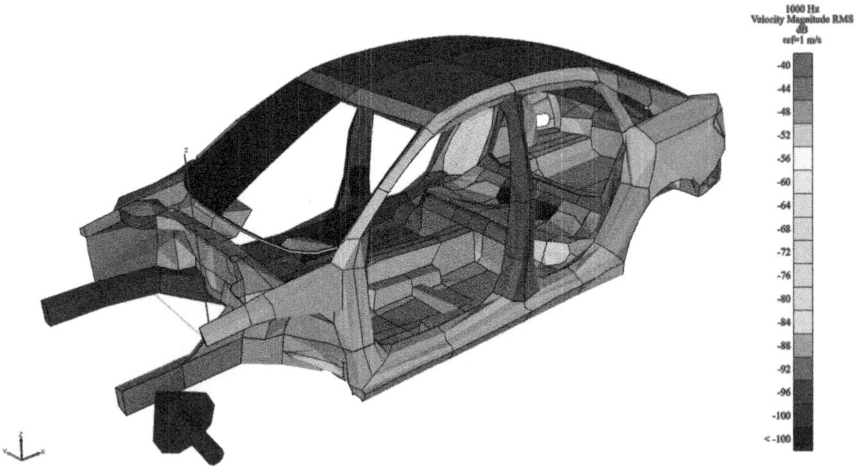

Figure 12.27 Plate subsystem configuration of the SEA carbody model with force excitation at the front longeron with $F_{rms=1}$ N. *Source:* Alexander Peiffer.

12.3.4.2 Carbody

The carbody structure (Figure 12.27) is modelled by several flat and curved plates that are in some cases quite small due to the given restrictions. Many parts in a car are treated with damping material so that the damping loss factors can be very high ($\eta \approx 20\%$). The error resulting from the small subsystems is therefore somehow limited; however, the subdivision of the carbody structure requires much experience to cope well with the realistic bending wave propagations along the carbody and over many subsystems.

12.3.4.3 Noise Control

The noise control of cars is done by a diversity of foams, fibers, and damping and absorption material. Each lay-up is considered in detail and with varying thickness because of the beaded steel plates. Every single plate that is treated with such material is modelled with the transfer matrix method as introduced in section 10.5.2. In Figure 12.28 the variety of shell properties and noise control treatments are denoted.

12.4 Trains

Vibroacoustic analysis of train structures in many ways resembles that of aircraft structures: the size is similar, and the structural lay-up and the materials applied display similarities (although the use of composites is pushed further for aircraft). Rail carbodies generally consist of a stiffened metal carbody structure, typically in steel or aluminum, with acoustically decoupled wall panels and walking floors making up the interior. All-in-all, the carbody constitutes a double wall structure with layers of thermal and acoustic insulation (fiberglass and foams) in the cavity. Recent body-shell concepts are based on GFRP or metal face sandwich. Like in aircraft, the walking floor

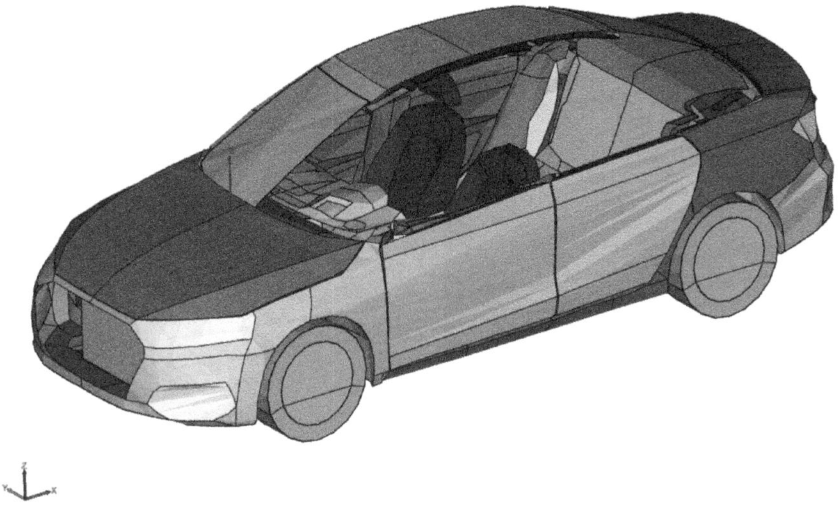

Figure 12.28 Car plate subsystems with different colors for different property and noise control treatment. *Source:* Alexander Peiffer.

structures are typically made using a sandwich material (honeycomb core or alike), but also wooden floors are common.

12.4.1 Structural Design

Steel carbodies are built in mild or stainless quality steels for which the sheeting is stiffened by beams in different layouts (Figure 12.29). For aluminum carbodies, wall, roof, and floors are normally made from extruded hollow profiles that are welded together (Figure 12.30). Aluminum carbodies can be made lighter than corresponding steel structures and can also be made to a high degree of pressure tightness, making them popular for high-speed trains. In addition, multi-material composite designs, often based on sandwich technologies, are increasingly being applied. These make use of the high stiffness properties of sandwich materials while providing smooth surfaces,

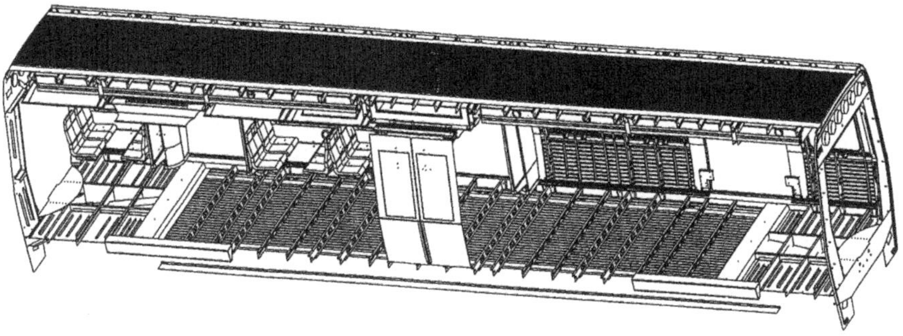

Figure 12.29 Structural layout of steel body-shell. Reproduced with permission from (Orrenius et al., 2003).

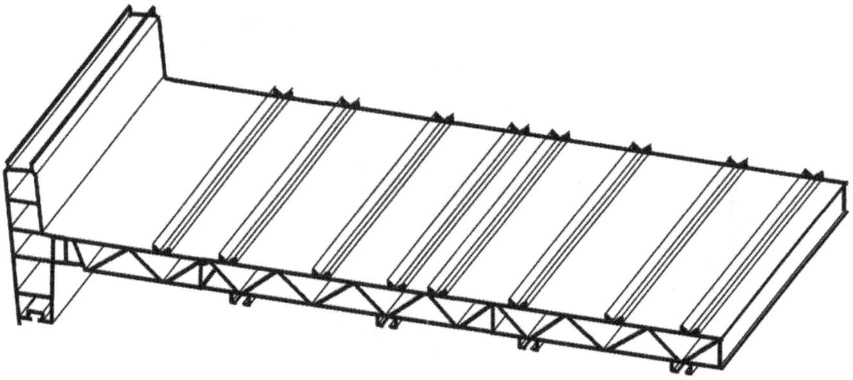

Figure 12.30 Structural layout of extruded body-shell section. Reproduced with permission from (Orrenius et al., 2003).

beneficial for the interior assembly and exterior aesthetics. Furthermore, GFRP sandwich is the most common material for cab-structures due to its flexibility in shaping the exterior.

12.4.2 Interior Design

To minimise transfer of acoustical energy to the inner floor and lining panels, these are normally designed to be acoustically decoupled from the structure. Floating floor arrangements are applied, as illustrated in Figure 12.31. The system illustrated has rubber springs with a conical shape, designed for a controlled system resonance independent of load. The walking floor may also be continuously supported on a foam or fiber mat.

12.4.3 Excitation and Transmission Paths

Noise sources on a rail vehicle can be categorized into two groups: (i) those associated with rail–wheel interaction, namely radiation and vibrations from wheel and rail and sleepers (Thompson et al., 2009) and (ii) those of vehicle components (Bistagninio et al., 2015). The latter group can be divided into subgroups with (a) components for which fans and other cooling devices are the main source and (b) components with a vibrating shell being the main source, e.g. due to electromagnetic forces (transformers, motors) or

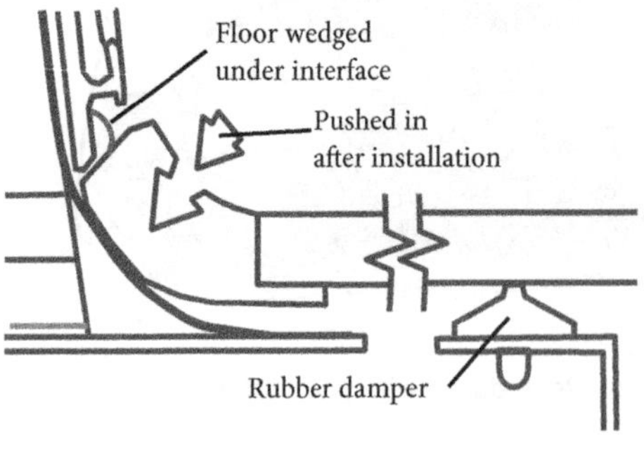

Figure 12.31 Floating floor arrangement. Reproduced with permission from (Orrenius et al., 2003).

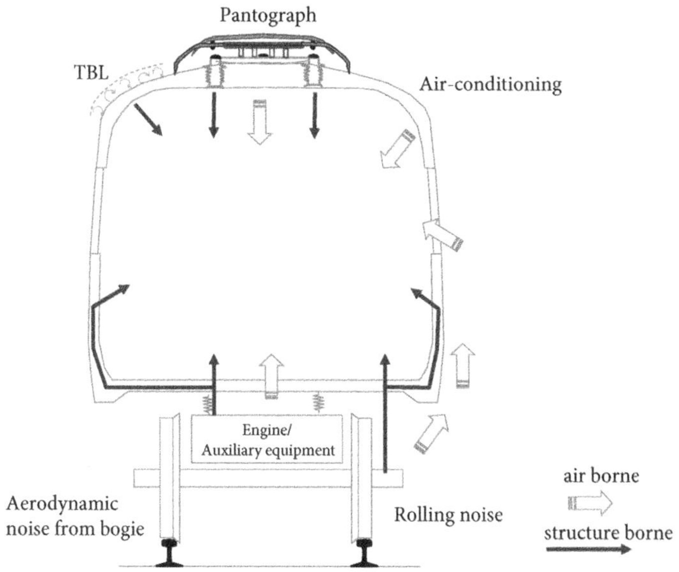

Figure 12.32 Vibroacoustic sources and transmission paths of rail car in operation. Reproduced with permission from Ulf Orrenius.

mechanical contact (gear boxes). At very high speeds components with aerodynamic sources (c) become increasingly important, e.g. bogies and pantographs (Thompson et al., 2009).

The main sources of a rail car in operation are depicted in Figure 12.32. Please note that the relative importance of the sources are strongly speed dependent. For a train at standstill, several auxiliary systems contribute to the interior levels: HVAC, transformers, electric motor, and diesel engine cooling fans (Orrenius et al., 2007). During acceleration (and retardation), drive system noise from gears and motors is typically the main source for interior noise. Such noise is generally of a tonal character and can therefore be very annoying. The transmission to the interior is rather dominated by the structural forces than the sound levels generated in the bogie (although an accelerating motor can be very loud).

At speeds above approximately 50 km/h, rolling noise typically becomes the main source. The excitation originates from imperfections on the rail head and wheels. For source modelling purposes these imperfections as well as the wheel and track properties must be described well (Thompson et al., 2009), which is typically handled within a dedicated software Twins (Jiang et al., 2011). For rail cars with under floor mounted diesel engines, structural vibration from the propulsion systems may exceed that of the rolling noise and is the main reason why electric trains provide a better acoustic comfort than diesel trains. Please note that for diesel engines, the main transmission path to the interior is typically via structural vibrations. At speeds above 250 km/h, low frequency aeroacoustic excitation becomes increasingly important. Such excitation is typically due to different objects interacting with the external flow, e.g. bogies, antennas, and the pantograph, rather than the TBL excitation as for a smooth aircraft fuselage. Note that locally, e.g. for passenger seats below the pantograph, aeroacoustic sources may dominate also at lower speeds.

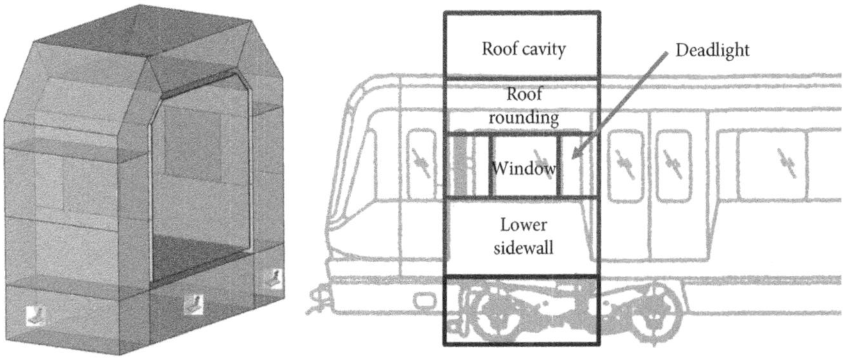

Figure 12.33 SEA model of a metro cross-section above the bogie from early transmission modelling work Stegemann (2002). *Source:* Alexander Peiffer.

12.4.4 Simulation Strategy

Being large enough to ensure sufficient modal densities and at a complexity level that both supports the use of a statistical model description and rather straightforward substructuring, train structures are generally suitable for SEA. From a passenger perspective the main sources are in the vicinity of the bogie. Therefore, vibroacoustic simulation models typically focus on the cross-section above the bogie. In addition, for high-speed trains, the driver's cab as well as the pantograph cross-sections, are critical. For a carbody cross-section of, say, three meters length, the size of the structure in combination with the broad frequency excitation, makes standard FE models inappropriate for audio frequency interior noise predictions and must be combined with high frequency methods like SEA or ray tracing.

The use of hybrid FEM/SEA models is helpful. Apart from being better suited for realistic excitation mechanisms, they offer a major improvement to standard SEA because of the possibility to simulate a structure containing both deterministic and statistical subsystems and thereby to cover a widened frequency range. In addition, including FEM subsystems much improves the chances to represent important excitation mechanisms, like aeroacoustics pressure fluctuations and structural forces from rolling sources and the traction system, with the accuracy needed for industrial predictions.

12.4.5 Applications to Rail Structures – Double Walls

The double wall model as outlined in Figure 12.34 is appropriate for most rail carbody structures. The model includes additional paths which describe the double leaf nonresonant system in addition to the mass law path, and is required to correctly represent the physics at and below the double wall resonance. The main double wall path T15 in Figure 12.34 corresponds exactly to equation (9.121). All other paths can be derived by modification of this formula. In case of low damping in the center cavity, the double wall junction must be used carefully, because it may lead to using the overall mass law path twice as shown by Wang (2015). However, real structures typically have lossy foams or fibers in the cavity, which simplifies the ananlysis.

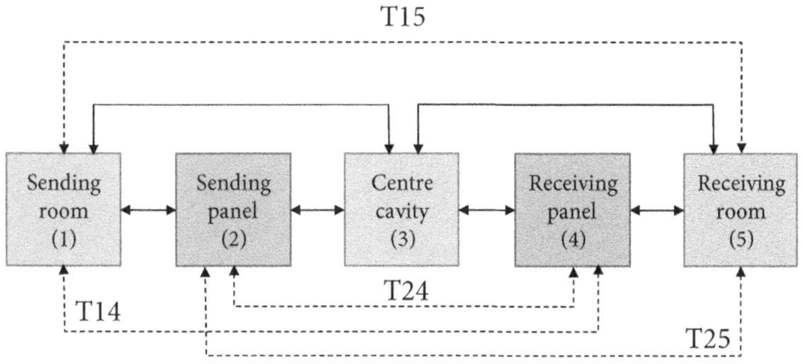

Figure 12.34 Double wall SEA block diagram with additional paths for double leaf calculation. *Source:* Alexander Peiffer.

Complicating factors are modelling and quantification of the coupling mechanisms between inner and outer walls: elastic elements, air-spring, edge conditions, etc; and also to correctly account for the vibroacoustic properties of the carbody parts, typically manufactured in rib-stiffened corrugated steel sheets or assembled from so-called extruded panels. These topics are further discussed by, e.g. Craik (1996); Orrenius et al. (2005); Cherif et al. (2017).

12.4.5.1 Carbody Elements from Extruded Profiles

As explained above, most carbodies are made in either steel or aluminum. In the following we will focus on extruded aluminum structures as they are rather challenging to model. Due to their poor acoustic insulation properties, they are also subject to extensive noise control treatments. Steel carbody designs much resemble those of aircraft fuselage structures as discussed in section 12.2.

SEA models for structures made from extruded profiles should account for the fact that the vibrational modes relevant for sound radiation can be divided into two groups: global and local modes (Orrenius et al., 2005; Li et al., 2021). This also applies to many other stiffened engineering structures like those of aerospace and ships, but the cut-on frequencies for the local modes are typically relatively higher for rail structures.

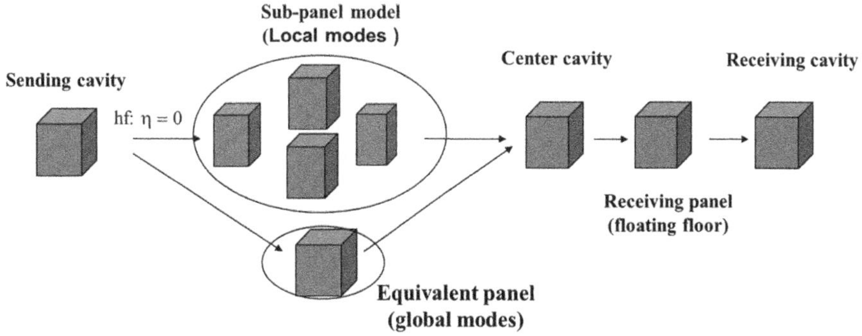

Figure 12.35 Schematic of energy flow paths for a train floor assembled from extruded panels. Reproduced with permission from Orrenius et al. (2005).

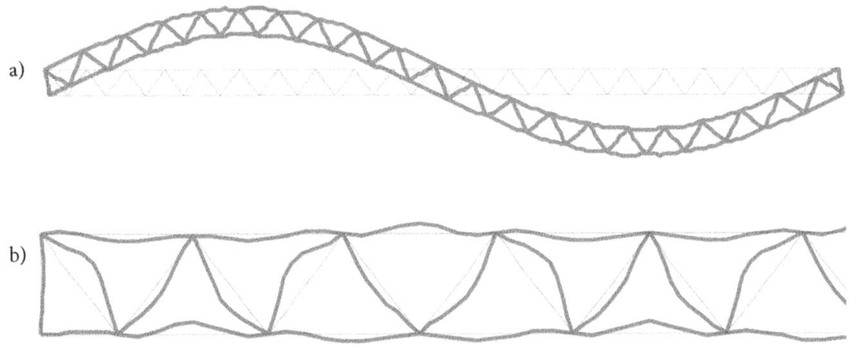

Figure 12.36 Example of local and global cross section modes determined from wave-guide finite elements. Reproduced with permission from Orrenius et al. (2005).

At low frequencies these profiles can be modelled by equivalent plate theory as their cross-sections do not deform when vibrating, see Figure 12.36a. At higher frequencies the individual subpanels start to vibrate independently, and the transmission is determined by the properties of these panels, see 12.36b. The frequency at which the subpanels start to have resonances is here referred to as the *local cut-on frequency*, typically at 400-500 Hz. Also important are the *coincidence frequencies* above which the structure radiates sound efficiently. For global modes, for which the panel structure vibrates as a whole, the coincidence frequency is typically between 130 and 170 Hz, while for the subpanel modes coincidence occurs around 4 kHz. The combined width of the two coincidence regions is one reason for poor sound reduction properties. See reference (Pang, 2004; Li et al., 2021) for examples of measured and calculated radiation efficiencies.

Resulting transmission losses for low and high frequency models are displayed in Figure 12.37, highlighting the need for including both modal groups as parallel subsystems in a model. The benefit of separating global and local modes is also clear

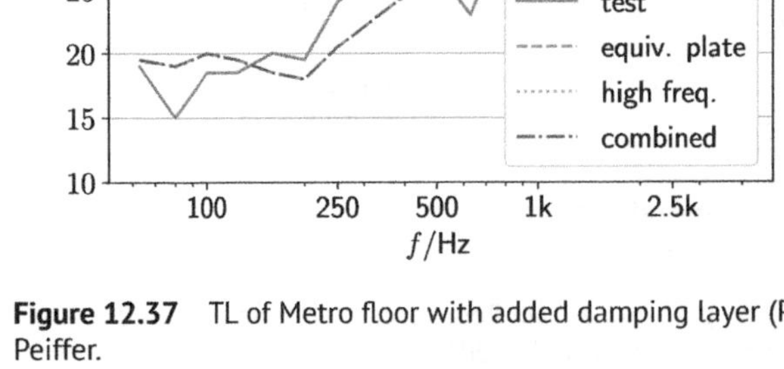

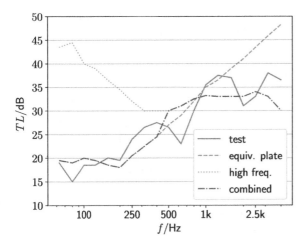

Figure 12.37 TL of Metro floor with added damping layer (Pang, 2004). *Source:* Alexander Peiffer.

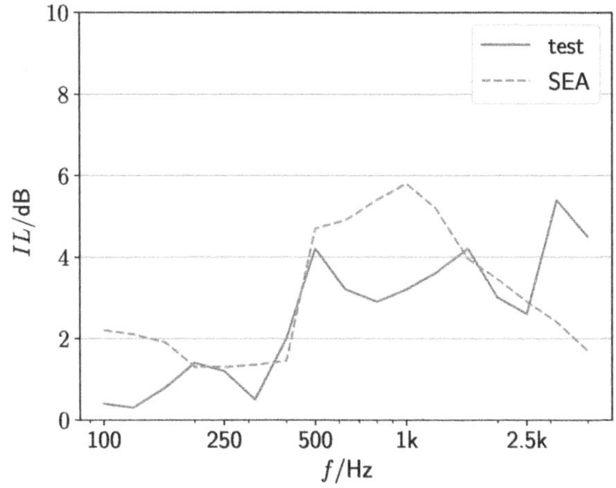

Figure 12.38 Insertion loss from application of damping layer measured and calculated (Orrenius et al., 2005). *Source:* Alexander Peiffer.

when examining calculated insertion loss of damping layers, as shown in Figure 12.38. Damping applied on the upper surface of the panel is mainly effective above the cut-on frequency of the local modes. Also, calculated panel radiation efficiencies and velocity level differences are useful indicators of model validity, see Orrenius et al. (2005) and Li et al. (2021) for comparasions of measured and caculated data.

Alternatively, periodic cell theory can be applied to derive an SEA model from a small structural FE model with periodical boundary conditions (Cotoni et al., 2008). Such procedures are successfully applied for extruded profiles by Orrenius et al. (2009a) and for aluminum aircraft structures (Orrenius et al., 2009b). In Wittsten (2016) the concept is applied to minimize the weight of an extruded train floor, subject to acoustical and structural constraints.

The periodic cell concept was also applied in work presented in (Orrenius et al., 2014) in which a complete floor assembly as depicted in Figure 12.40 was analyzed. Airborne transmission via the cavity is not included in the FE model and was added separately from using a standard double wall SEA model but with the springs disconnected, and the total transmission coefficient is determined as $\tau_{\text{total}} = \tau_{\text{air}} + \tau_{\text{struct}}$ where the airborne and structure borne transmission are denoted τ_{air} and τ_{struct}, respectively. For the transmission through the extruded panel, the fluid structure coupling was determined by input data from running the periodic cell model without the top floor. The

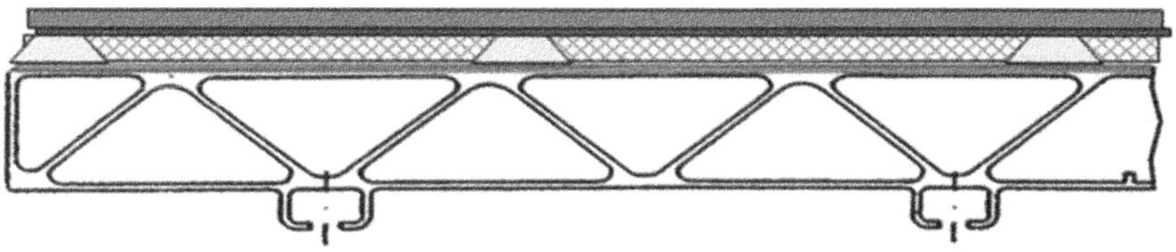

Figure 12.39 Schematic of floating floor arrangements with the walking floor resting on rubber mount. The air cavity is filled with foam material. Reproduced with permission from Ulf Orrenius.

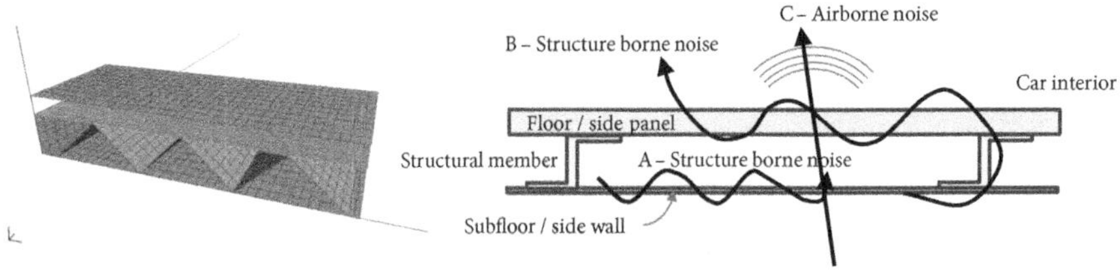

Figure 12.40 Left: Periodical cell model of the train floor. Right: Structural and airborne transmission. *Source:* Alexander Peiffer.

Metawell inner floor was modelled as a sandwich panel with equivalent core, and the rubber spring elements were modelled as lossy linear springs.

In Figure 12.41 the total transmission loss calculated is plotted together with the airborne and structure borne paths indicated. The airborne path dominates at low frequencies whereas the structural path dominates at high frequencies. Also, the measured transmission loss is displayed. The match between measured and calculated results is reasonable throughout the frequency range, although the SEA double wall model grossly underestimates the low frequency TL. Please note that the double wall junction functionality applied here accounts for indirect coupling between the sending and receiving cavity as well as between sending cavity and receiving panel and vice versa, see Figure 12.34.

For the assembly analyzed, the direct coupling between the two panels via connecting joints and constrained air was so strong that it dominates the transmission into the receiving panel. This suggests that additional absorption in the center cavity, e.g. by means of absorption material, or damping material applied on the extruded panel will not increase the transmission loss much unless this transmission is significantly reduced. The model thus provides a physical explanation to discouraging results from measurements on such noise control treatments. Instead, for increased

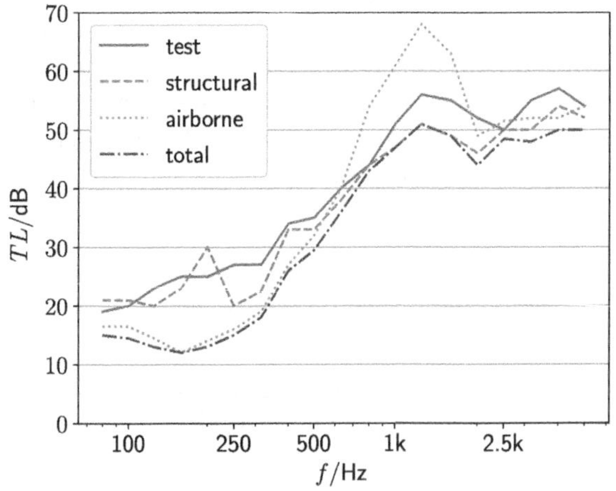

Figure 12.41 Structural vs. airborne transmission loss: Calculated and measured results. *Source:* Alexander Peiffer.

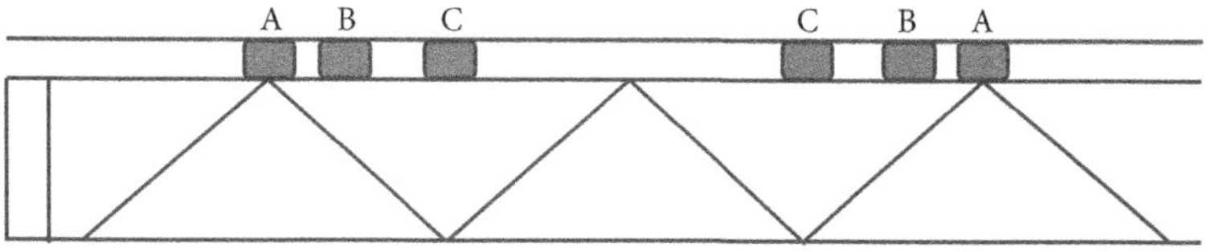

Figure 12.42 Cross-section of FE model with location of spring elements indicated: A, nominal; B, 20% offset; C, 50% offset. *Source:* Alexander Peiffer.

sound and vibration insulation, the direct coupling between the panels should preferably be relaxed, e.g. by increasing the distance between the walls, or using softer rubber elements. Typically, the spring elements are mounted on top of the web intersections of the extruded panels, the location with highest input impedance, position A in Figure 12.42. However, for practical reasons, such mounting is not always possible. A parameter study was undertaken, in which the spring elements were offset from the web intersection of the extruded panel by 20% and 50% of the division between the webs.

Transmission loss with different element locations is shown in Figure 12.43. It is notable that already 20% offset significantly reduces the transmission loss. One reason is that with offset springs, the lossy rubber elements are less deflected, and the modal loss factors of the inner floor modes are significantly reduced.

12.4.6 Carbody Sections – High Speed Applications

At higher speeds the source ranking drastically changes with the onset of aeroacoustic sources. For exterior noise, aeroacoustics becomes significant above speeds of 300–320 km/h. As the speed dependence is typically $50\log(v)$, rather than $30\log(v)$ as for the rolling sources (Thompson et al., 2009), they become dominant at higher speeds. For interior noise, aeroacoustics may locally be significant also at lower speeds, e.g for driver's cabs and for areas below the pantograph.

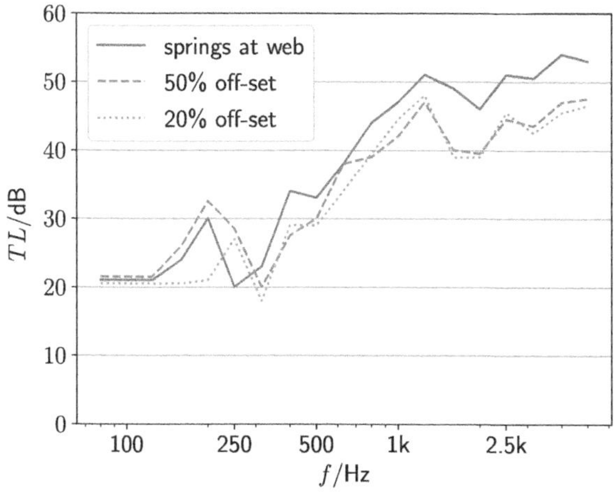

Figure 12.43 Transmission loss with different spring element locations. *Source:* Alexander Peiffer.

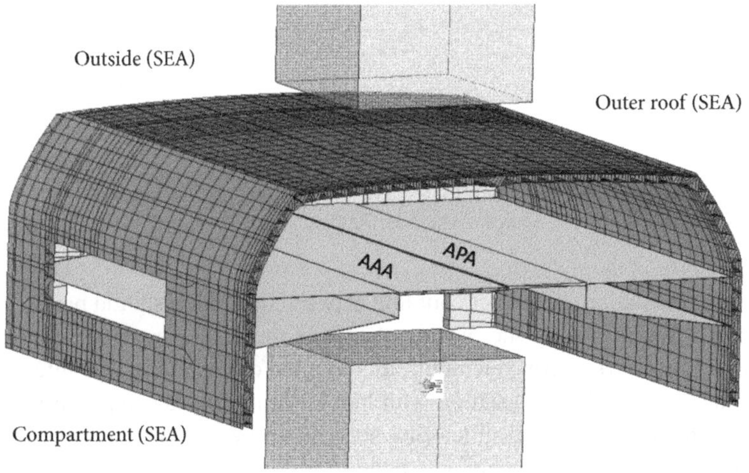

Figure 12.44 FEM/SEA model of a train roof structure. The duct cavities and the roof spacing cavity are hidden. "AAA" is aluminum–air–aluminum sandwich, and "APA" is aluminum–plywood–aluminum sandwich. Reproduced with permission from Orrenius and Kunkell (2011).

12.4.6.1 Roof Structures

Train roof structures are critical at high speeds when high noise and vibration levels are generated from pantographs and antennas etc. In Figure 12.44 an FEM/SEA model of a roof structure is depicted (Orrenius and Kunkell, 2011). The lower SEA cavity represents the passenger compartment, coupled to the inner ceiling and walls. The upper cavity represents the outside and was coupled to the roof. Generally, the aluminum roof structure was modelled in FE whereas the interiors, including the fluid cavities, were modelled using SEA subsystems. The ceiling was made from a sandwich with 12 mm plywood core covered with 1 mm aluminum face panels. However, the core was not homogeneous but contained a grid of rectangular $\approx 0.4 \times 0.4\ \text{m}^2$ cut-outs, where the ceiling essentially consisted of an aluminum double shell, see (Orrenius and Kunkell, 2011)). As the first resonance frequency of the AAA-panels in the cut-outs is quite low, 45 Hz assuming simply supported boundaries, it was decided to model all cut-outs with one single SEA system and the rest of the sandwich with another SEA subsystem. The constrained air in the cut-out cavity was modelled as a linear spring. In addition, SEA subsystems representing the ventilation airducts were added. Thus, in the model, three parallel SEA systems were connecting the compartment cavity to the roof cavity above the ceiling (not shown in the figure).

In Figure 12.45 calculated input powers via different paths into the roof cavity are displayed. Missing data are due to too few modes in the band. The aluminum–air–aluminum sandwich path dominates the transmission. The calculated transmission loss is shown to match that measured according to ISO 15186-2 by scanning of the outer roof with an intensity probe while exciting the interior using an omnidirectional loudspeaker. The model has been used for parametric studies to assess the effect of noise control treatment applied on the ceiling structure (Orrenius and Kunkell, 2011).

The pantograph becomes a very important source at high speeds and in this section, results from a pantograph noise transmission study are presented. The pantograph

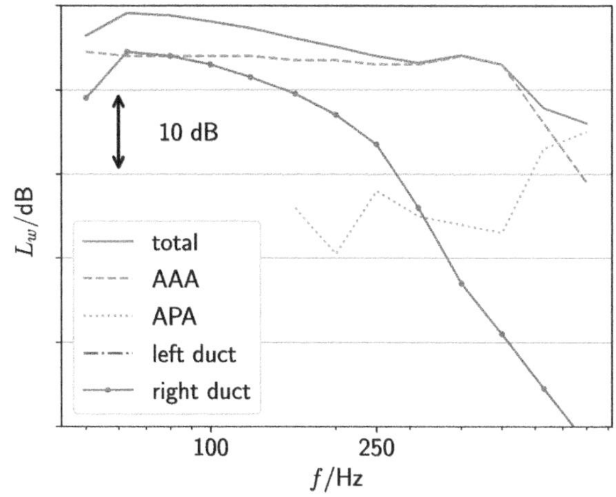

Figure 12.45 Input powers to roof cavity (Orrenius and Kunkell, 2011). *Source:* Alexander Peiffer.

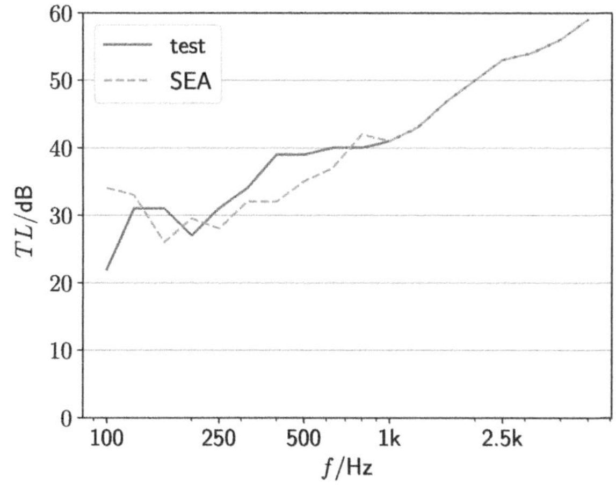

Figure 12.46 Measured and simulated transmission loss of roof (Orrenius and Kunkell, 2011). *Source:* Alexander Peiffer.

analyzed is mounted on vibration isolators at three connection points in a roof recess; see Figures 12.32 and 12.47. Due to the aerodynamic forces and the contact with the electrical overhead line, vibrations are generated and transmitted via point forces on the roof. In the study presented, forces from wind tunnel measurements were applied (Brick et al., 2011) with operational pantograph (not folded) but without the influence of the overhead lines.

A hybrid FEM/SEA model was created for the high-speed train roof. A computational fluid dynamics (CFD) model was used to determine the pressure fluctuations in the recess. In Figure 12.47 the model is displayed with the positions of the pantograph loads indicated as well as the CFD mesh. The CFD pressure field spectra were then projected onto the modal basis, representing the FE part of the roof structure. For this

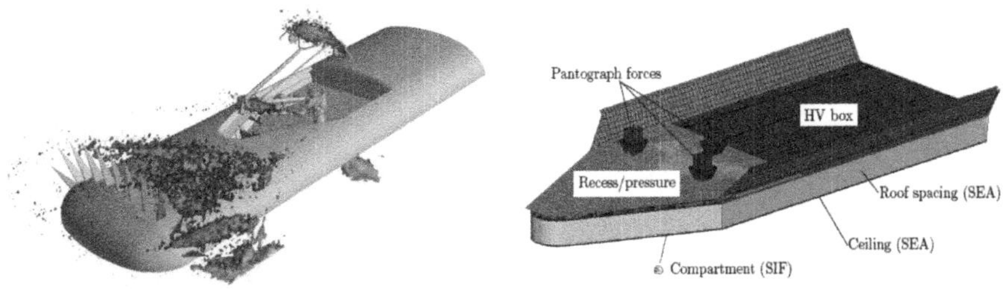

Figure 12.47 Left: Aeroacoustic simulation of pantograph in recess. (Right) FEM/SEA model of pantograph recess with CFD determined pressure field and structural forces. Reproduced with permission from (Orrenius and Kunkell, 2011).

construction the design of the interior ceiling is made in a flat thin GRP panel, thus more suitable to SEA than the ceiling described above. The roof spacing cavity is filled with an absorbing material, here modelled according to Delany and Bazley (1970) with the material described simply by its flow resistivity and density. Figure 12.48 shows the power input into the compartment for the three types of excitations: the pantograph structural forces, the pressure field in the recess, and (for reference) a turbulent boundary layer applied to the entire carbody circumference.

Combination of experimental data for the dynamic forces on the pantograph feet with pressure data from compressible and nonstationary CFD is a practical approach that serves its purpose well in view of the difficulty in predicting the dynamic forces. The calculations were carried out at a speed of 350 km/h whereas the windtunnel measurements were made at 280 km/h. Accordingly, the dynamic forces were scaled according to the speed dependence determined at the measurements (Orrenius and Kunkell, 2011).

The CFD pressure data available were determined for the recess only without the pantograph. To determine the effect of the uplifted pantograph, a correction term was

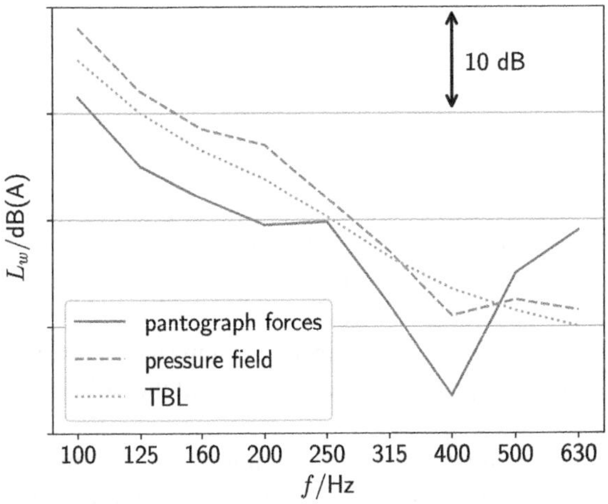

Figure 12.48 Power input to compartment at 350 km/h (Orrenius and Kunkell, 2011). *Source:* Alexander Peiffer.

applied based on measured wall pressure difference in the cavity with and without the uplifted pantograph. However, the spectral details were not considered for this correction which underestimates, e.g. the tonality associated with Strouhal effects. Alternatively, CFD calculations with the uplifted pantograph could have been used. In this case the pressure field would need to be complemented by the acoustic excitation of the pantograph, e.g. by using Lighthill's analogy (Lighthill, 1953).

A benefit of a coupled FEM/SEA model in this context is that parametric studies of the ceiling design including noise control treatments can readily be made without updating the FE model. Transmission path analysis is also straightforward.

12.4.6.2 Cab Structures

For cab structures, modal models based on FE can be solved up to a few kHz but need to be combined with either ray tracing or SEA for the interior cavity (Kirchner et al., 2012). Also the cab interior trim panels in, eg, honeycomb, plastic, or foam-covered sheet metal, are typically suitable for SEA modelling (or TMM). A Hybrid FEM/SEA approach is useful to analyze aeroacoustics noise with dominating power below 1 kHz, for which the receiving structures, like the floor above the bogie and the wind screen, are modelled with FE. Also, structural sources like vibrations transferred via bogie coupling elements, can be analyzed in this way. Like for the pantograph example above, vibration velocities measured at the mounting point of the coupling elements can be used together with FE calculated impedances to determine input powers. In Figure 12.49 two FEM/SEA models are displayed for low and high frequency analysis of a high-speed train cab. The low frequency model ($\approx 0.05 - 0.6$ kHz) contains mainly FE subsystems (335 000 elements), with interior panels and air cavities modelled with SEA subsystems. The floor pressure load is taken from CFD calculations and the structural forces on the yaw damper mounting point from in-situ testing. The high frequency

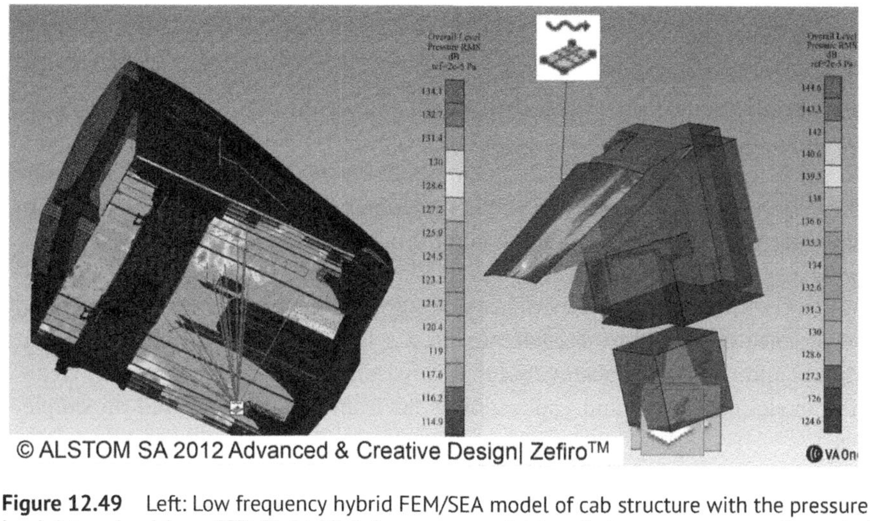

Figure 12.49 Left: Low frequency hybrid FEM/SEA model of cab structure with the pressure load determined from CFD. Right: High frequency model for wind screen aeroacoustic analysis. Reproduced with permission ALSTOM Holdings SA, 46 rue Albert Dhalenne 93000 St Ouen-sur-Seine, France. Alstom Group is the owner of all copyrights on these images herebefore listed.

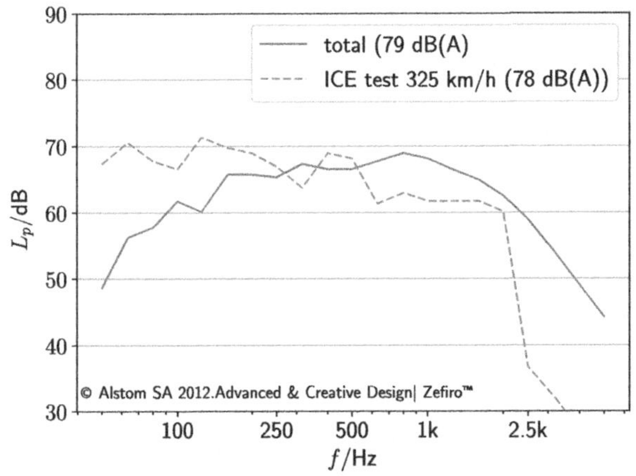

Figure 12.50 Calculated and measured sound levels in the cab cavity at 320 km/h. The measurements are from a different train but at the same speed. Reproduced with permission ALSTOM Holdings SA, 46 rue Albert Dhalenne 93000 St Ouen-sur-Seine, France. Alstom Group is the owner of all copyrights on these images herebefore listed.

model ($\approx 0.2-2$ kHz) is mainly made up of SEA subsystems, with only the windscreen modelled in FE. In this particular case the airborne transmission through the floor was taken from test data.

In 12.50, calculated cab sound levels are displayed together with data measured from a different high-speed cab at the same speed. Transfer path analysis was made to reveal the major transmission paths, which is highly useful in the design process when assessing the effect of various control measures. In this case it was shown that an increase in speed will mainly increase the low frequency transmission from the bogie area (aeroacoustics).

12.5 Summary

The aim of the presented test cases is to highlight different aspects and issues of vibroacoustic simulation. In the industrial context the task of simulation is mostly to support the design process in the early development phase but also to find the best solutions in later project phases. This later period is sometimes a '*fire fighting*' phase but even then the simulation can support the test and the development of countermeasures for a specific acoustic problem. However, the creation of a precise simulation model is a demanding task that requires experience and knowledge of the simulation method. The experience is mandatory because we need a feeling for the correct subsystem description and the level of abstraction that is reasonable. It hopefully became clear, that the model generation and application is far from being done based on simple recipes.

Bibliography

Jean-F. Allard and Noureddine Atalla. *Propagation of Sound in Porous Media*. Wiley, second edition, 2009. ISBN 978-0-470-74661-5.

W.V. Bhat. Flight test measurement of exterior turbulent boundary layer pressure fluctuations on Boeing model 737 airplane. *Journal of Sound and Vibration*, 14(4): 439–457, February 1971. ISSN 0022460X.

M. A. Biot. Generalized Theory of Acoustic Propagation in Porous Dissipative Media. *The Journal of the Acoustical Society of America*, 34(9A):1254–1264, September 1962. ISSN 0001-4966.

A. Bistagninio, G. Squicciarini, Ulf Orrenius, E. Bongini, M. Starnberg, R. Cordero, J. Sapena, and D. J. Thompson. Acoustical Source Modelling for Rolling Stock Vehicles: The Modeller's Point of View, Proceedings Euronoise 2015,. In *Euronoise*, pages 1991–1996, 2015.

Michael Blundell and Damian Harty. *Multibody Systems Approach to Vehicle Dynamics*. Automotive Engineering. Elsevier Butterworth-Heinemann, Amsterdam Heidelberg, transferred to digital print edition, 2010. ISBN 978-0-7506-5112-7.

M. Bouhaj, O. von Estorff, and A. Peiffer. An approach for the assessment of the statistical aspects of the SEA coupling loss factors and the vibrational energy transmission in complex aircraft structures: Experimental investigation and methods benchmark. *Journal of Sound and Vibration*, 403:152–172, September 2017. ISSN 0022460X.

Heike Brick, Torsten Kohrs, Ennes Sarradj, and Geyer. Noise from high-speed trains: Experimental determination of the noise radiation of the Pantograph. In *Proceedings Forum Acousticum*, Aalborg, Denmark, July 2011.

Arnaud Charpentier, Prasanth Sreedhar, Julio Cordioli, and Kazuki Fukui. Modeling process and validation of Hybrid FE-SEA method to structure-borne noise paths in a trimmed automotive vehicle. In *Proceedings 2008 SAE BRASIL Noise and Vibration Conference*, March 2008.

Raef Cherif, Andrew Wareing, and Noureddine Atalla. Evaluation of a hybrid TMM-SEA method for prediction of sound transmission loss through mechanically coupled aircraft double-walls. *Applied Acoustics*, 117:132–140, February 2017. ISSN 0003682X.

J.A. Cockburn and J.E. Robertson. Vibration response of spacecraft shrouds to in-flight fluctuating pressures. *Journal of Sound and Vibration*, 33(4):399–425, April 1974.

G. M. Corcos. Resolution of Pressure in Turbulence. *The Journal of the Acoustical Society of America*, 35(2):192–199, 1963.

V. Cotoni, R.S. Langley, and P.J. Shorter. A statistical energy analysis subsystem formulation using finite element and periodic structure theory. *Journal of Sound and Vibration*, 318(4-5):1077–1108, December 2008. ISSN 0022460X.

Vincent Cotoni, Robin Langley, and Phil Shorter. A Statistical Energy Analysis Subsystem Formulation using Finite Element and Periodic Structure Theory. In *Proceedings NOVEM 2009*, page 12, Oxford, England, April 2009.

Robert J M Craik. SEA for buildings. *Noise and Vibration Worldwide*, 27(6):13–, June 1996. ISSN 0957-4565.

Peter Davidsson. *Structure-Acoustic Analysis; Finite Element Modelling and Reduction Methods*. PhD thesis, Lund University, Lund, Sweden, August 2004.

Evan B. Davis. By Air by SEA. In *Noise-Con 2004*, pages 12–17, Baltimore, Maryland, July 2004.

Carlos de Matos. *Simulation des Schalldurchgangs einer Flugzeugwand*. Diploma Thesis, TU-Hamburg Harburg, Hamburg, Germany, February 2008.

ME Delany and EN Bazley. Acoustical properties of fibrous absorbent materials. *Applied acoustics*, 3(2):105–116, 1970.

B.M. Efimtsov and L.A. Lazarev. Forced vibrations of plates and cylindrical shells with regular orthogonal system of stiffeners. *Journal of Sound and Vibration*, 327(1–2):41–54, October 2009. ISSN 0022-460X.

Javier García de Jalón and Eduardo Bayo. *Kinematic and Dynamic Simulation of Multibody Systems The Real-Time Challenge*. 1994. ISBN 978-1-4612-2600-0.

Klaus Genuit, editor. *Sound-Engineering im Automobilbereich: Methoden zur Messung und Auswertung von Geräuschen und Schwingungen*. Springer, Heidelberg, 2010. ISBN 978-3-642-01415-4.

Michael Goody. Empirical Spectral Model of Surface Pressure Fluctuations. *AIAA Journal*, 42(9):1788–1794, September 2004. ISSN 0001-1452, 1533-385X.

Shijie Jiang, P. Meehan, David Thompson, and C Jones. Railway rolling noise prediction under European conditions. *Australian Acoustical SocietyConference 2011, Acoustics 2011: Breaking New Ground*, page 8, January 2011.

Karl-Richard Kirchner, F. Brännström, Ulf Orrenius, and R. Hallez. aeroacoustic noise generation and transmission modelling for a high-speed train driver's cab. In *Proceedings Internoise 2012*, New York, USA, 2012.

Alexander Klabes, Christina Appel, Michaela Herr, and Sören Callsen. Fuselage Excitation During Cruise Flight Conditions: A New CFD Based Pressure Point Spectra Model. In *Proceedings*, page 12, Hamburg, Germany, August 2016.

L. Leylekian, M. Lebrun, and P. Lempereur. An Overview of Aircraft Noise Reduction Technologies. *Aerospace Lab*, (7):15, June 2014.

Hui Li, Giacomo Squicciarini, David Thompson, Jungsoo Ryue, Xinbiao Xiao, Dan Yao, and Junlin Chen. A modelling approach for noise transmission through extruded panels in railway vehicles. *Journal of Sound and Vibration*, 502:116095, June 2021. ISSN 0022460X.

M.J. Lighthill. On sound generated aerodynamically I. General theory. *Proceedings of the Royal Society A: Mathematical, Physical and Engineering Science*, 221:24, 1953.

B.R. Mace and P.J. Shorter. Energy flow models from finite element analysis. *Journal of Sound and Vibration*, 233(3):369–389, June 2000. ISSN 0022-460X.

Gideon Maidanik. Response of Ribbed Panels to Reverberant Acoustic Fields. *The Journal of the Acoustical Society of America*, 34(6):809–826, June 1962. ISSN 0001-4966.

Laurent Maxit, Marion Berton, Christian Audoly, and Daniel Juvé. Discussion About Different Methods for Introducing the Turbulent Boundary Layer Excitation in Vibroacoustic Models. In Elena Ciappi, Sergio De Rosa, Francesco Franco, Jean-Louis Guyader, and Stephen A. Hambric, editors, *Flinovia - Flow Induced Noise and Vibration Issues and Aspects*, pages 249–278. Springer International Publishing, Cham, 2015. ISBN 978-3-319-09712-1.

Vinod G. Mengle, Ulrich W. Ganz, Eric Nesbitt, Eric J. Bultemeier, and Russell H. Thomas. Flight Test Results for Uniquely Tailored Propulsion-Airframe Aeroacoustic Chevrons: Shockcell Noise. In *Proceedings*, volume 2439, page 17. American Institute of Aeronautics and Astronautics (AIAA), 2006.

U. Orrenius, V. Cotoni, and A. Wareing. Analysis of sound transmission through periodic structures typical for railway car bodies and aircraft fuselages. *INTER-NOISE and NOISE-CON Congress and Conference Proceedings*, 2009(8):785–796, 2009a. ISSN 0736-2935.

Ulf Orrenius and H. Kunkell. Sound Transmission through a High-Speed Train Roof. In *Proceedings ICSV 2011*, Rio, Brasil, 2011.

Ulf Orrenius, E. Lundberg, and Bert Stegemann. Noise Control Elements in Train Design: An Overview of Solutions for Carbody Floors. In *Proceedings ICSV 2013*, Stockholm, Sweden, 2003.

Ulf Orrenius, Y.Y. Pang, and Bert Stegemann. Acoustic modelling of extruded profiles for railway cars. In *Proceedings NOVEM 2005*, St. Raphael, France, 2005.

Ulf Orrenius, Siv Leth, and Anders Frid. Noise reduction at urban hot-spots by vehicle noise control. In *Proceedings IWRN 2007*, München, Germany, 2007.

Ulf Orrenius, A. Waering, and Vincent Cotoni. Analysis of sound transmission through aircraft fuselages excited by turbulent boundary layer or diffuse acoustic pressure fields. In *Proceedings NOVEM 2009*, Oxford, England, 2009b.

Ulf Orrenius, Hao Liu, Andrew Wareing, Svante Finnveden, and Vincent Cotoni. Wave modelling in predictive vibro-acoustics: Applications to rail vehicles and aircraft. *Wave Motion*, 51(4):635–649, June 2014. ISSN 01652125. doi: 10.1016/j.wavemoti.2013.11.007.

Dan Palumbo. Determining correlation and coherence lengths in turbulent boundary layer flight data. *Journal of Sound and Vibration*, 331(16):3721–3737, April 2012. ISSN 0022-460X.

Y.Y. Pang. Modelling acoustic properties of truss-like periodic panels: Application to extruded aluminium profiles for railway structures. Master's thesis, KTH Stockholm, Stockholm, Sweden, 2004.

Alexander Peiffer. Comparison of the computational expense of hybrid FEM/SEA calculation. In *Proceedings NOVEM 2012*, Sorrento, Italy, April 2012.

Alexander Peiffer. Full frequency vibro-acoustic simulation in the aeronautics industry. In *Proceedings*, Leuven, Belgium, September 2016.

Alexander Peiffer and Uwe Christian Mueller. Review of Efficient Methods for the Computation of Transmission Loss of Plates with Inhomogeneous Material Properties and Curvature Under Turbulent Boundary Layer Excitation. In Elena Ciappi, Sergio De Rosa, Francesco Franco, Jean-Louis Guyader, Stephen A. Hambric, Randolph Chi Kin Leung, and Amanda D. Hanford, editors, *Flinovia—Flow Induced Noise and Vibration Issues and Aspects-II*, pages 233–251. Springer International Publishing, Cham, 2019. ISBN 978-3-319-76779-6 978-3-319-76780-2.

Alexander Peiffer and Zhiyi Wang. SEA Simulation einer Flugzeugseitenwand und Korrelation zu Testdaten. In *Fortschritte der Akustik*, page 4, Nürnberg, Germany, April 2015.

Alexander Peiffer, Stephan Brühl, and Daniel Redmann. Hybrid Modelleing of Random Excitation of Shell Structures. In *Proceedings NOVEM 2009*, Oxford, England, April 2009.

Alexander Peiffer, Mahjoub Mezni, and Clemens Moeser. Hybrid Modelling of Deterministically Loaded Subsystems. In *Proceedings 18th International Congress on Sound and Vibration*, page 8, Rio, Brasil, July 2011.

Alexander Peiffer, Clemens Moeser, and Arno Röder. Transmission loss modelling of double wall structures using hybrid simulation. In *Fortschritte Der Akustik*, pages 1161–1162, Merano, March 2013.

Alexander Peiffer, Clemens Moeser, Koen De Langhe, and Robin Boeykens. SIMULATING SOUND TRANSMISSION LOSS THROUGH AIRCRAFT FUSELAGE PANELS: AN UPDATE ON RECENT TECHNOLOGY EVOLUTIONS. In *Proceedings of NAFEMS 2015*, page 20, San Diego, June 2015.

Alexander Schell and Vincent Cotoni. Prediction of Interior Noise in a Sedan Due to Exterior Flow. *SAE International Journal of Passenger Cars - Mechanical Systems*, 8(3): 1090–1096, June 2015. ISSN 1946-4002.

Bert Stegemann. Development and validation of a vibroacoustic model of a metro rail car using Statistical Energy Analysis (SEA). Master's thesis, TU-Berlin, Berlin, 2002.

Stephan Tewes and Alexander Peiffer. Abschlussbericht RAST. Technical Report BMBF-FKZ-03CL09A-Abschlussbericht-RAST, EADS IW SP-NV, Hamburg, Germany, December 2013.

D. J. Thompson, Chris Jones, and Pierre-Etienne Gautier. *Railway Noise and Vibration: Mechanisms, Modelling and Means of Control*. Elsevier, Amsterdam ; Boston, 1st ed edition, 2009. ISBN 978-0-08-045147-3.

Joe Venor and Burghardt, Marc. SEA MODELLING AND VALIDATION OF SOUND-FIELDS AND SEALS IN AUTOMOBILES. In *Conference Proceedings*, page 7, Potsdam, Germany, 2005.

Joe Venor, A. Müller, and Silje Nintzel. SEA MODELLING AND VALIDATION OF STRUCTURE AND AIRBORNE INTERIOR NOISE IN AN AUTOMOBILE. In *Proceedings Vibro-Acoustic Users Conference Europe*, page 8, Leuven, Belgium, January 2003.

Xu Wang, editor. *Vehicle Noise and Vibration Refinement*. Woodhead Publishing in Mechanical Engineering. CRC Press, Boca Raton, 2010. ISBN 978-1-84569-497-5.

Zhiyi Wang. Correlation between SEA Simulation and Test Results for a Double Wall Structure. Master's thesis, TU-München, München, Germany, June 2015.

Christian Weineisen. Modellierung einer Flugzeugstruktur mit der Statistischen-Energieanalyse. Master's thesis, Technische Universität München (TUM), München, Germany, March 2014.

J. Wittsten. Optimization of Train Floors with Acoustic and Structural Constraints. Master's thesis, Chalmers University of Technology, Chalmers, Sweden, 2016.

Peter Zeller, editor. *Handbuch Fahrzeugakustik: Grundlagen, Auslegung, Berechnung, Versuch ; mit 43 Tabellen*. ATZ/MTZ-Fachbuch. Vieweg + Teubner, Wiesbaden, 1. aufl edition, 2009. ISBN 978-3-8348-9322-2.

13

Conclusions and Outlook

13.1 Conclusions

This book outlines many aspects of vibroacoustic simulation. We made the journey from lumped elements to distributed systems, from deterministic and random over to hybrid systems models. The idea was to present the full frequency portfolio of simulation methods but based on simple cases that are comprehensive but of practical relevance. It is clear that we had to stay with manageable systems and examples for clarity. Even the industrial cases in Chapter 12 can only provide some aspects of complex models.

The goal was to provide a comprising description of deterministic and random methods as far as the link between them, the hybrid FEM/SEA theory. To the author's experience, most acousticians are exclusively focussing on one method. Or even worse, they claim that either deterministic or random methods are the best.

Hopefully after this lecture it becomes clear that there is no good or bad method. There is only an appropriate frequency range and sometimes – when the conditions of hybrid FEM/SEA methods are predominant – both approaches are required within one model.

However, for *simple* systems constructed by plates, cavities, and noise control treatment a complete *toolbox* is provided to perform simulation on such systems over the full audible frequency range. The code and libraries of the examples can be found on the author's website: www.docpeiffer.com.

13.2 What Comes Next?

This textbook can cover only a part of the extensive subject of vibroacoustic simulation, and the author would like to propose to the reader some literature and topics for further reading and studies. The list is by no means complete but should give suggestions to either develop an interest for the ongoing research in the field or for more details of more precise material and system descriptions.

13.3 Experimental Methods

Experiments and tests are mandatory for every acoustician. They provide the link to the real world and validate the models that are used. There is sometimes a ridiculous

Vibroacoustic Simulation: An Introduction to Statistical Energy Analysis and Hybrid Methods,
First Edition. Alexander Peiffer.

competition between people doing simulation or experiments that does not make any sense. Experimental people claim to be more practical than theorists, and the simulation people may feel superior due to the fact that they can manage complex simulation tools of mathematical equations.

It is the author's conviction that both groups require a deep knowledge of the other's competence to become a true acoustic expert. An experimentalist who is just comparing curves without any idea about the theoretical reason for the differences may make the wrong design decisions. A numerical model helps a lot to understand the vibroacoustic system well and to draw the right conclusions. On the other hand, a theorist that is never visiting a lab will not develop any feeling for practical problems and limitations. She or he may also miss important source and transmission phenomena in the model. In addition modern test analysis methods are so advanced and complicated that experimental people also require a deep theoretical knowledge.

This section provides some literature proposals of advanced experimental methods that are required to provide test results that can be used in model updates.

13.3.1 Transfer Path Analysis

The transfer path analysis (TPA) is based on the measurement of the frequency transfer function of vibroacoustic systems. Hence, the dynamics of the systems is reduced to specific points of interest: mount connection, passenger ear positions, or excited locations. This transfer function is measured for multiple inputs (forces, moments, and acoustic volume strength) and responses (acceleration, displacement, and pressure). A good overview about the history and state of the art can be found, for example, in (van der Seijs et al., 2016).

The system is subdivided into subsystems that are connected via specific point shaped connections so that the transfer functions of the main paths are included. Those transfer functions are similar to the matrix (1.198) as introduced in section 1.7.

Both response and excitation are measured. For structural excitation this is done with piezoelectric transducers that measure the force generated by impact hammers or shakers and accelerometers for the response. Instead of accelerometers a scanning laser vibrometer can be used. Acoustic sources with given source strength $\boldsymbol{Q}$ are realized by

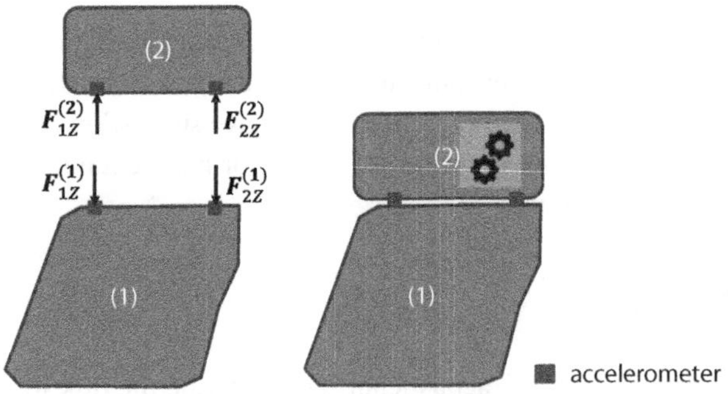

Figure 13.1 Principle of classical transfer path analysis. a) transfer matrix test of cut-free systems, b) operational test. *Source:* Alexander Peiffer.

special loudspeaker systems sometimes connected to long tubes in order to establish a point source. The pressure response is determined by microphones.

The determination of the transfer functions is performed using broadband signals (white noise, sweeps, or the given hammer impact history) and by applying equation (1.194). Hammer measurements are usually averaged over several impacts.

The transfer function of the subsystem is then inverted to get the stiffness or mobility matrix of the subsystems of interest. Especially this inversion is the critical point in TPA because noisy signals and test errors may lead to badly conditioned matrices and erroneous results. Many papers deal with methods to increase the robustness of this inversion by using pseudo inverses and additional reference sensors to create overdetermined systems.

In combination with an operational analysis, the TPA determines the forces occurring at the interface connections. From the forces and the local stiffness, the power input to the system can be calculated, and transfer paths are rated in terms of significance. The precision of these and power forces depends strongly on the quality of the matrix inversion and the test result in general.

When the systems become random, phase and magnitude of the system responses become uncertain. In this case the TPA can no longer be applied.

13.3.2 Experimental Modal Analysis

The mode shapes of structural or fluid systems can be measured directly. The tests and the related analysis methods are called experimental modal analysis (EMA). The system of interest is either equipped with a sufficient number of sensors to capture the modal shape and excited at one or a few excitation points. Instead, few response sensors and a large number of excitation positions can also be used. Due to reciprocity both methods provide the same result.

The are several methods to analyze measured frequency response functions and to extract the modal properties. A comprising overview about the existing methods is given by He and Fu (2001). The measured mode shapes can be used in modal frequency analysis similar to numerics, or the shapes are used as experimental reference. In this case the simulated modes are compared to the experimental shapes.

The major challenge is to estimate clean transfer functions for precise mode shapes, in particular for higher order modes. One advanced method is to estimate these functions by a best fit to a fraction of polynomial functions named PolyMAX parameter estimation (Peeters et al., 2004).

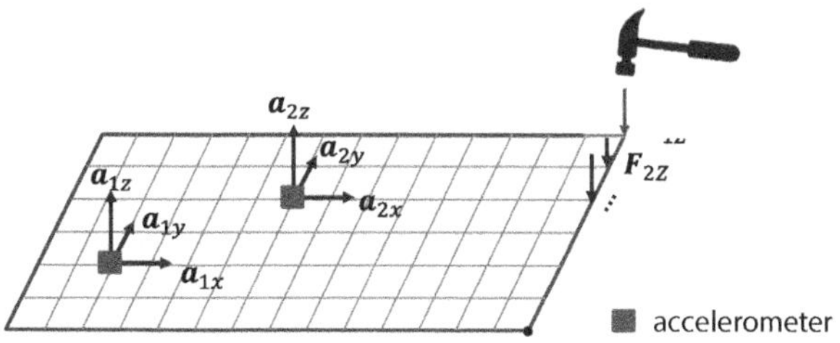

Figure 13.2 Typical setup of experimental modal analysis.
Source: Alexander Peiffer.

13.3.3 Correlation Between Test and Simulation

It is common practice that test and simulation results are compared based on point responses, e.g. pressure at driver's ear from simulation and test. This might work for low frequencies and deterministic systems but may be misleading for complex systems. It is much better to compare the system responses for many positions distributed over this full system.

It does not make much sense to compare hundreds of response curves in one diagram. Thus, specific criteria are introduced to compare shapes, for example shapes of modes or frequency responses. Here, the most used criteria for mode comparison is given, the so called modal assurance criteria (MAC):

$$MAC(\{\Psi_{test}\},\{\Psi_{sim}\}) = \frac{\left|\{\Psi_{test}\}^H \{\Psi_{sim}\}\right|^2}{\left(\{\Psi_{test}\}^H \{\Psi_{test}\}\right)\left(\{\Psi_{sim}\}^H \{\Psi_{sim}\}\right)} \tag{13.1}$$

Under the condition of normal modes and perfect agreement the MAC is supposed to be orthogonal, hence $MAC = 1$ for equal mode shapes and $MAC = 0$ for different shapes. For the global agreement of the dynamic behavior the modal frequencies must also be equal.

The application of this criterion is state of the art in model updating methods and included in many test software packages. Nevertheless, there are many practical pitfalls. Note, that structural modes are mass normalized as given by equation (1.119) and that the mass matrix is usually not available in experimental modal analysis. So, for systems with very irregular system and mass distribution, this must be considered. This affects the orthogonality of the modes; thus, different modes don't have zero MAC. See Allemang (2001) for more details. In addition the shape might be undersampled, meaning that the selected sensors are not dense enough to catch all details of the high frequency shapes.

There are several approaches besides engineering judgement to overcome such issues. One option is to check the so called AUTOMAC first. This is the MAC when the modes are correlated to themselves. If the result is the unit matrix, the selected test mesh and the mass distribution are supposed to be correct. A further issue is the case when modes have the same frequency and thus cannot be identified as distinct modes in the test. Obviously, only deterministic systems can be compared by this criterion.

When it comes to correlation of an SEA simulation with tests, things are getting more complicated. The SEA result is the average result of an ensemble of systems. Thus, curve comparison of single tests are reasonable when the systems are clearly random, as it is the case, for example, in the academic test cases of Chapter 10.

For the correlation of SEA models to tests, the system responses must be measured at many locations to get a good spatial average. Theoretically, even an ensemble of systems should be tested, which is not possible in most cases. Especially, single point response comparison are definitely no solution. However, when automatic test systems such as laser scanning systems or robots are used, a certain average can be achieved, and the comparison of physical units or the energy is possible. The investigation of the structure borne wave propagation along the aircraft fuselage in Chapter 12 is one example of this practice (Weineisen, 2014). A detailed correlation of the fuselage sidewall to SEA models is dealt with in Wang (2015).

13.3.4 Experimental or Virtual SEA

The experimental SEA can be considered as a transfer path analysis of random systems. The concept of virtual SEA is to measure the input power (from force and acceleration) at several excitation points on each subsystem and the velocity or pressure response on several locations of the connected and responding subsystems. The energy of such systems is estimated from (6.116) and (6.118). For the latter equation assumptions for the mass distribution must be made.

We start from the nonsymmetric version of the SEA equation (6.102), so the the modal density is not required. When we assume that the power introduced into the nth subsystem is given by $\Pi_{in}^{(n)}$ and the energy response of the mth subsystem due to this excitation is denoted by E_{mn} we get the following equation system for three systems

$$\begin{bmatrix} E_{11} & E_{12} & E_{13} \\ E_{21} & E_{22} & E_{23} \\ E_{31} & E_{32} & E_{33} \end{bmatrix} \begin{bmatrix} \eta_{11}+\eta_{12}+\eta_{13} & -\eta_{21} & -\eta_{31} \\ -\eta_{12} & \eta_{22}+\eta_{21}+\eta_{23} & -\eta_{33} \\ -\eta_{13} & \eta_{23} & \eta_{33}+\eta_{31}+\eta_{32} \end{bmatrix}$$

$$= \frac{1}{\omega} \begin{bmatrix} \Pi_{in}^{(1)} & & \\ & \Pi_{in}^{(2)} & \\ & & \Pi_{in}^{(3)} \end{bmatrix} \tag{13.2}$$

The subsystem energy is derived with (6.116) or (6.118). This can become a complicated task for irregular shapes and mass distribution, e.g. a ribbed plate of a fuselage panel (Bouhaj et al., 2017).

In equation (13.2) there are nine equations for nine parameters when we write the power load and energy matrices in one column. This system of equations is solved for η_{ij}. There are several investigations on experimental SEA, for example Lalor (1987, 1990, 1997), continued by the investigations on practical systems from Borello (2009, 2015) and Borello and Gagliardini (2007). Similar to the transfer path analysis, the determination of the coupling loss factors require an invertible matrix that is not always provided by tests. One key measure to overcome this is the method of averaging. Lalor proposed calculating the coupling loss factor from every measurement and average the coupling loss factors. Other authors average all excitation positions over one subsystem. Bouhaj et al. (2017) propose a statistical method to bring more robustness into the matrix inversion. In any case, virtual experimental SEA can be a promising tool to create SEA models from test data.

In addition the same method can be applied using FE simulations. In this case the introduced power and the energy estimation is determined from simulations. The biggest advantage here is, that the amount of sensors is nearly unlimited. The global analysis procedure is exactly the same. Such studies can be used to investigate, for example, very complicated junctions where no simplified equations for the radiation stiffness are available.

Note that the general decision how the substructuring is made is the duty of the experimentalist. Hence, also experimental SEA requires decisions about the SEA model set-up namely on the separation of the system into subsystems. This is further evidence for the fact that experimental people need theoretical knowledge.

13.4 Further Reading on Simulation

13.4.1 Advances in SEA and Hybrid FEM/SEA Methods

Further reading on SEA is provided by Le Bot (2015) who introduces many aspects of the statistical concept of SEA and constitutes the link to statistical physics. Langley (2007, 2016) developed an extended version of the diffuse field reciprocity (7.19) based on a modal approach and for subsystems that don't carry a perfectly diffuse wave field. An interesting field of research is the automatic recognition of components of a hybrid FEM/SEA model as developed by Langley and Kovalewski (2012).

Some research projects aimed at developing or collecting methods for the "mid frequency" range. One example is the Marie Curie project "MID-Frequency"– CAE-Methodologies for Mid-Frequency-Analysis in Vibration and Acoustics (Atak et al., 2012). A text book on current research in vibroacoustic simulation and many practical examples was edited by Hambric et al. (2016).

13.5 Energy Flow Method and Influence Coefficient

The work from Shorter (1998) and Mace (2005) provides a more sophisticated approach to determine the coupling loss factor using modal analysis of the full system and subsystems. In this case no power injection is required. Many software tools have implemented this method. Similar to virtual SEA the energy flow method (EFM) is a powerful tool to investigate complicated subsystems and junctions.

In this process of subsystem selection, it is also very important to know what is not an SEA subsystem. Tewes and Peiffer (2007) showed that the inclusion of deterministic subsystems into the SEA power balance leads to incorrect coupling loss factors in the experimental SEA process.

13.5.1 More Realistic Systems

In the initial chapters of this book we stayed with canonical and simple systems, such as homogenous plates and cavities. They were perfect for the exemplification of the method. Unfortunately, reality is neither canonical nor simple. Thus, a next step is to include more realistic but usually also complicated subsystems into the determination of modal density, wavenumber, and damping. Typical candidates are specific lay-ups, for example composite or sandwich plates. Integrated structures can be corrugated or beaded plates, rib-stiffened plates, or extruded profiles as shown in Chapter 12.

In many cases an analytical formulation of such materials or shapes does not make sense. The required equations may become more complicated and computationally even more demanding than numerical models. However, some more advanced descriptions of subsystems are available that can be managed with reasonable effort. Note that the thin shell is not correct for high frequencies, too. Here, thick plate theory must be applied, introduced, for example, in Cremer et al. (2005).

13.5.2 Anisotropic Material

All examples in this book are using isotropic materials, meaning that the mechanical or acoustical properties are independent from the orientation. The mechanical strength and superiority of carbon fiber or glass fiber reinforced plastics comes from the fact that highest strength is designed into the direction of the highest load. Hence, those materials must be anisotropic to fulfill this specification. Anisotropic materials are used in plates, shells, or beams. So, they are mostly used for example in plate theory.

13.5.3 Porous Elastic Material

In section 9.3.4 the limp fiber model was introduced, which is a useful description for porous material without frame dynamics. One reason for using this model was that the theory is not too complex. For some foam materials that are used in the automotive industry, for example as firewall mass-spring noise control treatment, the Biot (1962) model must be used. The consideration of such layers makes special layer formulations necessary plus interface conditions. The topic of porous materials is a research subject of its own and still an ongoing field especially in the experimental field. The book from Allard and Atalla (2009) gives a comprehensive treatment of this topic. There are many recent formulations (Horoshenkov et al., 2019; Lafarge et al., 1997) of the thermal and viscous dissipation during the wave propagation in the porous frame showing that the topic is still a field of basic research.

The simulation of coupled fluid structure models in combination with porous finite elements is still in the process of being implemented as standard procedure.

13.5.4 Composite Material

In section 3.7 we dealt with the in-plane forces (3.148) and moments (3.185). The linear laminate theory is well established in structural lightweight design and specific equations for bending and in-plane waves are elaborated, for example, in Baker and Scott (2016). Basically the behavior of a plate wave doesn't change; the main difference to isotropic plates is that stiffness ratios between bending and in-plane strain can be different. In addition, the line moments and stresses can be coupled in case of unsymmetrical lay-ups.

13.5.5 Sandwich

More optimised plate structures in terms of lightweight design are sandwich plates that put most of the strength in the skin layer. The task of the very lightweight core material is to keep the distance between both skins. The dynamics of sandwich panels is special because the skin layers decouple from a certain frequency, so sandwich panels may be statically stiff and dynamically soft. This was shown by Kurtze and Watters (1959) and further detailed in (Moore and Lyon, 1991). This effect of decoupling depends mainly on the shear modulus of the core material. Note that due to the high stiffness and low weight, sandwich structures have the potential to become a vibroacoustic disaster because of a low and extended coincidence frequency range. However, by careful selection of the shear stiffness, this can be mitigated by dynamical decoupling.

In addition sandwich plates with heavy layers can constitute double walls due to the symmetric mode of wave propagation. In (Moore and Lyon, 1991) the transfer impedance for symmetric and nonsymmetric modes is given and can be used in the transfer matrix method to calculate the transmission loss.

13.5.6 Shell Theory

In this book we dealt with flat plates. In most engineering applications shells with curvature are used. There are several detailed textbooks dealing exclusively with shells of isotropic material, for example the books from Leissa (1993) and Ventsel and Krauthammer (2001). Special shapes often used in analytical SEA are cylindrical shells, also called singly curved, or spherical shells called doubly curved. A treatment of natural modes of cylindrical (and ribbed) shells is given by Szechenyi (1971). The work of Koval deals with the transmission loss and extends this treatment from cylinders of isotropic material (Koval, 1976) including pre-stress from internal pressure (aircraft fuselage) to composite cylinders.

13.5.7 Wave Finite Element Method (WFE)

The wave finite element method can be considered as a bridge between FEM and SEA in that sense that small FE models are used to calculate the parameters of SEA. Please note that this doesn't mean the EFM or virtual SEA methods. The idea of the WFE is to take a representative unit cell and apply periodic boundary conditions. With these assumptions quantities such as speed of sound, wavenumber, and damping can be calculated from this unit cell.

This is a perfect method for simple systems as plates (Manconi et al., 2013) but especially for periodic systems such as rib-stiffened shells or extruded profiles (Cotoni et al., 2008; Orrenius et al., 2014).

In addition the WFE can be extended to the calculation of coupling loss factors as shown by Mitrou et al. (2017). This further opens the path to a systematic derivation of the coupling loss factor.

13.5.8 The High Frequency Limit

In section 6.3.1.5 it was shown that there is an upper limit for SEA. This is the case when the dissipation along the wave path is so large that a diffuse field cannot exist, because the waves are damped out and reverberation is not possible. The frequency limit for systems with the maximum length L with this condition is

$$\omega_{SEAmax} = \frac{c_{gr}}{\eta L}. \tag{13.3}$$

This is the domain of methods that include the energy distribution in the systems. One option to take care of this is ray tracing, also called geometrical acoustics, as presented in (Pierce, 1991; Kuttruff, 2014). Examples for the application of ray tracing in the SEA context are presented by Gardner and Macarios (2014).

A different approach is introduced by Tanner. He developed an element based ray tracing method called dynamic energy analysis (DEA). In Tanner et al. (2012) the method is applied to plate systems.

13.6 Vibroacoustics Simulation Software

All examples in this book are coded in python except those from Chapter 12. This led to an open source simulation software prototype for SEA and hybrid simulation that is presented at the author's website www.docpeiffer.com. The author would highly appreciate further contributions from readers by creating their own test cases based on the example scripts. Further extensions to the above mentioned more advanced methods are intended in the near future. Contributions from the acoustic community are very welcome.

Bibliography

Jean-F. Allard and Noureddine Atalla. *Propagation of Sound in Porous Media*. Wiley, second edition, 2009. ISBN 978-0-470-74661-5.

Randall J. Allemang. The Modal Assurance Criterion (MAC): Twenty Years of Use and Abuse. In *Proceedings*, 2001.

Onur Atak, Bert Pluymers, Wim Desmet, and Katholieke Universiteit te Leuven (1970-). *MID-Frequency- CAE Methodologies for MID-Frequency Analysis in Vibration and Acoustics*. 2012. ISBN 978-94-6018-523-6.

A. A. Baker and Murray L. Scott, editors. *Composite Materials for Aircraft Structures*. AIAA Education Series. AIAA/American Institute of Aeronautics and Astronautics, Inc, Reston, Virginia, third edition edition, 2016. ISBN 978-1-62410-326-1.

M. A. Biot. Generalized Theory of Acoustic Propagation in Porous Dissipative Media. *The Journal of the Acoustical Society of America*, 34(9A):1254–1264, September 1962. ISSN 0001-4966.

Gerard Borello. Analysis of Vibroacoustic Systems using Virtual SEA, June 2009.

Gerard Borello. SEA+ for Aerospace Applications, 2015.

Gerard Borello and L. Gagliardini. Virtual SEA: Towards an industrial process. *SAE Technical Paper*, 2007-0123-02, 2007.

M. Bouhaj, O. von Estorff, and A. Peiffer. An approach for the assessment of the statistical aspects of the SEA coupling loss factors and the vibrational energy transmission in complex aircraft structures: Experimental investigation and methods benchmark. *Journal of Sound and Vibration*, 403:152–172, September 2017. ISSN 0022460X.

V. Cotoni, R.S. Langley, and P.J. Shorter. A statistical energy analysis subsystem formulation using finite element and periodic structure theory. *Journal of Sound and Vibration*, 318(4-5):1077–1108, December 2008. ISSN 0022460X.

Lothar Cremer, Manfred Heckl, and Björn Petersson. *Structure-Borne Sound: Structural Vibrations and Sound Radiation at Audio Frequencies*. Springer Verlag, Berlin, Germany, 3rd edition edition, December 2005. ISBN 978-3-540-26514-6.

Bryce Gardner and Tiago Macarios. Combining Ray Tracing and SEA to Predict Speech Transmissibility. In *Proceedings 8th International Styrian Noise, Vibratoin & Harshness Congress: The European Noise Conference*, June 2014.

Stephen A. Hambric, S. H. Sung, and D. J. Nefske, editors. *Engineering Vibroacoustic Analysis: Methods and Applications*. Wiley, Chichester, West Sussex, United Kingdom, 2016. ISBN 978-1-119-95344-9.

Jimin He and Zhi-Fang Fu. *Modal Analysis*. Butterworth-Heinemann, Oxford ; Boston, 2001. ISBN 978-0-7506-5079-3.

Kirill V. Horoshenkov, Alistair Hurrell, and Jean-Philippe Groby. A three-parameter analytical model for the acoustical properties of porous media. *The Journal of the Acoustical Society of America*, 145(4):2512–2517, April 2019. ISSN 0001-4966.

L.R. Koval. On sound transmission into a thin cylindrical shell under “flight conditions”. *Journal of Sound and Vibration*, 48(2):265–275, September 1976. ISSN 0022460X.

G. Kurtze and B. G. Watters. New Wall Design for High Transmission Loss or High Damping. *The Journal of the Acoustical Society of America*, 31(6):739–748, June 1959. ISSN 0001-4966.

Heinrich Kuttruff. *Room Acoustics, Fifth Edition*. 2014. ISBN 978-1-4822-6645-0 978-0-203-87637-4.

Denis Lafarge, Pavel Lemarinier, Jean F. Allard, and Viggo Tarnow. Dynamic compressibility of air in porous structures at audible frequencies. *The Journal of the Acoustical Society of America*, 102(4):1995–2006, October 1997. ISSN 0001-4966.

N. Lalor. The Measurement of SEA Loss Factors on a fully assembled Structure. Technical Report ISVR Technical Report Np. 150, Southampton, U.K., August 1987.

N. Lalor. Practical Considerations for the Measurement of Internal and Coupling Loss Factors on Complex Structures. Technical Report ISVR Technical Report Np. 182, Southampton, U.K., June 1990.

N. Lalor. The practical implementation of SEA. In *Proceedings of the IUTAM Symposium*, pages 257–268, Southampton, U.K., July 1997. Kluwer Academic Publishers.

R. S. Langley. On the diffuse field reciprocity relationship and vibrational energy variance in a random subsystem at high frequencies. *The Journal of the Acoustical Society of America*, 121(2):913–921, 2007.

Robin S. Langley and Alice Kovalewski. Automatic recognition of the components of a hybrid fe-sea model from a finite element model. In *Proceedings NOVEM 2012*, page 11, Sorrento, Italy, April 2012.

R.S. Langley. On the statistical properties of random causal frequency response functions. *Journal of Sound and Vibration*, 361:159–175, January 2016. ISSN 0022460X.

A. Le Bot. *Foundation of Statistical Energy Analysis in Vibroacoustics*. Oxford University Press, Oxford, United Kingdom ; New York, NY, first edition edition, 2015. ISBN 978-0-19-872923-5.

Arthur W. Leissa. *Vibration of Shells*. American Inst. of Physics, Woodbury, NY, 1993. ISBN 978-1-56396-293-6.

B.R. Mace. Statistical energy analysis: Coupling loss factors, indirect coupling and system modes. *Journal of Sound and Vibration*, 279(1–2):141–170, January 2005. ISSN 0022-460X.

Elisabetta Manconi, Brian R. Mace, and Rinaldo Garziera. The loss-factor of pre-stressed laminated curved panels and cylinders using a wave and finite element method. *Journal of Sound and Vibration*, 332(7):1704–1711, 2013. ISSN 0022-460X.

Giannoula Mitrou, Neil Ferguson, and Jamil Renno. Wave transmission through two-dimensional structures by the hybrid FE/WFE approach. *Journal of Sound and Vibration*, 389:484–501, February 2017. ISSN 0022460X.

J. A. Moore and R. H. Lyon. Sound transmission loss characteristics of sandwich panel constructions. *The Journal of the Acoustical Society of America*, 89(2):777–791, February 1991. ISSN 0001-4966.

Ulf Orrenius, Hao Liu, Andrew Wareing, Svante Finnveden, and Vincent Cotoni. Wave modelling in predictive vibro-acoustics: Applications to rail vehicles and aircraft. *Wave Motion*, 51(4):635–649, June 2014. ISSN 01652125.

Bart Peeters, Herman Van der Auwerer, Patrick Guillaume, and Jan Leuridan. The PolyMAX frequency-domain method: A new standard for modal parameter estimation? *Shock and Vibration*, 11:395–409, 2004. ISSN 1070-9622/04.

Allan D. Pierce. *Acoustics - An Introduction to Its Physical Principles and Applications*. Acoustical Society of America (ASA), Woodbury, New York 11797,U.S.A., one thousand, nine hundred eighty-ninth edition, 1991. ISBN 0-88318-612-8.

P. J. Shorter. *Combining Finite Elements and Statistical Energy Analysis*. PhD thesis, University of Auckland, Aucklang, Newzealand, July 1998.

E. Szechenyi. Approximate methods for the determination of the natural frequencies of stiffened and curved plates. *Journal of Sound and Vibration*, 14(3):401–418, 1971.

Gregor Tanner, David J. Chappell, Dominik Löchel, and N. Sondergard. A DEA approach for solving multi-mode large scale vibro-acoustic wave problems in the mid-to-high frequency regime. In *Proceedings ISMA 2012*, Leuven, Belgium, September 2012.

Stephan Tewes and Alexander Peiffer. EFM-Modellierung einer Flugzeugdoppelwandstruktur. In *Fortschritte der Akustik*, pages 903–904, Stuttgart, March 2007.

Maarten V. van der Seijs, Dennis de Klerk, and Daniel J. Rixen. General framework for transfer path analysis: History, theory and classification of techniques. *Mechanical Systems and Signal Processing*, 68–69:217–244, February 2016. ISSN 08883270.

Eduard Ventsel and Theodor Krauthammer. *Thin Plates and Shells: Theory: Analysis, and Applications*. CRC Press, August 2001. ISBN 978-0-8247-0575-6.

Zhiyi Wang. Correlation between SEA Simulation and Test Results for a Double Wall Structure. Master's thesis, TU-München, München, Germany, June 2015.

Christian Weineisen. Modellierung einer Flugzeugstruktur mit der Statistischen-Energieanalyse. Master's thesis, Technische Universität München (TUM), München, Germany, March 2014.

Appendix A

Basic Mathematics

A.1 Fourier Analysis

We will start our introduction to the Fourier analysis with the series representation of periodic signals. The next step is moving towards the Fourier transform by the application of a limit process. We start with the analytical formulation of the theory: i.e. all signals are continuous, and we will later switch to digital signals in section A.2.

A.1.1 Fourier Series

Imagine a function $f(t)$ of time with periodicity T, the fundamental frequency $f_0 = 1/T$ and angular frequency $\omega_0 = 2\pi/T$

$$f(t) = f(t+T) \tag{A.1}$$

This function shall be synthesized by a series of sine and cosine functions, reading

$$f(t) = a_0 + \sum_{n=1}^{\infty} (b_n \cos(n\omega_0 t) + c_n \sin(n\omega_0 t)) \tag{A.2}$$

If (A.2) is integrated over the time interval $t \in [-T/2; T/2]$ and using the orthogonality of the sine and cosine functions, we get for the coefficients

$$a_0 = \frac{1}{T}\int_{-T/2}^{T/2} f(t)dt \tag{A.3}$$

$$b_n = \frac{1}{T}\int_{-T/2}^{T/2} f(t)\cos(n\omega_0 t)dt \tag{A.4}$$

$$c_n = \frac{1}{T}\int_{-T/2}^{T/2} f(t)\sin(n\omega_0 t)dt \tag{A.5}$$

The first coefficient a_0 is the mean value of the signal, the others are the cosine (b_n) and sine (c_n) coefficients . In order to illustrate this harmonic synthesis, we consider a rectangular periodic function

Vibroacoustic Simulation: An Introduction to Statistical Energy Analysis and Hybrid Methods,
First Edition. Alexander Peiffer.

$$f(t) = \begin{cases} 1 & \text{for } 0 \leq \text{t} \mod \text{T} < \text{T}/2 \\ -1 & \text{for T}/2 \leq \text{t} \mod \text{T} < \text{T} \end{cases} \tag{A.6}$$

Obviously the mean is zero ($a_0 = 0$), and this function is an odd function $f(t) = -f(-t)$; thus the coefficients b_n corresponding to the even cosine function will vanish. The remaining coefficients c_n read

$$c_n = \begin{cases} \frac{4}{\pi n} & \text{for n} = 1, 3, 5, ... \\ 0 & \text{otherwise} \end{cases} \tag{A.7}$$

In Figure A.1 the function and several Fourier series with an increasing number of coefficients are shown. We see that the function is better and better represented, but due to the infinite slope of the rectangular function, an infinite number of Fourier coefficients would be required for a perfect match.

An entirely equivalent representation can also be formulated by a series of complex Euler functions, this time running from $-\infty$ to ∞

$$f(t) = \sum_{n=-\infty}^{\infty} \boldsymbol{a}_n e^{jn\omega_0 t} \tag{A.8}$$

The coefficients a_n for $n > 0$ can be expressed in terms of the series coefficient from (A.2)

$$\boldsymbol{a}_n = \frac{1}{2}(b_n - jc_n) n > 0 \tag{A.9a}$$

$$\boldsymbol{a}_n = \frac{1}{2}(b_n + jc_n) n < 0 \tag{A.9b}$$

It is quite illustrative to link these expressions to a harmonic signal with angular frequency $\omega_0 = 2\pi f$

$$f_n(t) = Re\left(\boldsymbol{A}_n e^{jn\omega_0 t}\right) = \frac{1}{2}\left(\boldsymbol{A}_n e^{jn\omega_0 t} + \boldsymbol{A}_n^* e^{-jn\omega_0 t}\right) \tag{A.10}$$

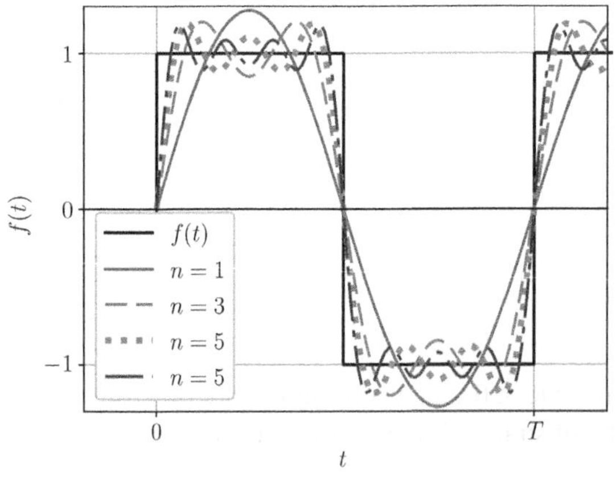

Figure A.1 Rectangular function and its Fourier series representation using up to four non-zero coefficients. *Source:* Alexander Peiffer.

Thus, for a given frequency $n\omega$ we get with (A.8)

$$f_n(t) = \boldsymbol{a}_n e^{jn\omega_0 t} + \boldsymbol{a}_{-n} e^{-jn\omega_0 t} \Rightarrow \boldsymbol{a}_n = \frac{\boldsymbol{A}_n}{2} \quad \boldsymbol{a}_{-n} = \frac{\boldsymbol{A}_n^*}{2} \tag{A.11}$$

An alternative and instructive derivation of this equation can be found by applying an integration over the period for (A.8) multiplied by $e^{-jm\omega_0 t}$, leading to

$$f(t)e^{jm\omega_0 t} = \sum_{n=-\infty}^{\infty} a_n e^{j(n-m)\omega_0 t} \tag{A.12}$$

If we integrate (A.12) over the period $[-T/2; T/2]$, we get

$$\int_{-T/2}^{T/2} f(t)e^{jm\omega_0 t} dt = \sum_{n=-\infty}^{\infty} \int_{-T/2}^{T/2} \boldsymbol{a}_n e^{j(n-m)\omega_0 t} dt. \tag{A.13}$$

We see that the right hand side is only non-zero for $m = n$. This is called the orthogonality property of the e-function in the integration, hence

$$\int_{-T/2}^{T/2} f(t)e^{jn\omega_0 t} dt = \boldsymbol{a}_n T \tag{A.14}$$

This shows in an instructive way how the orthogonality relation is used to derive an alternative expression for a_n to (A.9a)-(A.9b)

$$\boldsymbol{a}_n = \frac{1}{T} \int_{-T/2}^{T/2} f(t)e^{jn\omega_0 t} dt \tag{A.15}$$

If we replace $e^x = \cos x + j \sin x$ the former Equations (A.2) and (A.8) as far as the related coefficients a_0,b_n and c_n can also be derived. The squared value of the periodic function would have the following consequence for Fourier series

$$\langle f^2(t) \rangle = \frac{1}{T} \int_{-T/2}^{T/2} f(t)^2 dt = \frac{1}{T} \int_{-T/2}^{T/2} f(t)f(t)^* dt \tag{A.16}$$

With (A.8) we get

$$\langle f^2(t) \rangle = \frac{1}{T} \int_{-T/2}^{T/2} \left(\sum_{n=-\infty}^{\infty} \boldsymbol{a}_n e^{jn\omega_0 t} \right) \left(\sum_{m=-\infty}^{\infty} \boldsymbol{a}_m^* e^{-jm\omega_0 t} \right) \tag{A.17}$$

Performing the integration and using the orthogonality relationship giving zero for $m \neq n$ we get

$$\langle f^2(t) \rangle = \sum_{n=-\infty}^{\infty} |\boldsymbol{a}_n|^2 \tag{A.18}$$

and the total squared signal power is equal to the sum of the square of all coefficients.

A.1.2 Fourier Transformation

For stationary periodic processes the Fourier series is the perfect tool to investigate the harmonic content of a signal. The frequency resolution – the difference between two frequencies – is restricted by $\Delta\omega = \omega_0 = \frac{2\pi}{T}$ because of the periodicity.

In practical applications many nonperiodic or transient signals occur. For the frequency analysis of such signals we require a finer resolution and a different approach. This can be achieved if we perform the limit process for T to infinity for the Fourier series

$$f(t) = \sum_{n=-\infty}^{\infty} \boldsymbol{a}_n e^{jn\omega_0 t} \quad \text{with} \quad \boldsymbol{a}_n = \lim_{T\to\infty} \frac{1}{T} \int_{-T/2}^{T/2} f(t) e^{-jn\omega_0 t} dt \tag{A.19}$$

Now we let $\Delta\omega \to 0$. Arranging (A.19) in such a way that we eliminate all T by $\frac{2\pi}{\Delta\omega}$ and ω_0 by $\Delta\omega$

$$f(t) = \sum_{n=-\infty}^{\infty} \frac{2\pi \boldsymbol{a}_n}{\Delta\omega} \underbrace{e^{jn\Delta\omega t}}_{\to e^{j\omega t}} \frac{\Delta\omega}{2\pi} \tag{A.20}$$

$$\underbrace{\frac{2\pi \boldsymbol{a}_n}{\Delta\omega}}_{\to F(\omega)} = \int_{-\infty}^{\infty} f(t) e^{-j\omega t} dt \tag{A.21}$$

where we have also used $\omega = n\omega_0$. Finally, we get the expression for the pair of Fourier transforms by performing the limit process $T \to \infty$ and $\Delta\omega \to 0$

$$f(t) = \frac{1}{2\pi} \int_{-\infty}^{\infty} \boldsymbol{F}(\omega) e^{j\omega t} d\omega \qquad = \mathcal{F}^{-1}\{\boldsymbol{F}(\omega)\} \tag{A.22a}$$

$$\boldsymbol{F}(\omega) = \int_{-\infty}^{\infty} f(t) e^{-j\omega t} dt \qquad = \mathcal{F}\{f(t)\} \tag{A.22b}$$

Obviously, the Fourier transform (FT) of a harmonic signal does not converge because the infinite integration requires $f(t) \to 0$ for large arguments. The Fourier transform is a mathematical tool to investigate transient or pulse shaped signals. In order to illustrate some applications of the Fourier transform we investigate a function of rectangular shape

$$f(t) = \begin{cases} A & \text{for} -\tau \leq \text{t} \leq \tau \\ 0 & \text{for t} < \tau \quad \text{t} > \tau \end{cases} \tag{A.23}$$

with the following Fourier transform

$$\boldsymbol{F}(\omega) = A \int_{-\tau}^{\tau} e^{-j\omega t} dt = \frac{A}{j\omega} \left(e^{j\omega\tau} - e^{-j\omega\tau}\right) = 2A\tau \left(\frac{\sin(\omega\tau)}{\omega\tau}\right) \tag{A.24}$$

A.1.3 Dirac Delta Function

Based on this rectangular pulse the δ-function is derived, which is a very powerful tool for the description of mechanical and acoustical systems. We use the rectangular pulse with slightly adjusted parameters

$$2A\tau = 1 \qquad \tau \to 0 \qquad f(t) \to \delta(t) \tag{A.25}$$

The delta function is not a function in the classical sense, because is has infinite values A at $t = 0$ due to $\tau \to 0$. But, it is very useful due to the properties that are

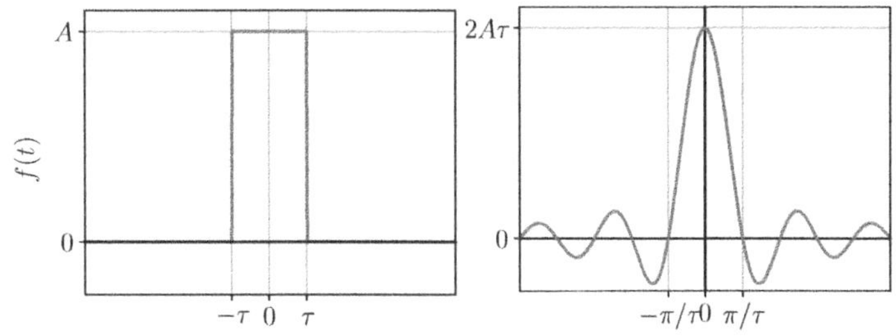

Figure A.2 Rectangular pulse function and its Fourier transform. *Source:* Alexander Peiffer.

presented next. If we integrate the delta function with the above derivation from the rectangular pulse, we find

$$\delta(t) = 0 \,\forall\, t \neq 0 \qquad \int_{-\infty}^{\infty} \delta(t)dt = 1 \tag{A.26}$$

If the delta function is centred at t_0, the same properties read

$$\delta(t - t_0) = 0 \,\forall\, t \neq t_0 \qquad \int_{-\infty}^{\infty} \delta(t - t_0)dt = 1 \tag{A.27}$$

The use of the delta function is given by its so-called *sifting* property that extracts any function if it occurs in the integral over a product with the delta function, such as

$$\int_{-\infty}^{\infty} f(t)\delta(t - t_0)dt = f(t_0)\int_{-\infty}^{\infty} \delta(t - t_0)dt = f(t_0) \tag{A.28}$$

The first term of the above equation is called the *convolution* of $f(t)$ and the δ-function. Using this the Fourier transform of the delta function reads as

$$\boldsymbol{F}(\omega) = \int_{-\infty}^{\infty} \delta(t - t_0)e^{-j\omega t}dt = e^{-j\omega t_0} \tag{A.29}$$

Thus, the frequency content of the delta function extends from negative to positive infinity with a constant value. On the other hand the inverse Fourier transform of the exponential function gives

$$\delta(t - t_0) = \frac{1}{2\pi}\int_{-\infty}^{\infty} e^{j\omega(t-t_0)}d\omega \tag{A.30}$$

Exchanging ω by ω_0 and $-t$ by t_0 provides the FT of the exponential function.

$$2\pi\delta(\omega - \omega_0) = \int_{-\infty}^{\infty} e^{j\omega_0 t}e^{-j\omega t}dt \tag{A.31}$$

$$e^{j\omega_0 t} = \frac{1}{2\pi}\int_{-\infty}^{\infty} 2\pi\delta(\omega - \omega_0)e^{j\omega t}d\omega \tag{A.32}$$

A.1.4 Signal Power

The signal power in the time domain is related to the FT similar to (A.18). With

$$\int_{-\infty}^{\infty} f^2(t)dt = \int_{-\infty}^{\infty} \left(\frac{1}{2\pi}\int_{-\infty}^{\infty} F(\omega_1)e^{j\omega_1 t}d\omega_1\right)\left(\frac{1}{2\pi}\int_{-\infty}^{\infty} F(\omega_2)e^{j\omega_2 t}d\omega_2\right)^* dt \tag{A.33}$$

and change of the integration order we get

$$\int_{-\infty}^{\infty} f^2(t)dt = \frac{1}{4\pi^2}\int_{-\infty}^{\infty}\int_{-\infty}^{\infty} F(\omega_1)F^*(\omega_2)\left(e^{j(\omega_1-\omega_2)t}dt\right)d\omega_1 d\omega_2 \tag{A.34}$$

Using (A.30) enables us to replace the parentheses by $2\pi\delta(\omega_1 - \omega_2)$, and then using the sifting property (A.28) gives

$$\int_{-\infty}^{\infty} f^2(t)dt = \frac{1}{2\pi}\int_{-\infty}^{\infty} F(\omega_1)F^*(\omega_1)d\omega_1 = \frac{1}{2\pi}\int_{-\infty}^{\infty} |F(\omega)|^2 d\omega \tag{A.35}$$

This is known as Parseval's formula. The total energy in the signal is associated with the integral over all frequency components. If we switch back to the frequency f and $d\omega = 2\pi df$ we get rid of the normalization by $1/2\pi$

$$\int_{-\infty}^{\infty} f^2(t)dt = \int_{-\infty}^{\infty} |F(f)|^2 df \tag{A.36}$$

A.1.5 Fourier Transform of Real Harmonic Signals

Even if the FT of harmonic signals will not provide a finite result, the delta function enables us to apply the FT even to harmonic signals. The cosine function is linked to the exponential function by

$$\cos(\omega_0 t) = \frac{1}{2}(e^{j\omega_0 t} + e^{-j\omega_0 t}) \tag{A.37}$$

Using (A.32) we get for the FT of the cosine function

$$\int_{-\infty}^{\infty} \cos(\omega_0 t)e^{j\omega t}dt = \pi\delta(\omega - \omega_0) + \pi\delta(\omega - \omega_0) \tag{A.38}$$

Applying the infinite integration interval of the FT to the cosine function leads consequently to infinity spectra in the frequency domain, here given by two symmetric delta functions with peaks at $\omega = \pm\omega_0$.

A.1.6 Useful Properties of the Fourier Transform

Working with Fourier transforms is much easier when specific relationships are used that link the time domain with the frequency domain. We start with the partial derivative with regard to time.

$$\frac{\partial f(t)}{\partial t} = \frac{\partial \mathcal{F}^{-1}\{\boldsymbol{F}(\omega)\}}{\partial t} = \frac{1}{2\pi}\int_{-\infty}^{\infty} [j\omega \boldsymbol{F}(\omega)]e^{j\omega t}d\omega = \mathcal{F}^{-1}\{j\omega \boldsymbol{F}(\omega)\} \tag{A.39}$$

Thus, the derivative in time domain corresponds to the multiplication by a factor of $j\omega$ in the frequency domain. This is the same factor that we would get if the time derivative is taken from harmonic signals (1.26) with time factor $e^{j\omega t}$. Both conventions are in agreement when we choose $e^{j\omega t}$ for the time dependence of the harmonic function and $e^{-j\omega t}$ for the exponential function in the Fourier transform.

Time reversal in the time domain means complex conjugate in frequency domain.

$$\mathcal{F}\{f(-t)\} = \int_{-\infty}^{\infty} f(-t)e^{-\omega t}dt \overset{t'=t}{=} \int_{-\infty}^{\infty} f(t')e^{\omega t'}dt' \tag{A.40}$$

$$\boldsymbol{F}(\omega) = \mathcal{F}\{f(t)\} \Rightarrow \boldsymbol{F}(\omega)^* = \mathcal{F}\{f(-t)\} \tag{A.41}$$

A further important theorem is the shift theorem which states

$$\int_{-\infty}^{\infty} f(t-\tau)e^{j\omega t}dt = \boldsymbol{F}(\omega)e^{j\omega\tau} \tag{A.42}$$

This can be easily proven by exchanging variables. A time delay τ creates a specific phase change for every frequency.

A very important relationship results from the FT of the convolution of two time signals. This is the infinite integral over the product with time delay τ for one function

$$\int_{-\infty}^{\infty} f(t-\tau)g(\tau)d\tau = f(t) * g(x) \tag{A.43}$$

The Fourier transform of the convolution gives:

$$\begin{aligned}
&\int_{-\infty}^{\infty}\left(\int_{-\infty}^{\infty} f(t-\tau)g(\tau)d\tau\right)e^{j\omega t}dt \\
&= \int_{-\infty}^{\infty}\left(\int_{-\infty}^{\infty} f(t-\tau)e^{j\omega t}dt\right)g(\tau)d\tau \\
&= \int_{-\infty}^{\infty}\left(\boldsymbol{F}(\omega)e^{j\omega\tau}\right)g(\tau)d\tau \\
&= \boldsymbol{F}(\omega)\int_{-\infty}^{\infty} g(\tau)e^{j\omega\tau}d\tau \\
&= \boldsymbol{F}(\omega)\boldsymbol{G}(\omega)
\end{aligned} \tag{A.44}$$

or

$$\mathcal{F}\{f(t) * g(t)\} = \mathcal{F}\{f(t)\}\mathcal{F}\{g(t)\} = \boldsymbol{F}(\omega)\boldsymbol{G}(\omega) \tag{A.45}$$

Thus, the Fourier transform of the convolution of two signals is the product of the Fourier transform of each function. It can be further shown that the inverse Fourier transform of a product of spectra is also a convolution in frequency domain.

$$\mathcal{F}\{f(t)g(t)\} = \boldsymbol{F}(\omega) * \boldsymbol{G}(\omega)) = \frac{1}{2\pi}\int_{-\infty}^{\infty} \boldsymbol{F}(\omega_1)\boldsymbol{G}(\omega-\omega_1)d\omega_1 \tag{A.46}$$

A.1.7 Fourier Transformation in Space

The Fourier presentation of time signals can be converted to the wave forms in space. We apply the pair of Fourier transformations from (A.22a) and (A.22b) to the space domain

$$f(x) = \frac{1}{2\pi}\int_{-\infty}^{\infty} F(k)e^{jkx}dk \qquad = \mathcal{F}^{-1}\{F(k\} \qquad \text{(A.47a)}$$

$$F(k) = \int_{-\infty}^{\infty} f(x)e^{-jkt}dx \qquad = \mathcal{F}\{f(x)\} \qquad \text{(A.47b)}$$

This can be further extended to multidimensional applications, for example two dimensional surfaces

$$f(x,y) = \frac{1}{4\pi^2}\int_{-\infty}^{\infty}\int_{-\infty}^{\infty} F(k_x,k_y)e^{j(k_x x+k_y y}dk_x dk_y \qquad = \mathcal{F}^{-1}\{F(k_x,k_y\} \qquad \text{(A.48a)}$$

$$F(k_x,k_y) = \int_{-\infty}^{\infty}\int_{-\infty}^{\infty} f(x,y)e^{-j(k_x x+k_y y)}dxdy \qquad = \mathcal{F}\{f(x,y)\} \qquad \text{(A.48b)}$$

A.2 Discrete Signal Analysis

Today, most acquisition systems and consumer electronics are based on digital systems. The digital representation of analog or continuous signals is advantageous, because the information can be easily stored and transmitted without losses. This is not the case for analog systems such as, for example, the transmission of frequency modulated radio signals. In addition the results of numerical simulation are also digital. However, the digital analysis and especially the spectral analysis of discrete signals leads to specific effects that must be carefully considered if you would like to avoid mistakes or misinterpretations. The most popular and well known effect is the under-sampling of high frequency data that creates artefacts, e.g. wheels in movies that rotate in the wrong direction. The impact of digital signal analysis is separated into two steps:

1. What is the impact of sampling?
2. What is the impact of limited signal length or limited number of samples?

We will not consider the discretization effect of sampling. Analog to digital (AD) converters cannot present the samples in a continuous way. However, due to the vast development of AD converters and digital systems, 24 bit is used even for consumer electronics. This corresponds to a dynamics larger than 16 million or 140 dB, which is often higher than the signal to noise ratio of the sensor–amplifier system.

A.2.1 Fourier Transform of Discrete Signals

For dealing with this phenomena, a mathematical formulation of the sampling process is required. The sampling can be represented by multiplying the continuous signal by a sum of delta functions

$$f[n] = f(n\Delta T) \equiv f_s(t) = f(t) \sum_{n=-\infty}^{\infty} \delta(t - n\Delta T) \tag{A.49}$$

called a delta comb. Here, ΔT is the sampling period. The rectangular bracket denotes the discrete argument n here. In order to derive the effect of this sampling process to the spectrum we apply the Fourier transformation (A.22a) to (A.49)

$$\boldsymbol{F}(e^{j\omega\Delta T}) = \int_{-\infty}^{\infty} \left(f(t) \sum_{n=-\infty}^{\infty} \delta(t - n\Delta T) \right) e^{-j\omega t} dt \tag{A.50}$$

and get with the sifting property

$$\boldsymbol{F}(e^{j\omega\Delta T}) = \sum_{n=-\infty}^{\infty} f[n] e^{-j\omega n\Delta T} \tag{A.51}$$

In Figure A.3 the results of Equation (A.49) are presented graphically. The continuous function is now represented by an infinite train of delta functions. The area of each delta peak at $n\Delta T$ equals the function value at this time according to Equation (A.28), and it is denoted by the length of the arrow.[1]

The sampling has a very peculiar effect on the Fourier transform. Note that the discrete FT has an argument of the form $e^{j\omega\Delta T}$. In order to understand the effect of sampling in the frequency domain, we interpret the delta comb as a periodic signal with period ΔT and the fundamental sampling frequency $\omega_s = \frac{2\pi}{\Delta T}$. The Fourier series in exponential form (A.8) is applied to the delta comb, giving

$$\sum_{n=-\infty}^{\infty} \delta(t - n\Delta t) = \sum_{m=-\infty}^{\infty} a_m e^{jm\omega_s t} \tag{A.52}$$

Figure A.3 Process of sampling by a delta comb function. *Source:* Alexander Peiffer.

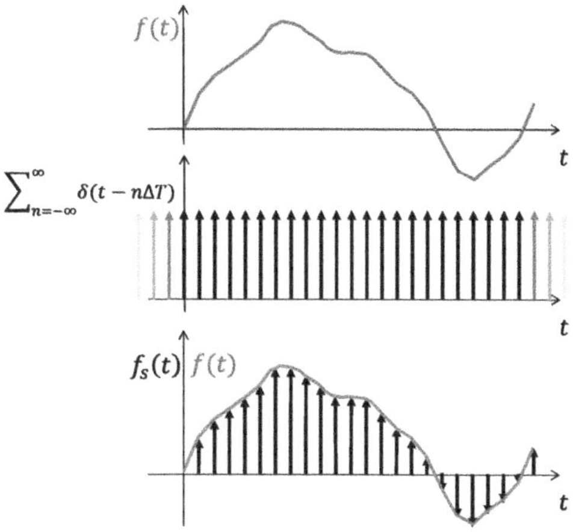

1 This is not the length of the delta peak, because this is infinite

In the Equation (A.15) for the complex Fourier coefficients a_m, there is only one delta peak in the integration interval.

$$a_m = \frac{1}{\Delta T} \int_{-\Delta T/2}^{\Delta T/2} \delta(t) e^{jm\omega_s t} dt = \frac{1}{\Delta T} \tag{A.53}$$

With these coefficients the Fourier series of the delta comb is given by

$$\sum_{n=-\infty}^{\infty} \delta(t - n\Delta T) = \sum_{m=-\infty}^{\infty} \frac{1}{\Delta T} e^{jm\omega_s t} \tag{A.54}$$

Entering this into (A.50)

$$\boldsymbol{F}(e^{j\omega\Delta T}) = \int_{-\infty}^{\infty} \left(f(t) \sum_{m=-\infty}^{\infty} \frac{1}{\Delta T} e^{jm\omega_s t} \right) e^{-j\omega t} dt \tag{A.55}$$

as far as rearranging sum and integration of the Fourier spectrum of the sampled signal leads to the following expression

$$\boldsymbol{F}(e^{j\omega\Delta T}) = \frac{1}{\Delta T} \sum_{m=-\infty}^{\infty} \left(\int_{-\infty}^{\infty} f(t) e^{-j(\omega - m\omega_s)t} dt \right) \tag{A.56}$$

This term in parentheses is the Fourier transform with frequency argument $\omega - m\omega_s$ in the exponential function, hence

$$\boldsymbol{F}(e^{j\omega\Delta T}) = \frac{1}{\Delta T} \sum_{m=-\infty}^{\infty} \boldsymbol{F}(\omega - m\omega_s) \tag{A.57}$$

Thus, the sampling of the function $f(t)$ leads to a periodic repetition of the spectrum $\boldsymbol{F}(\omega)$ at $n\omega_s$ with $n \in \mathbb{Z}$. This effect is illustrated in Figure A.4. The periodic summation of the original spectrum requires a band limited spectrum. If the original spectrum has contributions above $\omega_s/2$ or below $-\omega_s/2$ there will be an overlap. When sampling continuous signals the sampling rate must be at least twice the maximum frequency content of the analog signal to avoid this effect called aliasing. In signal acquisition systems low pass filters make sure that there is no spectral content in the critical range.

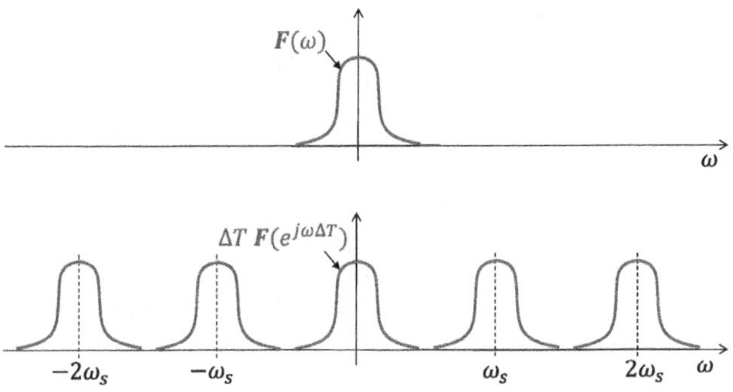

Figure A.4 The Fourier transform of the sampled signal with its periodically repeated continuous spectrum at intervals of ω_s. *Source:* Alexander Peiffer.

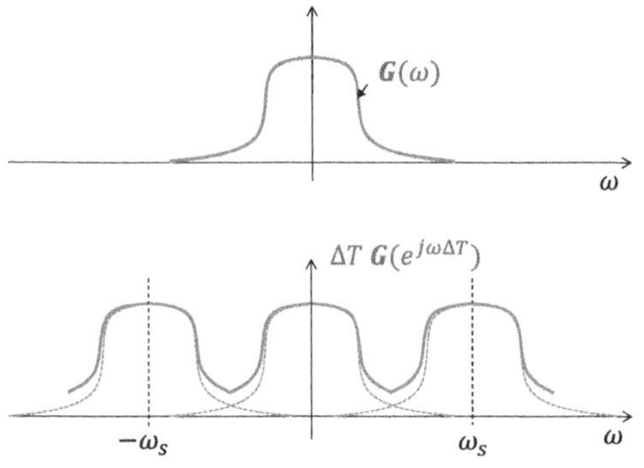

Figure A.5 The Fourier transform of a sampled signal with overlapping spectra due to spectral content above the maximum sampling frequency. *Source:* Alexander Peiffer.

In Figure A.5 the spectrum of the sampled signal is shown when the analog signal has frequency contributions above the maximum frequency $\omega_s/2$.

We can conclude on the first question: What is the effect of sampling? The sampling process gives a discrete set of data values approximating the continuous signal. The frequency domain of the original spectrum is converted into a periodic spectrum with frequency interval $\omega_s = 2\pi/\Delta T$.

A.2.2 The Discrete Fourier Transform

As digital signals will not contain an unlimited number of samples, the infinite sum has to be replaced by a sum over N samples, and we get the discrete Fourier transform (DFT).

$$\hat{F}(e^{j\omega\Delta T}) = \sum_{n=0}^{N-1} f[n]e^{j\omega n\Delta t} \tag{A.58}$$

From this formula all continuous values of ω can be calculated. But, there is evidence that a spectrum derived from N samples cannot contain more information as is given by these N samples. So, we have to select N points in the spectrum. A natural choice for this is to select N samples in the spectrum between 0 and ω_s

$$\omega_k = k\Delta\omega = k\frac{2\pi}{N\Delta T} \text{ with } k = 0 \dots N-1 \tag{A.59}$$

and the discrete Fourier transform reads

$$F[k] = \sum_{n=0}^{N-1} f[n]e^{j2\pi nk/N} \tag{A.60}$$

It can be proven (see e.g. Oppenheim et al., 1999) that we can exactly recover the samples $f[n]$ by

$$f[n] = \frac{1}{N}\sum_{k=0}^{N-1} F[k]e^{j2\pi nk/N} \quad \forall \quad 0 \leq n < N-1 \tag{A.61}$$

To conclude, the finite number of samples (and not the sampling procedure) causes a finite number of samples in the Discrete Fourier transform. In numerical tools the above formulas of the DFT are implemented in a numerically more efficient way: the

Fast Fourier transform (FFT). A special algorithm is implemented that recursively calculates the DFT but only for $N = 2^M$. Newer developments allow for arbitrary lengths by using the FFTW algorithm (Frigo and Johnson, 2005). We don't care about the details, because most signal analysis packages in Matlab or NumPy provide methods for calculating the FFT efficiently.

A.2.3 Windowing

In principle we have all the means to investigate time signals and their spectral content numerically. There is one remaining issue that is worth mentioning. The spectrum is sampled at discrete values. What happens when we analyze a harmonic signal of frequency ω that is, for example, between two sampling points ω_k and ω_{k+1}?

In order to illustrate this, we switch back to continuous signals. The finite number of samples of the DFT can be interpreted as a rectangular window function. So we multiply the time signal with a function $w(t)$.

$$f_w(t) = f(t)w(t) \tag{A.62}$$

According to (A.46) this results in a convolution in frequency domain

$$F_w(\omega) = \frac{1}{2\pi}\int_{-\infty}^{\infty} F(\omega_1)W(\omega - \omega_1)d\omega_1 \tag{A.63}$$

In other words, the spectrum of a windowed function is the convolution of the FT with the FT of the window function. The FT of a rectangular window that is defined as follows

$$w_r(t) = \begin{cases} 1 & -\frac{T_w}{2} \le t \le \frac{T_w}{2} \\ 0 & t < -\frac{T_w}{2}; t > \frac{T_w}{2} \end{cases} \tag{A.64}$$

has the FT in the form of a sinc function

$$W(\omega) = T_w \frac{\sin(\omega T_w/2)}{\omega T_w/2} = T_w sinc(\omega T_w/2) \tag{A.65}$$

Consider now the spectrum of a cosine function (A.38) that consists of two symmetric peaks. When convoluted with the FT of a window function, the spectrum looks as shown in Figure A.6 where the FT of the window appears at the positions of the delta function. The peak of the window function is given by

$$W(0) = \int_{-\infty}^{\infty} w(t)dt \tag{A.66}$$

being T_w for the rectangular window. So, the amplitude of the cosine function can be derived from the peak value if the maximum is divided by $W(0)/\pi$.

In Figure A.6b the shape of the windows FT is shown, and the sampling of the related discrete FT is denoted by blue dots. The sampling occurs exactly at the zeros of the sinc function. Thus, when the frequencies of the cosine function fit exactly to one spectral sampling value, you get a single peak. What happens if this is not the case? In this case the sampling occurs at the sides of the window peak that is very steep so that the amplitude is underestimated. This is called spectral leakage.

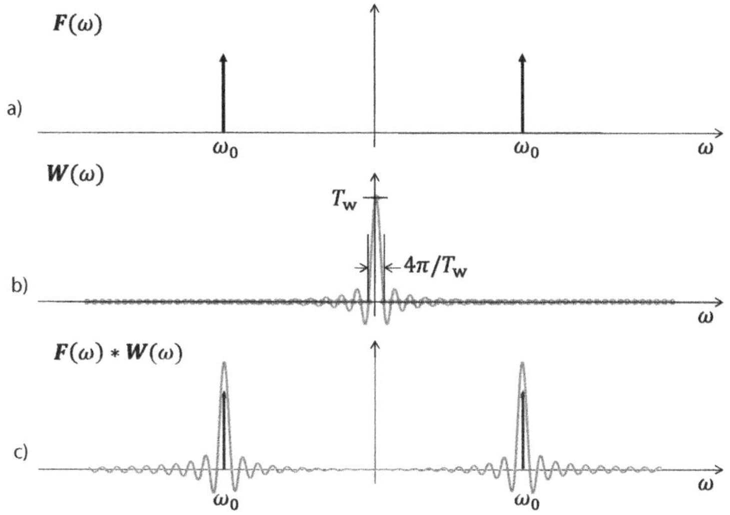

Figure A.6 Effect of windowing on the spectrum of the cosine function.
Source: Alexander Peiffer.

For avoiding this there are two options:

1. We increase the spectral sampling rate by extending the time interval artificially by adding zero values to the time signals, the so-called zero-padding.
2. We use a window function with a broader peak in the frequency domain to mitigate the leakage effect.

An often used window function is the Hanning window given by

$$w_{hann}(t) = \begin{cases} \cos^2(\frac{\pi t}{T_w}) & -\frac{T_w}{2} \leq t \leq \frac{T_w}{2} \\ 0 & t < -\frac{T_w}{2}; t > \frac{T_w}{2} \end{cases} \tag{A.67}$$

The effect of this broader window is sketched in Figure A.7. The leakage is reduced, but a certain underestimation of the amplitude is still possible.

Mitigating spectral leakage requires compromises; either you need fine spectral resolution, but then you accept high leakage. When choosing a broad peak you reduce the leakage but you loose spectral resolution. A further option is to take a higher number of samples or increase the time length T_w of the window. However, this is not always possible.

The sample interval in the spectrum is $\Delta\omega = \frac{2\pi}{N\Delta T} = \frac{2\pi}{T_w}$, and this doesn't depend on the sampling rate of the time signal but on the size of the time window $T_w = N\Delta T$. The highest considerable frequency sample is $\omega_s/2 = \pi/\Delta T$. The sampling rate influences the width of the spectrum or the maximum frequency value that can be considered.

A.3 Coordinate Transformation of Discrete Equation of Motion

The conversion between coordinate systems is a useful means to reduce the size of the numerical problem, for example to change from global coordinates to local dynamic coordinates with an easier formulation of the dynamics.

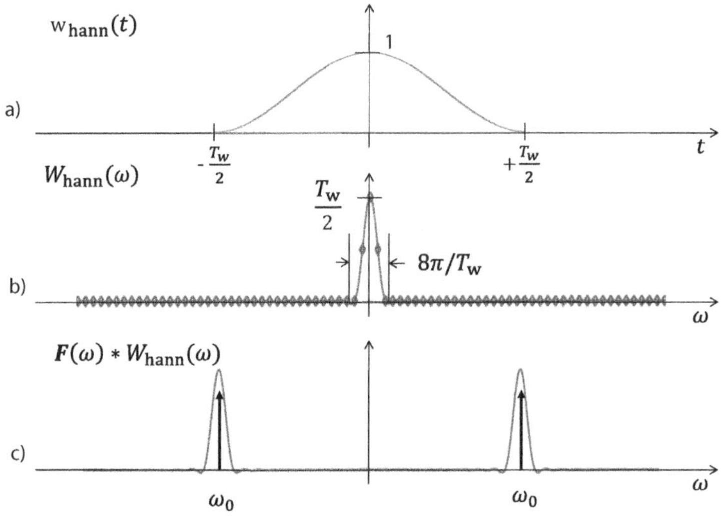

Figure A.7 Cosine function spectrum with Hanning windowing function. *Source:* Alexander Peiffer.

The equation of motion in discrete form

$$[D]\{q\} = \{f\} \tag{A.68}$$

shall be converted into the coordinates q' defined by

$$\{q\} = [T]\{q'\} \text{ and } \{q'\} = [T]^{-1}\{q\} \tag{A.69}$$

Here, $[T]$ is the matrix of new base vectors, e.g. the mode shapes. Entering the left hand side of (A.69) into (A.68) reads

$$[D][T]\{q'\} = \{f\} \tag{A.70}$$

and multiplying from the left with $[T]^{-1}$ gives the final converted form:

$$[T]^{-1}[D][T]\{q'\} = [T]^{-1}\{f\} \tag{A.71}$$

$$[D']\{q'\} = \{f'\} \tag{A.72}$$

So, the matrix and force are converted into the new system by

$$[D'] = [T]^{-1}[D][T] \qquad \{f'\} = [T]^{-1}\{f\} \tag{A.73}$$

Note that from Equation (A.69), the matrix must be invertible and therefore consist of linear independent vectors.

Bibliography

M. Frigo and S. G. Johnson The Design and Implementation of FFTW3. *Proceedings of the IEEE*, 93(2):216–231, 2005. ISSN 0018-9219.

Alan V. Oppenheim, Ronald W. Schafer, and John R. Buck. *Discrete-Time Signal Processing*. Prentice Hall Signal Processing Series. Prentice-Hall, Upper Saddle River, NJ, second edition, internat. ed edition, 1999. ISBN 978-0-13-083443-0.

Appendix B

Specific Solutions

B.1 Second Moments of Area

The bending wave behavior of beams is determined by the cross-section of the beam. We saw in Equations (3.94)–(3.95) that those quantities are a result of an area integration for the infinitesimally small area times the lever related to the neutral axis. There are some definitions and rules that are mandatory to deal with beam dynamics. Figure B.1 shows a typical cross section from building construction. The I or double T-beam. The second moments of area for the axis through the centroid and thus neutral axes are

$$I'_{yy} = \int_A z'^2 dy'dz' \tag{B.1}$$

$$I'_{zz} = \int_A y'^2 dy'dz' \tag{B.2}$$

$$I'_{yz} = I_{zy} = \int_A y'z'dy'dz' \tag{B.3}$$

Here, the first two terms are called the second moment of area with respect to the y' and z' axes, respectively. The last term is called the product moment of area. There is a further moment linked to the torsion of the beam, the polar moment of area J_{xx}

$$J_{xx} = \int_A r^2 dy'dz' \tag{B.4}$$

The polar moment is related to the area moment by the perpendicular axis theorem:

$$J_{xx} = \int_A r^2 dy'dz' = \int_A (y'^2 + z'^2)dy'dz' = I_{yy} + I_{zz} \tag{B.5}$$

If, for instance, a beam is made up of several standard sections or it is mounted on a plate that defines the neutral axis, the transformation to different axes is required. The moments of area regarding the axes y_0 and z_0 are:

$$I_{yy} = Az_0^2 + I_{yy'} \tag{B.6}$$

$$I_{zz} = Ay_0^2 + I_{zz'} \tag{B.7}$$

$$I_{yz} = Az_0y_0 + I_{yz'} \tag{B.8}$$

Vibroacoustic Simulation: An Introduction to Statistical Energy Analysis and Hybrid Methods, First Edition. Alexander Peiffer.

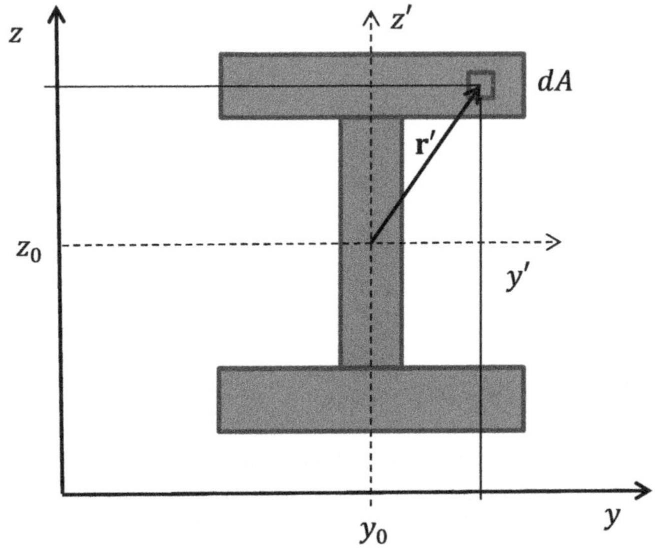

Figure B.1 Cross section of I-beam with axes x', z' coinciding with the centroid. *Source:* Alexander Peiffer.

In addition one might rotate the coordinate system around the center by α leading to the new axes y'', z''. The updated moments are

$$I_{yy''} = \frac{1}{2}(I_{yy'} + I_{zz'}) + \frac{1}{2}(I_{yy'} - I_{zz'})\cos 2\alpha - I_{yz'}\sin 2\alpha \tag{B.9}$$

$$I_{zz''} = \frac{1}{2}(I_{yy'} + I_{zz'}) - \frac{1}{2}(I_{yy'} - I_{zz'})\cos 2\alpha + I_{yz'}\sin 2\alpha \tag{B.10}$$

$$I_{yz''} = \frac{1}{2}(I_{yy'} - I_{zz'})\sin 2\alpha + J_{yz'}\cos 2\alpha \tag{B.11}$$

From Equation (B.11) it follows that you can always find an angle α with $I_{yz''} = 0$

$$\tan 2\alpha = \frac{2I_{yz'}}{I_{zz'} - I_{yy'}} \tag{B.12}$$

Those axes that fulfill this requirement are called the principle axes of the cross-section. For symmetric sections they coincide with the axes of symmetry.

B.2 Wave Transmission

In contrast to the hybrid coupling loss factor formulation from section 7.3 based on the diffuse field reciprocity, the wave based theory relies on the transmission and reflection of plane waves. This theory was derived by Langley and Heron (1990) for plate edges and from Langley and Shorter (2003) for point connections. For plane acoustic waves a similar approach was shown in section 2.6 for the introduction of the wave transmission coefficient. The coupling loss factor can equivalently be derived from the transmission coefficient. For one- and two-dimensional junctions, the diffuse field transmission must then be calculated by averaging over the appropriate angles.

For plates we have four wave types, and the transmission and reflection is rather complex and must be expressed by several transmission coefficients. For arbitrary line junctions the extension to plane wave transmission based on the theory shown in section 8.2.5.1 is achieved by extending the solution of pure radiating waves in Equations (8.115)–(8.118) by selecting one specific incoming wave from

$$\begin{Bmatrix} \boldsymbol{u} \\ \boldsymbol{v} \\ \boldsymbol{w} \\ \boldsymbol{\beta}_x \end{Bmatrix} = \begin{bmatrix} k_x & j\mu_S & 0 & 0 \\ j\mu_L & -k_x & 0 & 0 \\ 0 & 0 & 1 & 1 \\ 0 & 0 & \mu_{B1} & \mu_{B2} \end{bmatrix} \begin{Bmatrix} \boldsymbol{\Psi}_L\, e^{\mu_L y} \\ \boldsymbol{\Psi}_S\, e^{\mu_S y} \\ \boldsymbol{\Psi}_{B1} e^{\mu_{B1} y} \\ \boldsymbol{\Psi}_{B2} e^{\mu_{B1} y} \end{Bmatrix} e^{-jk_x x} \tag{B.13}$$

For $y = 0$ we get the edge formulation as in Equation (8.133). In short form the out-going wave field of (B.13) reads:

$$\{\boldsymbol{q}(x,y)\} = [\boldsymbol{T}_\Psi]\{\boldsymbol{\Psi}_{\text{out}}(x,y)\} \tag{B.14}$$

Applying the differential equations of the free edge (8.108a)–(8.108d) to the above solution of four radiating waves gives the equations of motion in wave amplitude coordinates as given by the block matrices from (8.120) and (8.123).

$$\begin{Bmatrix} \boldsymbol{F}'_x \\ \boldsymbol{F}'_y \\ \boldsymbol{F}'_z \\ \boldsymbol{M}'_x \end{Bmatrix} = [\boldsymbol{S}'_{\text{dir}}] \begin{Bmatrix} \boldsymbol{\Psi}_L \\ \boldsymbol{\Psi}_S \\ \boldsymbol{\Psi}_{B1} \\ \boldsymbol{\Psi}_{B2} \end{Bmatrix} \tag{B.15}$$

$$[\boldsymbol{S}'_{dir}] = \dots$$

$$\begin{bmatrix} -2S\mu_L k_x & jS(k_S^2 - 2k_x^2) & 0 & 0 \\ jS(k_S^2 - 2k_x^2) & -Sk_x\mu_S & 0 & 0 \\ 0 & 0 & B(\mu_{B1}^3 - (2-\nu)\mu_{B1}k_x^2) & B(\mu_{B2}^3 - (2-\nu)\mu_{B2}k_x^2) \\ 0 & 0 & B(\nu k_x^2 - \mu_{B1}^2) & B(\nu k_x^2 - \mu_{B2}^2) \end{bmatrix}$$

The inverse wave transformation matrix provides the radiation stiffness in edge coordinates

$$\{\boldsymbol{F}'_e\} = [\boldsymbol{S}'_{\text{dir}}][\boldsymbol{T}_\Psi]^{-1}\{\boldsymbol{q}_{e,\text{out}}\} = [\boldsymbol{D}'_{dir}]\{\boldsymbol{q}_{e,\text{out}}\} \tag{B.16}$$

For plates with purely out-going waves, the above equation is valid. Imagine a plate that carries out-going waves and one incoming wave of the form:

$$\{\boldsymbol{q}_{\text{in},L}(x,y)\} = \Psi_{L,in} \begin{Bmatrix} k_x \\ j\mu_{\text{in},L} \\ 0 \\ 0 \end{Bmatrix} e^{\mu_{\text{in},L} y} e^{-jk_x x} \tag{B.17}$$

$$\{\boldsymbol{q}_{\text{in},S}(x,y)\} = \Psi_{\text{in},S} \begin{Bmatrix} j\mu_{\text{in},S} \\ -k_x \\ 0 \\ 0 \end{Bmatrix} e^{\mu_{\text{in},S} y} e^{-jk_x x} \tag{B.18}$$

$$\left\{\boldsymbol{q}_{\text{in},B}(x,y)\right\} = \Psi_{\text{in},B2}\begin{Bmatrix} 0 \\ 0 \\ 1 \\ \mu_{\text{in},B2} \end{Bmatrix} e^{\mu_{\text{in},B2}y}e^{-jk_xx} \tag{B.19}$$

For the incoming waves the evanescent bending wave $B1$ can be omitted because it cannot propagate. The incoming waves change in sign and are related to the outgoing by

$$\mu_L = -\mu_{\text{in},L} \qquad \mu_S = -\mu_{\text{in},S} \qquad \mu_{B2} = -\mu_{\text{in},B2}$$

The wave field on the plate carrying both waves, one incoming and all out-going waves, can be described by:

$$\{\boldsymbol{q}(x,y)\} = [\boldsymbol{T}_\Psi]\{\boldsymbol{\Psi}_{\text{out}}(x,y)\} + \{\boldsymbol{q}_{\text{in}}(x,y)\} \tag{B.20}$$

Applying boundary conditions (8.108a)–(8.108d) on the full wave field gives for the specific edge force

$$\{\boldsymbol{F}'_e\} = [\boldsymbol{S}'_{\text{dir}}]\{\boldsymbol{q}_\Psi\} + \{\boldsymbol{F}'_{\text{in}}\} = [\boldsymbol{S}'_{\text{dir}}][\boldsymbol{T}_\Psi]^{-1}\left\{\boldsymbol{q}_{e,out}\right\} + \{\boldsymbol{F}'_{\text{in}}\} \tag{B.21}$$

$$= [\boldsymbol{D}'_{\text{dir}}]\left\{\boldsymbol{q}_{e,out}\right\} + \{\boldsymbol{F}'_{\text{in}}\} \tag{B.22}$$

The edge motion is a combination of both waves. Hence,

$$\{\boldsymbol{q}_e\} = \left\{\boldsymbol{q}_{e,out}\right\} + \left\{\boldsymbol{q}_{e,in}\right\}$$

and using this gives

$$\{\boldsymbol{F}'_e\} = [\boldsymbol{D}'_{\text{dir}}]\left(\{\boldsymbol{q}_e\} - \left\{\boldsymbol{q}_{e,in}\right\}\right) + \{\boldsymbol{F}'_{\text{in}}\} \tag{B.23}$$

$$= [\boldsymbol{D}'_{\text{dir}}]\{\boldsymbol{q}_e\} - [\boldsymbol{D}'_{\text{dir}}]\left\{\boldsymbol{q}_{e,in}\right\} + \{\boldsymbol{F}'_{\text{in}}\} \tag{B.24}$$

When no external forces are acting on the edge we may rewrite

$$[\boldsymbol{D}'_{\text{dir}}]\{\boldsymbol{q}_e\} = [\boldsymbol{D}'_{\text{dir}}]\left\{\boldsymbol{q}_{e,in}\right\} - \{\boldsymbol{F}'_{\text{in}}\} \tag{B.25}$$

So, the right hand side of the above equation may be interpreted as the blocked force generated by the incoming wave

$$\{\boldsymbol{F}'_{b,in}\} = [\boldsymbol{D}'_{\text{dir}}]\left\{\boldsymbol{q}_{e,in}\right\} - \{\boldsymbol{F}'_{\text{in}}\} \tag{B.26}$$

If we assume the mth plate carrys the incoming wave, the global edge motion of the full junction follows from this force transformed into the global edge system

$$\{\boldsymbol{F}'_{b0,in}\} = [T^{(m)}]\{\boldsymbol{F}'_{bm,in}\} \tag{B.27}$$

acting on the total stiffness of the full line junction (8.129)

$$[\boldsymbol{D}'_{\text{tot}}]_{e0}\{\boldsymbol{q}_{e0}\} = \{\boldsymbol{F}'_{b0,in}\} \tag{B.28}$$

The solution of (B.28) provides the edge motion in global coordinates. Rotating into the coordinate system of the nth plate and applying the wave transformation matrix provides the amplitude of each wave:

$$\{\boldsymbol{q}_{\Psi n}\} = \left[\boldsymbol{T}_{\Psi}^{(n)}\right]\left[T^{(n)}\right]\{\boldsymbol{q}_{e0}\} \tag{B.29}$$

For the determination of the transmission coefficient we must calculate the radiated and irradiated power. Due to the infinite length of the line, this must be a length intensity. In addition, it is weighted by the angle φ_m to consider only the component orthogonal to the line edge.

$$\Pi'_{m,j} = E'' c_{\mathrm{gr},mv} \sin(\varphi_v^{(m)}) \tag{B.30}$$

With m denoting the subsystem, v the wave type (L, S, B). For example $c_{\mathrm{gr},mv}$ is the group velocity of the v-wave of the mth plate. The energy density follows from the area density $M'' = \rho_0 h$ and the displacement amplitudes with[1] $E'' = \frac{1}{2}\rho_0 h \hat{\Psi}^2$

Note that due to the formulation of the in-plane wave propagation, Ψ_v is not the displacement amplitude in these cases. The amplitudes are $k_L\Psi_L$, $k_S\Psi_S$ and Ψ_{B2} for longitudinal, shear and bending waves, respectively.

Finally the intensities for each wave type are

$$\Pi'_L = \frac{1}{2}\rho_0 h\omega^3 k_L \Psi_L^2 \sin\vartheta_L \tag{B.31}$$

$$\Pi'_S = \frac{1}{2}\rho_0 h\omega^3 k_S \Psi_S^2 \sin\vartheta_S \tag{B.32}$$

$$\Pi'_B = \rho_0 h\omega^3 / k_B \Psi_B^2 \sin\vartheta_B \tag{B.33}$$

The incoming wave and angle drives the wavenumber k_x; the transmitted angles follow from the transmitted wavenumbers

$$\cos\varphi_v^{(m)} = k_x / k_{mv} \tag{B.34}$$

Finally the transmission coefficient follows from the ratio of transmitted to irradiated intensity

$$\tau_{mj,nk} = \frac{\Pi'_{nk}(k_x)}{\Pi'_{mj}(k_x)} \tag{B.35}$$

When $k_x > k_{mv}$ there will be no solution for $\varphi_v^{(m)}$ and $\tau = 0$. So, not all wave types are available for all angles. For example in Figure B.2 the full angle of the shear wave leads to a minimal angle that does not start at zero, because longitudinal waves occur first when the projected wavenumber k_x is equal to or smaller than kL.

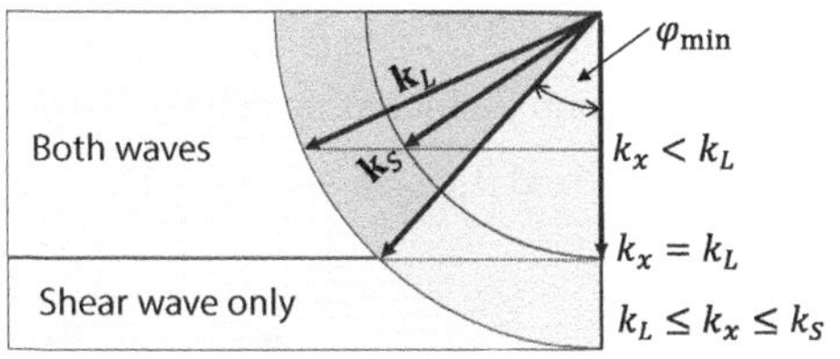

Figure B.2 Angle range for shear wave with given 0–90° range of longitudinal wave. *Source:* Alexander Peiffer.

1 Keep in mind that for all quantities, the figures for the relevant plate must be chosen. For better readability the subsystem and wave indexing are not used in all equations.

The wave transmission theory has one advantage: it is also valid for strong coupling, for example a transmission factor of $\tau = 1$ when similar plates are coupled together at an angle of 180°. The diffuse field reciprocity requires a diffuse field, which is hard to achieve with fully coupled and therefore fully absorbing boundary conditions.

B.2.1 The Blocked Forces Interpretation

The following term was described in Section B.2 as the blocked force

$$\{\boldsymbol{F}'_b\} = [\boldsymbol{D}'_{\text{dir}}]\{\boldsymbol{q}_{e,in}\} - \{\boldsymbol{F}'_{\text{in}}\} \tag{B.36}$$

It is worth illustrating this fact in more detail. The incoming wave generates the force $\boldsymbol{F}'_{\text{in}}$ given by the boundary conditions of the edge and an edge motion $\boldsymbol{q}_{e,in}$. In order to achieve the bocked boundary condition, a force must act in such a way on the boundary that it excites an out-going that matches exactly the incoming wave with negative sign, thus:

$$\{\boldsymbol{q}_{e,out}\} = -\{\boldsymbol{q}_{e,in}\} \tag{B.37}$$

This force is given by

$$\{\boldsymbol{F}_{e,out}\} = [\boldsymbol{D}'_{\text{dir}}]\{\boldsymbol{q}_{e,out}\} = -[\boldsymbol{D}'_{\text{dir}}]\{\boldsymbol{q}_{e,in}\} \tag{B.38}$$

So in total, the edge must act with force $\{\boldsymbol{F}'_{\text{in}}\} - [\boldsymbol{D}'_{\text{dir}}]\{\boldsymbol{q}_{e,in}\}$ in order to block the edge motion of an incoming wave. From the force balance follows that the incoming and reflected wave acts with the force as shown in Equation (B.26).

There are further consequences. First, for the calculation of waves that are reflected in one sense but *transmitted* into a different wave type but the same plate, the reflected blocked force components must by subtracted. The easiest way to do so is to subtract $\boldsymbol{q}_{e,in}$ from the global solution in Equation (B.28). Second, for the realization of the blocked force of one wave type, the in-plane wave requires both in-plane waves. Thus, there is an inert exchange of energy between both subystems that breaks the rule for the diffuse field reciprocity.

In the remainder of Section B.2, the blocked forces for each wave type are derived in detail. According to the diffuse field reciprocity (7.19), this blocked force is also given by the free field radiation stiffness of each wave type. This separation into waves is clear for bending waves, but not for the in-plane waves where the edge motion usually generates a combination of both wave types, shear and longitudinal waves.

B.2.2 Bending Waves

The wave solution for incoming bending waves relies on the positive and real wavenumber solution k_B and $\mu_{\text{in},B2} = +\sqrt{k_x^2 - k_B^2} = -\mu_{B2}$. Using this and setting the evanescent

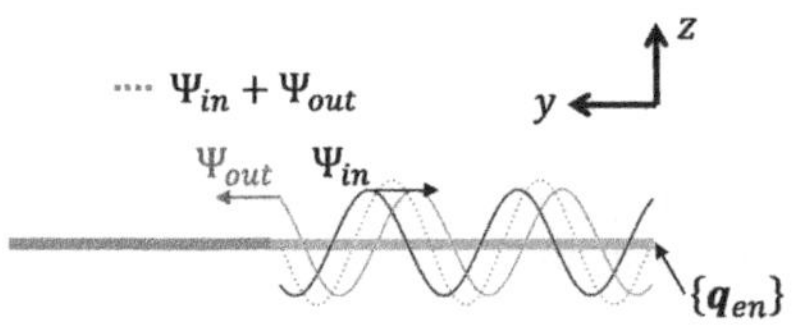

Figure B.3 In-coming and out-going waves assuring the blocked condition at boundary $y = 0$. *Source:* Alexander Peiffer.

wave $\boldsymbol{\Psi}_{B1} = 0$, the incoming wave reads as

$$\boldsymbol{w} = \boldsymbol{\Psi}_{B2} e^{-\mu_{B2} y} e^{-jk_x x + j\omega t} \tag{B.39}$$

$$\beta_x = \mu_{B2} \boldsymbol{\Psi}_{B2} e^{-\mu_{B2} y} e^{-jk_x x + j\omega t} \tag{B.40}$$

With $y = 0$ we get the edge displacement:

$$\left\{\boldsymbol{q}_{e,in}\right\} = \begin{Bmatrix} \boldsymbol{w}_{e,in} \\ \beta_{x,e,in} \end{Bmatrix} = \boldsymbol{\Psi}_{B2} \begin{Bmatrix} 1 \\ -\mu_{B2} \end{Bmatrix} \tag{B.41}$$

and for the traction force vector

$$\left\{\boldsymbol{F}'_{e,in}\right\} = -\boldsymbol{\Psi}_{B2} B \begin{Bmatrix} \mu_{B2}^3 - (2-\nu) k_x^2 \mu_{B2} \\ \mu_{B2}^2 - \nu k_x^2 \end{Bmatrix} \tag{B.42}$$

The reaction to the edge displacement follows from the radiation stiffness times the edge motion

$$\left[\boldsymbol{D}_{e,\mathrm{dir}}\right]\left\{\boldsymbol{q}_{e,in}\right\} = \boldsymbol{\Psi}_{B2} B \begin{Bmatrix} 2\mu_{B1}\mu_{B2}^2 + \mu_{B1}^2\mu_{B2} + \nu k_x^2 \mu_{B2} \\ -2\mu_{B1}\mu_{B2} - \nu k_x^2 - \mu_{B2}^2 \end{Bmatrix} \tag{B.43}$$

Using these equations in (B.36) gives the blocked reverberant force due to bending waves

$$\left\{\boldsymbol{F}'_{bB}\right\} = 2B\boldsymbol{\Psi}_{B2} \begin{Bmatrix} \mu_{B1}\mu_{B2}^2 + \mu_{B1}^2\mu_{B2} \\ -\mu_{B1}\mu_{B2} - \mu_{B2}^2 \end{Bmatrix} \tag{B.44}$$

For further consideration it is better to replace the propagation constants $\mu_{B1/2}$ by wavenumber expressions

$$\mu_{B1} = -k_{yB1} = -\sqrt{k_x^2 + k_B^2} \qquad \mu_{B1} = -jk_{yB2} = -j\sqrt{k_B^2 - k_x^2} \tag{B.45}$$

Using the this convention

$$\left\{\boldsymbol{F}'_{bB}\right\} = 2B\boldsymbol{\Psi}_{B2} \begin{Bmatrix} k_{yB1}k_{yB2}^2 - jk_{yB1}^2 k_{yB2} \\ -jk_{yB1}k_{yB2} + k_{yB2}^2 \end{Bmatrix} \tag{B.46}$$

leads to the cross spectral density matrix:

$$\left[S_{ff}^*\right]_B'' = \left\langle \boldsymbol{F}'\boldsymbol{F}'^H \right\rangle_E = 8B^2\boldsymbol{\Psi}_{B2}^2 k_{yB2}^2 k_B^2 \begin{bmatrix} k_{yB1}^2 & k_{yB1} \\ k_{yB1} & 1 \end{bmatrix} \tag{B.47}$$

This cross spectrum is per square length, as the force is a specific force per length.

For the consideration of the reciprocity, the modal density of plates and expression for the total energy is required. As explained in section 8.2.4.1, the plane wave formulation with the projected wavenumber as degree of freedom has to be treated as a one-dimensional system.

$$n_{1D}(\omega) = \frac{L}{2\pi c_{\mathrm{gr}}} \tag{B.48}$$

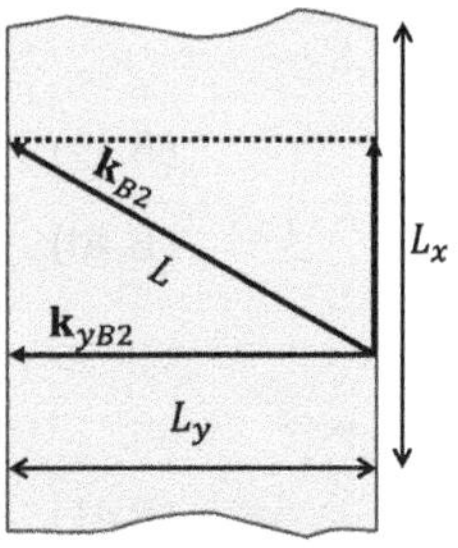

Figure B.4 Effective subsystem length for plane waves in two-dimensional systems.

From Figure B.4 it follows that the effective length L follows from L_y and the wavenumber ratio.

$$\frac{L}{L_y} = \frac{k_B}{k_{yB2}} \tag{B.49}$$

Together with $c_{gr} = 2c_B$, we get the modal density

$$n_{1D}(\omega) = \frac{L_y k_B^2}{4\pi k_{yB2}\omega} \tag{B.50}$$

The energy in the plate subsystem is given by

$$E = \frac{\omega^2}{2}\mathbf{\Psi}_{B2}^2 \rho_0 h A \text{ with } A = L_x L_y \tag{B.51}$$

In this case the energy length density is required, thus

$$E' = \frac{E}{L_x} = \frac{\omega^2}{2}\mathbf{\Psi}_{B2}^2 \rho_0 h L_y \tag{B.52}$$

For the determination of the radiation stiffness, the same convention of the propagation constants (B.45) is used

$$Im\left[\mathbf{D}'_{\text{dir}}\right]_B = BIm\begin{bmatrix} -\mu_{B1}^2\mu_{B2} + \mu_{B1}\mu_{B2}^2 & \mu_{B1}\mu_{B2} - \nu k_x^2 \\ \mu_{B1}\mu_{B2} + \nu k_x^2 & -\mu_{B1} + \mu_{B2} \end{bmatrix} = Bk_{yB2}\begin{bmatrix} k_{B1}^2 & k_{B1} \\ k_{B1} & k_{B1} \end{bmatrix}.$$

With the modified length specific diffuse field reciprocity (7.19) we get finally:

$$\begin{aligned} \left[S_{ff}\right]''_B &= \frac{4E'}{\pi\omega n_{1D}(\omega)} Im\left[\mathbf{D}'_{\text{dir}}\right]_B \\ &= 8\mathbf{\Psi}_{B2}^2 k_{yB2} k_B^2 \underbrace{\frac{\rho_0 h\omega^2}{k_B^4}}_{=B} Im\left[\mathbf{D}'_{\text{dir}}\right]_B \\ &= 8B^2\mathbf{\Psi}_{B2}^2 k_{yB2}^2 k_B^2 \begin{bmatrix} k_{yB1}^2 & k_{yB1} \\ k_{yB1} & 1 \end{bmatrix} \end{aligned} \tag{B.53}$$

This proves the diffuse field reciprocity in case of bending waves of homogeneous plates.

B.2.3 Longitudinal Waves

The wave solution for incoming longitudinal waves relies on the positive solution $\mu_{\text{in},L} = +\sqrt{k_x^2 - k_L^2} = -\mu_L$. Using this the incoming wave reads as:

$$\boldsymbol{u} = \boldsymbol{\Psi}_L k_x e^{-\mu_L y} e^{-jk_x x + j\omega t} \tag{B.54}$$

$$\boldsymbol{v} = \boldsymbol{\Psi}_L - j\mu_L e^{-\mu_L y} e^{-jk_x x + j\omega t} \tag{B.55}$$

with $y = 0$ we get the edge displacement

$$\left\{\boldsymbol{q}_{e,in}\right\} = \begin{Bmatrix} \boldsymbol{u}_{e,in} \\ \boldsymbol{v}_{e,in} \end{Bmatrix} = \boldsymbol{\Psi}_L \begin{Bmatrix} k_x \\ -j\mu_L \end{Bmatrix} \tag{B.56}$$

and for the traction force vector

$$\{\boldsymbol{F}'_{e,in}\} = -\boldsymbol{\Psi}_L S \begin{Bmatrix} -2k_x\mu_L \\ j(2k_x^2 - k_x^2) \end{Bmatrix} \tag{B.57}$$

The reaction to the edge displacement follows from the radiation stiffness times the edge motion

$$\left[\boldsymbol{D}_{e,dir}\right]\left\{\boldsymbol{q}_{e,in}\right\} = \boldsymbol{\Psi}_L S \begin{Bmatrix} 2k_x\mu_L\left(1 - \frac{k_s^2}{k_x^2 - \mu_S\mu_L}\right) \\ -j\left[2k_x^2 + k_s^2\left(1 - \frac{2k_x^2}{k_x^2 - \mu_S\mu_L}\right)\right] \end{Bmatrix} \tag{B.58}$$

Using these equations in (B.36) gives the blocked force due to longitudinal waves

$$\{\boldsymbol{F}'_{bL}\} = \frac{2S\boldsymbol{\Psi}_L k_s^2}{k_x^2 - \mu_S\mu_L} \begin{Bmatrix} -k_x\mu_L \\ j\mu_S\mu_L \end{Bmatrix} \tag{B.59}$$

For further consideration the propagation constants $\mu_{L/S}$ are replaced by wavenumber expressions

$$\mu_L = -jk_{yL} = -j\sqrt{k_L^2 - k_x^2} \qquad \mu_S = -jk_{yS} = -j\sqrt{k_S^2 - k_x^2} \tag{B.60}$$

Using this convention

$$\{\boldsymbol{F}'_{bL}\} = \frac{2S\boldsymbol{\Psi}_L k_s^2}{k_x^2 - \mu_S\mu_L} \begin{Bmatrix} +jk_x k_{yL} \\ -jk_{yS}k_{yL} \end{Bmatrix} \tag{B.61}$$

leads to the cross spectral density matrix

$$\left[S_{ff}^*\right]_L'' = \left\langle \boldsymbol{F}'\boldsymbol{F}'^H \right\rangle_E = \frac{4S^2\boldsymbol{\Psi}_L^2 k_s^4 k_{yL}^2}{(k_x^2 + k_{yS}k_{yL})^2} \begin{bmatrix} k_x^2 & -k_{yS}k_x \\ -k_{yS}k_x & k_{yS}^2 \end{bmatrix} \tag{B.62}$$

With the effective length

$$\frac{L}{L_y} = \frac{k_L}{k_{yL}} \tag{B.63}$$

and the fact that for L and S-waves, the group velocity equals the phase velocity $c_{gr} = c_L$, we get the modal density

$$n_{1D}(\omega) = \frac{L_y k_L^2}{2\pi k_{yL}\omega} \tag{B.64}$$

The energy in the plate subsystem is given by

$$E = \frac{1}{2}(\Psi_L k_L)^2 \rho_0 h A \text{ with } A = L_x L_y \tag{B.65}$$

Note that the amplitude for in-plane waves is given as a product including the wavenumber. So, the longitudinal wave amplitude displacement is given by $\hat{\Psi}_L k_L$. The energy length density reads as

$$E' = \frac{E}{L_x} = \frac{1}{2}(\Psi_L k_L)^2 \rho_0 h L_y \tag{B.66}$$

The diffuse field reciprocity requires the following expression to relate the radiation stiffness to the cross spectrum

$$\frac{4E'}{\pi \omega n_{1D}(\omega)} = 4\rho_0 h \Psi_L^2 \omega^2 k_{yL}$$

In the above equation the mass per area can be replaced using the expression for the wavenumber (3.167)

$$k_L^2 = \frac{\omega^2(1+\nu)(1-\nu)\rho_0}{E} = \frac{\omega^2(1-\nu)\rho_0 h}{2S}$$

With the identity $k_s^2 = 2k_L^2/(1-\nu)$ we get

$$\frac{4E'}{\pi \omega n_{1D}(\omega)} = \frac{4S\, 2k_L^2 \Psi_L^2 k_{yL}}{(1-\nu)} = 4S\Psi_L^2 k_s^2 k_{yL} \tag{B.67}$$

With the modified length specific diffuse field reciprocity (7.19) we get finally

$$\begin{aligned} Im\left[\boldsymbol{D}'_{\text{dir}}\right]_L &= \frac{\left[S_{ff}\right]_L}{4S\Psi_L^2 k_s^2 k_{yL}} \\ &= \frac{S k_s^2 k_{yL}}{(k_x^2 + k_{yS}k_{yS})^2} \begin{bmatrix} k_x^2 & -k_{yS}k_x \\ -k_{yS}k_x & k_{yS}^2 \end{bmatrix} \end{aligned} \tag{B.68}$$

This matrix can be used as imaginary of the radiation stiffness, but this is not derived from diffuse field reciprocity. The radiation stiffness is defined by a coupled motion of both waves, making it impossible to identify a pure longitudinal wave radiation.

B.2.4 Shear Waves

Following the similar procedure as in section B.2.3 with an incoming shear wave, the blocked force reads

$$\{F'_{bS}\} = \frac{2S\Psi_S k_s^2}{k_x^2 - \mu_S \mu_L} \begin{Bmatrix} j\mu_S \mu_L \\ k_x \mu_S \end{Bmatrix} \tag{B.69}$$

As illustrated in Figure B.2, there is an angle or wavenumber range where both propagation constants are complex, but also a range where μ_L is real and only μ_S remains imaginary. For $k_x \leq k_L$ we use (B.60)

$$\{\boldsymbol{F}'_{bS}\} = \frac{2S\boldsymbol{\Psi}_S k_S^2}{k_x^2 - \mu_S\mu_L} \begin{Bmatrix} -jk_{yS}k_{yL} \\ -jk_x k_{yS} \end{Bmatrix} \tag{B.70}$$

giving the following blocked force matrix

$$\left[S^*_{ff}\right]''_S = \left\langle \boldsymbol{F}'\boldsymbol{F}'^H \right\rangle_E = \frac{4S^2\boldsymbol{\Psi}_S^2 k_S^4 k_{yS}^2}{(k_x^2 + k_{yS}k_{yL})^2} \begin{bmatrix} k_{yL}^2 & k_{yL}k_x \\ k_{yL}k_x & k_x^2 \end{bmatrix} \tag{B.71}$$

Applying shear wave quantities to the one-dimensional modal density and energy gives

$$\frac{4E'}{\pi\omega n_{1D}(\omega)} = 4\rho_0 h \boldsymbol{\Psi}_S^2 \omega^2 k_{yS} \tag{B.72}$$

Using $\omega^2\rho_0 h = k_S^2 S$ we get

$$\begin{aligned} Im\left[\boldsymbol{D}'_{\text{dir}}\right]_S &= \frac{\left[S_{ff}\right]_S}{4S\boldsymbol{\Psi}_S^2 k_S^2 k_{yS}} \\ &= \frac{Sk_S^2 k_{yS}}{(k_x^2 + k_{yS}k_{yS})^2} \begin{bmatrix} k_{yL}^2 & k_{yL}k_x \\ k_{yL}k_x & k_x^2 \end{bmatrix} \end{aligned} \tag{B.73}$$

In the wavenumber range $k_L < k_x \leq k_S$ only k_S remains real, and we keep the longitudinal propagation constant

$$\{\boldsymbol{F}'_{bS}\} = \frac{2S\boldsymbol{\Psi}_S k_S^2}{k_x^2 + jk_{yS}\mu_L} \begin{Bmatrix} \mu_L k_{yS} \\ -jk_x k_{yS} \end{Bmatrix} \tag{B.74}$$

The cross correlation provides

$$\left[S^*_{ff}\right]''_{S2} = \frac{4S^2\boldsymbol{\Psi}_S^2 k_S^4 k_{yS}^2}{k_x^4 + k_{yS}^2\mu_L^2} \begin{bmatrix} \mu_L^2 & j\mu_L k_x \\ -j\mu_L k_x & k_x^2 \end{bmatrix} \tag{B.75}$$

We note that the diffuse field reciprocity is impossible to fulfill here, because the matrix contains imaginary components. However, if we ignore this fact, and with the diffuse field reciprocity factor (B.72), we get

$$\begin{aligned} Im\left[\boldsymbol{D}'_{\text{dir}}\right]_{S2} &\neq \frac{\left[S_{ff}\right]''_{S2}}{4S\boldsymbol{\Psi}_S^2 k_S^2 k_{yS}} \\ &= \frac{Sk_S^2 k_{yS}}{k_x^4 + k_{yS}^2\mu_L^2} \begin{bmatrix} \mu_L^2 & j\mu_L k_x \\ -j\mu_L k_x & k_x^2 \end{bmatrix} \end{aligned} \tag{B.76}$$

B.2.5 In-plane Waves

Both waves can be combined by adding both radiation imaginary parts of the radiation stiffnesses (B.68) and (B.73)

$$
\begin{aligned}
Im[\mathbf{D}'_{\text{dir}}]_{SL} &= Im[\mathbf{D}'_{\text{dir}}]_L + Im[\mathbf{D}'_{\text{dir}}]_S \\
&= \frac{Sk_s^2}{(k_x^2 + k_{yS}k_{yL})^2}\begin{bmatrix} k_{yL}k_x^2 + k_{yS}k_{yL}^2 & -k_{yS}k_{yL}k_x + k_{yS}k_{yL}k_x \\ -k_{yS}k_{yL}k_x + k_{yS}k_{yL}k_x & k_{yL}k_{yS}^2 + k_{yS}k_x^2 \end{bmatrix} \\
&= \frac{S}{k_x^2 + k_{yS}k_{yL}}\begin{bmatrix} k_{yL}k_s^2 & 0 \\ 0 & k_{yS}k_s^2 \end{bmatrix} \qquad \text{(B.77)}
\end{aligned}
$$

In contrast to this the imaginary of the in-plane stiffness matrix (8.121) reads

$$
Im[\mathbf{D}'_{\text{dir}}]_{SL} = \frac{S}{k_x^2 + k_{yS}k_{yL}}\begin{bmatrix} k_{yL}k_S^2 & -k_x(-2k_{yS}k_{yS} + k_S^2 - 2k_x^2) \\ asym & k_{yS}k_S^2 \end{bmatrix} \qquad \text{(B.78)}
$$

So, except the off diagonal components, the blocked force cross correlation is equal to the imaginary radiation stiffness. This equation is valid for $k_x \leq k_L$; for $k_L < k_x \leq k_S$, Equation (B.76) must be used as there is no propagating longitudinal wave. For the comparison to the combined in-plane stiffness matrix (8.121), we use $\mu_S = -jk_{yS}$ and keep μ_L getting

$$
[\mathbf{D}'_{\text{dir}}]_{SL} = \frac{S}{k_x^2 - jk_{yS}\mu_L}\begin{bmatrix} -\mu_L k_S^2 & -k_x(2k_{yS}\mu_L + j(k_S^2 - 2k_x^2)) \\ asym & jk_{yS}k_S^2 \end{bmatrix} \qquad \text{(B.79)}
$$

Extension of the denominator and extracting the imaginary part gives

$$
Im[\mathbf{D}'_{\text{dir}}]_{SL} = \frac{S}{k_x^4 + k_{yS}^2\mu_L^2}\begin{bmatrix} k_{yS}\mu_L^2 k_S^2 & -k_x(k_x^2 k_S^2 + 2(k_x^4 - \mu_S^2 k_{yS}^2)) \\ asym & k_{yS}k_x^2 k_S^2 \end{bmatrix} \qquad \text{(B.80)}
$$

Besides the off diagonals the equation corresponds also to the result from the blocked forces cross correlation (B.76). Thus, the radiation stiffness including both in-plane waves can be used for the calculation of the cross correlation function from the diffuse field reciprocity.

B.3 Conversion Formulas of Transfer Matrix

The transfer matrix method is similar to the two pole theory in electronics. In the application of this theory, there are many different representations of the same system by different matrices. In this book the typical and most used cases are the transfer matrix and the dynamic stiffness matrix. Thus, the most important conversion formulas and some derivations are given.

B.3.1 Derivation of Stiffness Matrix from Transfer Matrix

The stiffness matrix of the noise control treatment can be calculated from the transfer matrix of the infinite layer (9.96)

$$\begin{Bmatrix} \boldsymbol{p}_1(k_x) \\ \boldsymbol{v}_{z1}(k_x) \end{Bmatrix} = \begin{bmatrix} T_{11}(k_x) & T_{12}(k_x) \\ T_{21}(k_x) & T_{22}(k_x) \end{bmatrix} \begin{Bmatrix} \boldsymbol{p}_2'(k_x) \\ \boldsymbol{v}_{z1}(k_x) \end{Bmatrix} = \begin{Bmatrix} -\boldsymbol{p}_2(k_x) \\ \boldsymbol{v}_{z1}(k_x) \end{Bmatrix}$$

due to the discussions in section 9.1.1. Reordering the variables links the transfer matrix to the impedance matrix by:

$$\frac{1}{\boldsymbol{T}_{21}} \begin{bmatrix} \boldsymbol{T}_{11} & -\det([\boldsymbol{T}]) \\ -1 & \boldsymbol{T}_{22} \end{bmatrix} \begin{Bmatrix} \boldsymbol{v}_{z1} \\ \boldsymbol{v}_{z2} \end{Bmatrix} = \begin{Bmatrix} \boldsymbol{p}_1 \\ \boldsymbol{p}_2 \end{Bmatrix} \quad \begin{bmatrix} \boldsymbol{Z}_{11} & \boldsymbol{Z}_{12} \\ \boldsymbol{Z}_{12} & \boldsymbol{Z}_{22} \end{bmatrix} = \frac{1}{\boldsymbol{T}_{21}} \begin{bmatrix} \boldsymbol{T}_{11} & -\det([\boldsymbol{T}]) \\ -1 & \boldsymbol{T}_{22} \end{bmatrix} \tag{B.81}$$

With $\boldsymbol{v}_{zi} = j\omega \boldsymbol{u}_i$ the final transformation to the stiffness matrix reads as:

$$[\boldsymbol{D}'']_T = \begin{bmatrix} \boldsymbol{D}''_{11} & \boldsymbol{D}''_{12} \\ \boldsymbol{D}''_{12} & \boldsymbol{D}''_{22} \end{bmatrix}_T = \frac{j\omega}{\boldsymbol{T}_{21}} \begin{bmatrix} \boldsymbol{T}_{11} & -\det([\boldsymbol{T}]) \\ -1 & \boldsymbol{T}_{22} \end{bmatrix} \tag{B.82}$$

Bibliography

R. S. Langley and P. J. Shorter. The wave transmission coefficients and coupling loss factors of point connected structures. *The Journal of the Acoustical Society of America*, 113(4): 1947–1964, 2003.

R.S. Langley and K.H. Heron. Elastic wave transmission through plate/beam junctions. *Journal of Sound and Vibration*, 143(2): 241–253, December 1990. ISSN 0022460X.

Appendix C

Symbols

Table C.1 Greek Symbols I.

α_s	Absorption coefficient
$\beta_x, \beta_y, \beta_z$	Positive rotation around x, y, and z-axis
$\Delta\omega$	Angular frequency bandwidth
ΔT	Sampling interval
α	Decay rate
α_x	Strain damping coefficient
$\delta(\cdot)$	Dirac delta function
δ_{BC}	Correction factor for certain boundary condition
ζ	Critical damping ratio
η	Damping loss factor
η_m, η_{mm}	Damping loss factor of mth system
η_{mn}	Coupling loss factor of system m to n
ϑ	Azimuthal angle
$\kappa = c_v/c_p$	Specific heat ratio
λ	Wavelength, first Lamé coefficient
μ	Frequency ratio, ratio factor
ν	Poisson ratio
ρ	Density
ρ_{fg}	Correlation between f and g
ϕ	Phase angle
φ	Polar angle
Φ	Velocity potential (chapter 2)
Φ_n	Normalized mode shape
Ψ_n	Mode shape
$\mathbf{\Psi}_{L/S}$	Amplitude of plate longitudinal and shear wave
$\mathbf{\Psi}_{B1/2}$	Amplitude of evanescent and propagating plate bending waves
Π	Power
Π_{in}	Input power
$\Pi_{i\to j}$	Transmitted power from subsystem i to j

Vibroacoustic Simulation: An Introduction to Statistical Energy Analysis and Hybrid Methods, First Edition. Alexander Peiffer.

Table C.2 Greek Symbols II.

σ	Stress
σ_{rad}	Radiation efficiency
σ'	Surface porosity
τ	Decay rate
τ_{mn}	Transmission coefficient
ϕ	Wave incident angle
$\zeta = c_v/c_{vc}$	Ratio of viscous to critical viscous damping
ω	Angular frequency
ω_0	Angular resonance frequency of undamped harmonic oscillator
ω_d	Angular resonance frequency of damped harmonic oscillator
ω_d	Angular frequency of maximum amplitude of damped harmonic oscillator
ω_n	Modal angular frequency of mode n
$\omega_s = 2\pi/\Delta t$	Angular sampling frequency

Table C.3 Capital latin Symbols I.

A	Surface, area
A_c	Cross section of beam or tube
$\boldsymbol{A}$	Complex amplitude
B	Plate bending stiffness
B_y, B_z	Beam bending stiffness for y and x-axis
$[C]$	Viscous damping matrix
$[\boldsymbol{D}]$	Dynamic stiffness matrix
E	Energy, Young's modulus
E_{kin}	Kinetic energy
E_{pot}	Potential energy
F	Force
$\boldsymbol{F}(\omega), \boldsymbol{G}(\omega)$	Functions in frequency domain
$\boldsymbol{F}$	Complex force amplitude
G	Shear modulus
$\boldsymbol{H}(\omega)$	Frequency response function
He$= kL$	Helmholtz number
$[I]$	Unit matrix
J	Torsional moment of rigidity
I_{ii}	Area inertial moment
K	Bulk modulus
$[K]$	Stiffness matrix

(Continued)

Table C.3 (Continued)

L	Length
L_j	Length of junction
L_p	Pressure level
L_v	Velocity level
L_j	Junction length
$[L]$	Nonsymmetric power SEA matrix
$[L']$	Symmetric SEA matrix
M	Mass (total)
$\boldsymbol{M}$	Moment vector
M_x	Moment in x-direction
$[M]$	Mass matrix
N	Mode count
$N'_{x/y}$	Force per length in plates
P	Perimeter
Q	Quality factor
$\boldsymbol{Q}$	Volume source strength
$Q'_{x/y}$	Shear stress per length in plates
R	Radius
$\boldsymbol{R}$	Reflection factor
$R_{fg}(t_1, t_2)$	Cross correlation function
S_{ff}	Power spectral density
S	Surface, section area
S_j	Surface of junction
$\boldsymbol{S}_{fg}$	Cross spectral density
T	Time interval
$\boldsymbol{T}$	Transmission factor
T_{mem}	Membrane tension
TL	Transmission loss
V	Volume
$\boldsymbol{W}(\omega)$	Spectrum of window function
$\boldsymbol{Y}$	Mobility
$\boldsymbol{Z} = \boldsymbol{F}/\boldsymbol{V}$	Mechanical impedance
$\boldsymbol{Z}_a = \boldsymbol{p}_{\text{surf}}/\boldsymbol{Q}_{\text{surf}}$	Radiation impedance

Table C.4 Small latin Symbols I.

a_n, b_n, c_n	Fourier coefficients
c	Wave speed
c_0	Wave speed in fluids
c_{gr}	Wave group speed
c_ϕ	Wave phase speed
c_v	Viscous damping
c_L	Longitudinal wave speed in solids
c_{LB}	Longitudinal wave speed in beams
c_{LP}	Longitudinal wave speed in plates
c_S	Shear wave speed
$c_{vc} = \sqrt{4mk_s}$	Critical viscous damping
$c_{v\text{eq}}$	Equivalent viscous damping
$\mathbf{d} = \left\{u, v, w\right\}^T$	Displacement vector
e	Energy density
e	Euler constant
f	Frequency
f_c	Critical frequency
f_q	General source term of acoustic wave equation
$f(t), g(t)$	Functions in time domain
j	Imaginary number
h	Thickness of a plate
k	Wavenumber
$\boldsymbol{k}$	Complex wavenumber
k_s	Spring stiffness
$n(\omega)$	Modal density
m	Mass
m'	Mass per unit length
m''	Mass per unit area
m_n	Modal mass
p	Acoustic pressure
q	Generalised displacement
q_s	Volume source strength density

(Continued)

Table C.4 (Continued)

r_s	Reflection coefficient
t	Time
u, v, w	Displacement in x, y, and z-direction
v_x, v_y, v_z	Velocity in x, y, and z-direction
u_r	Displacement at resonance
v_{x0}	Initial velocity in x-direction
$\boldsymbol{v}_x$	Complex velocity amplitude in x-direction
$w(t)$	Window function in signal analysis
z_0	Characteristic acoustic impedance
$\boldsymbol{z} = \boldsymbol{p}/\boldsymbol{v}$	Acoustical impedance
$\boldsymbol{z}_a = \boldsymbol{p}_{\text{surf}}/\boldsymbol{v}_{\text{surf}}$	Acoustic radiation impedance

Table C.5 Other Symbols..

$Im\boldsymbol{Z}$	Imaginary part of $\boldsymbol{Z}$
$Re\boldsymbol{Z}$	Real part of $\boldsymbol{Z}$
∞	Infinity
$\langle\cdot\rangle$	Average operator
$\langle\cdot\rangle_T$	Average over time operator
$\langle\cdot\rangle_E$	Ensemble average operator
$\langle\cdot\rangle_V$	Average over volume operator
$\langle\cdot\rangle_S$	Average over area operator
$\langle\cdot\rangle_L$	Average over length operator
$E[\cdot]$	Expected value
$\mathcal{F}\{\cdot\}$	Fourier transform
$\mathcal{F}^{-1}\{\cdot\}$	Inverse Fourier transform
$\hat{\cdot}$	Real valued amplitude
$\begin{Bmatrix}\cdot\end{Bmatrix}$	Column vector (of coefficients)
$\mathbf{r}$	Vector, geometrical
$\begin{bmatrix}\cdot\end{bmatrix}$	Matrix
$(\cdot)'$	Denoted quantity per length
$(\cdot)''$	Denoted quantity per area
$\dot{x}$	Time derivative
$\ddot{x}$	Double time derivative

Table C.6 Indexes.

i,j	Index for degrees of freedom
i	Index for inner degrees of freedom
s	Index for surface degrees of freedom
jXY	Index for junction degrees of freedom connected to system X and Y
m,n	Index for systems
(m)	Index for system when subindex used by, e.g. other degrees of freedom
-1	Inverse of a matrix
T	Transpose of a vector
H	Hermitian of a matrix or vector
$-H$	Hermitian of the inverse of a matrix
ff	Cross correlation of force degrees of freedom
uu	Cross correlation of general displacement degrees of freedom
in	Denoting input into a system
out	Denoting output leaving a system
ref	Denoting reference values
refl	Denoting reflection
trans	Cross correlation of force degrees of freedom
∞	Index denoting quantities of infinite or semi infinite systems

Index

Vibroacoustic Simulation: An Introduction to Statistical Energy Analysis and Hybrid Methods,
First Edition. Alexander Peiffer.

d

e

f

g

h

i

j

k

l

m

Printed and bound by CPI Group (UK) Ltd, Croydon, CR0 4YY

16/04/2025

14658426-0005